Stereochemical Analysis of Alicyclic Compounds by C-13 NMR Spectroscopy

Stereochemical Analysis of Alicyclic Compounds by C-13 NMR Spectroscopy

James K. Whitesell and Mark A. Minton

London   New York
CHAPMAN AND HALL

First published in 1987 by Chapman and Hall Ltd
11 New Fetter Lane, London EC4P 4EE
Published in the USA by Chapman and Hall
29 West 35th Street, New York, NY 10001

© 1987 Chapman and Hall Ltd

Printed in Great Britain at the
University Press, Cambridge

ISBN 0 412 29550 4

British Library Cataloguing in Publication Data

Whitesell, J.K.
Stereochemical analysis of alicyclic
compounds by C-13 NMR spectroscopy.
1. Chemistry, Physical organic   2. Nuclear
magnetic resonance spectroscopy   3. Carbon
—Isotopes
I. Title   II. Minton, M.A.
547.3′0877   QD476

ISBN 0-412-29550-4

Library of Congress Cataloging in Publication Data

Whitesell, J.K. (James K.)
Stereochemical analysis of alicyclic compounds by
C-13 NMR spectroscopy.
Bibliography: p.
1. Nuclear magnetic resonance spectroscopy.
2. Carbon—Isotopes—Analysis.   3. Stereochemistry.
4. Alicyclic compounds—Analysis.
I. Minton, M.A. (Mark A.)
II. Title.
QD96.N8W47   1987   547′.5046   86-34312

ISBN 0-412-29550-4

Contents

Preface

Through numerous conversations with other synthetic chemists it became apparent that the great power of carbon nuclear magnetic resonance was being significantly underutilized. In our own work we have found that ^{13}C spectroscopy is a more powerful tool than ^{1}H NMR spectroscopy, especially for probing subtle stereochemical questions in complicated systems. This is especially true in five membered ring compounds where ^{1}H NMR is at a particular disadvantage. The two techniques can be used independently to solve the same question– that of stereochemistry – but they do so in different ways. Advantage can be taken in ^{1}H NMR of a relatively consistent relationship between stereochemical orientation and coupling constants between vicinal protons, while in ^{13}C NMR it is the correlation between spatial relationships of non-hydrogen, γ substituents and their effect on chemical shift that can be used to assign stereochemistry.

It was also clear that the use of ^{13}C NMR required a different approach to problem solving than that typically used with ^{1}H NMR. While the latter technique could be employed with a very general approach (e.g., the Karplus equation), ^{13}C NMR would, at least for the immediate future, require a relatively extensive set of model systems from which the consequences of stereochemical changes could be derived for any given carbon framework. Indeed, our own research efforts required such a set of spectral data for model, monosubstituted systems and served as the impetus for the collection of the extensive data that will be found in Chapters 5–18. With this extensive data available in computer searchable form, we are beginning to assemble the equivalent of a Karplus relationship for γ substituent effects in ^{13}C NMR and hope to make these findings available in the not too distant future. However, the origin of such effects is not at all well defined and the observed spectral shifts may be caused by the coincidence of several relatively independent factors.

We have exercised extreme care both in compiling and in reproducing the chemical shift data in the numerous tables to be found in the later chapters of this work. The process involved considerably more than the lifting of chemical shift data from the literature as the spectral information for each compound was checked and in numerous cases we have made reassignments so that the collection of data for a particular ring system would be self-consistent. In addition, the tables were printed directly from our computer data bank of this information and photographically reproduced for this manuscript, avoiding the inevitable errors that would occur in typesetting such an extensive collection of numbers. We hope that in this way we have been able to assemble a ^{13}C NMR data set that is not only more extensive but, and perhaps more importantly, more accurate than any currently available.

James K. Whitesell
Mark A. Minton

Introduction

The dramatic changes that have taken place in synthetic chemistry are a direct result of improved instrumentation for structural analysis of complex organic molecules. As instrumentation evolved from ultraviolet to infrared to proton nuclear magnetic resonance spectroscopy, it was the synthetic community that was first to develop and then to exploit the full power of each new technique. This does not appear to be the case with the latest tool, ^{13}C NMR spectroscopy, which has yet to find significant usage by synthetic chemists other than as a rather complicated means for counting carbons. The power of the technique is far greater than as a replacement for elemental analysis and indeed can be used to solve complicated structural problems more readily than less straightforward methods such as 2-D ^1H NMR techniques.

Effective use of ^{13}C NMR requires a relatively extensive collection of model systems that can be used to predict the effect of substituents in many situations. The tables of spectral data to be found in Chapters 5 through 18 represent as complete a collection of spectra of monosubstituted, bicyclic carbon compounds as was practical to assemble. Two presentations are provided: structural representations of spectral data for compounds bearing a single functionality; and complete compilations for mono- as well as poly-functionalized systems.

The data compilation for the monosubstituted compounds include, in addition to the chemical shift, the shift - of - shift ($\Delta\delta$) value that represents the difference between the observed shift and the shift for the same carbon of the unfunctionalized, parent system. These numbers are in parentheses. For most of the parent bicyclic systems, there are several different sets of values reported in the literature, although we have listed in the table only that set which we consider to be the most useful (e.g., that from the more recent literature taken as a dilute solution in deuterochloroform). In the calculation of these $\Delta\delta$ values, however, we have used for each compound the set of parent shifts obtained under comparable conditions, if available, since it was felt that shift differences would vary less from solvent to solvent (and with other factors such as concentration and temperature) than would the actual shifts themselves. Thus, there is no ready mechanism for the user of these tables to check the validity of the $\Delta\delta$ values that we have provided. However, these numbers were calculated semiautomatically by computer from data bank information and should be without error.

In the interest of a clear presentation, we have not provided all of the information relating to each spectrum that might be desired. For instance, it might be useful in certain circumstances to know what solvent was used for a given spectrum. To permit access to this information we have provided literature citations for all compounds. In many cases, several authors reported slightly different data for the same compound and again we have chosen the set of values which we believe will be most useful in a general sense, and that citation is listed first in the sequence of references. On occasion we found it necessary to combine two different sets of data. For instance, multiplicity data obtained in a relatively unusual solvent was required to assign spectral shift data obtained in deuterochloroform. In these cases the critical references are listed first, separated by a + sign.

While the collection of bicyclic systems is nearly complete, there is one obvious omission - the hydroazulenes or bicyclo[5.3.0]decanes. It was felt that the magnitude of effort required to make a reasonably complete collection of data in this case would not be repaid because of the relatively great conformational freedom present in this ring system.

For approximately 5% of the spectra, the assignments reported in the literature were sufficiently inconsistent with the remainder of the data collected that we felt justified in making our own shift correlations. In the majority of cases, these 'corrections' represent nothing more than hindsight on our part since we had the advantage of the more recent data when analyzing older literature. We have corrected "obvious" character transpositions, such as 43.4 for 34.4, in a limited number of cases, again using the total collection of data to insure that such changes were made only with the most compelling justification.

The first four chapters should serve as sufficient introduction to carbon NMR spectroscopy so that someone well-versed in proton spectroscopy will be able to take full advantage of the data in the subsequent chapters. However, we make no pretense as to providing a meaningful background for development of the subject of nuclear magnetic resonance in general and the beginner to this field is referred to any one of the numerous and excellent texts available on this subject.

Chapter 1

Fundamental Effects on Carbon Shifts

The power of carbon NMR spectroscopy for the solution of structural problems in organic chemistry can be attributed mainly to a single phenomenon – that, in most cases, the total effect on chemical shifts due to the presence of a large number and variety of substituents can be predicted to be the sum of the effects of the individual groups. It is this "substituent additivity" that sets carbon spectroscopy distinctly apart from proton spectroscopy. For example, the chemical shift values for the indicated carbons in ethane, propane, butane, and pentane progress in the order: 6.5, 16.3, 24.8, 34.2 δ[1].

$$CH_3\text{-}CH_3 \qquad CH_3\text{-}CH_2\text{-}CH_3 \qquad CH_3\text{-}CH_2\text{-}CH_2\text{-}CH_3$$

$$CH_3\text{-}CH_2\text{-}CH_2\text{-}CH_2\text{-}CH_3$$

The increments as the carbon chain grows, 9.8, 8.5, 9.4 δ, are relatively constant and represent the addition of either an α or a β substitutent. It is indeed remarkable that the effect of each additional carbon atom as an α or β substituent remains reasonably constant and unaffected by the nature and number of other groups. Yet using a value of +8.0 for the effect of an α or β carbon atom leads to predictions of 8.0, 16.0, 24.0, and 32.0 δ for the examples above with the largest error (2.2 δ) less than 1% of the normal range for carbon absorption. The additivity of shift-of-shift ($\Delta\delta$) effects is found to hold for most substituent groups with the exceptions limited to highly polarizable functionalities such as alkyl halides (especially bromides and iodides) and carbonyl groups and to situations where there is steric interaction between the substituents. Thus, with a relatively simple set of $\Delta\delta$ effects it is usually possible to predict the absorptions of most or all of the carbons in a complex structure with sufficient accuracy so that a correlation between the observed shifts and carbons can be made.

The value of carbon NMR spectroscopy is further enhanced by the observation, noted above as an exception, that there is a perturbation in the $\Delta\delta$ effect of substituents that are sterically interacting. In the majority of cases this will be limited to those situations where groups are γ to one another and the magnitude can be directly correlated with the spatial relationship between these groups. The effect is greatest when a γ substituent and the carbon under investigation are closest together. In total, then, the shift of a carbon can be predicted based primarily upon the nature and number of α and β substituents with comparatively smaller changes resulting from the stereochemical orientations of γ substituents.

1.1 Simple examples

The major focus of this book will be on the use of carbon NMR spectroscopy for the assessment of configurational and conformational stereochemical features of a molecule. In order to glean such information from spectral data it will be necessary to establish first what chemical shift would be anticipated based solely upon the α and β substituents. The difference between such predictions and observed spectra can then be attributed to the nature and spatial orientation of the γ substituents, thus providing key information regarding the relative stereochemical orientation of groups in the molecule. For example, from a comparison of the carbon spectra for exo- and endo-2-methylnorbornane with each other and with that for the unsubstituted parent system (Figure 1-1), it can be seen that the methyl group exerts a large upfield shift on C-7 of the exo isomer while it

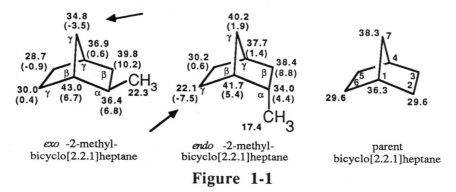

exo -2-methyl-
bicyclo[2.2.1]heptane

endo -2-methyl-
bicyclo[2.2.1]heptane

parent
bicyclo[2.2.1]heptane

Figure 1-1

is C-6 of the endo isomer that is similarly affected. In each case these represent the nearest carbons that are γ to the methyl group. The difference in chemical shifts for these two carbons between the two isomers is quite large, while those for the carbons of the two isomers that are either α or β to the methyl group (1, 2, and 3) differ to a much smaller degree. The methyl group is also affected by this interaction, and is shifted further upfield in the endo isomer where the γ carbon (C-6) is also shifted upfield more than C-7 is in the exo isomer.

The observations made above for the bicyclo[2.2.1]heptane systems are not unique. It is generally observed that carbon absorptions are shifted upfield by steric interactions with γ substituents as can be seen by comparision of the Δδ values for the stereoisomeric pairs of 6- and 2-methylbicyclo[3.2.1]octanes (Figure 1-2).

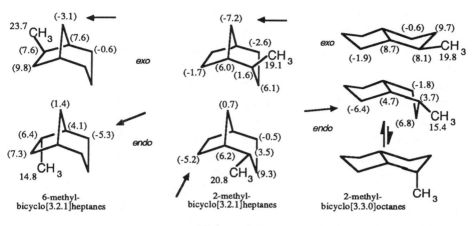

6-methyl-
bicyclo[3.2.1]heptanes

2-methyl-
bicyclo[3.2.1]heptanes

2-methyl-
bicyclo[3.3.0]octanes

Figure 1-2

The absorption for C-8 in both *exo* isomers is consistently upfield of the corresponding carbons in the *endo* isomers, while in the latter compounds, it is on the one hand C-4 and on the other C-7 that is affected by the methyl group. Care must be taken in making such analyses in systems where alternate diastereomers adopt different conformations. Thus, in the case of the 2-methyl - bicyclo[3.3.0]octanes, the shift of C-3 in the endo isomer is upfield compared to that of the exo diastereomer, presumably because the conformation of the former is such that C-3 and C-8 are sterically interacting, as depicted. In many of the cases with conformational freedom, analysis of carbon spectral data can provide not only configurational but conformational information as well.

Not only are the carbons that undergo a steric interaction with a γ substitutent shifted upfield but the same is true for the two carbons that are between these groups. This effect can be seen most easily in the shifts of C-3 in the isomeric 2-methylbicyclo[3.2.1]octanes. There is a gauche butane-like interaction between the methyl group and C-4 only in the exo isomer, and the shift of C-3 for this isomer is 3.2 δ upfield from that in the endo diastereomer. This "internal" gauche butane effect can be quite useful because the magnitude for a carbon within a given system remains relatively constant with changes in the nature of the substituents directly involved in the steric interaction and can be observed even when these latter groups are not carbon.

Absorptions must be assigned to the individual carbons before such an analysis of subtle, stereochemical effects can be made. It is best to make spectral assignments by analogy with closely related models with very similar substituents. As this is not always possible, a collection of anticipated effects for the α and β substituents is essential.

1.2   Substituent Shift effects

There are many "systems" in the literature that provide relatively complicated tables listing the magnitude of the anticipated $\Delta\delta$ effects for substituents under a variety of situations. Unfortunately, the provisions included in these systems for the subtle variations caused by minor changes make them cumbersome to use. The system that we are currently using was first suggested by Dr Ben Shoulders of the University of Texas at Austin, and although we have made modifications and refinements, the simplicity and ease of use of this system are a result of his efforts. However, no simple system of this kind can deal with spatial relationships in a general way and thus will not be reliable in complicated systems; the number and nature of the effects are simply too large and their magnitudes are not fully independent of each other. Nonetheless, it is essential that a correlation of absorptions with carbons be made and we rely heavily on the following method for this purpose. Once these fundamental assignments have been accomplished, correlations with model systems that represent as closely as possible the actual system involved can be used to make quite accurate predictions about stereochemical relationships.

Table 1-1 on the following page represents a set of shift - of - shift, or $\Delta\delta$, effects that can be used to approximate the anticipated chemical shift of carbons in a variety of situations. The values listed invariably are compromises between the obvious need to make accurate predictions and the desire to have an easily remembered (and therefore useful) set of effects that can be used to make initial assignments of the observed shifts in a carbon spectrum. Once such a correlation has been made, then specific model compounds (to be found in the latter chapters of this book) can be used to provide much more precise assessments of shift effects in a specific situation.

In all cases the $\Delta\delta$ values listed represent the effect to be anticipated through the addition of a particular substituent atom or functional group to a carbon under consideration. As an example, the methyl group of acetone has one α and two β substituents (3*8.0 = 24.0) and has present the α carbonyl group (+7.5) for an anticipated shift of 31.5 (observed 30.6). With more complicated systems, and especially in situations where model compounds are available, it will be found more convenient to assess a *total* effect anticipated for a given substituent.

General Effects on Carbon NMR Shifts

$$\underline{C}-\alpha-\beta-\gamma$$

For any sp^3 carbon, add to 0.0 for each:
    α or β substituent   8.0

In addition, add for each:

	1°	2°	3°
α oxygen substituent		38.0	
α nitrogen substituent		22.5	
α trans C=C		2.5	
α cis C=C		-2.5	
α ester or acid		-2.5	
α ketone		7.5	
α aldehyde		15.0	
α chlorine	22.0	30.5	41.0
α bromine	10.5	24.0	variable
α iodine		variable	
γ carbon		-2.0	
γ oxygen		-5.0	

For 3° carbons, for each:
    β substituent   -1.5

For 4° carbons, for each:
    β substituent   -3.5

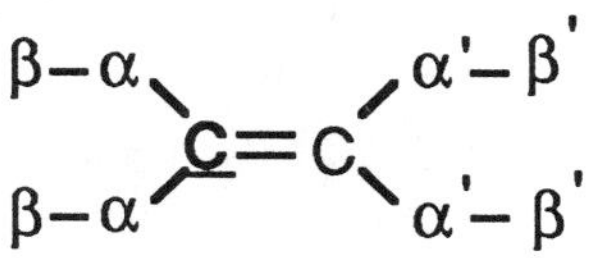

$$\beta-\alpha \diagdown \quad \diagup \alpha'-\beta'$$
$$\underline{C}=C$$
$$\beta-\alpha \diagup \quad \diagdown \alpha'-\beta'$$

For an sp^2 carbon of a C=C bond, add to 121.0 for each:
    α or β carbon substituent   8.0
    α' carbon substituent   -8.0
    cis   -1.0

Table 1-1

The application of these rules to simple systems provides amazingly accurate predictions. Thus, for the hydrocarbon below, C1 has one α and one β carbon (+16.0), and two γ carbons (-4.0) for a prediction of 12.0 (obs. 11.6). Similarly, C3 has three α and two β carbons (40.0) and is 3° with two β substituents (-3.0) for a prediction of 37.0 (obs. 36.6), and C2 has two α and two β carbons (32.0) and one γ carbon (-2.0) - pred. 30.0 (obs. 29.5). Similarly, the C3 methyl group has one α and two β and two γ substituents (24.0 - 4.0 = 20.0, 18.9 obs).

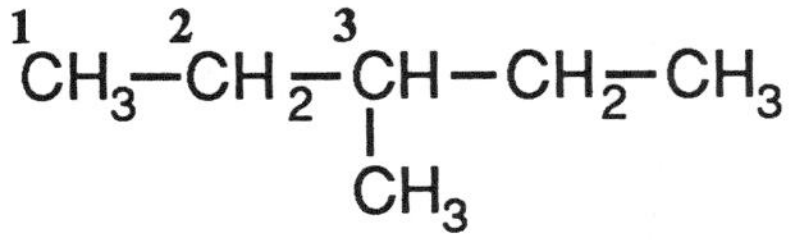

$$\overset{1}{CH_3}-\overset{2}{CH_2}-\overset{3}{CH}-CH_2-CH_3$$
$$\underset{CH_3}{|}$$

It may be noticed that the $\Delta\delta$ effect value in Table 1-1 for each α and β substituent carbon atom is 8.0, a value that is somewhat smaller than that required for correlation of the spectra of the simple, linear hydrocarbons ethane through pentane mentioned previously where the average of the spectral shift effect of each additional α and β carbon is 9.2 δ. The $\Delta\delta$ values for the various substituents in Table 1-1 are those that provide a "best fit" with the more complicated systems where real questions of structure and/or stereochemistry arise. This difference also highlights the fact that the effects of substituents are not completely independent of each other and in general, the greater the number of substituents, the smaller the effect of each.

This system works well for functionally substituted systems, for example, the alcohol below:

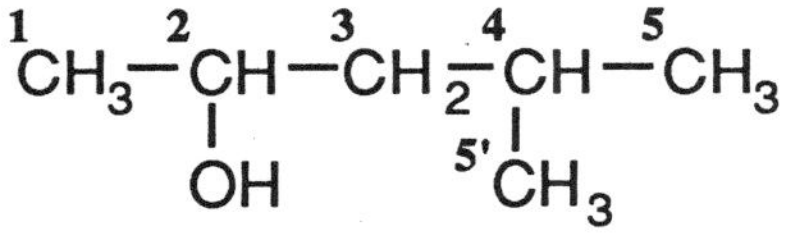

$$\overset{1}{CH_3}-\overset{2}{CH}-\overset{3}{CH_2}-\overset{4}{CH}-\overset{5}{CH_3}$$
$$\underset{OH}{|}\qquad\underset{\overset{5'}{CH_3}}{|}$$

	α/β	α O	γ	3°		(observed,[2])
C1	3*8.0		-2.0		=	22.0 (23.9)
C2	4*8.0	+38.0	2*-2.0	-1.5	=	64.5 (65.8)
C3	6*8.0				=	48.0 (48.8)
C4	4*8.0		-5.0 -2.0	-1.5	=	23.5 (24.9)
C5	3*8.0		-2.0		=	22.0 (23.2 & 22.5)

Corrections are made in both examples for the presence of γ substituents based on the values given in Table 1-1. The origins of these effects are not well understood, and as will become clear in Chapter 2, the magnitude varies with the spatial relationship of the two atoms, reaching a maximum when they are in the equivalent of an eclipsed butane relationship. However, the effect is not simply spatial, and can still represent a significant shift effect when the γ group is anti, especially with more electronegative substituents such as oxygen. The $\Delta\delta$ effects given in Table 1–1 represent "average" values that will apply to acyclic systems with typical conformational populations. As will be developed in Chapter 2, for the purposes of determining spatial relationships a predicted shift is first made by ignoring γ effects. The difference then between such predictions and observed spectral shifts can be attributed to γ interactions. An indication of the sensitivity of this effect to subtle influences can be seen by the observation of two distinct signals for the diastereotopic methyl groups (5 and 5') in the alcohol shown above.

The treatment of polar and polarizable functional groups is more complicated as each situation becomes somewhat unique. Indeed, the absorption for carbonyl carbons is quite variable, responding in the same compound to factors such as solvent and even concentration. There are two methods for treating the perturbation caused by the presence of a carbonyl group: either as a correction to be added to the normal effect of the functional group substituents or as a change expected from the conversion of a methylene to a carbonyl group. The latter method is more generally useful, especially for cyclic ketones and lactones, since it is quite often the case that shift values can be found for a model system that is structurally quite similar to that under consideration. In addition, there are small but real changes in the magnitudes of the effects with ring size. In general, the latter method is easier (and avoids, for instance, the question of how many β substituents there are for each ring carbon of a cyclobutane), and becomes nearly essential with more complicated mono- and polycyclic systems where many of the carbons are in unique situations.

References

1. Wiberg, K.B., Pratt, W.E. and Bailey, W.F., *J. Org. Chem.* **45**, 4936 (1980).
2. *Sadtler Standard Carbon-13 NMR Spectra*, Sadtler Research Laboratories, Philadelphia, Pa 19104 (1976-1982).

Chapter 2

Mono- and Bicyclic Systems

The concept of shift additivity that was developed in Chapter 1 with acyclic systems can be extended to cover cyclic arrays as well. Unfortunately, each ring system has unique features and a relatively simple but "universal" set of $\Delta\delta$ values such as that found in Table 1-1 will be too general to provide accurate predictions for mono- and especially polycyclic systems. It is for this reason that we have gathered the extensive sets of carbon shift data that will be found in the following chapters that individually cover virtually all of the bicyclic ring systems that are commonly used in synthesis. Thus, while the $\Delta\delta$ values of substituents differ from one system to another and even between positions on a given framework, the majority of these values will be found in this book. In addition, these $\Delta\delta$ values can be applied directly to tri- and tetracyclic systems by an analysis that treats each local area as part of bicyclic subunit.

The discussion that follows in this chapter will cover the variations to be expected between acyclic and cyclic systems since a general understanding of these differences will provide a basis for dealing with the changes to be found amongst the various bicyclic ringsystems.

2.1 Monocyclic systems

The carbon shift data for a number of monofunctionalized, monocyclic compounds are illustrated on the following pages (Figure 2-1), along with the shift values for the appropriate, parent hydrocarbons. Notice that the shifts for the ring carbons of the parents do not correspond to those predicted by using the system detailed in Chapter 1.

For example, in cyclopentane each carbon has 2 α and 2 β substituent carbons and thus all would be predicted to absorb at 32.0, not 26.2 as observed (a rationalization of this effect will be provided in Chapter 3). There are two ways to deal with the upfield shift for cyclic systems: either as a correction (eg. –5.8 for a carbon in a five-membered ring) or by using the value for the appropriate hydrocarbon as a base to which the effects of all substituents other than the ring atoms are added. As the complexity of the systems increases, it becomes increasingly easy to use the latter approach, and it is highly recommended that this method be used consistently. Notice also that the shift - of - shift values that result from the introduction of the various functionalities vary with the size of the ring. Thus, it should be clear that no simple system such as that detailed in Chapter 1 can be universally applicable. Nonetheless, it is possible to arrive at reasonable predictions for cyclic systems in the absence of γ interactions by using Table 1-1 adjusted for the shift for the parent ring.

Figure 2-1

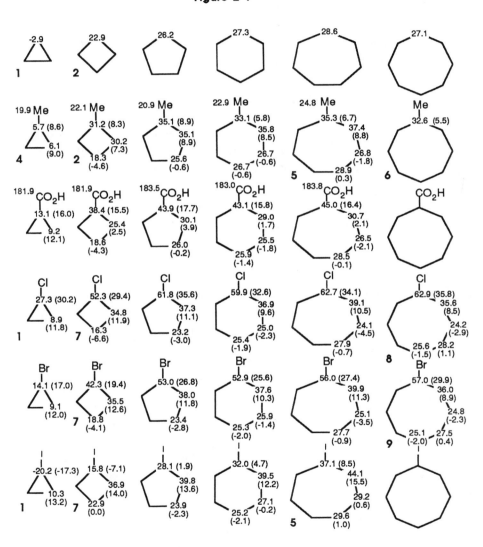

Figure 2-1 (cont.)

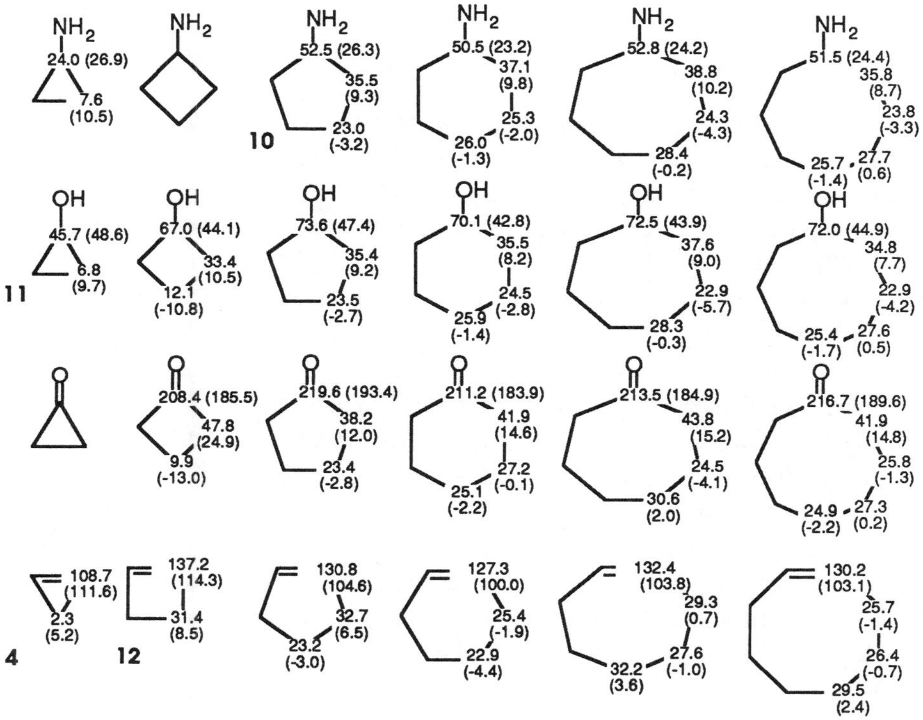

All references are (3) unless otherwise indicated by boldfaced number to lower left.
Where shift data are not provided, the information is currently unavailable in the literature.

As an example, carbons 1 and 2 of methylcyclohexane would be predicted to be found as follows:

	C-1	C-2	C-3
cyclohexane	27.3	27.3	27.3
α or β subst.	+8.0	+8.0	
3°, β correction	-3.0	——	——
	32.3	35.3	27.3
observed	33.1	35.8	26.7

2.2 Bicyclic systems

The shift values for the common bicyclic systems (as well as for adamantane) are provided on the following page (Figure 2-2). Perhaps the most noticeable differences between the various systems are in the bridgehead carbons - varying from 9.8 to 46.9. In part this large spread results from the different ring sizes involved, but even within those systems that involve only five- and six-membered rings, the spread is substantial (24.0 to 46.9). In addition, it can be seen that there are large differences between the trans and cis isomers, with the shift of the bridgehead carbons of the latter consistently upfield. This observation is as would be expected, since there are gauche interactions in the cis isomers that are not present in the trans isomers.

Reasonable predictions for the shift values for trans-decalin can be based on the simple substituent effects outlined in Chapter 1 and the stereochemical considerations mentioned above. Thus, the bridgehead carbons would be predicted to absorb at 45.3 (27.3, + 24.0 for three additional β substituents, and -6.0 for a 3° carbon with four β substituents, 43.7 obs.). Similarly, the prediction for C-2 would be 35.3 (27.3 + 8.0), 34.3 observed, and for C-3 would be 27.3 (26.9 obs.) since there are no additional α or β substituents relative to cyclohexane and the γ carbons are anti, not gauche. There is a small upfield shift, typically 0.5 for an anti, γ carbon, but in most cases this effect is too small to be of significant use in stereochemical predictions. The situation with cis-decalin is more complicated because of the superposition of two equal energy conformations. Nonetheless, the shifts observed are understandable once this factor is combined with stereochemical considerations to be detailed in Chapter 3.

The shift - of - shift effect of substituents is not constant from one system to another. Perhaps the most dramatic variations are seen in the effect of a bridgehead substituent on the opposite bridgehead in bridged bicyclic systems (Table 2-1). It is thus imperative, especially with either polarizable or more electronegative substituents, to use substituent effects derived from the system under investigation. Tables 2-2 through 2-7 on the following pages list in order of magnitude the Δδ effects of various substituents on some of the more common bicyclic systems, and should serve as useful introductions to the effects to be expected from various substituents.

Fused and Bridged Bicyclic Systems

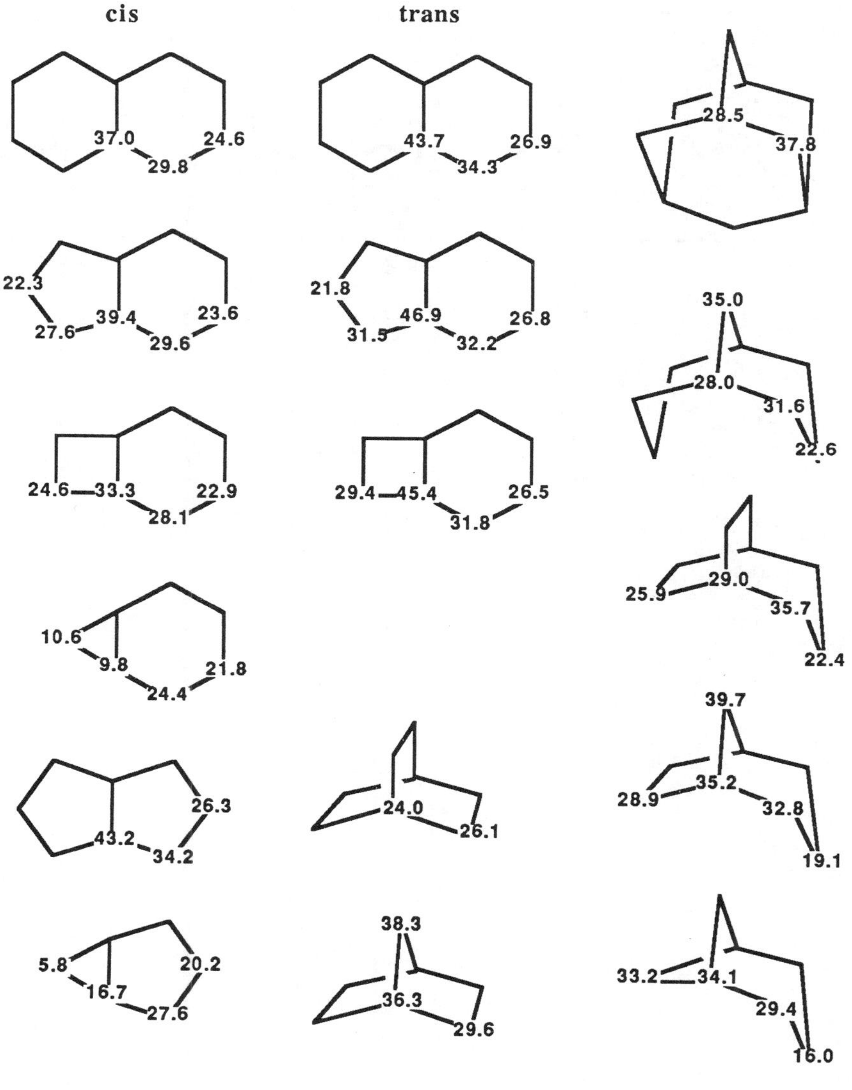

cis   trans

Figure 2-2

Bridgehead - Bridgehead Effects

Table 2-1

X	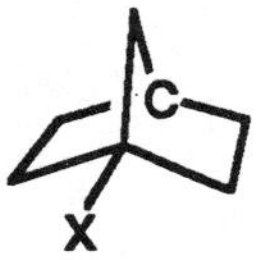	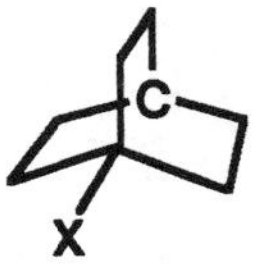	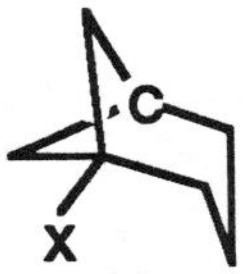	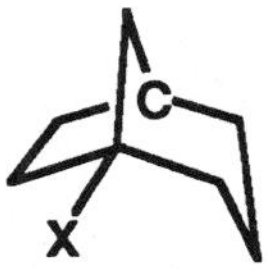	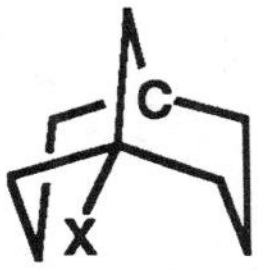
F	−3.3	0.0	−9.2	-	+5.5
OH	−1.6	−0.1	-	-	+4.2
Cl	−1.6	−1.0	-	-	+4.9
Br	−1.6	−1.6	-	-	+5.4
I	−1.8	−2.8	-	-	+5.3
Me	+1.4	+0.7	−4.0	+0.8	-
Ph	+0.9	+0.6	-	-	-
CO_2H	+1.4	+0.6	-	-	−0.2

Table 2-2

Substituent	α C-1	Substituent	β C-2	Substituent	γ C-3
1-F	67.6	1-I	13.2	1-I	2.3
1-OMe	51.3	1-Br	10.2	1-Br	1.8
1-OEt	50.3	1-Cl	8.6	1-Me	1.4
1-OH	46.4	1-Ph	7.4	1-N=)2[Z]	1.2
1-N=)2[E]	46.3	1-Me	6.9	1-Ph-Me	1.2
1-N=)2[Z]	45.1	1-OH	5.6	1-Et	1.1
1-Cl	33.5	1-Ph-Me	4.8	1-Cl	1.1
1-NHAc	26.2	1-SnMe3	4.5	1-Ph	1.1
1-Br	25.8	1-Et	4.1	1-CH2OH	0.9
1-Ph-Me	16.3	1-NHAc	3.9	1-SnMe3	0.9
1-COOMe	15.8	1-COOMe	3.3	1-N=)2[E]	0.5
1-COOH	15.7	1-COOH	3.1	1-OH	0.5
1-Ph	14.9	1-N=)2[Z]	2.7	1-COOMe	0.2
1-CH2OH	13.8	1-N=)2[E]	2.7	1-COOH	0.2
1-Et	11.6	1-F	2.4	1-NHAc	0.1
1-Me	7.3	1-CH2OH	2.2	1-OMe	0.1
1-I	1.3	1-OEt	1.6	1-F	0.0
1-SnMe3	−2.8	1-OMe	1.2	1-OEt	−0.1

Substituent	γ C-4	Substituent	β C-7
1-Me	1.4	1-I	12.4
1-COOH	1.4	1-Br	9.8
1-CH2OH	1.0	1-N=)2[Z]	9.5
1-Ph	0.9	1-Cl	8.4
1-Et	0.7	1-Me	6.8
1-SnMe3	−0.2	1-Ph-Me	5.5
1-N=)2[E]	−0.3	1-OH	5.5
1-Ph-Me	−0.7	1-Ph	4.4
1-COOMe	−0.8	1-COOH	3.9
1-NHAc	−1.1	1-COOMe	3.9
1-Br	−1.6	1-SnMe3	3.7
1-OH	−1.6	1-N=)2[E]	3.7
1-Cl	−1.6	1-NHAc	3.3
1-I	−1.8	1-F	2.7
1-N=)2[Z]	−2.0	1-CH2OH	2.3
1-OMe	−2.5	1-OEt	2.0
1-OEt	−3.0	1-OMe	1.6
1-F	−3.3	1-Et	−3.1

Table 2-2 (cont.)

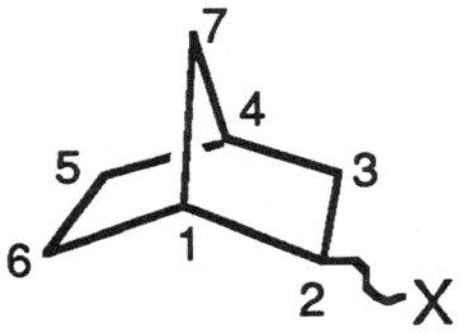

Substituent	β C-1	Substituent	β C-1	Substituent	α C-2	Substituent	α C-2
X2-I	11.4	N2-SnMe3	4.2	X2-F	65.9	X2-PMe2	15.0
X2-Br	10.1	N2-OAc	3.9	X2-OTs	55.7	X2-PSMe2	14.9
X2-Cl	9.6	X2-SnMe3	3.8	X2-OMe	54.2	X2-P(OMe)2	14.7
N2-Ph-Cl	9.2	N2-PSMe2	3.7	N2-OMs	53.0	X2-Et	14.5
X2-NH2	8.9	N2-)2	3.6	N2-OMe	51.9	N2-PMe2	13.7
N2-I	8.8	N2-Et	3.5	X2-OAc	47.9	N2-PSMe2	13.3
X2-OH	7.9	X2-OMe	3.4	X2-OCHO	47.4	X2-POMe2	13.2
N2-Br	7.5	N2-CN	3.4	N2-OAc	45.5	N2-P(OMe)2	13.1
X2-OTMS	7.5	N2-NH3Cl	3.2	X2-OH	45.1	N2-Ph-Cl	12.9
N2-Cl	7.2	N2-PCl2	3.1	X2-OTMS	43.8	N2-CH2OH	12.9
N2-NH2	6.8	N2-OMe	2.9	N2-OH	43.3	N2-Et	12.7
X2-Me	6.7	N2-POMe2	2.8	N2-OTMS	41.6	N2-POMe2	12.2
N2-OH	6.2	N2-PO(OMe)2	2.6	N2-O-3N	32.4	N2-CH2NH3Cl	10.0
N2-OTMS	5.8	X2-PCl2	2.2	X2-Cl	32.1	X2-CH2CH2OH	8.7
X2-OTs	5.8	X2-PSMe2	2.1	N2-Cl	31.0	X2-PO(OPh)2	8.7
X2-OCHO	5.6	X2-CH2OH	2.0	X2-POCl2	25.4	X2-CH2COOH	8.3
X2-CN	5.5	X2-PMe2	1.9	X2-NH2	25.3	X2-PO(OMe)2	8.2
N2-Me	5.4	X2-PMe3I	1.8	X2-PCl2	24.9	N2-CX	7.5
X2-F	5.3	X2-PO(OMe)2	1.8	N2-PCl2	23.8	X2-CH2COMe	7.2
X2-OAc	5.3	X2-PO(OPh)2	1.8	N2-Br	23.7	X2-Me	6.8
X2-CH2TMS	5.3	N2-PMe2	1.7	X2-Br	23.5	X2-PMe3I	6.2
N2-OMS	5.2	N2-CH2OH	1.6	N2-NH2	23.3	X2-CH2CH2Br	5.9
X2-CH2COMe	5.0	N2-O-3N	1.4	N2-NH3Cl	22.3	N2-Me	4.4
X2-CH2CH2OH	5.0	X2-POMe2	1.4	X2-O-3X	21.4	N2-I	2.8
X2-CH2COOH	4.9	X2-O-3X	0.5	X2-COOMe	16.9	X2-CN	1.0
X2-COOMe	4.8	N2-CH2NH3Cl	0.3	X2-COOH	16.9	N2-CN	0.1
X2-COOH	4.7	X2-POCl2	0.0	N2-COOH	16.5	X2-I	0.0
X2-Et	4.7	N2-CX	-0.3	N2-)2	16.5	N2-SnMe3	−1.1
X2-CH2CH2Br	4.5			N2-COOMe	16.4	X2-SnMe3	−2.1
N2-COOMe	4.3			X2-CH2OH	15.4		
N2-COOH	4.3			X2-CH2TMS	15.2		

Table 2-2 (cont.)

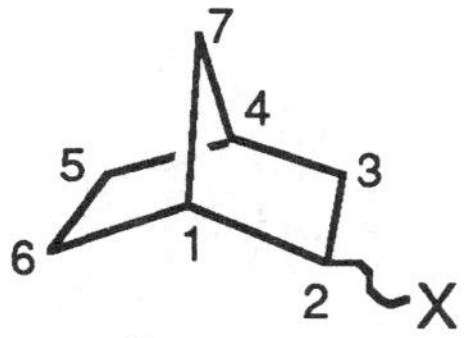

Substituent	β C-3	Substituent	β C-3	Substituent	γ C-4	Substituent	γ C-4
N2-O-3N	32.4	N2-OMe	7.4	N2-CX	3.6	N2-CH2OH	0.4
X2-O-3X	21.4	N2-OAc	7.2	N2-CH2NH3Cl	1.5	X2-(CH2)2OH	0.3
X2-I	15.2	X2-CN	6.3	N2-O-3N	1.4	N2-SnMe3	0.3
X2-Br	14.2	N2-CN	5.5	X2-I	1.4	X2-(CH2)2Br	0.2
N2-I	14.1	X2-SnMe3	5.2	N2-OTMS	1.4	N2-OMS	0.2
X2-Cl	13.8	N2-CH2NH3Cl	4.9	N2-Me	1.4	N2-OAc	0.2
X2-OH	12.6	X2-COOMe	4.7	N2-Ph-Cl	1.3	N2-CN	0.2
X2-NH2	12.4	N2-NH3Cl	4.6	X2-CH2COMe	1.3	N2-NH3Cl	0.2
N2-Br	11.8	X2-COOH	4.6	N2-NH2	1.2	N2-OMe	0.1
X2-OTMS	11.6	X2-CH2OH	4.6	N2-PSMe2	1.2	X2-PSMe2	0.1
N2-Cl	11.2	N2-PMe2	4.2	X2-SnMe3	1.1	X2-PMe3I	0.1
N2-Ph-Cl	11.0	X2-PMe2	4.1	N2-I	1.0	X2-Cl	0.0
N2-NH2	10.5	N2-CH2OH	4.1	N2-Et	1.0	X2-PO(OMe)2	0.0
X2-(CH2)2Br	10.4	N2-SnMe3	4.0	N2-)2	1.0	X2-POMe2	0.0
X2-Me	10.2	N2-PMe3I	4.0	N2-OH	0.9	X2-COOMe	−0.1
X2-OTs	10.1	N2-PCl2	3.7	N2-COOMe	0.9	X2-CH2OH	−0.1
X2-OAc	10.1	X2-PCl2	3.5	N2-PO(OMe)2	0.9	X2-COOH	−0.2
X2-OCHO	10.1	X2-POCl2	3.0	X2-CH2TMS	0.8	X2-CN	−0.3
X2-F	9.9	X2-PSMe2	2.8	N2-COOH	0.8	X2-PO(OPh)2	−0.3
N2-OH	9.8	N2-COOMe	2.5	X2-Br	0.7	X2-NH2	−0.4
X2-OMe	9.6	X2-PMe3I	2.5	X2-PCl2	0.7	X2-OCHO	−0.5
X2-(CH2)2OH	9.0	X2-PO(OPh)2	2.3	N2-Br	0.6	X2-PMe2	−0.6
N2-OTMS	8.8	X2-PO(OMe)2	2.3	N2-Cl	0.6	X2-OTMS	−0.8
X2-CH2COOH	8.8	N2-COOH	2.2	X2-Me	0.6	X2-OAc	−0.8
N2-CX	8.8	X2-POMe2	1.7	X2-CH2COOH	0.6	X2-OH	−0.9
N2-Me	8.8	N2-PO(OMe)2	1.3	X2-O-3X	0.5	X2-OTs	−0.9
X2-CH2TMS	8.7	N2-PSMe2	0.9	X2-POCl2	0.5	X2-OMe	−1.8
X2-CH2COMe	8.6	N2-)2	0.8	X2-Et	0.5	X2-F	−2.1
X2-Et	8.6	X2-P(OMe)2	0.8				
N2-Et	7.7	N2-POMe2	0.1				
N2-OMS	7.4						

Table 2-2 (cont.)

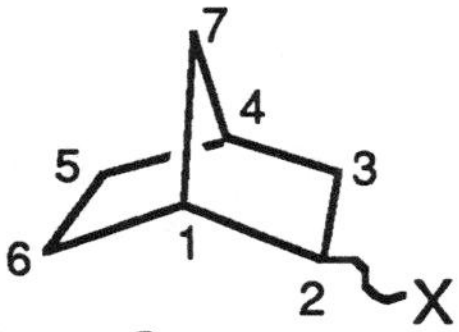

Substituent	δ C-5	Substituent	δ C-5	Substituent	γ C-6	Substituent	γ C-6
N2-PMe3I	1.9	N2-PMe2	−0.7	X2-SnMe3	3.8	N2-PO(OMe)2	−4.2
N2-Et	0.8	X2-COOMe	−0.8	X2-PSMe2	2.8	X2-O-3X	−4.3
N2-)2	0.8	X2-CH2COMe	−0.9	X2-POMe2	2.5	N2-POMe2	−4.5
N2-Me	0.6	X2-CH2TMS	−0.9	X2-PMe3I	2.3	N2-COOMe	−4.6
N2-NH2	0.6	X2-Me	−0.9	X2-PO(OMe)2	2.2	N2-COOH	−4.7
N2-Ph-Cl	0.5	X2-CH2COOH	−0.9	X2-P(OMe)2	1.9	N2-PCl2	−4.8
N2-CH2OH	0.5	X2-(CH2)2Br	−0.9	X2-PO(OPh)2	1.9	N2-CN	−4.9
N2-SnMe3	0.5	X2-COOH	−0.9	X2-PCl2	1.6	X2-OTMS	−4.9
N2-OTMS	0.3	X2-OTMS	−1.0	X2-POCl2	1.4	N2-Br	−5.2
N2-I	0.3	X2-PCl2	−1.1	X2-PMe2	1.1	X2-OCHO	−5.2
N2-OH	0.3	X2-PO(OMe)2	−1.1	N2-CX	1.0	X2-OAc	−5.2
N2-OMe	0.1	X2-OMe	−1.1	X2-Et	0.8	X2-OH	−5.2
N2-CH2NH3Cl	0.0	X2-I	−1.2	X2-(CH2)2OH	0.6	N2-PSMe2	−5.3
N2-P(OMe)2	0.0	X2-PSMe2	−1.2	X2-CH2TMS	0.5	X2-OMe	−5.3
N2-PCl2	−0.1	X2-POMe2	−1.2	N2-SnMe3	0.4	N2-PMe2	−5.7
X2-P(OMe)2	−0.1	X2-NH2	−1.2	X2-CH2OH	0.4	X2-OTs	−5.7
X2-SnMe3	−0.2	X2-OCHO	−1.3	X2-CH2COMe	0.4	N2-)2	−6.4
N2-Br	−0.2	X2-OAc	−1.3	X2-Me	0.4	N2-Ph-Cl	−6.8
N2-Cl	−0.2	X2-PO(OPh)2	−1.4	X2-(CH2)2Br	0.4	N2-CH2OH	−7.0
N2-OMS	−0.3	X2-PMe2	−1.4	X2-CH2COOH	0.3	N2-Et	−7.2
N2-POMe2	−0.3	X2-OH	−1.5	X2-COOMe	0.0	N2-CH2NH3Cl	−7.2
N2-COOMe	−0.4	X2-Br	−1.6	X2-COOH	−0.1	N2-Cl	−7.4
N2-PO(OMe)2	−0.4	X2-Cl	−1.6	N2-I	−0.9	N2-Me	−7.5
N2-OAc	−0.4	X2-CN	−1.6	N2-PMe3I	−1.2	X2-F	−7.7
N2-COOH	−0.5	X2-POCl2	−1.7	X2-I	−1.5	N2-NH3Cl	−8.1
X2-CH2OH	−0.5	X2-OTs	−1.7	X2-CN	−1.5	N2-OMS	−8.8
X2-Et	−0.5	X2-F	−2.0	X2-Br	−2.2	N2-OAc	−8.8
X2-(CH2)2OH	−0.6	X2-PMe3I	−2.2	X2-NH2	−3.1	N2-NH2	−9.5
N2-NH3Cl	−0.6	N2-O-3N	−4.1	X2-Cl	−3.1	N2-OTMS	−9.5
N2-PSMe2	−0.7	X2-O-3X	−4.3	N2-P(OMe)2	−3.8	N2-OH	−9.6
N2-CN	−0.7	N2-CX	−7.0	N2-O-3N	−4.1	N2-OMe	−9.6

Table 2-2 (cont.)

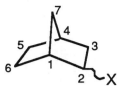

Substituent	γ C-7	Substituent	γ C-7
N2-O-3N	12.1	X2-PO(OMe)2	−1.2
N2-)2	4.8	X2-PO(OPh)2	−1.3
N2-PMe3I	4.6	X2-PCl2	−1.3
N2-PSMe2	3.6	X2-CN	−1.3
N2-POMe2	3.1	X2-PSMe2	−1.4
N2-CX	2.5	N2-OMe	−1.4
N2-P(OMe)2	2.4	X2-COOH	−1.7
N2-PO(OMe)2	2.4	X2-COOMe	−1.8
N2-SnMe3	2.3	N2-I	−2.0
N2-PCl2	2.2	X2-I	−2.3
N2-COOMe	2.0	X2-PMe2	−2.3
N2-COOH	2.0	X2-Br	−2.8
N2-Me	1.9	X2-OTMS	−2.8
N2-Et	1.8	X2-(CH2)2OH	−2.9
N2-OTMS	1.7	X2-Et	−2.9
N2-CH2OH	1.5	X2-CH2COMe	−2.9
N2-PMe2	1.2	X2-OCHO	−2.9
N2-Br	0.9	X2-CH2OH	−3.0
X2-POCl2	0.7	X2-OAc	−3.0
N2-NH2	0.3	X2-CH2COOH	−3.1
X2-SnMe3	0.2	X2-OMe	−3.2
N2-NH3Cl	0.1	X2-Cl	−3.3
N2-CN	0.0	X2-OTs	−3.3
N2-CH2NH3Cl	−0.2	X2-CH2TMS	−3.4
N2-Cl	−0.4	X2-Me	−3.5
X2-PMe3I	−0.6	X2-F	−3.8
X2-(CH2)2Br	−0.7	X2-OH	−3.9
N2-OH	−0.7	N2-Ph-Cl	−3.9
X2-POMe2	−1.0	X2-NH2	−4.4
N2-OMS	−1.1	X2-O-3X	−12.0
N2-OAc	−1.1		

Table 2-2 (cont.)

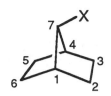

	β	γ-gauche		γ-anti		α	
Substituents	C-1	Substituents	C-2	Substituents	C-5	Substituents	C-7
7-I	7.4	7-SnMe3	0.5	7-PMe2	2.1	7-F	60.7
7-Br	6.2	7-COOMe	−1.5	7-PCl2	1.9	7-OH	41.4
7-PCl2	5.9	7-I	−1.6	7-SnMe3	1.8	7-OTMS	39.9
7-Cl	5.8	7-COOH	−1.6	7-CH2OH	1.4	7-Cl	27.8
7-CH2TMS	4.9	7-Ph	−1.7	7-Ph	1.3	7-PCl2	27.1
7-Me	4.2	7-CH2OH	−1.8	7-Me	0.9	7-Br	19.8
7-OH	4.1	7-PMe2	−2.0	7-CH2TMS	0.7	7-CH2OH	18.7
7-SnMe3	4.0	7-PCl2	−2.1	7-)2	0.5	7-PMe2	16.6
7-Ph	3.8	7-)2	−2.4	7-COOMe	0.3	7-Ph	15.9
7-OTMS	3.4	7-CH2TMS	−2.5	7-COOH	0.3	7-COOH	15.5
7-PMe2	3.1	7-Br	−2.6	7-Cl	−2.5	7-COOMe	15.4
7-COOH	2.9	7-OH	−2.8	7-Br	−2.6	7-)2	8.8
7-COOMe	2.9	7-Me	−2.9	7-OH	−2.6	7-CH2TMS	7.2
7-F	1.9	7-OTMS	−3.0	7-OTMS	−2.6	7-Me	5.6
7-)2	1.9	7-Cl	−3.2	7-I	−3.0	7-SnMe3	0.5
7-CH2OH	1.4	7-F	−3.4	7-F	−4.7	7-I	−2.9

Table 2-3

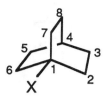

α		β		γ		δ	
Substituent	C-1	Substituent	C-2	Substituent	C-3	Substituent	C-4
1-F	69.8	1-I	14.2	1-I	3.5	1-Me	0.7
1-OAc	56.0	1-Br	11.3	1-Br	2.8	1-Ph	0.6
1-OMe	49.0	1-Cl	9.9	1-Cl	1.9	1-COOH	0.6
1-OH	44.5	1-Me	7.3	1-F	1.1	1-CH2OH	0.5
1-Cl	43.6	1-OH	6.5	1-SnMe3	1.0	1-COMe	0.1
1-Br	40.7	1-Ph	6.1	1-Me	0.7	1-F	0.0
1-I	23.2	1-F	4.9	1-Ph	0.5	1-OMe	0.0
1-COMe	20.5	1-SnMe3	3.7	1-COOH	−0.1	1-OH	−0.1
1-COOH	14.4	1-COOH	2.6	1-OH	−0.2	1-OAc	−0.3
1-Ph	10.1	1-OAc	2.4	1-OMe	−0.3	1-SnMe3	−0.7
1-CH2OH	8.3	1-OMe	2.1	1-OAc	−0.5	1-Cl	−1.0
1-Me	3.4	1-CH2OH	1.5	1-CH2OH	−0.5	1-Br	−1.6
1-SnMe3	−2.6	1-COMe	1.4	1-COMe	−0.6	1-I	−2.8

Table 2-3 (cont.)

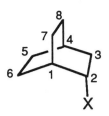

β		α		β		γ	
Substituent	C-1	Substituent	C-2	Substituent	C-3	Substituent	C-4
M2-OH	7.7	M2-OH	43.4	M2-OH	11.4	M2-Et	1.5
M2-Me	6.4	M2-COMe	23.4	M2-Me	9.7	M2-Me	1.2
M2-Et	4.7	M2-CHO	23.1	M2-Et	8.6	M2-OH	0.8
M2-COOH	3.5	M2-*CHOHMe	17.7	M2-*CHOHMe	4.5	M2-CH2OH	0.4
M2-COOEt	3.3	M2-*CHOHMe	17.7	M2-*CHOHMe	4.3	M2-*CHOHMe	0.0
M2-COMe	2.3	M2-COOH	15.8	M2-CH2OH	4.1	M2-*CHOHMe	–0.3
M2-*CHOHMe	1.8	M2-COOEt	15.5	M2-COOH	1.9	M2-COOH	–0.3
M2-CHO	1.2	M2-Et	12.4	M2-COOEt	1.8	M2-COOEt	–0.5
M2-CH2OH	0.8	M2-CH2OH	12.3	M2-COMe	–0.2	M2-CHO	–0.5
M2-*CHOHMe	0.2	M2-Me	4.3	M2-CHO	–1.4	M2-COMe	–0.6

(* diastereomeric pairs)
For an explanation of the stereochemical descriptors M and G see Chapter 6.

δ		γ		γ		δ	
Substituents	C-5	Substituents	C-6	Substituents	C-7	Substituents	C-8
M2-Et	0.6	E2-=CHMe	1.2	M2-Et	1.9	E2-=CHMe	0.1
M2-CH=CH2	0.3	2-=CH2	0.7	M2-Me	1.4	Z2-=CHMe	0.1
Z2-=CHMe	0.1	Z2-=CHMe	0.3	E2-=CHMe	1.2	2-=CH2	–0.1
E2-=CHMe	0.1	E2-=NOH	–0.8	M2-CH=CH2	1.1	M2-Et	–0.2
M2-Me	0.1	2-=O	–2.7	M2-CHOHMe	1.0	M2-CH=CH2	–0.4
M2-CH2OH	0.0	M2-COOH	–4.2	M2-CH2OH	0.9	M2-CH2OH	–0.5
2-=CH2	–0.1	M2-CHO	–4.5	2-=CH2	0.7	M2-Me	–0.9
M2-CHOHMe	–0.4	M2-COOEt	–4.5	Z2-=CHMe	0.3	M2-COOH	–1.0
M2-OH	–0.5	M2-Et	–4.9	M2-CHOHMe	0.3	M2-CHO	–1.1
M2-CHOHMe	–0.5	M2-CH=CH2	–4.9	M2-COOH	0.2	E2-=NOH	–1.1
M2-COOH	–0.8	M2-COMe	–5.1	M2-CHO	–0.1	M2-COOEt	–1.2
E2-=NOH	–1.1	M2-CH2OH	–5.4	M2-COOEt	–0.1	M2-CHOHMe	–1.3
M2-CHO	–1.1	M2-CHOHMe	–5.5	M2-COMe	–0.2	2-=O	–1.3
M2-COOEt	–1.1	M2-Me	–5.7	E2-=NOH	–0.8	M2-COMe	–1.6
2-=O	–1.3	M2-CHOHMe	–5.8	M2-OH	–2.3	M2-OH	–1.6
M2-COMe	–1.5	M2-OH	–7.5	2-=O	–2.7	M2-CHOHMe	–2.0

Table 2-4

α		β		γ	
Substituent	C-1	Substituent	C-2	Substituent	C-3
1-OH	47.7	1-OH	8.0	1-BMe2	0.1
1-SMe	43.2	1-Me	7.7	1-COOH	0.0
1-CMe2OH	18.9	1-SMe	6.9	1-CMe2OH	−0.1
1-COOH	16.6	1-CH2CH2COMe	5.0	1-OH	−0.2
1-CH2CH2COMe	10.0	1-COOH	3.9	1-SMe	−0.3
1-Me	6.6	1-CMe2OH	3.8	1-Me	−0.4
		1-BMe2	2.7	1-CH2CH2COMe	−0.4

γ		β	
Substituent	C-4	Substituent	C-5
1-BMe2	0.8	1-OH	8.8
1-CMe2OH	0.7	1-Me	7.7
1-Me	0.3	1-SMe	7.7
1-CH2CH2COMe	0.1	1-CH2CH2COMe	6.7
1-SMe	−0.1	1-COOH	6.6
1-COOH	−0.2	1-BMe2	2.7
1-OH	−0.5	1-CMe2OH	1.8

Table 2-4 (cont.)

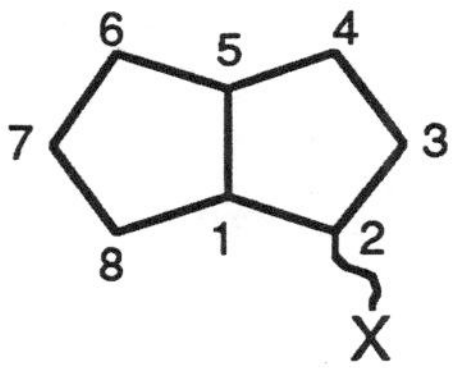

Substituent	β C-1	Substituent	α C-2	Substituent	β C-3	Substituent	γ C-4
X2-OH	9.3	2-=O	189.0	2-=O	11.7	X2-Me	−0.6
2-=O	9.0	2-=CH2	124.6	X2-Me	9.7	X2-COOMe	−0.6
X2-Me	8.7	N2-OMs	51.1	X2-OH	8.0	X2-COOH	−0.6
X2-OAc	6.9	X2-OAc	48.3	2-=CH2	7.9	X2-CH2OH	−0.9
X2-COOH	4.9	X2-OH	45.7	N2-OH	7.8	X2-CHO	−1.0
X2-COOMe	4.9	N2-OAc	43.6	N2-Me	6.8	N2-CH2OH	−1.5
2-=CH2	4.7	N2-OH	41.1	N2-OMs	6.2	N2-Me	−1.8
N2-Me	4.7	X2-CHO	25.9	X2-OAc	5.2	N2-COOMe	−2.1
N2-OH	4.3	N2-CHO	22.5	X2-COOMe	5.2	N2-CHO	−2.1
X2-CH2OH	3.3	X2-COOH	17.8	N2-OAc	5.1	N2-COOH	−2.2
N2-OMs	3.3	X2-COOMe	17.8	X2-COOH	5.1	2-=CH2	−2.4
N2-COOMe	2.7	X2-CH2OH	16.3	X2-CH2OH	4.5	X2-OAc	−2.4
N2-COOH	2.6	N2-COOH	14.7	N2-COOMe	3.8	X2-OH	−2.5
N2-OAc	2.5	N2-COOMe	14.6	N2-COOH	3.7	N2-OH	−4.9
N2-CH2OH	1.9	N2-CH2OH	12.2	N2-CHO	3.1	N2-OAc	−4.9
N2-CHO	1.5	X2-Me	8.1	X2-CHO	1.7	N2-OMs	−5.3
X2-CHO	1.2	N2-Me	3.7	N2-CH2OH	1.6	2-=O	−7.7

Substituent	γ C-5	Substituent	δ C-6	Substituent	δ C-7	Substituent	γ C-8
X2-COOH	0.5	N2-Me	1.5	N2-Me	1.5	2-=CH2	−0.7
X2-CHO	0.5	N2-COOH	1.0	N2-CH2OH	1.2	X2-CH2OH	−1.0
N2-CHO	0.5	N2-CH2OH	1.0	N2-OMs	1.2	X2-COOH	−1.0
2-=CH2	0.5	N2-COOMe	0.9	N2-OAc	1.1	X2-COOMe	−1.0
X2-COOMe	0.4	N2-OH	0.9	X2-OAc	0.6	X2-CHO	−1.1
X2-Me	0.2	N2-OAc	0.5	N2-COOMe	0.5	X2-Me	−1.9
X2-CH2OH	0.2	N2-OMs	0.4	N2-OH	0.3	X2-OAc	−3.5
N2-Me	0.0	N2-CHO	0.1	X2-OH	0.3	X2-OH	−3.8
N2-COOMe	0.0	X2-OAc	0.1	N2-COOH	0.2	2-=O	−4.3
N2-COOH	−0.1	X2-OH	0.1	2-=CH2	0.2	N2-CHO	−6.3
N2-CH2OH	−0.3	X2-Me	−0.2	2-=O	0.1	N2-Me	−6.4
N2-OH	−0.6	X2-CH2OH	−0.5	X2-CHO	−0.3	N2-OH	−6.4
N2-OAc	−0.7	X2-COOH	−0.6	X2-CH2OH	−0.5	N2-OMs	−6.5
N2-OMs	−0.9	X2-CHO	−0.6	X2-COOH	−0.7	N2-OAc	−6.6
X2-OAc	−1.0	X2-COOMe	−0.6	X2-COOMe	−0.7	N2-COOMe	−6.6
X2-OH	−1.3	2-=CH2	−0.7	X2-Me	−0.8	N2-COOH	−6.6
2-=O	−2.1	2-=O	−0.7	N2-CHO	−1.5	N2-CH2OH	−6.7

Table 2-4 (cont.)

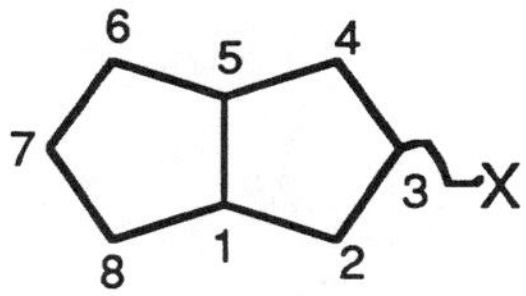

	γ		β		α
Substituent	C-1	Substituent	C-2	Substituent	C-3
X3-OAc	−1.8	3-=O	10.5	3-=O	194.6
X3-OH	−2.2	X3-OH	8.5	N3-OBs	57.9
N3-OAc	−2.8	N3-OH	8.2	X3-OAc	52.0
N3-OH	−3.0	X3-OAc	5.6	N3-OAc	50.3
N3-OBs	−3.4	N3-OBs	4.9	X3-OH	48.5
3-=O	−3.5	N3-OAc	4.7	N3-OH	47.5

	δ		ε
Substituent	C-6	Substituent	C-7
X3-OAc	−0.3	X3-OAc	−0.5
X3-OH	−0.3	X3-OH	−0.5
N3-OAc	−0.4	3-=O	−0.7
N3-OH	−0.5	N3-OAc	−0.9
3-=O	−0.7	N3-OBs	−1.1
N3-OBs	−0.7	N3-OH	−1.2

Table 2-5

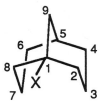

α		β		γ		δ	
Substituent	C-1	Substituent	C-2	Substituent	C3	Substituent	C4
1-F	64.7	1-I	15.3	1-I	3.8	1-CH2OH	−0.5
1-OMe	45.2	1-Br	12.0	1-Br	2.8	1-OMe	−0.9
1-Cl	42.6	1-Cl	10.5	1-Cl	1.8	1-COOMe	−1.0
1-Br	41.6	1-OH	7.9	1-F	0.7	1-F	−1.1
1-OH	41.0	1-F	5.7	1-OMe	0.7	1-CH2OTs	−1.1
1-I	27.4	1-OMe	3.5	1-OH	0.4	1-OH	−1.4
1-COOMe	13.0	1-COOMe	1.8	1-COOMe	−0.5	1-COOH	−1.5
1-COOH	12.8	1-CH2OH	1.5	1-CH2OH	−0.6	1-Br	−2.0
1-CH2OH	6.4	1-COOH	1.3	1-COOH	−0.8	1-Cl	−2.0
1-CH2OTs	5.0	1-CH2OTs	1.0	1-CH2OTs	−1.3	1-I	−2.1

γ		β	
Substituent	C5	Substituent	C9
1-F	5.5	1-I	15.2
1-Br	5.4	1-Br	12.1
1-I	5.3	1-Cl	11.0
1-Cl	4.9	1-OH	8.4
1-OH	4.2	1-F	6.5
1-OMe	3.8	1-OMe	4.5
1-CH2OH	0.6	1-CH2OH	1.4
1-COOMe	0.4	1-COOMe	1.3
1-CH2OTs	−0.2	1-CH2OTs	0.9
1-COOH	−0.2	1-COOH	0.8

Table 2-5 (cont.)

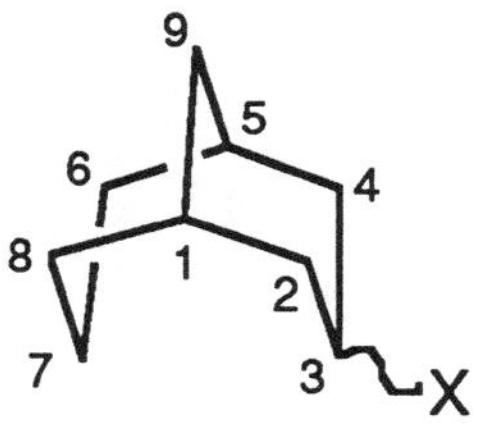

γ		β		α	
Substituent	C-1	Substituent	C-2	Substituent	C-3
X3-Br	3.6	3-=O	15.6	3-=O	190.6
X3-OH	2.9	X3-Br	12.7	3-=CH2	127.6
X3-Cl	2.7	X3-Cl	11.8	N3-OMe	51.4
3-=O	2.3	X3-OH	10.1	X3-OCOPh	48.4
3-=CH2	1.8	X3-Me	10.1	N3-OAC	44.5
X3-OCOPh	1.5	3-=CH2	9.1	X3-OH	44.1
X3-Me	1.2	X3-OCOPh	5.8	N3-OH	42.9
X3-COPh	−0.2	N3-OH	4.5	X3-Cl	35.0
X3-COOMe	−0.5	N3-Me	4.4	X3-Br	28.2
X3-C(Me)2OH	−1.0	X3-COPh	2.7	X3-C(Me)2OH	20.6
N3-OMe	−1.2	X3-COOMe	2.4	X3-COPh	18.8
N3-OAC	−1.3	N3-OMe	1.4	N3-C(Me)2OH	17.7
N3-OH	−1.3	N3-OAC	1.2	X3-COOMe	16.5
N3-Me	−1.8	X3-C(Me)2OH	0.6	N3-COOEt	13.5
N3-C(Me)2OH	−2.6	N3-COOEt	−2.4	N3-COOMe	13.4
N3-COOEt	−3.0	N3-COOMe	−2.5	X3-Me	5.9
N3-COOMe	−3.0	N3-C(Me)2OH	−4.4	N3-Me	3.1

δ		ε		δ	
Substituent	C-6	Substituent	C-7	Substituent	C-9
N3-Me	2.7	X3-Me	2.3	X3-Me	0.4
N3-C(Me)2OH	2.1	X3-Cl	0.3	X3-OH	−0.3
N3-OH	1.6	X3-Br	0.3	3-=CH2	−0.3
N3-COOMe	1.5	X3-OH	0.1	X3-OCOPh	−0.5
N3-OAC	1.4	X3-COPh	0.0	X3-C(Me)2OH	−0.5
N3-COOEt	1.4	X3-OCOPh	−0.1	X3-COPh	−0.7
N3-OMe	1.3	X3-C(Me)2OH	−0.3	X3-Cl	−0.8
3-=CH2	1.2	X3-COOMe	−0.5	X3-Br	−0.9
3-=O	0.4	3-=CH2	−2.5	X3-COOMe	−0.9
X3-Me	0.0	3-=O	−4.4	3-=O	−2.3
X3-C(Me)2OH	−0.1	N3-OAC	−6.0	N3-OMe	−4.2
X3-COPh	−0.5	N3-OMe	−6.0	N3-OAC	−4.8
X3-OH	−0.7	N3-OH	−6.2	N3-Me	−5.3
X3-COOMe	−0.7	N3-Me	−6.6	N3-OH	−5.6
X3-OCOPh	−0.8	N3-C(Me)2OH	−6.6	N3-COOMe	−5.9
X3-Cl	−1.0	N3-COOMe	−6.6	N3-COOEt	−5.9
X3-Br	−1.2	N3-COOEt	−6.6	N3-C(Me)2OH	−6.1

Table 2-6

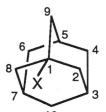

Substituents	α C-1	Substituents	β C2	Substituents	γ C-3	Substituents	δ C4
1-F	63.9	1-I	14.5	1-I	4.4	1-iPr	−0.2
1-NO2	56.0	1-Br	11.6	1-Br	4.1	1-(1-AD)	−0.3
1-OMe	43.4	1-Cl	9.9	1-Cl	3.2	1-Bu	−0.4
1-N=NCF3	42.9	1-SH	9.8	1-F	2.9	1-Pr	−0.4
1-OTMS	42.6	1-NH2	8.3	1-OTMS	2.3	1-SnMe3	−0.4
1-Cl	39.7	1-OTMS	8.0	1-OH	2.3	1-Et	−0.4
1-OH	39.4	1-OH	7.5	1-OMe	2.2	1-CH2Cl	−0.4
1-Br	38.0	1-Me	6.8	1-SH	1.6	1-CH2COMe	−0.5
1-N=C=S	29.9	1-N=C=S	6.0	1-SS(1-AD)	1.6	1-iBu	−0.5
1-NH3Cl	24.7	1-C≡CTMS	5.5	1-NH2	1.3	1-CH2CH2OH	−0.6
1-NHAc-	23.1	1-SS(1-AD)	5.4	1-NH3Cl	1.1	1-CMe=[E]NOH	−0.6
1-I	21.7	1-Pr	5.4	1-NO2	1.0	1-CH2OH	−0.6
1-SS(1-AD)	18.8	1-iBu	5.3	1-NHAc-	0.9	1-CH2F	−0.7
1-NH2	18.6	1-Ph	5.2	1-N=C=S	0.8	1-CH=CH2	−0.8
1-COMe	17.0	1-CH2COMe	5.2	1-CH2COMe	0.7	1-Me	−0.9
1-COOEt	12.1	1-CH2CH2OH	5.1	1-N=NCF3	0.6	1-CH2COOH	−1.0
1-COOH	11.9	1-F	4.8	1-Ph	0.5	1-OMe	−1.1
1-CMe=[E]NOH	11.3	1-Bu	4.8	1-Bu	0.5	1-Ph	−1.1
1-C(Me)2OH	10.1	1-CH2COOH	4.6	1-iPr	0.5	1-CH2I	−1.1
1-iPr	7.9	1-CH2I	4.3	1-(1-AD)	0.5	1-CH2Br	−1.1
1-(1-AD)	7.8	1-Et	4.3	1-Me	0.4	1-C≡CTMS	−1.2
1-SH	7.7	1-CH2CN	4.1	1-Pr	0.4	1-COOEt	−1.3
1-Ph	7.6	1-CH=CH2	4.0	1-iBu	0.4	1-NHAc-	−1.4
1-CH=CH2	7.0	1-SnMe3	3.8	1-SnMe3	0.4	1-CH2CN	−1.4
1-CH2OH	6.0	1-NHAc-	3.7	1-CH2CH2OH	0.3	1-N=NCF3	−1.5
1-CH2Br	5.1	1-OMe	3.4	1-CH2COOH	0.3	1-SS(1-AD)	−1.6
1-CH2Cl	5.0	1-NH3Cl	2.9	1-Et	0.3	1-NH2	−1.6
1-CH2COMe	5.0	1-CH2Br	2.9	1-CH2I	0.2	1-COOH	−1.6
1-iBu	4.7	1-NO2	2.7	1-CMe=[E]NOH	0.2	1-OH	−1.7
1-CH2COOH	4.3	1-N=NCF3	2.0	1-CH2Cl	0.1	1-C(Me)2OH	−1.7
1-Pr	4.0	1-CMe=[E]NOH	2.0	1-C(Me)2OH	0.1	1-OTMS	−1.8
1-CH2I	3.9	1-CH2Cl	1.8	1-CH=CH2	0.0	1-SH	−1.9
1-Bu	3.8	1-iPr	1.6	1-C≡CTMS	−0.1	1-F	−2.0
1-Et	3.8	1-CH2OH	1.3	1-CH2CN	−0.1	1-Br	−2.1
1-CH2CH2OH	3.5	1-CN	1.3	1-CH2Br	−0.1	1-COMe	−2.1
1-C≡CTMS	2.2	1-COOEt	1.0	1-CH2OH	−0.3	1-Cl	−2.2
1-Me	1.2	1-COOH	0.5	1-COOEt	−0.4	1-NH3Cl	−2.2
1-CN	1.0	1-CH2F	0.5	1-CH2F	−0.5	1-N=C=S	−2.2
1-CH2F	0.1	1-COMe	−0.5	1-COOH	−0.7	1-I	−2.3
1-SnMe3	−0.7	1-C(Me)2OH	−0.7	1-COMe	−1.3	1-NO2	−2.6
		1-(1-AD)	−2.5	1-CN	−2.1	1-CN	−2.9

Table 2-6 (cont.)

Substituents	β C2	Substituents	γ C-3	Substituents	δ C4	Substituents	β C-1
1-I	14.5	1-I	4.4	1-iPr	−0.2	2-=S	28.3
1-Br	11.6	1-Br	4.1	1-(1-AD)	−0.3	2-=O	18.4
1-Cl	9.9	1-Cl	3.2	1-Bu	−0.4	2-=CH2	10.8
1-SH	9.8	1-F	2.9	1-Pr	−0.4	2-=C(CN)2	10.3
1-NH2	8.3	1-OTMS	2.3	1-SnMe3	−0.4	2-I	9.1
1-OTMS	8.0	1-OH	2.3	1-Et	−0.4	2-Br	8.0
1-OH	7.5	1-OMe	2.2	1-CH2Cl	−0.4	2-Cl	7.5
1-Me	6.8	1-SH	1.6	1-CH2COMe	−0.5	2-SH	7.2
1-N=C=S	6.0	1-SS(1-AD)	1.6	1-iBu	−0.5	2-OTMS	6.8
1-C≡CTMS	5.5	1-NH2	1.3	1-CH2CH2OH	−0.6	2-NH2	6.3
1-SS(1-AD)	5.4	1-NH3Cl	1.1	1-CMe=[E]NOH	−0.6	2-OH	6.2
1-Pr	5.4	1-NO2	1.0	1-CH2OH	−0.6	2-Me	5.5
1-iBu	5.3	1-NHAc-	0.9	1-CH2F	−0.7	2-SEt	5.0
1-Ph	5.2	1-N=C=S	0.8	1-CH=CH2	−0.8	2-NH(2-AD)	4.7
1-CH2COMe	5.2	1-CH2COMe	0.7	1-Me	−0.9	2-OMs	4.5
1-CH2CH2OH	5.1	1-N=NCF3	0.6	1-CH2COOH	−1.0	2-Et	4.4
1-F	4.8	1-Ph	0.5	1-OMe	−1.1	2-F	4.4
1-Bu	4.8	1-Bu	0.5	1-Ph	−1.1	2-SMe	4.2
1-CH2COOH	4.6	1-iPr	0.5	1-CH2I	−1.1	2-SnPh3	3.9
1-CH2I	4.3	1-(1-AD)	0.5	1-CH2Br	−1.1	2-=AD	3.7
1-Et	4.3	1-Me	0.4	1-C≡CTMS	−1.2	2-SnMe3	3.6
1-CH2CN	4.1	1-Pr	0.4	1-COOEt	−1.3	2-NHAc-	3.6
1-CH=CH2	4.0	1-iBu	0.4	1-NHAc-	−1.4	2-NHCOPh	3.5
1-SnMe3	3.8	1-SnMe3	0.4	1-CH2CN	−1.4	2-OAc-	3.5
1-NHAc-	3.7	1-CH2CH2OH	0.3	1-N=NCF3	−1.5	2-Pr	3.4
1-OMe	3.4	1-CH2COOH	0.3	1-SS(1-AD)	−1.6	2-N3	3.4
1-NH3Cl	2.9	1-Et	0.3	1-NH2	−1.6	2-N(CH2)4	3.2
1-CH2Br	2.9	1-CH2I	0.2	1-COOH	−1.6	2-=CHMe	3.1
1-NO2	2.7	1-CMe=[E]NOH	0.2	1-OH	−1.7	2-OMe	3.0
1-N=NCF3	2.0	1-CH2Cl	0.1	1-C(Me)2OH	−1.7	2-Ph	2.6
1-CMe=[E]NOH	2.0	1-C(Me)2OH	0.1	1-OTMS	−1.8	2-NH3Cl	1.7
1-CH2Cl	1.8	1-CH=CH2	0.0	1-SH	−1.9	2-CN	1.5
1-iPr	1.6	1-C≡CTMS	−0.1	1-F	−2.0	2-COOMe	1.3
1-CH2OH	1.3	1-CH2CN	−0.1	1-Br	−2.1	2-NMe2	1.3
1-CN	1.3	1-CH2Br	−0.1	1-COMe	−2.1	2-=NOMe	1.1
1-COOEt	1.0	1-CH2OH	−0.3	1-Cl	−2.2	2-SiMe3	1.0
1-COOH	0.5	1-COOEt	−0.4	1-NH3Cl	−2.2	2-COOH	1.0
1-CH2F	0.5	1-CH2F	−0.5	1-N=C=S	−2.2	2-CH2OH	1.0
1-COMe	−0.5	1-COOH	−0.7	1-I	−2.3	2-N(CH2)5	0.8
1-C(Me)2OH	−0.7	1-COMe	−1.3	1-NO2	−2.6	2-CH2OTs	0.4
1-(1-AD)	−2.5	1-CN	−2.1	1-CN	−2.9	2-(2-AD)	−0.4

Table 2-6 (cont.)

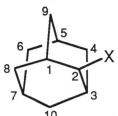

Substituents	ε C-6	Substituents	δ C-7	Substituents	γ C-8
2-Et	1.9	2-Et	1.1	2-SiMe3	3.5
2-Me	0.9	2-=AD	0.4	2-SnMe3	3.1
2-CH2OH	0.6	2-=CH2	0.1	2-SnPh3	2.9
2-(2-AD)	0.6	2-Me	−0.2	2-=S	2.9
2-SnMe3	0.4	2-CH2OH	−0.2	2-Et	2.7
2-I	0.4	2-SnMe3	−0.2	2-=CH2	2.1
2-NH(2-AD)	0.3	2-SiMe3	−0.2	2-=CCN2	1.9
2-N(CH2)4	0.3	2-SnPh3	−0.3	2-=AD	1.8
2-N(CH2)5	0.2	2-Pr	−0.3	2-Me	1.8
2-SnPh3	0.2	2-(2-AD)	−0.5	2-(2-AD)	1.8
2-SMe	0.2	2-NH(2-AD)	−0.5	2-Pr	1.6
2-SEt	0.2	2-=NOMe	−0.5	2-CH2OH	1.4
2-Br	0.2	2-SMe	−0.6	2-Ph	1.4
2-SiMe3	0.1	2-Ph	−0.7	2-=O	1.4
2-Ph	0.1	2-COOMe	−0.7	2-=NOMe	1.3
2-CH2OTs	0.1	2-N(CH2)4	−0.8	2-SMe	1.1
2-SH	0.1	2-N(CH2)5	−0.8	2-SEt	1.1
2-Cl	0.1	2-SEt	−0.8	2-SH	1.1
2-NMe2	0.1	2-CH2OTs	−0.8	2-I	1.0
2-OH	−0.1	2-OMe	−0.9	2-Br	1.0
2-OTMS	−0.1	2-=O	−0.9	2-CH2OTs	0.6
2-OMe	−0.1	2-COOH	−1.0	2-COOMe	0.5
2-NHAc-	−0.1	2-NMe2	−1.2	2-Cl	0.5
2-COOMe	−0.2	2-OH	−1.2	2-COOH	0.4
2-NHCOPh	−0.2	2-NHAc-	−1.2	2-NH(2-AD)	0.0
2-=AD	−0.3	2-F	−1.2	2-NMe2	0.0
2-=CH2	−0.3	2-OTMS	−1.3	2-N(CH2)5	−0.2
2-OAc-	−0.3	2-OAc-	−1.3	2-N(CH2)4	−0.5
2-NH2	−0.3	2-=S	−1.3	2-NHAc-	−0.5
2-COOH	−0.4	2-=CCN2	−1.3	2-NH2	−0.5
2-N3	−0.4	2-I	−1.4	2-NHCOPh	−0.6
2-F	−0.5	2-SH	−1.4	2-N3	−1.0
2-OMs	−0.7	2-NHCOPh	−1.4	2-OMe	−1.2
2-NH3Cl	−0.8	2-N3	−1.4	2-OH	−1.2
2-=NOMe	−1.1	2-Br	−1.5	2-OTMS	−1.3
2-=O	−1.5	2-Cl	−1.5	2-OAc-	−1.3
2-=S	−1.6	2-NH2	−1.5	2-NH3Cl	−1.3
2-CN	−1.6	2-OMs	−1.9	2-OMs	−1.5
2-=CCN2	−2.0	2-NH3Cl	−1.9	2-CN	−1.8
2-Pr	−2.8	2-CN	−2.2	2-F	−2.0

Table 2-7

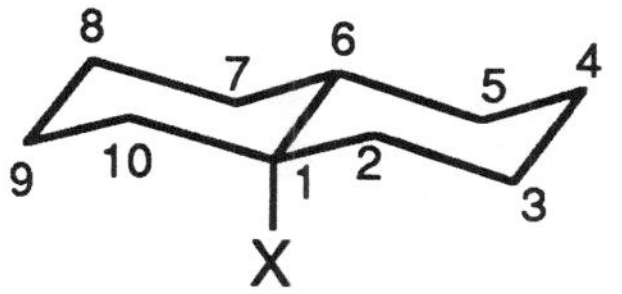

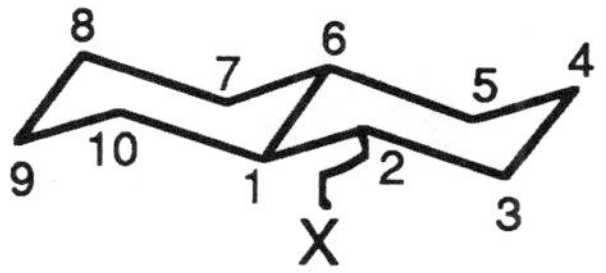

Substituents	α C-1	Substituents	β C-2	Substituents	γ C-3
1-OH	26.5	1-I	11.4	1-I	−2.0
1-F	49.5	1-Br	8.9	1-COOH	−3.3
1-Cl	33.6	1-Me	7.8	1-Br	−3.8
1-Br	36.8	1-Cl	7.8	1-Cl	−4.7
1-I	31.8	1-OH	5.6	1-Me	−4.8
1-COOH	5.1	1-COOH	4.5	1-F	−4.8
1-Me	−9.8	1-F	3.3	1-OH	−5.2

Substituents	δ C-4	Substituents	γ C-5	Substituents	β C-6
1-Me	0.3	1-I	−1.8	1-I	5.0
1-COOH	−0.1	1-Br	−4.0	1-Br	4.1
1-I	−0.4	1-COOH	−4.9	1-Cl	3.1
1-Br	−0.5	1-Me	−5.0	1-Me	2.1
1-Cl	−0.5	1-Cl	−5.2	1-COOH	2.0
1-F	−0.5	1-F	−5.2	1-OH	0.6
1-OH	−0.5	1-OH	−5.6	1-F	0.1

Substituents	β C-1	Substituents	α C-2	Substituents	β C-3
E2-OH	6.7	E2-OH	40.3	E2-Me	9.5
E2-Me	6.0	A2-OH	36.1	E2-OH	8.9
A2-OH	3.7	E2-NMe2	33.1	A2-Me	7.5
A2-Me	2.8	E2-COOMe	15.9	A2-OH	7.4
E2-NMe2	2.2	A2-COOMe	10.2	E2-COOMe	3.4
A2-COOMe	1.0	E2-Me	3.4	E2-NMe2	2.7
E2-COOMe	0.9	A2-Me	−1.3	A2-COOMe	2.5

Substituents	γ C-4	Substituents	δ C-5	Substituents	γ C-6
E2-Me	−0.5	A2-Me	0.7	E2-Me	−0.7
E2-COOMe	−1.5	A2-COOMe	0.5	E2-NMe2	−0.7
E2-NMe2	−1.7	E2-Me	0.3	E2-COOMe	−1.6
E2-OH	−2.8	E2-NMe2	0.1	E2-OH	−2.5
A2-COOMe	−5.1	A2-OH	−0.5	A2-COOMe	−7.4
A2-Me	−6.2	E2-OH	−0.6	A2-Me	−7.9
A2-OH	−6.9	E2-COOMe	−0.8	A2-OH	−8.1

Table 2-7 (cont.)

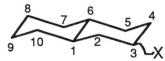

	γ		β		α
Substituents	C-1	Substituents	C-2	Substituents	C-3
E3-Me	−0.6	E3-Me	8.7	E3-OH	43.3
E3-COOMe	−1.3	E3-OH	8.7	A3-OH	39.7
E3-OH	−2.5	A3-OH	6.1	A3-NMe2	35.0
A3-COOMe	−4.2	A3-Me	5.6	E3-COOMe	15.7
A3-Me	−6.7	A3-NMe2	2.8	A3-COOMe	12.2
A3-NMe2	−7.0	E3-COOMe	2.0	E3-Me	6.0
A3-OH	−7.3	A3-COOMe	0.3	A3-Me	1.5

	β		γ		δ
Substituents	C-4	Substituents	C-5	Substituents	C-6
E3-OH	8.7	E3-Me	−0.4	A3-NMe2	0.5
E3-Me	8.6	E3-COOMe	−1.2	A3-Me	0.5
A3-OH	7.0	E3-OH	−2.3	E3-COOMe	−0.3
A3-Me	5.2	A3-COOMe	−4.1	A3-OH	−0.5
E3-COOMe	2.2	A3-NMe2	−4.8	E3-Me	−0.6
A3-NMe2	1.1	A3-Me	−6.3	A3-COOMe	−0.6
A3-COOMe	0.4	A3-OH	−6.7	E3-OH	−1.4

References

1. Crecely, K.M., Crecely, R.W. and Goldstein, J.H., *J. Phys. Chem.* **74**, 2680 (1970).

2. Eliel, E.L. and Pietrusiewicz, K.M., *Org. Magn. Reson.* **13**, 193 (1980).

3. *Sadtler Standard Carbon-13 NMR Spectra*, Sadtler Research Laboratories, Philadelphia, Pa 19104 (1976-1982).

4. Gunther, H. and Seel, H., *Org. Magn. Reson.* **8**, 299 (1976).

5. Thomas, F. and Schneider, H.-J., *Ann. Univ. Sarav., Math.-Naturwis. Fak.* **16**, 53 (1981).

6. Casanova, J., Zahra, J.-P. and Waegell, B., *Org. Magn. Reson.* **13**, 299 (1980).

7. Wiberg, K.B., Pratt, W.E. and Bailey, W.F., *J. Org. Chem.* **45**, 4936 (1980).

8. Subbotin, O.A. and Sergeev, N.M., *J. Struct. Chem.* **16**, 838 (1975).

9. Kleinpeter, E., Haufe, G. and Borsdorf, R., *J. Prakt. Chem.* **322**, 125 (1980).

10. Schneider, H.-J., Nguyen-ba, N. and Thomas, F., *Tetrahedron* **38**, 2327 (1982).

11. Kalabin, G.A., Kushnarev, D.F., Shostakovskii, S.M. and Voropaeva, T.K., *Bull. Acad. Sci. USSR, Div. Chem. Sci.* **24**, 2346 (1975).

12. Dorman, D.E., Jautelat, M. and Roberts, J.D., *J. Org. Chem.* **36**, 2757 (1971).

Chapter 3

Stereochemical Considerations

One of the most powerful features of ^{13}C NMR spectroscopy is the sensitivity of chemical shifts to the steric environment about each carbon. While steric effects are also important in proton spectroscopy, the superposition of both bimolecular and intramolecular interactions makes partitioning of these effects difficult. On the other hand, the chemical shift of carbons (except those bearing polarizable groups such as iodides or carbonyl groups) are relatively independent of such external factors as concentration and solvent. Thus, changes in shifts due to steric interactions can be attributed to intramolecular factors and provide a wealth of information concerning stereochemistry, both configurational and conformational. So long as bond angles do not vary significantly, the spatial relationship of a carbon and its α or β substituents does not vary with stereochemistry (three points define a plane) and thus changes in carbon chemical shifts with stereochemistry can be catalogued as caused by γ, δ and more remote groups. In general, it is only the γ groups that are gauche to a carbon that cause significant perturbations.

3.1 Gauche γ effects

The introduction of a second methyl group γ to the methyl of methylcyclohexane (Figure 3-1, next page) results in an upfield shift of the methyl group of –2.5 and represents an example of quite general perturbations due to gauche butane-like interactions. A similar shift of –2.7 is observed upon introduction of a second methyl group to equatorial 3-methyl-trans-decalin (Figure 3-2).

A somewhat larger upfield shift (–4.1) is observed upon in the introduction of a γ oxygen in trans-2-methylcyclohexanol and it is generally the case that shifts due to the presence of a γ oxygen are larger than observed with a γ carbon. The upfield shift of the carbon experiencing a gauche butane-like interaction with a γ substituent is believed to be caused by the overlap of electron density of hydrogen atoms on the terminal atoms, resulting in a shift of σ-bond electron density toward the two terminal atoms of the four atom system.

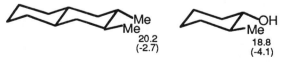

Figure 3-2

Figure 3-1

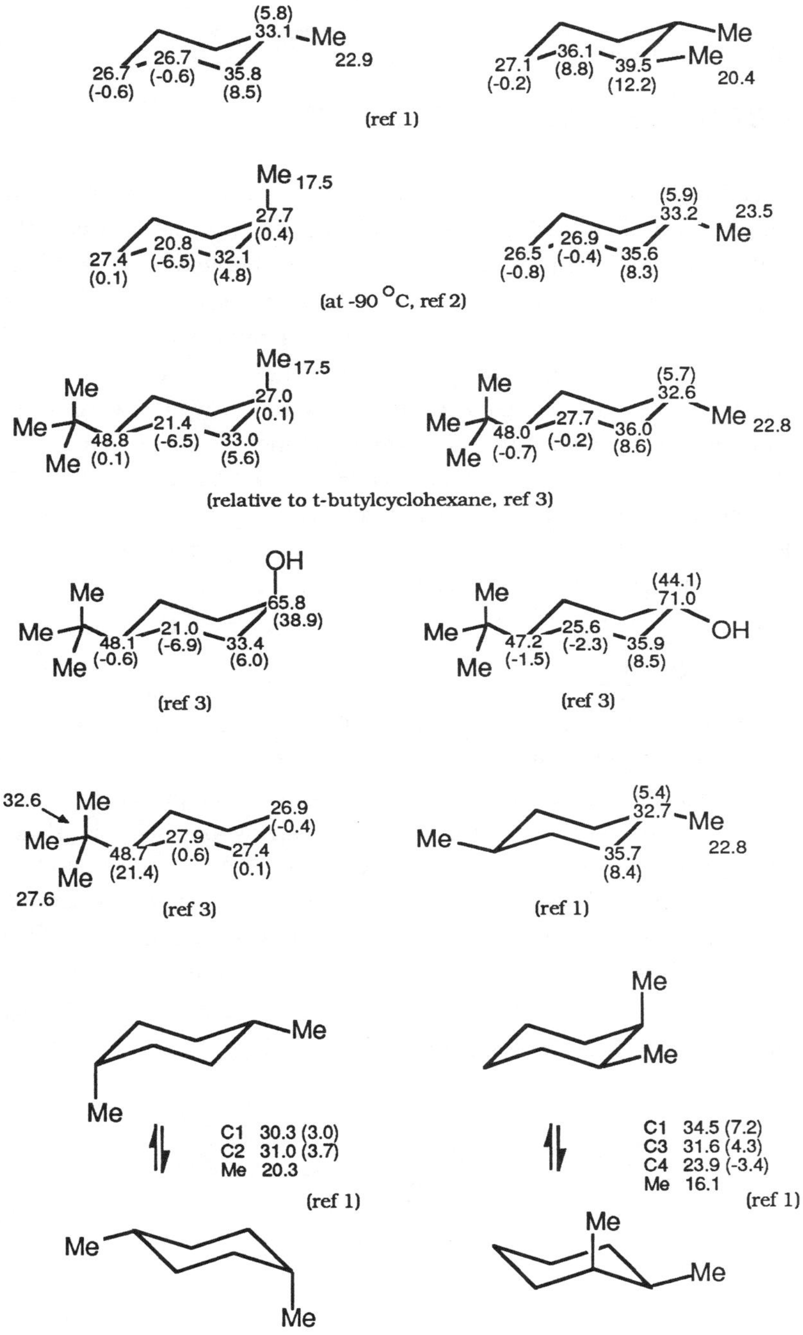

(ref 1)

(at -90 °C, ref 2)

(relative to t-butylcyclohexane, ref 3)

(ref 3)   (ref 3)

(ref 3)   (ref 1)

C1 30.3 (3.0)
C2 31.0 (3.7)
Me 20.3
(ref 1)

C1 34.5 (7.2)
C3 31.6 (4.3)
C4 23.9 (-3.4)
Me 16.1
(ref 1)

(all shift-of-shift values are relative to cyclohexane at 27.3 unless otherwise indicated)

The effects resulting from the stereochemical change from equatorial to axial methylcyclohexane are also quite dramatic, with the ring carbons 1, 2, and 6 shifted upfield in addition to the methyl group and ring carbons 3 and 5. Axially substituted cyclohexanes actually represent rather complicated examples of the gauche butane effect. The large upfield shift of the methyl group (–6.0) is the result of two gauche butane interactions, with both ring carbons 3 and 5 (Figure 3-3). While each of these latter carbons is only gauche to the methyl group, there appears to be a synergistic effect in the interaction of three hydrogens, one from the methyl group and the two axial hydrogens on C-3 and C-5 because the former is flanked by the latter. The resulting upfield shifts for C-3 and C-5 are thus quite large (–6.1). A number of other examples of substituted cyclohexanes are shown in Figure 3-1.

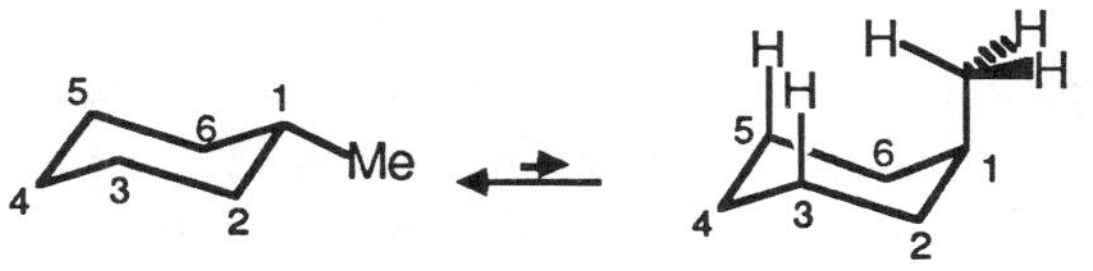

![Figure 3-3 structure]

Figure 3-3

Both simple and cooperative γ interactions are present in axial 2-methyl-trans-decalin (Figure 3.4), with large (and unequal) upfield shifts for carbons 4 and 6 (–6.2 and –7.9), and a smaller shift for carbon 10 (–2.8) that is involved in a simple interaction with the methyl group. In general, the effect on a carbon due to each, non-cooperative gauche butane-like interaction is between –2.5 and –3.0 when both atoms are carbon and somewhat larger when oxygen is involved. A value of –3.0 for the upfield shift due to the presence of a gauche carbon will be used here. The combination of effects, for instance three interactions of the methyl group in axial 2-methyldecalin (–9.5), appears to be nearly linear.

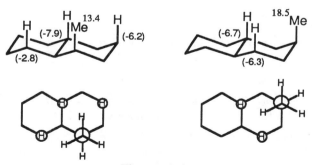

Figure 3-4

In axial methylcyclohexane, the carbon bearing the methyl group (C-1) as well as the C-2 and C-6 ring methylene carbons are also shifted upfield (–5.5 and –3.5) compared to the equatorial isomer. We attribute this latter effect to the change in the number of anti, hydrogen – hydrogen interactions where the loss of each pair results in an upfield shift of approximately 2.5 for each of the carbons involved. Such an effect is seen quite dramatically in comparing the shift of the central carbon of pentane (34.2, 2 α and 2 β substituents) where there are four such interactions (Figure 3-5), with that of cyclohexane where the number of α and β substituents is unchanged but the absorption has shifted upfield 6.9 through the loss of two such interactions.

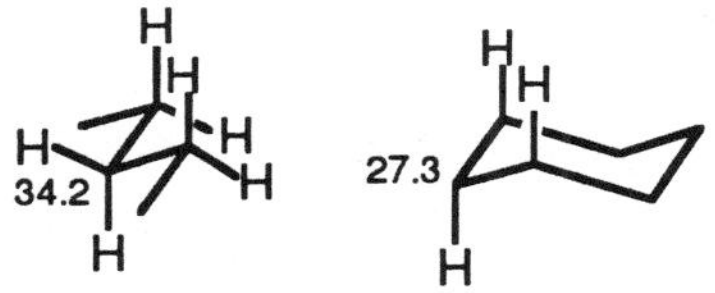

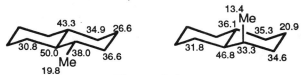

Figure 3-5

These two effects are both critically dependent upon stereochemical relationships and thus can be used to assign both configuration as well as conformation. In so far as the system under consideration is favored in a staggered conformational arrangement, the magnitudes noted above for these effects will be applicable. For instance, the methyl groups and C-1 and C-2 of trans-1,2-dimethylcyclohexane can be treated in the following fashion. Each of the ring carbons is derived from cyclohexane (27.3) with one additional α and β substituent (+16.0). However, the change from 2° to 3° requires a correction of –4.5 (–1.5 for each β) for a predicted shift of 38.8 (39.5 observed). On the other hand, the methyl groups are similar to that in equatorial methylcyclohexane except that each has a new γ substituent that represents a gauche interaction. Thus, the predicted shift would be 19.9 (22.9 – 3.0) which is quite close to the observed shift of 20.4.

Figure 3-6

Consider also the epimeric 2-methyl-trans-decalins (Figure 3-6). In the equatorial isomer the methyl group experiences a gauche interaction with C-10 (22.9 – 3.0 = 19.9 predicted, 19.8 observed) although the introduction of this substituent onto the parent does not change the number of anti-hydrogen interactions. Thus, C-1 is predicted at 50.2 (43.7 + 8.0 – 1.5), 50.0 observed; C-2 at 37.8 (34.3 + 8.0 – 4.5 for 2° to 3° with 3 β), observed 38.0; C-3 at 34.9 (26.9 +8.0), 36.6 observed; and C-10 at 31.3 (34.3 – 3.0 for one gauche), 30.8 observed. Noting the differences between the axial and equatorial isomers, C-1 of the former should be shifted –2.5 for the loss of one H–H interaction, (50.0 – 2.5) 47.5 versus 46.8 observed. Carbon-2 has lost two such interactions (38.0 – 5.0) 33.0 predicted, 33.3 observed, C-3 has lost one H–H (36.6 – 2.5) 34.1 versus 34.6 observed, and C-10 should be unchanged (30.8 predicted, 31.8 observed). The methyl group in the axial isomer now has three gauche interactions (with C-4, C-6, and C-10) as compared to only one in the equatorial isomer and would be predicted to be at 13.8 (19.8 – 2*3.0), 13.4 observed. The introduction of an α substituent usually entails an additional trans H–H interaction, but this effect is incorporated into the α effect of +8.0.

The situation becomes more complicated with the cis isomer of 1,2-dimethylcyclohexane because of the presence of two equal energy (though mirror image) conformations. Thus, the shift of each methyl group is the result of equal contributions from arrangements where the methyl is axial with three gauche interactions, and where it is equatorial, with only one gauche interaction. Each methyl group experiences an average of two gauche interactions (versus none in methylcyclohexane) and would thus be predicted to occur at 17.5 (23.5 – 2*3.0), 16.1 observed. On the other hand, the ring carbons bearing the methyl groups have lost, on average, 1.5 H–H interactions as compared with trans-1,2-dimethylcyclohexane and would be predicted to occur at 35.8 (39.5 –1.5*2.5), 34.5 observed. Each of the ring carbons C-4 and C-5 experiences a gauche interaction with the methyl groups in only one of the two conformations and thus can be estimated to be the average of the values for axial and equatorial methylcyclohexane, 23.9, the same as observed. However, this estimation ignores the small but consistent upfield shift of approximately –0.5 for the carbon across the ring from that bearing an equatorial substituent but the effect here is trivial since C-4 and C-5 each experience this interaction in only one of the two conformations. It should be clear at this point that ^{13}C NMR spectroscopy can provide a wealth of information concerning both configurational and conformational relationships.

3.2 γ Eclipsed interactions

The bias imposed by the bridging present in bicyclo[2.2.1]heptanes and in bicyclo[2.2.2]octanes favors boat, or near boat-like conformations (Figure 3-7). Thus, the C-2 to C-3 (and C-5 to C-6) bond in norbornanes is held rigidly eclipsed. Substituents on adjacent pairs of carbons are thus held in unusual spatial relationships, with cis exo and cis endo groups having a dihedral angle of 0° and trans groups separated by 120°. The relationships of groups on C-1 and C-2 are also atypical, with angles of approximately 42° with exo and 77° with endo substituents. The effect of substituents upon each other follow the pattern expected, that is the smaller the angle, the larger the upfield shift due to the presence of a γ substituent. The effect on C-2 and C-3 when both bear cis related groups is quite large, and may be unrelated to that noted above for gauche interactions resulting from the removal of anti hydrogens.

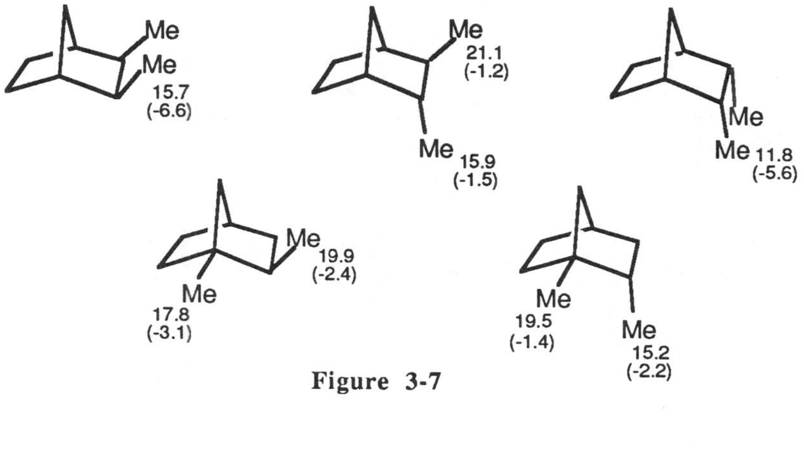

Figure 3-7

3.3 γ Anti interactions

In the majority of situations, a substituent to a carbon that is γ but anti effects only a modest upfield shift. When the substituent is carbon, the shift is usually less than 1.0, and typically only 0.5. The effect is larger with oxygen (–2.5), and thus there is a consistent trend from hydrogen to carbon to oxygen based upon the electronegativity of the substituent, presumably reflecting an increased, inductive withdrawal of electron density by the more electronegative substitutents. Notice, however, that the *direction* of the effect can not be explained by a direct, through bond interaction. It is likely that the effect is transmitted indirectly through the hydrogen on the carbon bearing the substituent. Thus, the effect on C-6 of an exo-2-halogen substituent in the bicyclo[2.2.1]heptanes is an increasing upfield shift with increasing electronegativity (Figure 3-8). On the other hand, the effect on C-3 of a C-1 substituent halogen atom is in the opposite direction (downfield) although the trend with electronegativity is the same.

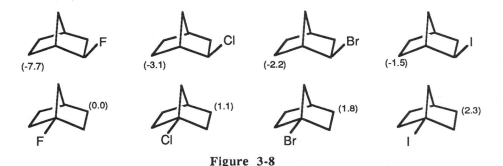

Figure 3-8

In most circumstances, the effect of two γ substituents that are on the same carbon is not additive, and it appears that the effect due to the combination of two geminal groups is somewhat less than the sum of the individual shift effects. Errors may be especially large where an anti oxygen functionality is one of the two substituents involved since, as mentioned above, it is likely that the relatively large effect of anti γ groups is transmitted through a hydrogen. The examples provided in Figure 3-9 are typical.

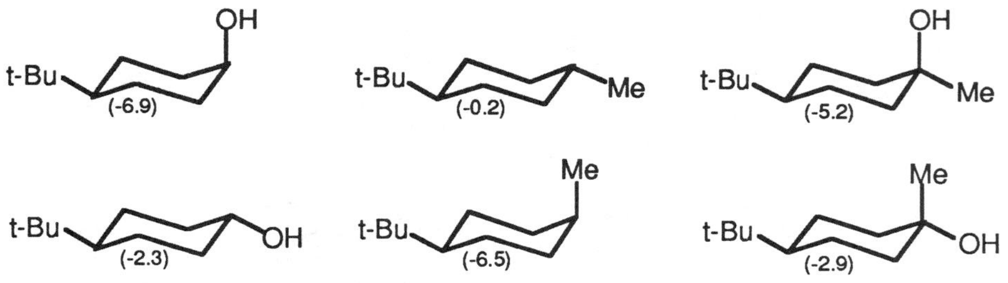

Figure 3-9

3.4 Enforced δ Interactions

The upfield shift observed for a carbon experiencing a gauche butane-like interaction was explained as being the result of hydrogen – hydrogen electron repulsion, shifting bonding electron density toward carbon. It would be anticipated then that if the hydrogen responsible for the shift of a carbon were replaced with a substituent, then this group would "insulate" the carbon and the γ effect would disappear. While such enforced δ interactions are not common because of the severe strain that they entail, those examples that are available (Figure 3-10) fit well with the above analysis. Thus, for two, 1,3-diaxial methyl groups on a six membered ring, the shift of the ring carbons bearing the methyl groups are best approximated by ignoring the presence of the group on the other ring carbon. Simple addition of the Δδ effects of the appropriate models would preserve the γ effects of both groups and result in significant error. Care must be taken in such cases. The three examples shown below clearly illustrate how the introduction of a γ substituent that interacts sterically with a group substituted on a carbon rather than with a hydrogen has little effect. Often the shift that is seen is slightly downfield.

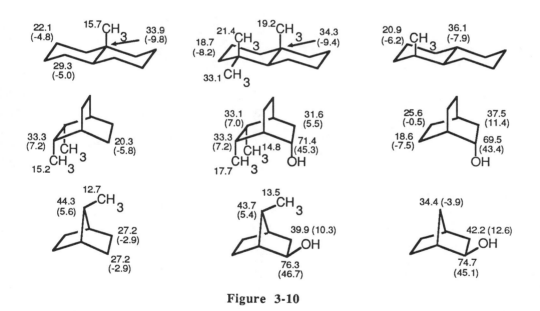

Figure 3-10

Finally, it must be noted, and without explanation, that the anticipated upfield shift of a carbon that gains a δ interaction is not observed and indeed the effect is the reverse and certainly not consistent with the explanation given for γ interactions. Fortunately, δ interactions are not common.

References

1. *Sadtler Standard Carbon-13 NMR Spectra*, Sadtler Research Laboratories, Philadelphia, Pa 19104 (1976-1982).

2. Whitesell, J.K. and Minton, M.A., unpublished results.

3. Schneider, H.-J. and Hoppen, V., *J. Org. Chem.* **43**, 3866 (1978).

Chapter 4

Multiple Substituent Shift Additivity

The relatively simple table of shift effects provided in Chapter 1 can be used to make approximate predictions of chemical shifts which in turn are useful in making first inspection assignments between carbons and absorptions. It is of course implicit in such a treatment that the magnitude of the shift–of–shift effect for a substituent be relatively independent of the presence of other groups. Thus, the shift of a given carbon might be considered to be the sum of the independent effects of all of the neighboring groups. This does indeed hold true in those situations where substituents do not directly interact with each other and where the conformation of the system is relatively constant. In theory it should be possible to arrive at predicted spectra for all possible isomers of an unknown compound and comparison of these with the observed values would narrow the possible candidates to one or at most two choices. However, it should be clear from the discussion in Chapter 2 that the magnitude of the shift effect of a substituent varies with the carbon framework to which it is attached, and therefore, no simple collection of generalized substituent effects will provide accurate predictions in all situations. Nonetheless, with appropriate spectral data for model compounds available, the total shift effect of several, noninteracting substituents can be predicted to be the sum of the $\Delta\delta$ effects for each individual substituent.

4.1   Examples

A simple example of how the total effect of more than one substituent can be predicted as the sum of the $\Delta\delta$ effects of the individual groups can be seen with the dimethyldecalin shown below. The $\Delta\delta$ effects of an axial methyl group on C-2 and that of an equatorial methyl group on C-4 are added to the observed shifts for the parent to obtain a predicted spectrum that is quite close to that observed, with an average error of 0.4 and with a maximum error of only 1.2. While many of the carbons are not perturbed significantly by either substituent (e.g. C-8 and C-9), large effects for both substituents must be combined for C-3 and C-4, yet, errors here are very small.

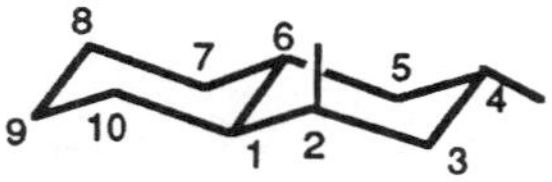

	C-1	C-2	C-3	C-4	C-5	C-6	C-7	C-8	C-9	C-10
a2-methyl	2.8	−1.3	7.5	−6.2	0.7	−7.9	0.7	0.0	0.4	−2.8
e4-methyl	−0.6	−0.4	8.6	6.0	8.7	−0.6	−0.2	−0.2	−0.2	−0.2
parent	43.7	34.3	26.9	26.9	34.3	43.7	34.3	26.9	26.9	34.3
predicted	45.9	32.6	43.0	26.7	43.7	35.2	34.8	26.7	27.1	31.3
observed	46.1	33.8	43.5	26.5	44.2	35.9	35.1	26.9	27.5	31.4
error	0.2	1.2	0.5	0.2	0.5	0.7	0.3	0.2	0.4	0.1

It should be mentioned here parenthetically that the $\Delta\delta$ values provided in the tables of monosubstituted systems in this book were derived by comparing the observed shifts with those obtained for the parent that we felt best represented the same experimental conditions. Thus, the $\Delta\delta$ values will often not correspond to the difference between the single set of absorption values reported for the parent (usually taken in deuterochloroform). They do reflect as accurately as possible the perturbation expected from a given substituent.

A somewhat more complex case is represented by the two, stereoisomeric bicyclo[3.3.0]octanes shown below. The $\Delta\delta$ values for each of the monosubstituted compounds (endo- or exo-2-methyl, endo- or exo-3-hydroxy, and exo-6-methyl) can be added to find a predicted total effect. These values can then be added to the shift values of the parent to arrive at a predicted absorption for each of the framework carbons. In this case the predictions are generally close to the observed spectrum so long as the substituents do not interact with each other either sterically or electronically.

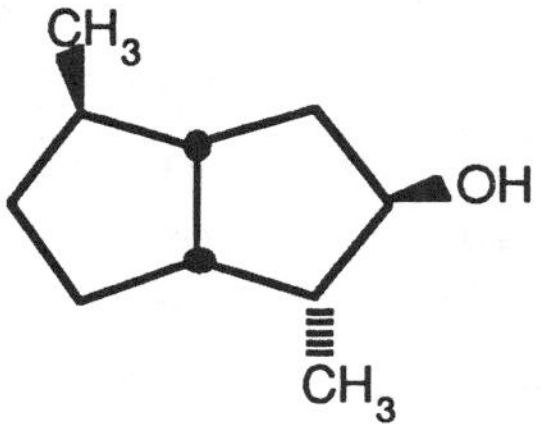

	C-1	C-2	C-3	C-4	C-5	C-6	C-7	C-8
endo-2-methyl	4.7	3.7	6.8	−1.8	0.0	1.5	1.5	−6.4
exo-3-hydroxy	−2.2	8.5	48.5	8.5	−2.2	−0.3	−0.5	−0.3
exo-6-methyl	0.2	−0.2	−0.8	−1.9	8.7	8.1	9.7	−0.6
parent	43.2	34.2	26.3	34.2	43.2	34.2	26.3	34.2
predicted	45.9	46.2	80.8	39.0	49.7	43.5	37.0	26.9
observed	44.3	46.8	77.4	40.1	46.8	44.1	35.8	28.1
error	1.6	−0.6	3.4	−1.1	2.9	−0.6	1.2	−1.2

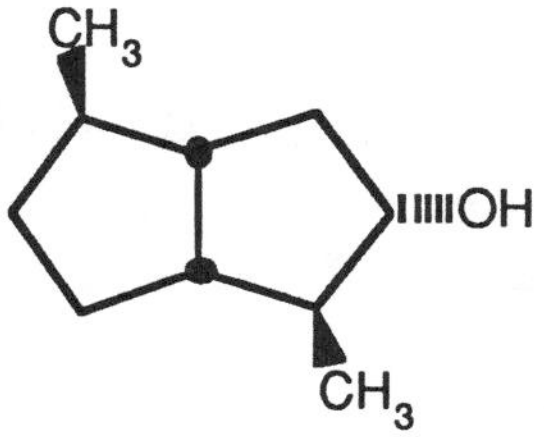

	C1	C2	C3	C4	C5	C6	C7	C8
exo-2-methyl	8.7	8.1	9.7	−0.6	0.2	−0.2	−0.8	−1.9
endo-3-hydroxy	−3.0	8.2	47.5	8.2	−3.0	−0.5	−1.2	−0.5
exo-6-methyl	0.2	−0.2	−0.8	−1.9	8.7	8.1	9.7	−0.6
parent	43.2	34.2	26.3	34.2	43.2	34.2	26.3	34.2
predicted	49.1	50.3	82.7	39.9	49.1	41.6	34.0	31.2
observed	48.6	49.2	80.6	40.8	46.9	42.5	34.8	30.6
error	0.5	1.1	2.1	−0.9	2.2	−0.9	−0.8	0.6

The two isomers used for the example given above were derived from an alkene by hydroboration – oxidation, and therefore it could be assumed that the methyl and adjacent hydroxyl group are trans in both diastereomers. However, it is instructive to analyze this system as if the stereochemical orientation of all three substituent groups were unknown and derive predictions for the anticipated chemical shifts of all eight possible diastereomers. The assignment of carbons and chemical shifts of absorptions is of course unknown at the outset of such an analysis and correlations must at first be attempted simply by a "best fit" approach (with due consideration for multiplicity). Listed below, in decreasing order of magnitude, are the predicted chemical shifts for all eight possible diastereomers.

obs	1	2	3	4	5	6	7	8	obs
X2-Me	X2-Me	N2-Me	X2-Me	N2-Me	N2-Me	X2-Me	X2-Me	N2-Me	N2-Me
N3-OH	X3-OH	N3-OH	N3-OH	X3-OH	N3-OH	X3-OH	N3-OH	X3-OH	X3-OH
X6-Me	X6-Me	N6-Me	N6-Me	N6-Me	X6-Me	N6-Me	X6-Me	X6-Me	X6-Me
80.6	83.7	82.1	85.0	83.1	79.8	86.0	82.7	80.8	77.4
49.2	50.6	47.6	52.0	47.9	48.9	52.3	50.3	49.7	46.8
48.6	49.9	44.9	48.9	45.7	45.9	49.7	49.1	46.2	46.8
46.9	49.9	44.9	45.1	45.7	45.1	45.9	49.1	45.9	44.3
42.5	41.8	38.9	37.2	39.1	43.3	37.4	41.6	43.5	44.1
40.8	40.2	34.2	35.4	34.5	38.7	35.7	39.9	39.0	40.1
34.8	34.7	33.4	31.1	34.1	36.3	31.8	34.0	37.0	35.8
30.6	31.4	25.5	30.0	25.7	26.7	30.2	31.2	26.9	28.1

					average errors in fit				
——>	1.4	3.2	3.0	2.9	1.7	3.0	1.1	1.6	
	3.2	3.0	4.2	3.0	1.3	4.5	2.9	1.6	<——

The observed absorptions for the exo-2, endo-3, exo-6 isomer do not fit well with isomers 2, 3, 4, and 6, leaving the choice among 1, 5, 7, and 8. On the other hand, the endo-2, exo-3, exo-6 isomer fits well only with isomers 5 and 8. However, to refine the choice further it is necessary to examine the shifts for the methyl group at C-2 to determine the orientations of this and the adjacent hydroxyl group. The absorption of this methyl would be expected to be shifted upfield as compared to the monosubstituted, endo- or exo-2-methyl system since the oxygen substituent at C-3 is gauche to this C-2 methyl. The effect of this relationship of these two groups can be estimated from the shift difference between the methyl groups of methylcyclopentane (21.4) and cis- (14.9) and trans- (19.5) 2-methylcyclopentanol. Thus, the addition of a cis- hydroxyl group results in an upfield shift of –6.5 for the methyl group while the methyl group is shifted only –1.9 in the trans alcohol. These effects can be applied to the observed shift values for the methyl groups of exo-2- and of endo-2-methylbicyclo[3.3.0]octane to arrive at the predicted shifts for the C-2 methyl given on the next page.

C-2 (high field) Methyl

obs	1	2	3	4	5	6	7	8	obs
X2-Me	X2-Me	N2-Me	X2-Me	N2-Me	N2-Me	X2-Me	X2-Me	N2-Me	N2-Me
N3-OH	X3-OH	N3-OH	N3-OH	X3-OH	N3-OH	X3-OH	N3-OH	X3-OH	X3-OH
X6-Me	X6-Me	N6-Me	N6-Me	N6-Me	X6-Me	N6-Me	X6-Me	X6-Me	X6-Me
16.6	13.3	8.9	17.9	13.5	8.9	13.3	17.9	13.5	12.7

The observed absorption at 16.6 for the exo, endo, exo isomer fits well only with diastereomers 3 and 7, while the methyl group absorption for the endo, exo, exo isomer is reasonably close to that predicted for isomers 1, 4, 6, and 8. When these conclusions are combined with the analysis above of the framework carbon atom absorptions, all but one diastereomer can be excluded for each of the observed compounds, leading to a reliable assignment of stereochemistry.

The most important point to note from the above numerical analysis is that the predicted spectral absorptions for the two isomers are quite different and thus a reliable assignment of stereochemistry could be made even if only one of the two isomers were available. In addition, just as it was necessary to correct for the influence of the hydroxy group on the C-2 methyl, the errors made in the predictions for the framework atoms can be accounted for, at least qualitatively, based on interactions between these substituents. The magnitude of the γ effect in exo-3-hydroxy - bicyclo[3.3.0]octane (–2.2) is at the low end of the normal range. It can therefore be presumed that the monosubstituted compound adopts a preferred conformation with the oxygen anti to the bridgeheads. Any change in the conformation of the bicyclic system caused by the additional substituents would thus be expected to increase the magnitude of the γ effect. While considerations such as these can not yet be quantified, it should be clear that the interpretations for poly- substituted systems based on total Δδ effects should be made with a clear understanding of those situations where substituent effects are not additive.

The following example (Figure 4-1) will demonstrate the use of smaller model compounds for the prediction of shifts for polycyclic systems where exact models are not available.

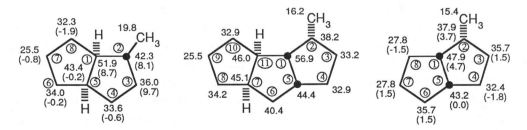

Figure 4-1

The absorptions for endo- and exo-2-methylbicyclo[3.3.0]octane can be extrapolated for use with the tricyclic system shown [1]. Indeed, since the only carbon substituents that contribute significantly are those that are either α or β or γ and gauche, the shifts for carbons 2-5 and 7-10 would be predicted to be unchanged from the appropriate bicyclic model. Agreement between these predictions and the observed spectrum is quite good, with the largest error only 1.7 and an average error of 0.6. Likewise, the shift of the methyl group (16.2 observed) would be expected to be quite similar to that for the endo-2-methyl model (15.4).

On the other hand, carbons 1, 6, and 11 must be dealt with individually because of their unique situation within the tricyclic framework. For C-1 the appropriate model carbon is C-1 of the endo bicyclic model (47.9), corrected for the presence of C-10 which represents an exo substituent appended onto a bicyclic framework consisting of the middle and right hand rings. This correction can be derived from the effect of the methyl group on C-1 in the exo model, 8.7 (51.9 – 43.2), resulting in the predicted shift of 56.6 (56.9 obs). The same derivation applies to C-11, except that here there is the additional perturbation due to the γ methyl group. This effect can be approximated by the effect of the methyl group on C-8 in the endo model (27.8 – 34.2 = –6.4), and thus C-11 would be predicted to be at 45.5 (47.9 + 8.7 – 6.4), also in good agreement with the observed value of 46.0. Carbon 6 can be modeled from the exo-methylbicyclo[3.3.0]octane with a correction for the addition of a β substituent (C-4). Since the appropriate shift effects of alkyl groups at the wingtip, or C-3, position of a bicyclooctane are not available, the value 8.7 derived above can be used, leading to a prediction of 42.3 (33.6 + 8.7), in reasonably good agreement with the observed value of 40.4. Note the less complicated approach of using 8.0 as the correction for an additional β substituent would still have produced excellent correspondence between predicted and observed shifts.

In the examples above the assumption was made that the total effect of several groups could be predicted simply as the sum of the individual groups' shift effects. While this process resulted in only relatively small errors, there are certain situations where the resulting errors are substantial and it is important to have a clear understanding of when and why shift additivity does not hold. The largest errors are found when groups interact either electronically or sterically and the sections that follow will deal with those situations.

4.2 Polarizable groups

In Chapter 1 it was noted that the shift effects of the halogens, and especially iodine, were variable because of the ease with which these groups could be polarized by neighboring groups as well as intermolecular interactions with solvent. In addition, groups with π-bonds, especially carbonyl groups, are easily polarized by interaction with electron donating or withdrawing groups as well as with other π systems. For instance, the β carbon to the carbonyl group in an enone will be found at lower field and the α carbon at higher field than would be expected because the electron withdrawing nature of the carbonyl group is transmitted through the carbon – carbon π system (Figure 4-2). This interaction is substantial and the resulting errors are quite large. In most situations the shifts of sp^2 hybridized carbons cannot be accurately predicted by shift additivity.

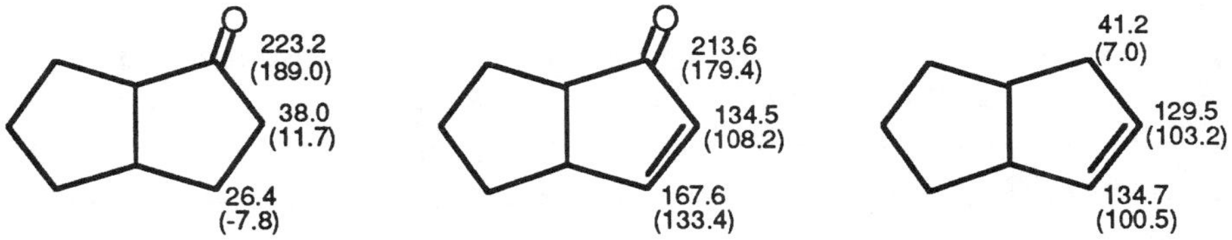

Figure 4-2

Even with less highly polarizable groups, there is still a significant error involved in ignoring interactions between groups. This is especially notable in cis-1,2-glycols and diacetates in the bicyclo[3.3.0]octane system (Figure 4-3, next page), where there is an apparent mutual diminishing of the inductive, α effect of oxygen such that a more accurate prediction is arrived at by ignoring the presence of the second oxygen as a β substituent. This effect is also seen in cis-1,2-dihydroxycyclopentane, but not in the trans isomer. The shift values for C-1 and C-2 in both cis- and trans-1,2-cyclohexanediol reflect more normal effects anticipated from the incorporation of an additional β substituent.

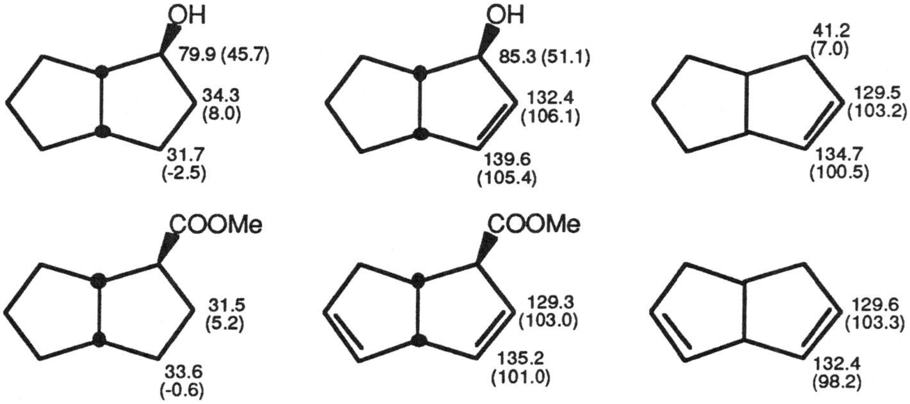

Figure 4-4

Other systems where substituent shift additivity fails include allylic alcohols and β,γ-unsaturated esters. The examples in Figure 4-4 illustrate these functional group relationships. The errors that would result from simple addition of substituent effects can be seen by inspection. In these cases, as well as with the enone example, the direction of the effect can be explained as being the result of the polarization of the π bond toward the electron withdrawing group.

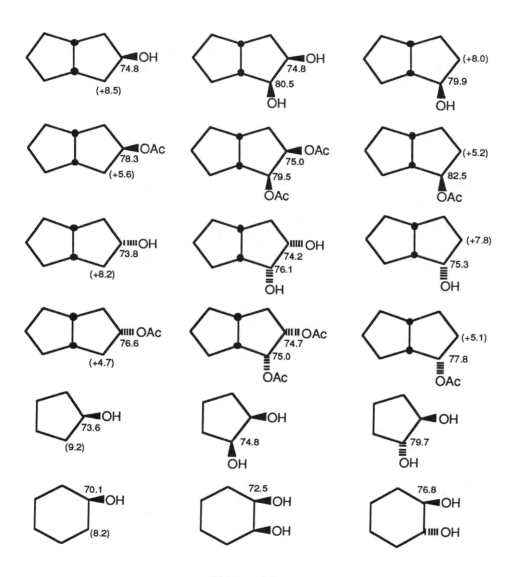

Figure 4-3

4.3   Conformationally mobile systems

In systems that can readily adopt two or more reasonable conformations, such as cis-decalins, a substituent can impart a significant bias for one arrangement. Thus, the spectral properties of exo- and endo-3-methyl-cis-decalin (Figure 4-5) will reflect in each case the almost exclusive participation of that conformation with the methyl group equatorial.

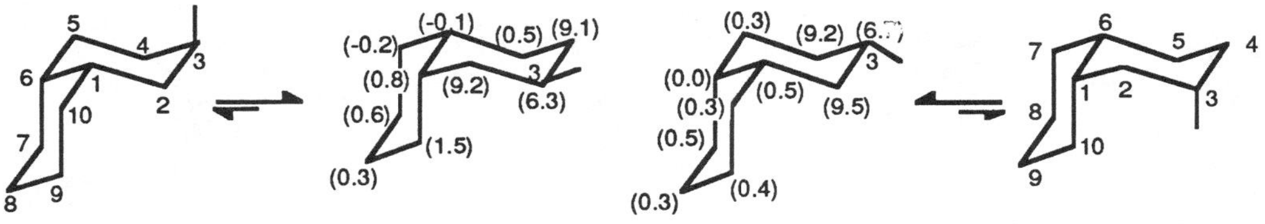

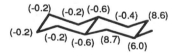

Figure 4-5

Calculating the shift–of–shift values for these two diastereomers using the values obtained from the room temperature averaged spectrum of the parent will lead to substantial errors. On the other hand, if the appropriate carbon shifts are taken from the low temperature spectrum for the conformationally frozen parent (as was done for the values shown in Figure 4-5), then the shift effects imparted by the equatorial methyl groups will be as anticipated, with values quite similar to those observed with the corresponding trans decalins (e.g. Figure 4-6).

Figure 4-6

When two or more groups which impart the same conformational bias are present, shift additivity will provide reasonable predictions. For example, the methyl groups in exo-3-exo-8-dimethyl-cis-decalin (next page) are either both equatorial or both axial, with the former arrangement obviously favored. Since this conformation preserves the spatial orientation of the methyl group in the monosubstituted model, then shift additivity will hold so long as the effect of the methyl group is assessed relative to the shifts of the conformationally fixed parent obtained at low temperature.

shifts relative to low temperature parent

	C-1	C-2	C-3	C-4	C-5	C-6	C-7	C-8	C-9	C-10
X3-methyl	0.8	9.2	6.3	9.1	0.5	−0.1	−0.2	0.6	0.3	1.5
X8-methyl	−0.1	−0.2	0.6	0.3	1.5	0.8	9.2	6.3	9.1	0.5
parent	36.1	32.6	21.0	27.2	25.8	36.1	32.6	21.0	27.2	25.8
predicted	36.8	41.6	27.9	36.6	27.8	36.8	41.6	27.9	36.6	27.8
observed	36.8	41.7	27.1	36.3	27.1	36.8	41.7	27.1	36.3	27.1
error	0.0	−0.1	0.8	0.3	0.7	0.0	−0.1	0.8	0.3	0.7

When two substituents are present that exert opposite effects, then it will not be possible to estimate the effects from the monosubstituted systems. For example with endo-3-methyl-endo-9-methyl-cis-decalin (Figure 4-7) the two conformations will be equally populated and there is no appropriate monosubstituted methyl decalin that bears an axial methyl group. This situation may be improved somewhat in the future by the availablity of additional spectral data for minor conformers obtained at low temperature. Nonetheless, predictions are still possible. For example, C-3 in the conformer with the methyl group equatorial would be predicted to be essentially unchanged from that observed at low temperature for endo-3-methyl-cis-decalin (33.6). In the conformer with this methyl axial, a prediction can be made by ignoring the presence of the δ interaction between this methyl and C-10 and using the $\Delta\delta$ value of −5.5 for C-1 of axial methylcyclohexane, for a predicted shift of 28.1. The average of these two values (since the conformers are equally populated), 30.8, is quite close to that observed (30.7).

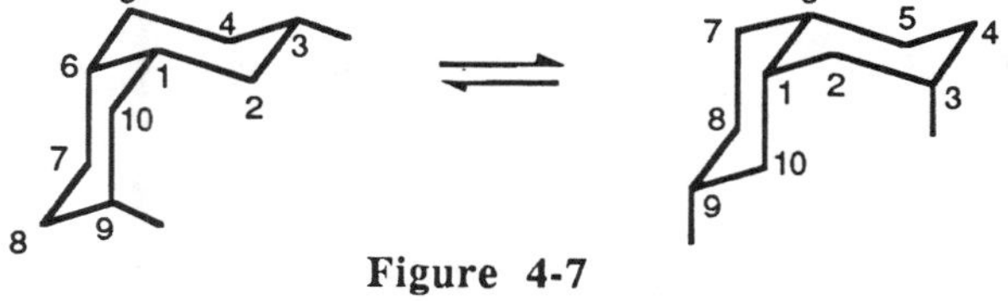

Figure 4-7

The same problem occurs with the bicyclo[3.3.1]nonane system when there is an endo substituent at both C-3 and at C-7 resulting in a double twist boat conformation. Since each of the model, monosubstituted compounds will be predominately in a boat – chair arrangement, they will not serve as good sources for $\Delta\delta$ effects for the disubstituted system and their use results in sizable errors for C-3 and C-7.

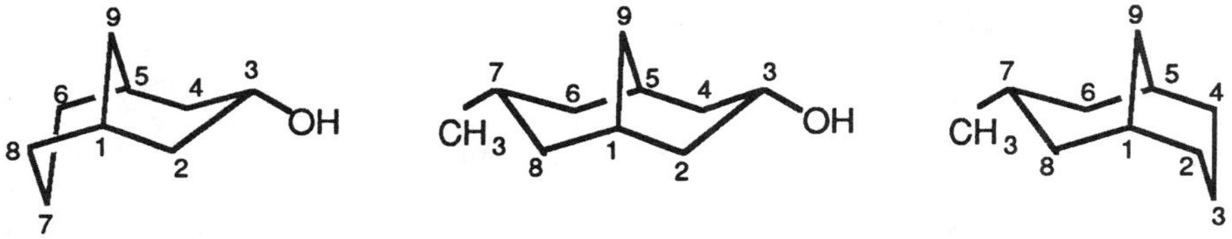

	C-1	C-2	C-3	C-4	C-5	C-6	C-7	C-8	C-9
N3-hydroxyl	–1.3	4.5	42.9	4.5	–1.3	1.6	–6.2	1.6	–5.6
N7-methyl	–1.8	2.7	–6.6	2.7	–1.8	4.4	3.1	4.4	–5.3
parent	28.0	31.6	22.6	31.6	28.0	31.6	22.6	31.6	35.0
predicted	24.9	38.8	58.9	38.8	24.9	37.6	19.5	37.6	24.1
observed	25.3	36.2	68.0	36.2	25.3	40.9	28.1	40.9	25.0
error	–0.4	2.6	–9.1	2.6	–0.4	–3.3	–8.6	–3.3	–0.9

While it is not possible to use models in the above case, the simple system developed in Chapter 1 is still applicable, so long as the stereochemical factors discussed in Chapter 3 are incorporated as well. Thus, in the above example, C-3 is in a six-membered ring (27.3) with one additional α oxygen substituent (+8.0 + 38.0) but is 3° with two β substituents (2*–1.5) and has a gauche interaction with a γ group (C-9, –3.0) for a predicted shift of 67.3 (68.0 obs). Similar treatment of C-7 leads to a predicted shift of 29.3 (28.1 obs).

References

1. Curran, D.P. and Rakiewicz, D.M., *Tetrahedron* **41**, 3943 (1985).

Chapter 5

Bicyclo[2.2.1]heptanes

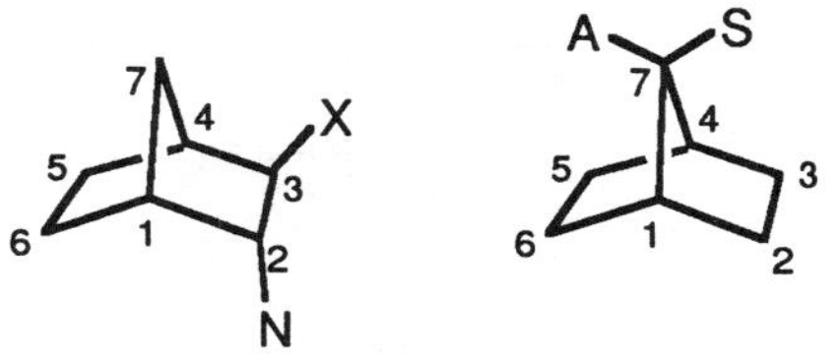

Nomenclature for this system is straightforward, with exo and endo referring to the orientation of substituents on carbons 2, 3, 5, and 6 relative to C-7, and syn and anti used for C-7 substituents. For 7-monosubstituted compounds, the numbering is defined so that the substituent will be syn. Relatively unusual dihedral angles are found about most of the bonds and thus caution should be exercised in applying $\Delta\delta$ effects observed here to other systems.

F
41.0 (2.7)
33.0 (-3.3) — 29.6 (0.0)
29.6 (0.0)
103.9 (67.6) — 32.0 (2.4)
32.0 (2.4)

Cl
46.8 (8.4)
34.8 (-1.6) — 30.9 (1.1)
30.9 (1.1)
69.9 (33.5) — 38.4 (8.6)
38.4 (8.6)

Br
48.2 (9.8)
34.8 (-1.6) — 31.6 (1.8)
31.6 (1.8)
62.2 (25.8) — 40.0 (10.2)
40.0 (10.2)

I
50.8 (12.4)
34.6 (-1.8) — 32.1 (2.3)
32.1 (2.3)
37.7 (1.3) — 43.0 (13.2)
43.0 (13.2)

OH
43.9 (5.5)
34.8 (-1.6) — 30.3 (.5)
30.3 (.5)
82.8 (46.4) — 35.4 (5.6)
35.4 (5.6)

OMe
40.0 (1.6)
33.9 (-2.5) — 29.9 (.1)
29.9 (.1)
87.7 (51.3) — 31.0 (1.2)
31.0 (1.2)

NHAc
41.7 (3.3)
35.3 (-1.1) — 29.9 (.1)
29.9 (.1)
62.6 (26.2) — 33.7 (3.9)
33.7 (3.9)

SnMe$_3$
42.0 (3.7)
36.1 (-.2) — 30.5 (.9)
30.5 (.9)
33.5 (-2.8) — 34.1 (4.5)
34.1 (4.5)

5.1 Saturated bicyclo [2.2.1] heptanes

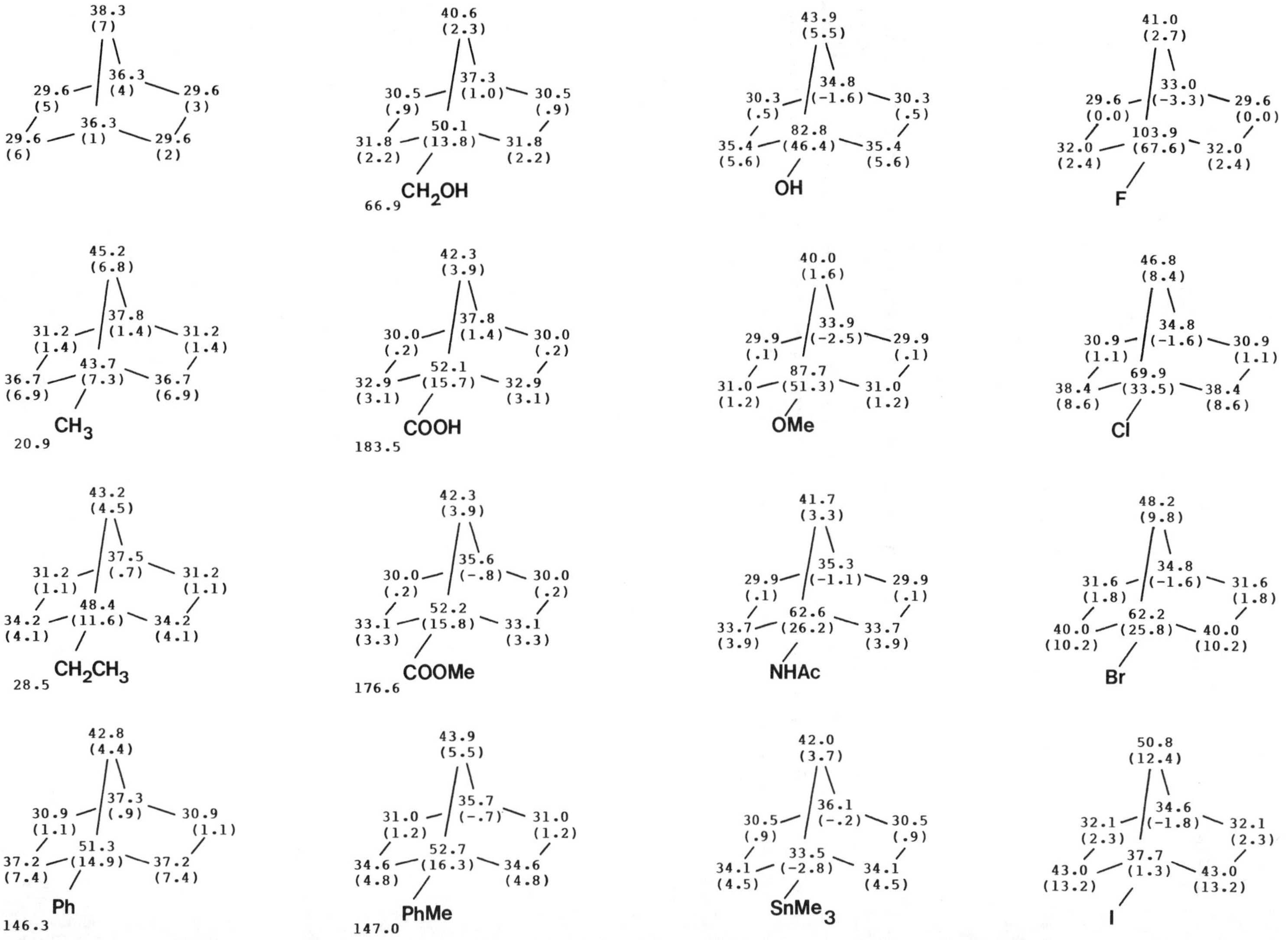

CH$_2$OH
40.6 (2.3)
37.3 (1.0) — 30.5 (.9)
30.5 (.9)
50.1 (13.8) — 31.8 (2.2)
31.8 (2.2)
66.9

COOH
42.3 (3.9)
37.8 (1.4) — 30.0 (.2)
30.0 (.2)
52.1 (15.7) — 32.9 (3.1)
32.9 (3.1)
183.5

COOMe
42.3 (3.9)
35.6 (-.8) — 30.0 (.2)
30.0 (.2)
52.2 (15.8) — 33.1 (3.3)
33.1 (3.3)
176.6

PhMe
43.9 (5.5)
35.7 (-.7) — 31.0 (1.2)
31.0 (1.2)
52.7 (16.3) — 34.6 (4.8)
34.6 (4.8)
147.0

CH$_3$
38.3 (7)
36.3 (4) — 29.6 (3)
29.6 (5) — 36.3 (1) — 29.6 (2)
29.6 (6)
20.9

CH$_2$CH$_3$
43.2 (4.5)
37.5 (.7) — 31.2 (1.1)
31.2 (1.1)
48.4 (11.6) — 34.2 (4.1)
34.2 (4.1)
28.5

Ph
42.8 (4.4)
37.3 (.9) — 30.9 (1.1)
30.9 (1.1)
51.3 (14.9) — 37.2 (7.4)
37.2 (7.4)
146.3

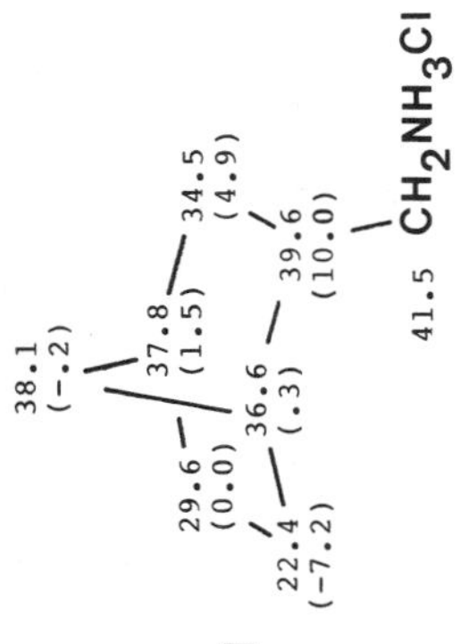

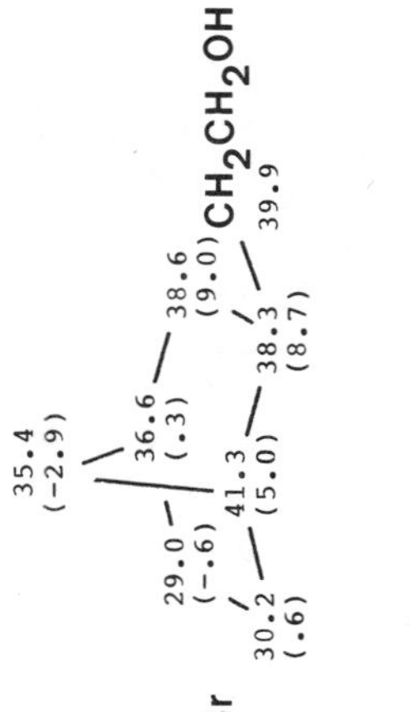

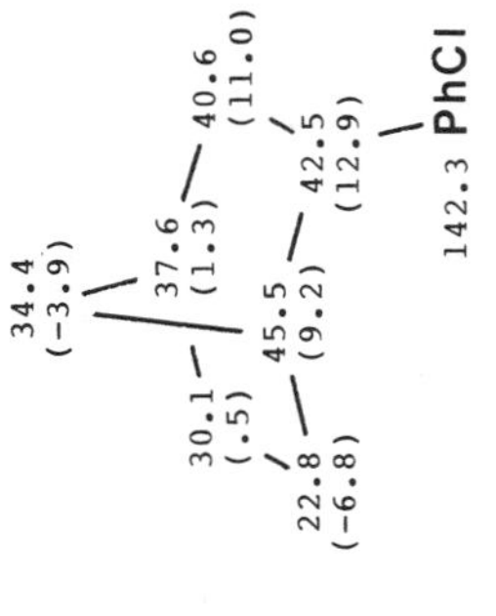

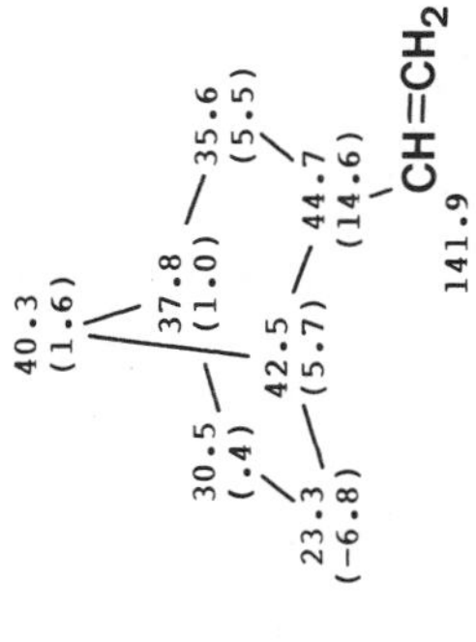

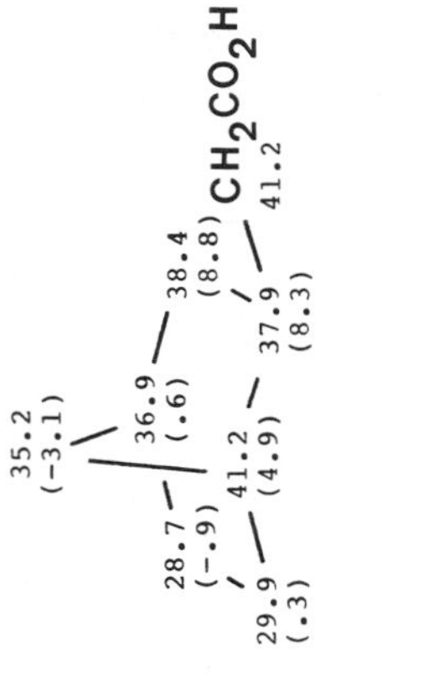

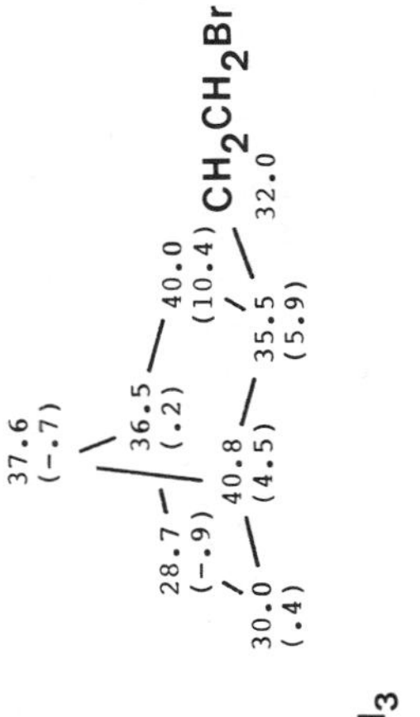

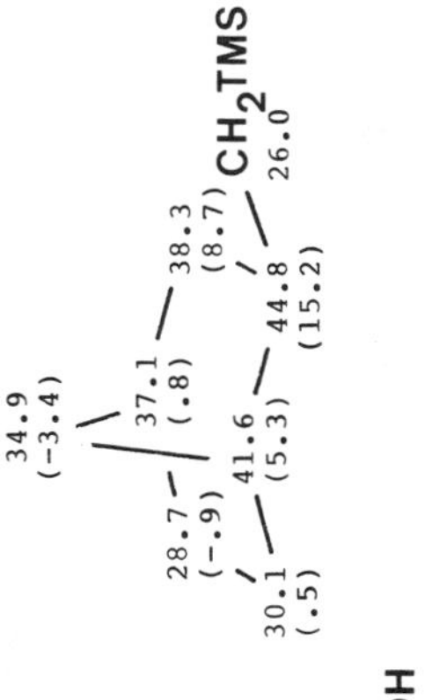

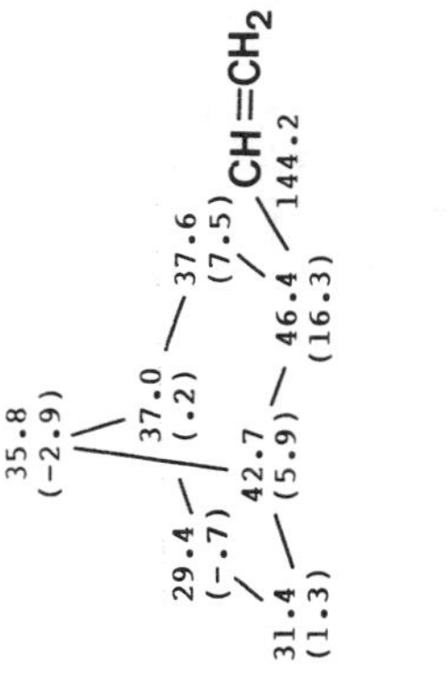

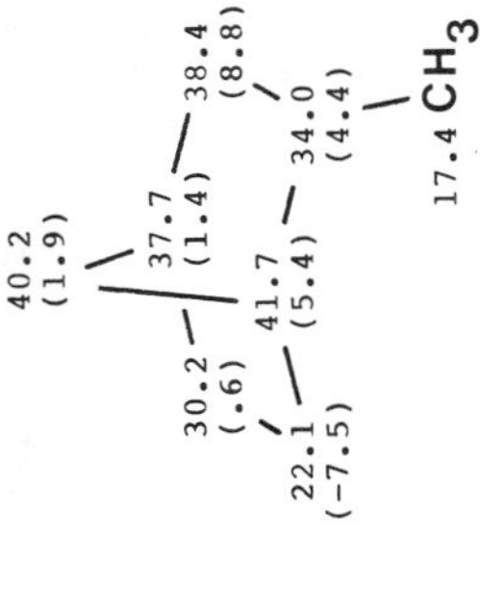

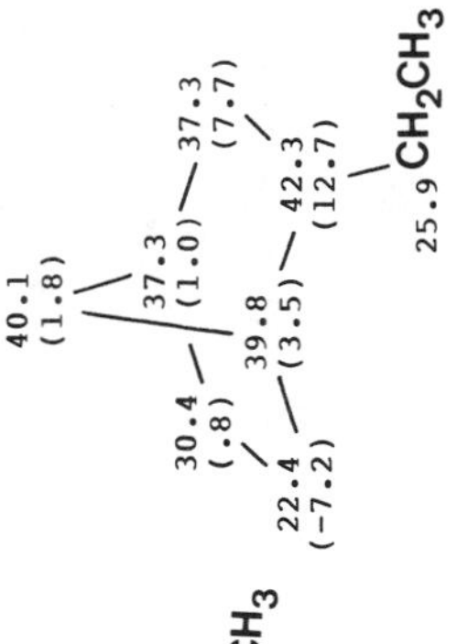

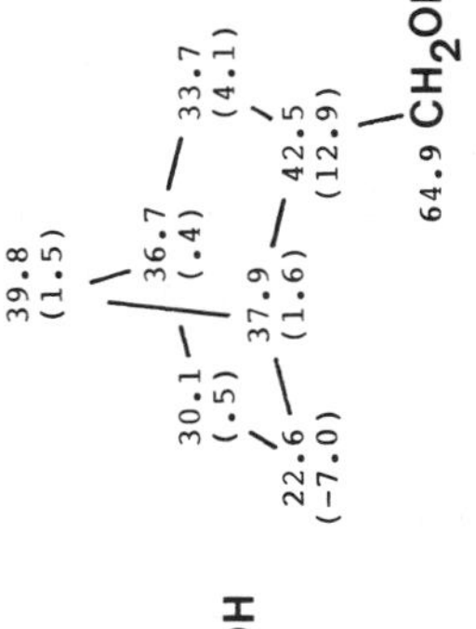

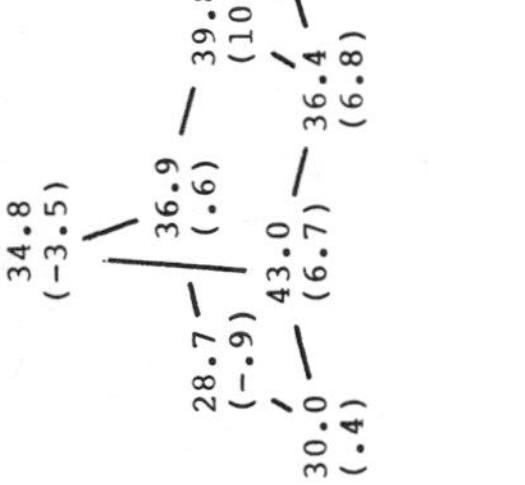

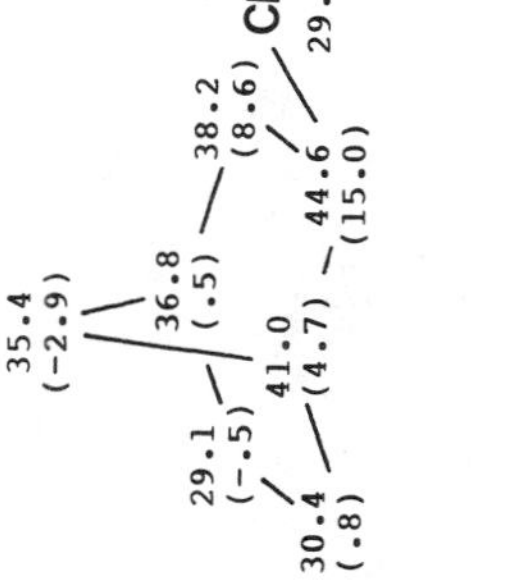

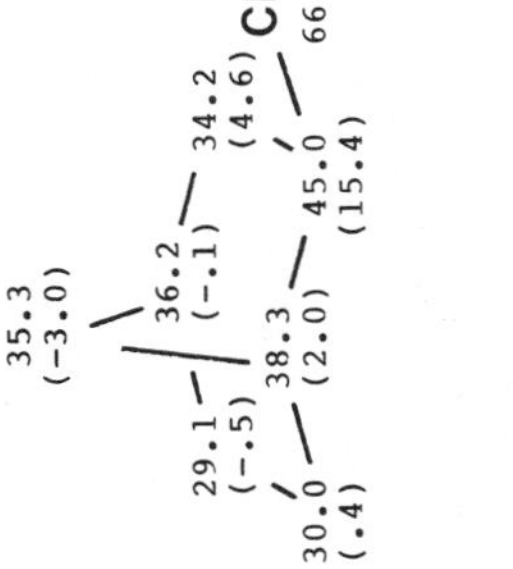

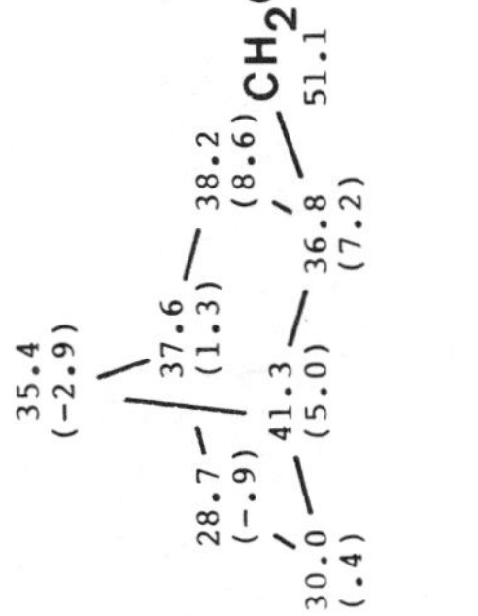

CH₂NH₃Cl 41.5
38.1 (-.2); 37.8 (1.5); 34.5 (4.9); 29.6 (0.0); 36.6 (.3); 39.6 (10.0); 22.4 (-7.2)

CH₂CH₂OH 39.9
35.4 (-2.9); 36.6 (.3); 38.6 (9.0); 29.0 (-.6); 41.3 (5.0); 38.3 (8.7); 30.2 (.6)

PhCl 142.3
34.4 (-3.9); 37.6 (1.3); 40.6 (11.0); 30.1 (.5); 45.5 (9.2); 42.5 (12.9); 22.8 (-6.8)

CH=CH₂ 141.9
40.3 (1.6); 37.8 (1.0); 35.6 (5.5); 30.5 (.4); 42.5 (5.7); 44.7 (14.6); 23.3 (-6.8)

CH₂CO₂H 41.2
35.2 (-3.1); 36.9 (.6); 38.4 (8.8); 28.7 (-.9); 41.2 (4.9); 37.9 (8.3); 29.9 (.3)

CH₂CH₂Br 32.0
37.6 (-.7); 36.5 (.2); 40.0 (10.4); 28.7 (-.9); 40.8 (4.5); 35.5 (5.9); 30.0 (.4)

CH₂TMS 26.0
34.9 (-3.4); 37.1 (.8); 38.3 (8.7); 28.7 (-.9); 41.6 (5.3); 44.8 (15.2); 30.1 (.5)

CH=CH₂ 144.2
35.8 (-2.9); 37.0 (.2); 37.6 (7.5); 29.4 (-.7); 42.7 (5.9); 46.4 (16.3); 31.4 (1.3)

CH₃ 17.4
40.2 (1.9); 37.7 (1.4); 38.4 (8.8); 30.2 (.6); 41.7 (5.4); 34.0 (4.4); 22.1 (-7.5)

CH₂CH₃ 25.9
40.1 (1.8); 37.3 (1.0); 37.3 (7.7); 30.4 (.8); 39.8 (3.5); 42.3 (12.7); 22.4 (-7.2)

CH₂OH 64.9
39.8 (1.5); 36.7 (.4); 33.7 (4.1); 30.1 (.5); 37.9 (1.6); 42.5 (12.9); 22.6 (-7.0)

CH₃ 22.3
34.8 (-3.5); 36.9 (.6); 39.8 (10.2); 28.7 (-.9); 43.0 (6.7); 36.4 (6.8); 30.0 (.4)

CH₂CH₃ 29.8
35.4 (-2.9); 36.8 (.5); 38.2 (8.6); 29.1 (-.5); 41.0 (4.7); 44.6 (15.0); 30.4 (.8)

CH₂OH 66.9
35.3 (-3.0); 36.2 (-.1); 34.2 (4.6); 29.1 (-.5); 38.3 (2.0); 45.0 (15.4); 30.0 (.4)

CH₂COMe 51.1
35.4 (-2.9); 37.6 (1.3); 38.2 (8.6); 28.7 (-.9); 41.3 (5.0); 36.8 (7.2); 30.0 (.4)

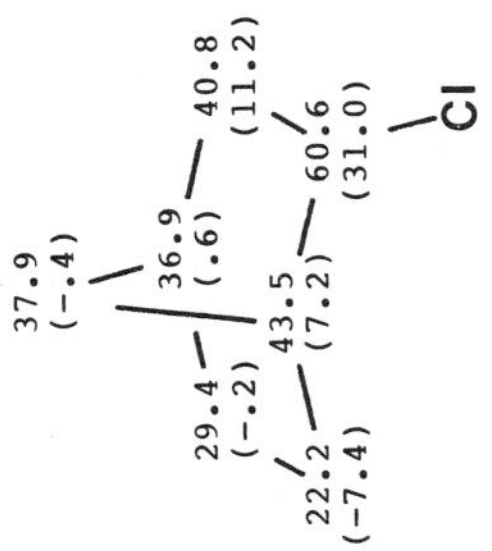

Cl
37.9 (-.4); 36.9 (.6); 40.8 (11.2); 60.6 (31.0); 29.4 (-.2); 43.5 (7.2); 22.2 (-7.4) — Cl

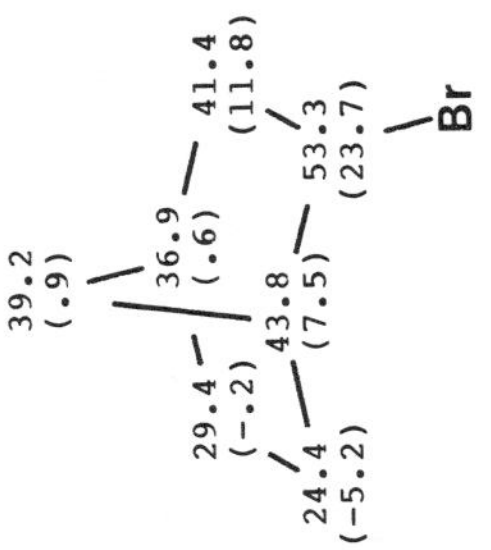

Br
39.2 (.9); 36.9 (.6); 41.4 (11.8); 53.3 (23.7); 29.4 (-.2); 43.8 (7.5); 24.4 (-5.2) — Br

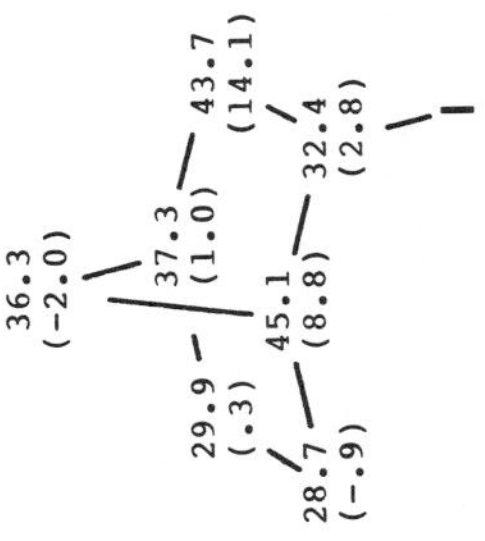

I
36.3 (-2.0); 37.3 (1.0); 43.7 (14.1); 32.4 (2.8); 29.9 (.3); 45.1 (8.8); 28.7 (-.9) — I

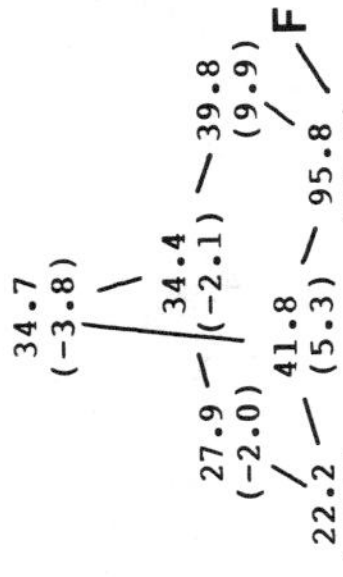

F
34.7 (-3.8); 34.4 (-2.1); 39.8 (9.9); 95.8 (65.9); 27.9 (-2.0); 41.8 (5.3); 22.2 (-7.7) — F

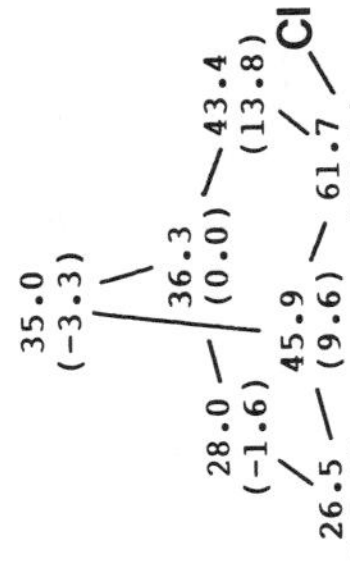

Cl
35.0 (-3.3); 36.3 (0.0); 43.4 (13.3); 61.7 (32.1); 28.0 (-1.6); 45.9 (9.6); 26.5 (-3.1) — Cl

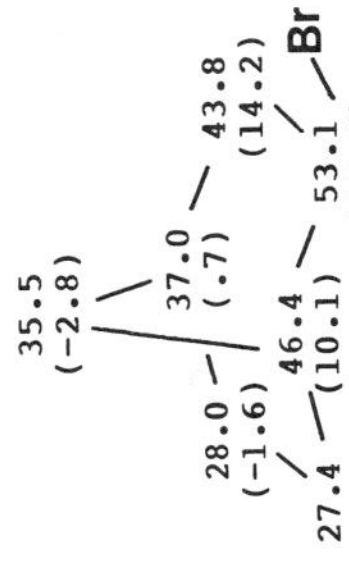

Br
35.5 (-2.8); 37.0 (.7); 43.8 (14.2); 53.1 (23.5); 28.0 (-1.6); 46.4 (10.1); 27.4 (-2.2) — Br

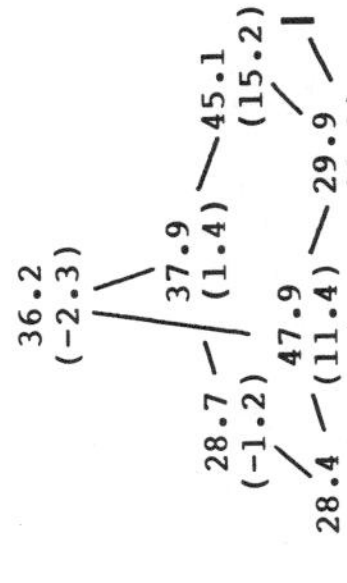

I
36.2 (-2.3); 37.9 (1.4); 45.1 (15.2); 29.9 (0.0); 28.7 (-1.2); 47.9 (11.4); 28.4 (-1.5) — I

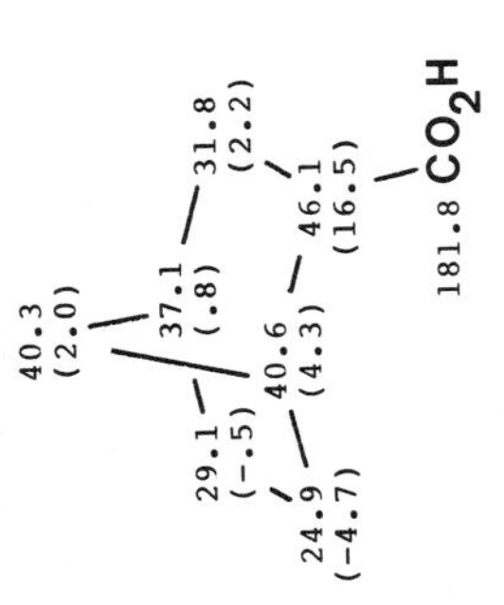

CO_2H 181.8
40.3 (2.0); 37.1 (.8); 31.8 (2.2); 29.1 (-.5); 46.1 (16.5); 40.6 (4.3); 24.9 (-4.7)

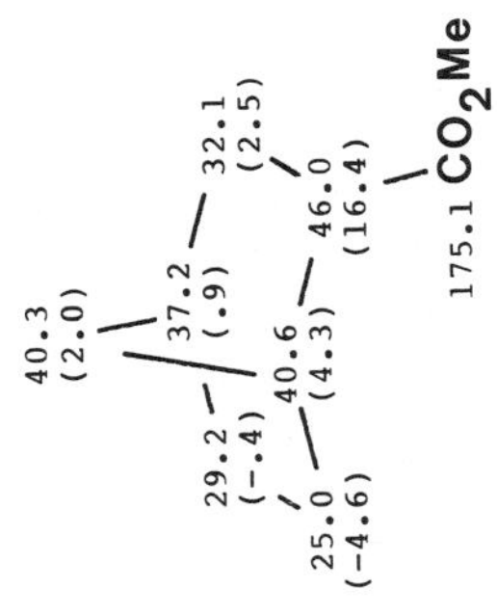

CO_2Me 175.1
40.3 (2.0); 37.2 (.9); 32.1 (2.5); 29.2 (-.4); 46.0 (16.4); 40.6 (4.3); 25.0 (-4.6)

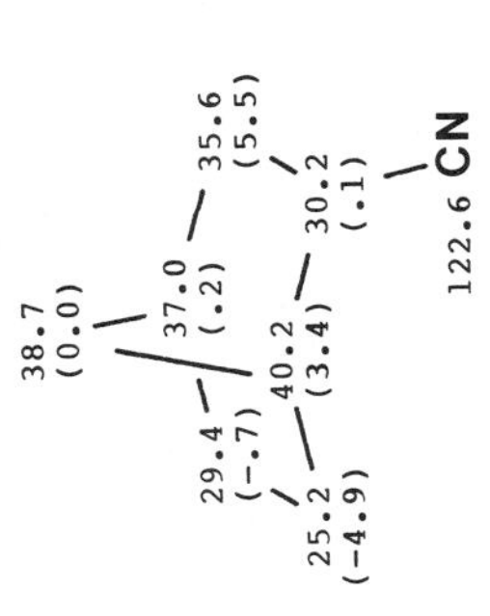

CN 122.6
38.7 (0.0); 37.0 (.2); 35.6 (5.5); 29.4 (-.7); 30.2 (.1); 40.2 (3.4); 25.2 (-4.9)

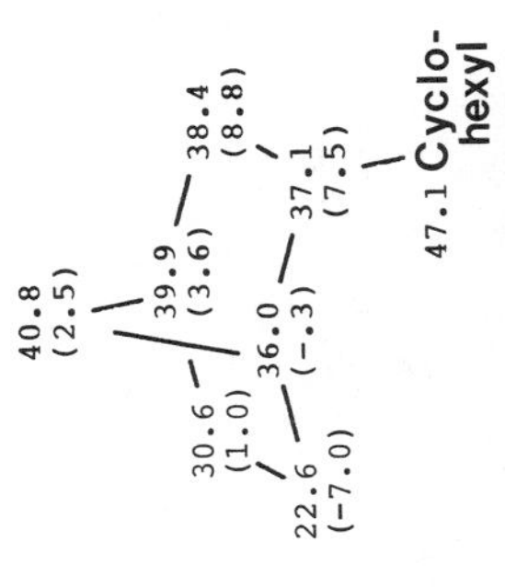

Cyclohexyl 47.1
40.8 (2.5); 39.9 (3.6); 38.4 (8.8); 30.6 (1.0); 37.1 (7.5); 36.0 (-.3); 22.6 (-7.0)

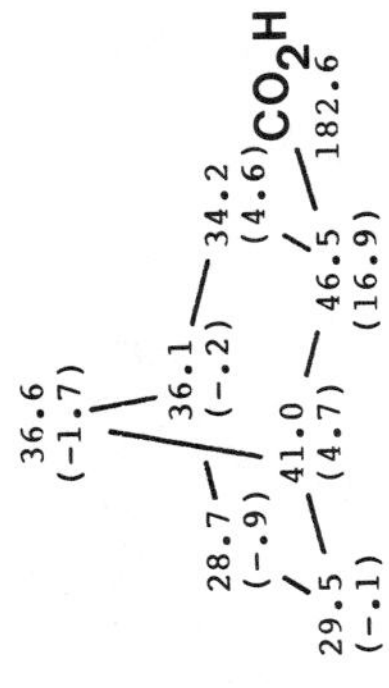

CO_2H 182.6
36.6 (-1.7); 36.1 (-.2); 34.2 (4.6); 28.7 (-.9); 46.5 (16.9); 41.0 (4.7); 29.5 (-.1)

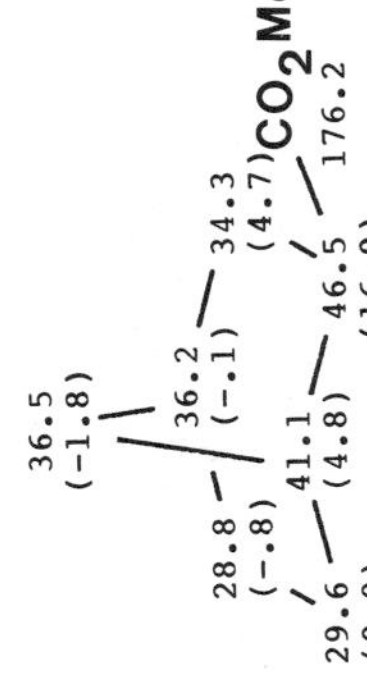

CO_2Me 176.2
36.5 (-1.8); 36.2 (-.1); 34.3 (4.7); 28.8 (-.8); 46.5 (16.9); 41.1 (4.8); 29.6 (0.0)

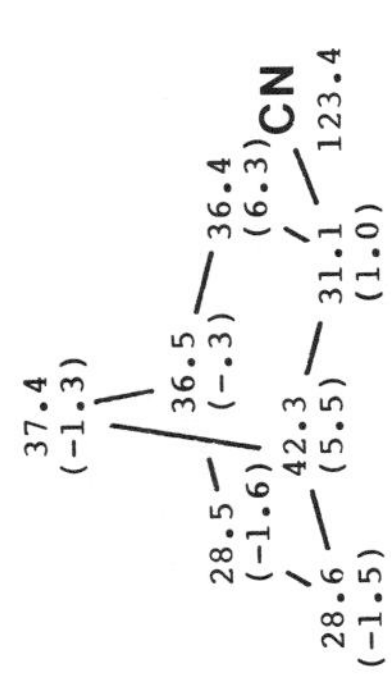

CN 123.4
37.4 (-1.3); 36.5 (-.3); 36.4 (6.3); 31.1 (1.0); 28.5 (-1.6); 42.3 (5.5); 28.6 (-1.5)

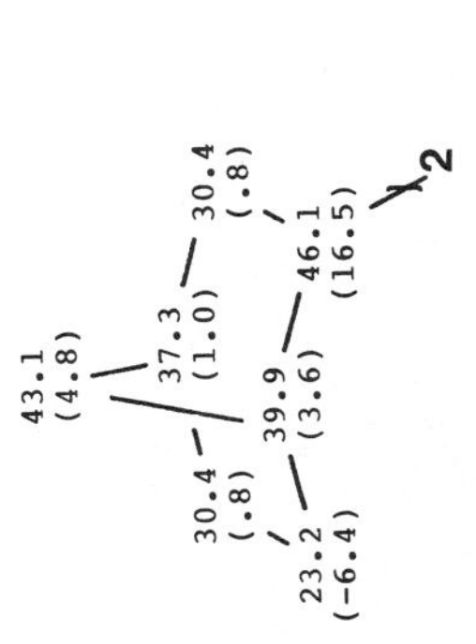

$_2$
43.1 (4.8); 37.3 (1.0); 30.4 (.8); 39.9 (3.6); 46.1 (16.5); 23.2 (-6.4); 30.4 (.8)

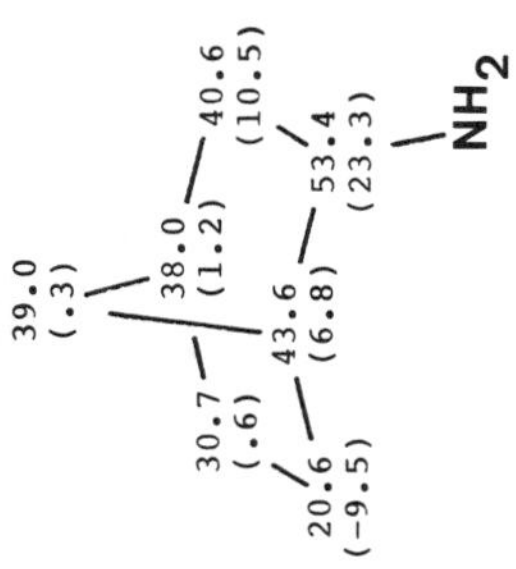

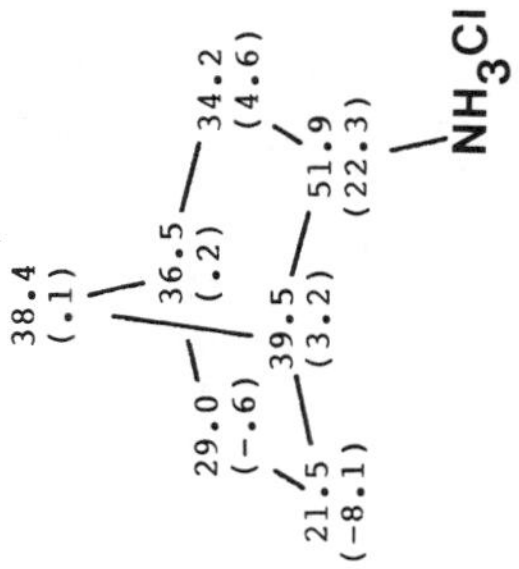

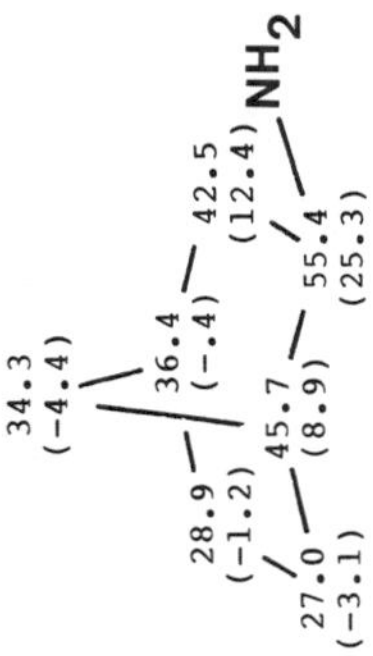

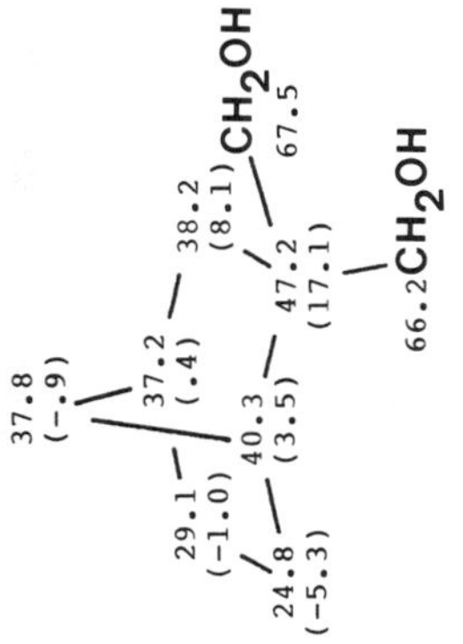

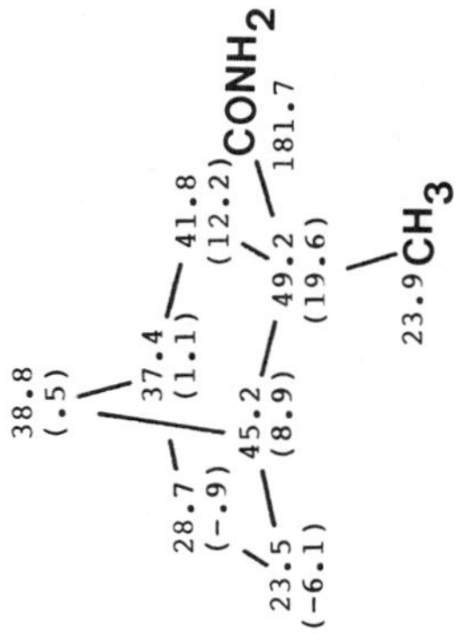

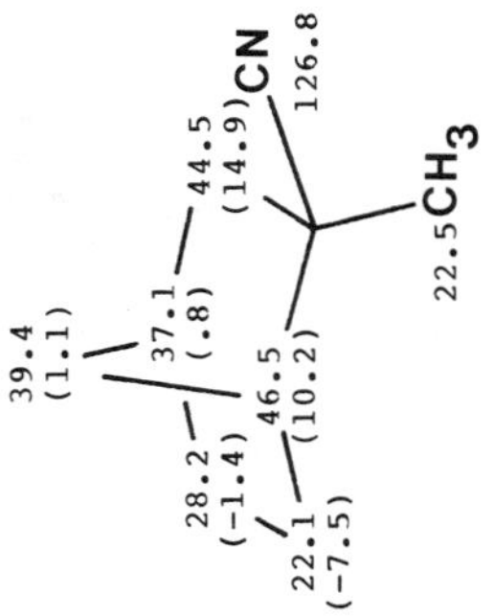

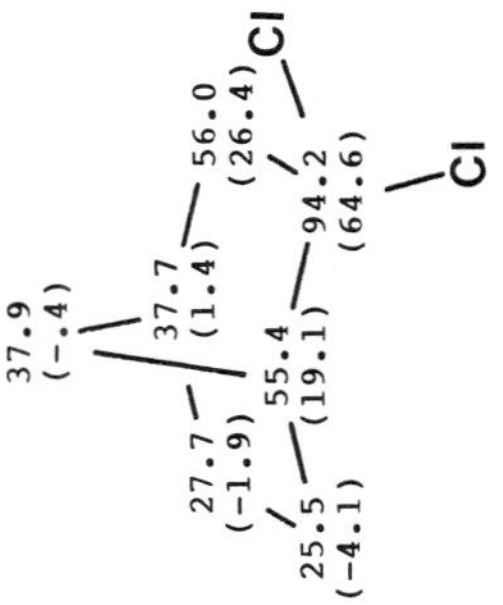

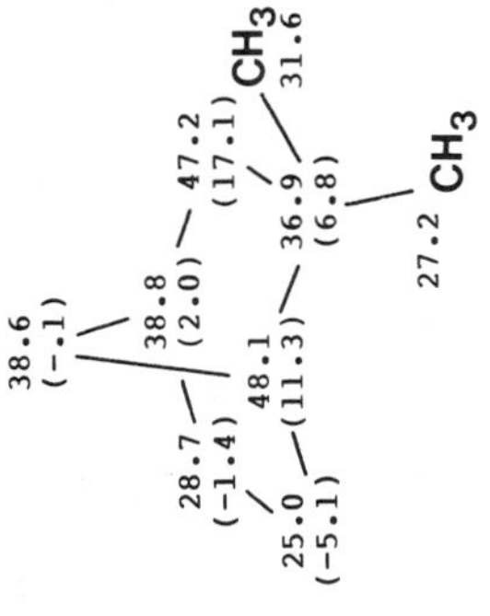

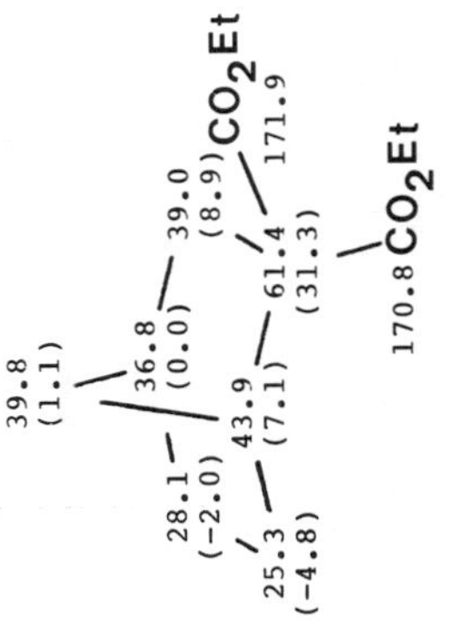

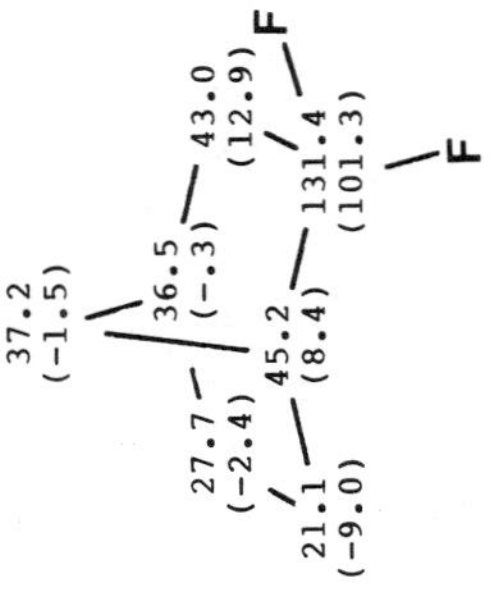

NH₂ derivative:
39.0 (.3) — 38.0 (1.2) — 40.6 (10.5) — 53.4 (23.3) — NH$_2$
30.7 (.6) — 43.6 (6.8)
20.6 (−9.5)

NH₃Cl derivative:
38.4 (.1) — 36.5 (.2) — 34.2 (4.6) — 51.9 (22.3) — NH$_3$Cl
29.0 (−.6) — 39.5 (3.2)
21.5 (−8.1)

NH₂ derivative (2):
34.3 (−4.4) — 36.4 (−.4) — 42.5 (12.4) — 55.4 (25.3) — NH$_2$
28.9 (−1.2) — 45.7 (8.9)
27.0 (−3.1)

CH₂OH derivative:
37.8 (−.9) — 37.2 (.4) — 38.2 (8.1) — CH$_2$OH 67.5
47.2 (17.1)
29.1 (−1.0) — 40.3 (3.5)
24.8 (−5.3)
66.2 CH$_2$OH

CONH₂ / CH₃ derivative:
38.8 (.5) — 37.4 (1.1) — 41.8 (12.2) — CONH$_2$ 181.7
49.2 (19.6)
28.7 (−.9) — 45.2 (8.9)
23.5 (−6.1)
23.9 CH$_3$

CN / CH₃ derivative:
39.4 (1.1) — 37.1 (.8) — 44.5 (14.9) — CN 126.8
46.5 (10.2)
28.2 (−1.4)
22.1 (−7.5)
22.5 CH$_3$

Cl / Cl derivative:
37.9 (−.4) — 37.7 (1.4) — 56.0 (26.4) — Cl
94.2 (64.6)
27.7 (−1.9) — 55.4 (19.1)
25.5 (−4.1)
Cl

CH₃ / CH₃ derivative:
38.6 (−.1) — 38.8 (2.0) — 47.2 (17.1) — CH$_3$ 31.6
36.9 (6.8)
28.7 (−1.4) — 48.1 (11.3)
25.0 (−5.1)
27.2 CH$_3$

CO₂Et / CO₂Et derivative:
39.8 (1.1) — 36.8 (0.0) — 39.0 (8.9) — CO$_2$Et 171.9
61.4 (31.3)
28.1 (−2.0) — 43.9 (7.1)
25.3 (−4.8)
170.8 CO$_2$Et

F / F derivative:
37.2 (−1.5) — 36.5 (−.3) — 43.0 (12.9) — F
131.4 (101.3)
27.7 (−2.4) — 45.2 (8.4)
21.1 (−9.0)
F

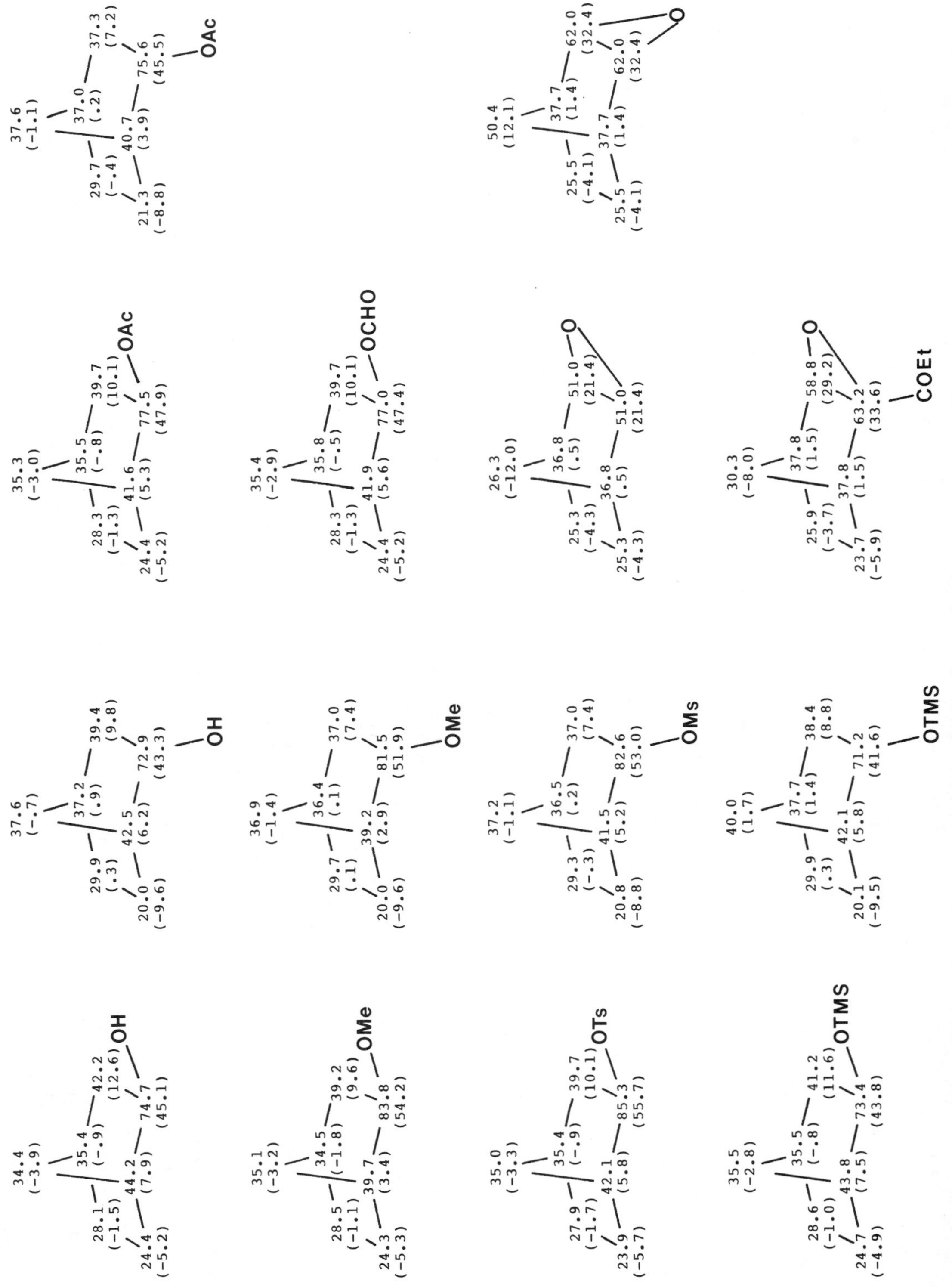

OAc
37.6 (−1.1)
37.0 (.2)
37.3 (7.2)
29.7 (−.4)
40.7 (3.9)
75.6 (45.5)
21.3 (−8.8)

O
50.4 (12.1)
37.7 (1.4)
62.0 (32.4)
25.5 (−4.1)
37.7 (1.4)
62.0 (32.4)
25.5 (−4.1)

OAc
35.3 (−3.0)
35.5 (−.8)
39.7 (10.1)
28.3 (−1.3)
41.6 (5.3)
77.5 (47.9)
24.4 (−5.2)

OCHO
35.4 (−2.9)
35.8 (−.5)
39.7 (10.1)
28.3 (−1.3)
41.9 (5.6)
77.0 (47.4)
24.4 (−5.2)

O
26.3 (−12.0)
36.8 (.5)
51.0 (21.4)
25.3 (−4.3)
36.8 (.5)
51.0 (21.4)
25.3 (−4.3)

COEt
30.3 (−8.0)
37.8 (1.5)
58.8 (29.2)
25.9 (−3.7)
37.8 (1.5)
63.2 (33.6)
23.7 (−5.9)

OH
37.6 (−.7)
37.2 (.9)
39.4 (9.8)
29.9 (.3)
42.5 (6.2)
72.9 (43.3)
20.0 (−9.6)

OMe
36.9 (−1.4)
36.4 (.1)
37.0 (7.4)
29.7 (.1)
39.2 (2.9)
81.5 (51.9)
20.0 (−9.6)

OMs
37.2 (−1.1)
36.5 (.2)
37.0 (7.4)
29.3 (−.3)
41.5 (5.2)
82.6 (53.0)
20.8 (−8.8)

OTMS
40.0 (1.7)
37.7 (1.4)
38.4 (8.8)
29.9 (.3)
42.1 (5.8)
71.2 (41.6)
20.1 (−9.5)

OH
34.4 (−3.9)
35.4 (−.9)
42.2 (12.6)
28.1 (−1.5)
44.2 (7.9)
74.7 (45.1)
24.4 (−5.2)

OMe
35.1 (−3.2)
34.5 (−1.8)
39.2 (9.6)
28.5 (−1.1)
39.7 (3.4)
83.8 (54.2)
24.3 (−5.3)

OTs
35.0 (−3.3)
35.4 (−.9)
39.7 (10.1)
27.9 (−1.7)
42.1 (5.8)
85.3 (55.7)
23.9 (−5.7)

OTMS
35.5 (−2.8)
35.5 (−.8)
41.2 (11.6)
28.6 (−1.0)
43.8 (7.5)
73.4 (43.8)
24.7 (−4.9)

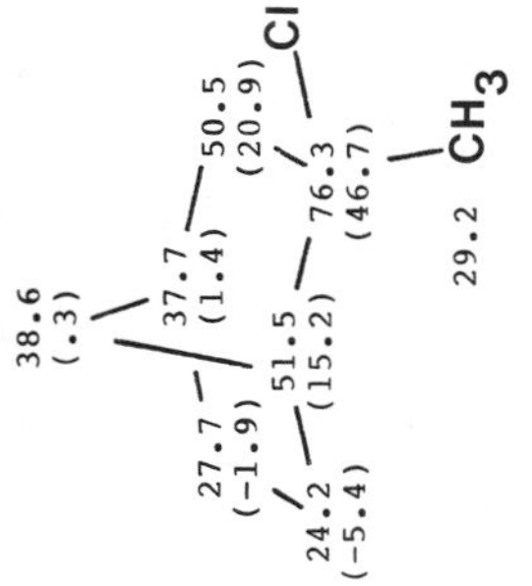

Structure 1 (Cl / CH₃):
38.6 (.3)
37.7 (1.4)
50.5 (20.9) — Cl
27.7 (−1.9)
51.5 (15.2)
76.3 (46.7)
24.2 (−5.4)
29.2 CH₃

Structure 2 (CH₃ / OH):
38.7 (.4)
37.3 (1.0)
47.0 (17.4)
30.4 CH₃
28.3 (−1.3)
48.4 (12.1)
77.2 (47.6) — OH
22.0 (−7.6)

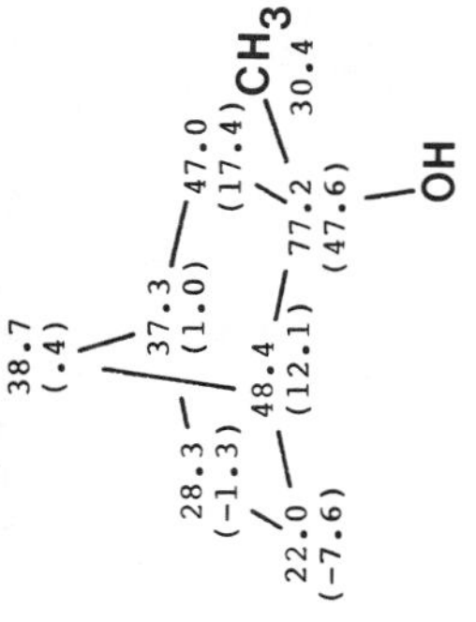

Structure 3 (CH₂CH₃ / OH):
38.8 (.1)
37.2 (.4)
44.6 (14.5)
34.8 CH₂CH₃
28.7 (−1.4)
46.9 (10.1)
79.4 (49.3) — OH
22.4 (−7.7)

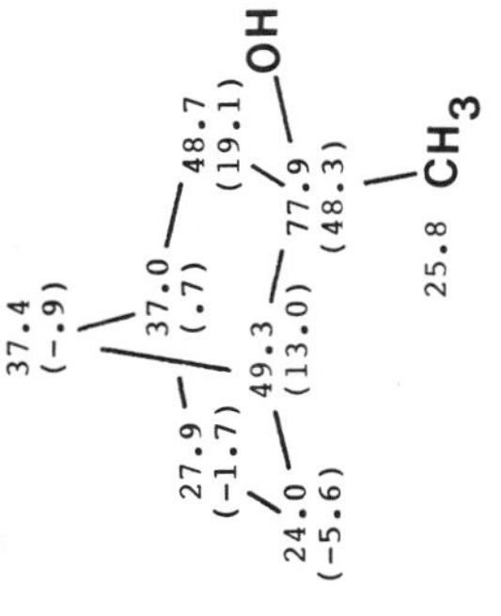

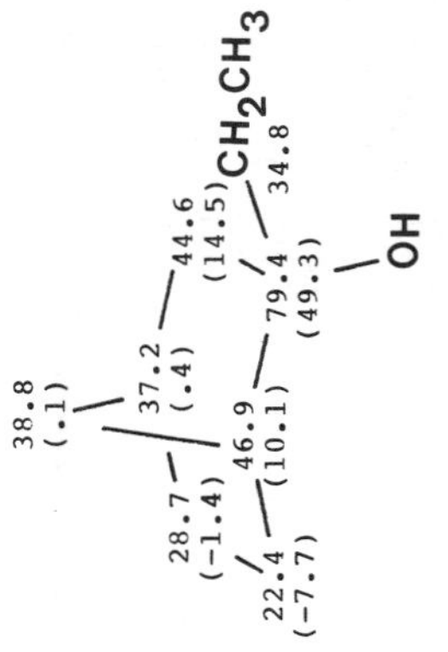

Structure 4 (CHMe₂ / OH):
38.6 (.3)
37.5 (1.2)
44.5 (14.9)
36.1 CHMe₂
28.3 (−1.3)
45.3 (9.0)
81.5 (51.9) — OH
22.8 (−6.8)

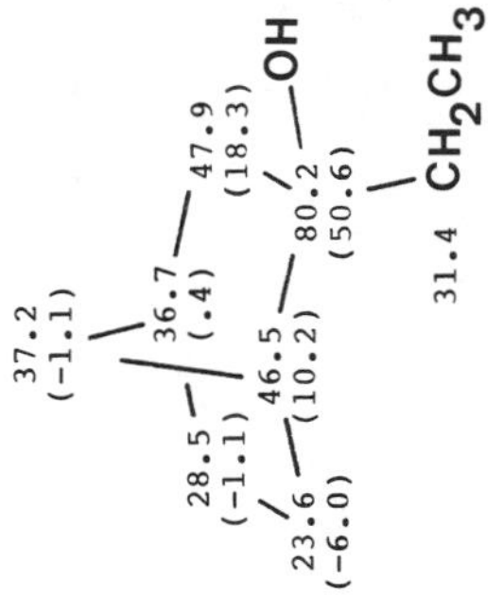

Structure 5 (Ph / OH):
38.8 (.5)
37.6 (1.3)
46.6 (17.0)
149.1 Ph
29.1 (−.5)
47.1 (10.8)
80.7 (51.1) — OH
22.3 (−7.3)

Structure 6 (OH / CH₃):
37.4 (−.9)
37.0 (.7)
48.7 (19.1) — OH
27.9 (−1.7)
49.3 (13.0)
77.9 (48.3)
24.0 (−5.6)
25.8 CH₃

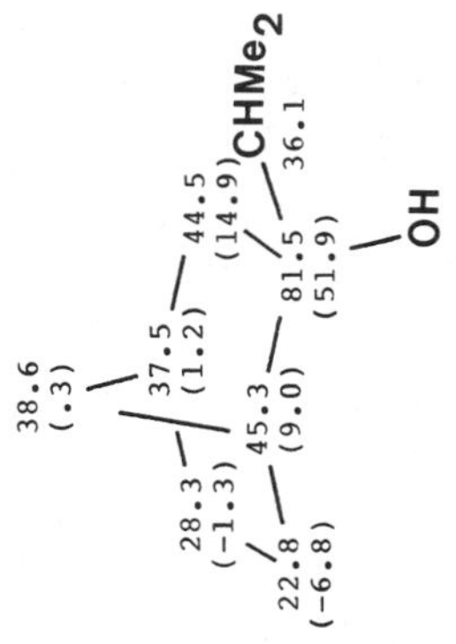

Structure 7 (OH / CH₂CH₃):
37.2 (−1.1)
36.7 (.4)
47.9 (18.3) — OH
28.5 (−1.1)
46.5 (10.2)
80.2 (50.6)
23.6 (−6.0)
31.4 CH₂CH₃

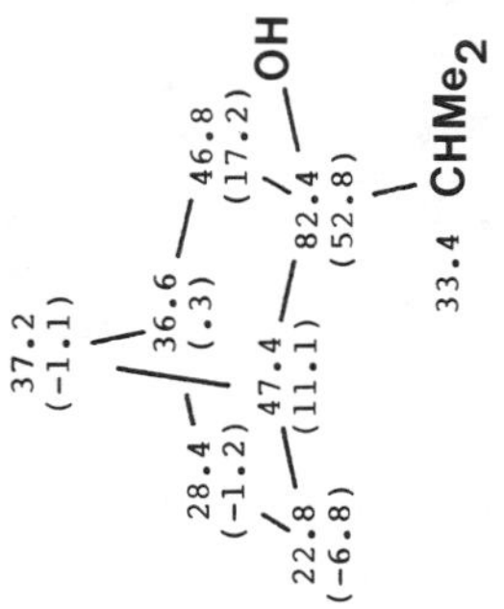

Structure 8 (OH / CHMe₂):
37.2 (−1.1)
36.6 (.3)
46.8 (17.2) — OH
28.4 (−1.2)
47.4 (11.1)
82.4 (52.8)
22.8 (−6.8)
33.4 CHMe₂

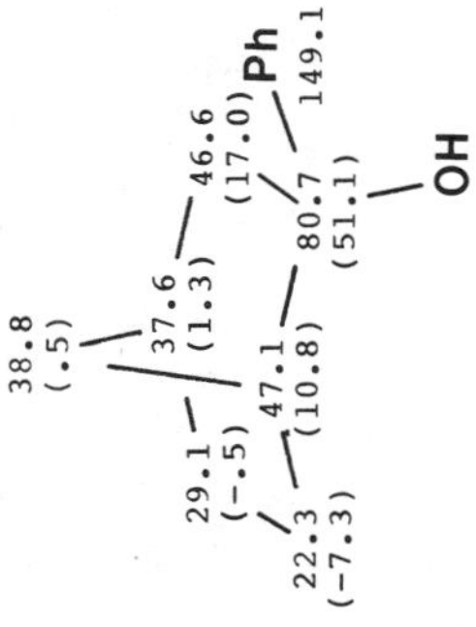

Structure 9 (CH₃ / OTMS):
39.0 (.7)
36.8 (.5)
45.9 (16.3) — CH₃
27.9 (−1.7)
47.0 (10.7)
72.8 (43.2)
21.5 (−8.1)
OTMS

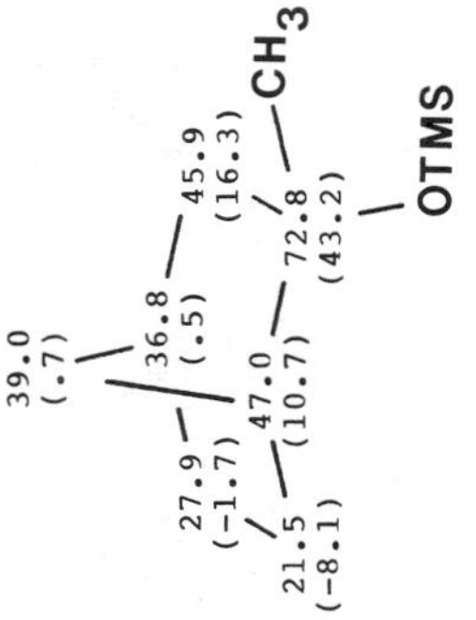

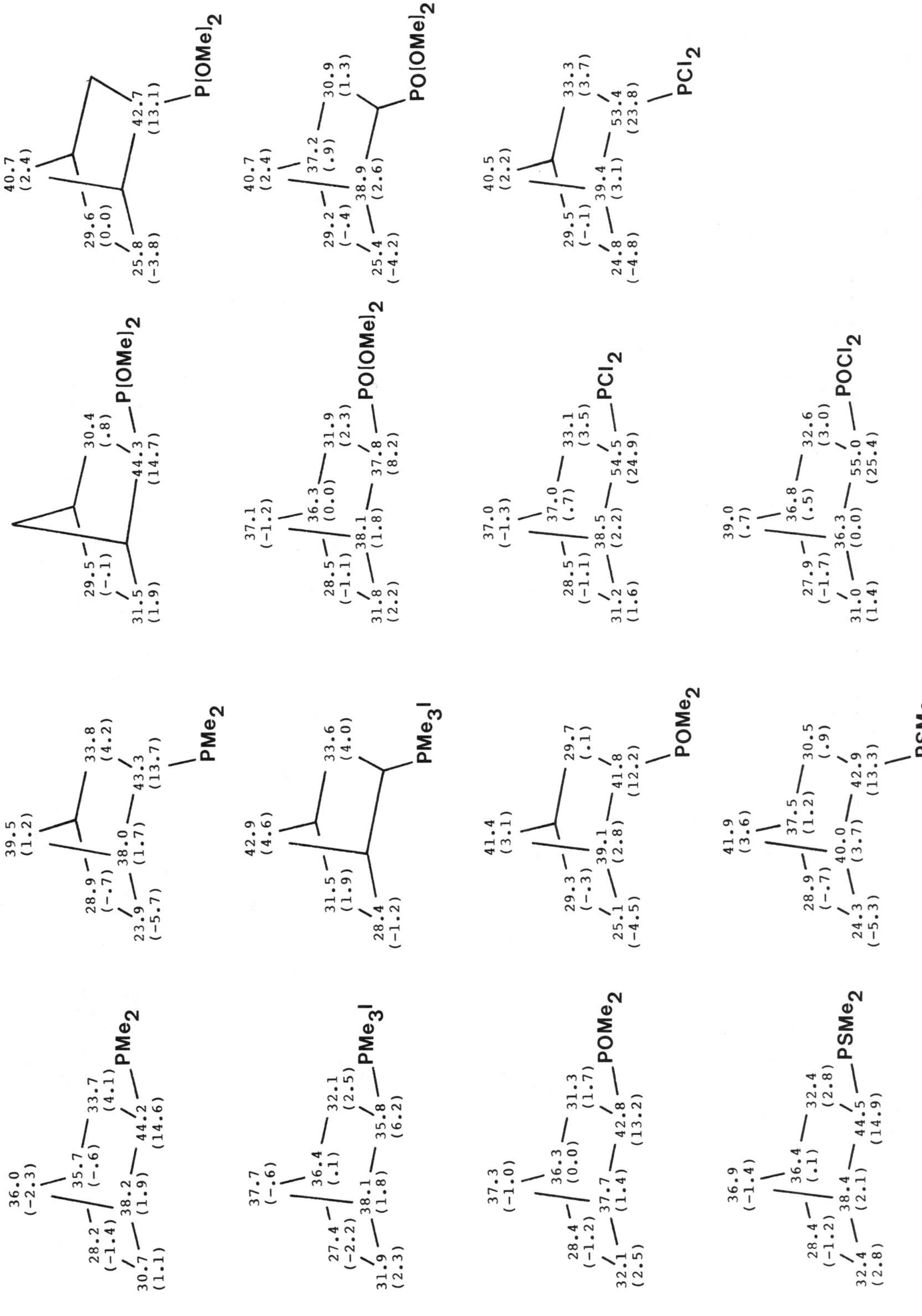

P(OMe)₂

40.7 (2.4)
42.7 (13.1)
29.6 (0.0)
25.8 (-3.8)

PO(OMe)₂

40.7 (2.4)
37.2 (.9)
30.9 (1.3)
29.2 (-.4)
38.9 (2.6)
25.4 (-4.2)

PCl₂

40.5 (2.2)
33.3 (3.7)
53.4 (23.8)
29.5 (-.1)
39.4 (3.1)
24.8 (-4.8)

P(OMe)₂

30.4 (.8)
44.3 (14.7)
29.5 (-.1)
31.5 (1.9)

PO(OMe)₂

37.1 (-1.2)
36.3 (0.0)
31.9 (2.3)
38.1 (1.8)
37.8 (8.2)
28.5 (-1.1)
31.8 (2.2)

PCl₂

37.0 (-1.3)
37.0 (.7)
33.1 (3.5)
38.5 (2.2)
54.5 (24.9)
28.5 (-1.1)
31.2 (1.6)

POCl₂

39.0 (.7)
36.8 (.5)
32.6 (3.0)
36.3 (0.0)
55.0 (25.4)
27.9 (-1.7)
31.0 (1.4)

PMe₂

39.5 (1.2)
33.8 (4.2)
43.3 (13.7)
28.9 (-.7)
38.0 (1.7)
23.9 (-5.7)

PMe₃¹

42.9 (4.6)
33.6 (4.0)
31.5 (1.9)
28.4 (-1.2)

POMe₂

41.4 (3.1)
29.7 (.1)
41.8 (12.2)
29.3 (-.3)
39.1 (2.8)
25.1 (-4.5)

PSMe₂

41.9 (3.6)
37.5 (1.2)
30.5 (.9)
40.0 (3.7)
42.9 (13.3)
28.9 (-.7)
24.3 (-5.3)

PMe₂

36.0 (-2.3)
35.7 (-.6)
33.7 (4.1)
38.2 (1.9)
44.2 (14.6)
28.2 (-1.4)
30.7 (1.1)

PMe₃¹

37.7 (-.6)
36.4 (.1)
32.1 (2.5)
38.1 (1.8)
35.8 (6.2)
27.4 (-2.2)
31.9 (2.3)

POMe₂

37.3 (-1.0)
36.3 (0.0)
31.3 (1.7)
37.7 (1.4)
42.8 (13.2)
28.4 (-1.2)
32.1 (2.5)

PSMe₂

36.9 (-1.4)
36.4 (.1)
32.4 (2.8)
38.4 (2.1)
44.5 (14.9)
28.4 (-1.2)
32.4 (2.8)

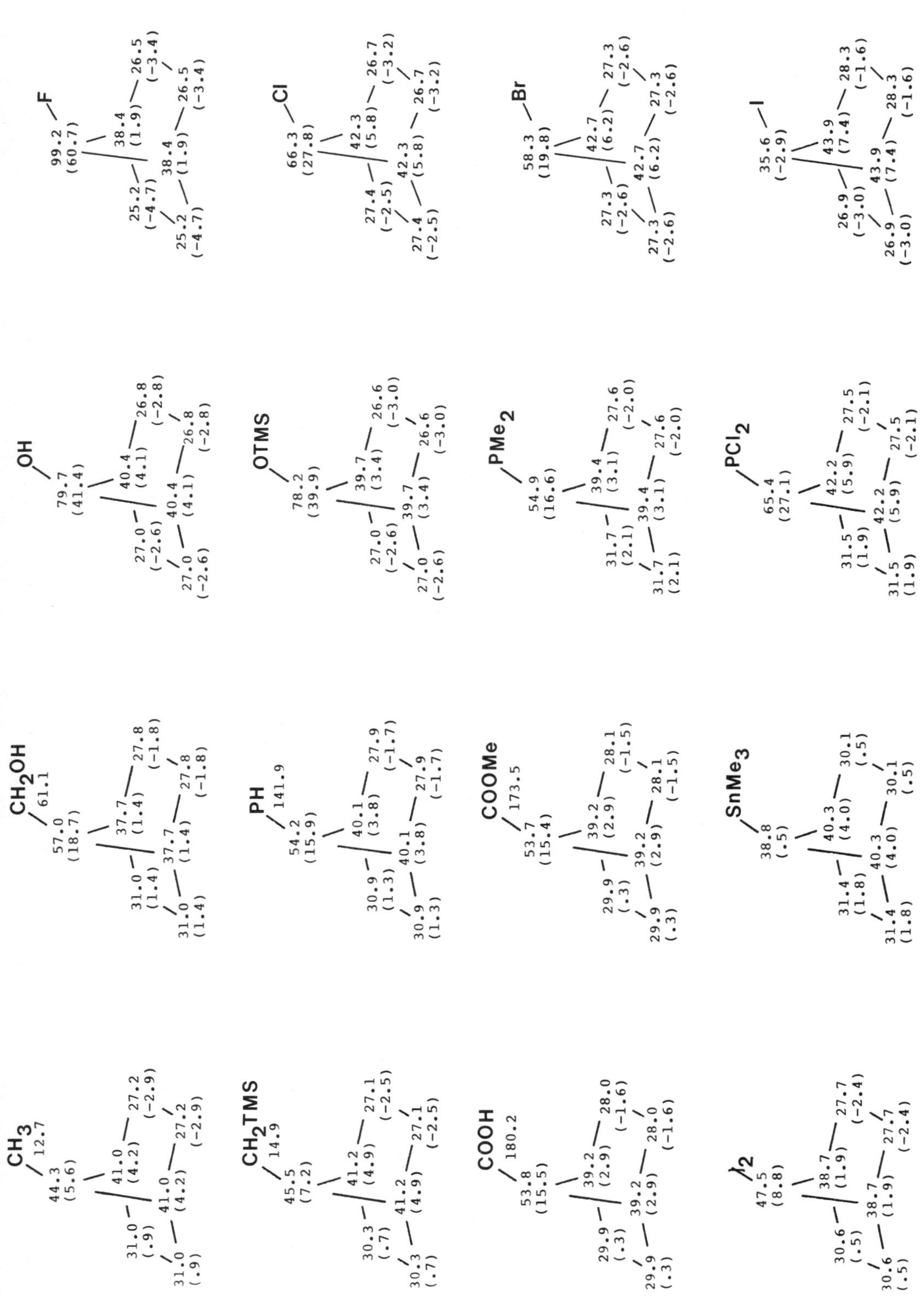

F
99.2 (60.7)
38.4 (1.9) — 26.5 (−3.4)
38.4 — 26.5 (−3.4)
25.2 (−4.7) — 26.5 (1.9)
25.2 (−4.7)

Cl
66.3 (27.8)
42.3 (5.8) — 26.7 (−3.2)
42.3 (5.8) — 26.7 (−3.2)
27.4 (−2.5) — 26.7 (5.8)
27.4 (−2.5)

Br
58.3 (19.8)
42.7 (6.2) — 27.3 (−2.6)
42.7 (6.2) — 27.3 (−2.6)
27.3 (−2.6) — 27.3 (6.2)
27.3 (−2.6)

I
35.6 (−2.9)
43.9 (7.4) — 28.3 (−1.6)
43.9 (7.4) — 28.3 (−1.6)
26.9 (−3.0) — 43.9 (7.4)
26.9 (−3.0)

OH
79.7 (41.4)
40.4 (4.1) — 26.8 (−2.8)
40.4 (4.1) — 26.8 (−2.8)
27.0 (−2.6) — 40.4 (4.1)
27.0 (−2.6)

OTMS
78.2 (39.9)
39.7 (3.4) — 26.6 (−3.0)
39.7 (3.4) — 26.6 (−3.0)
27.0 (−2.6) — 39.7 (3.4)
27.0 (−2.6)

PMe₂
54.9 (16.6)
39.4 (3.1) — 27.6 (−2.0)
39.4 (3.1) — 27.6 (−2.0)
31.7 (2.1) — 39.4 (3.1)
31.7 (2.1)

PCl₂
65.4 (27.1)
42.2 (5.9) — 27.5 (−2.1)
42.2 (5.9) — 27.5 (−2.1)
31.5 (1.9) — 42.2 (5.9)
31.5 (1.9)

CH₂OH
61.1
57.0 (18.7)
37.7 (1.4) — 27.8 (−1.8)
37.7 (1.4) — 27.8 (−1.8)
31.0 (1.4) — 37.7 (1.4)
31.0 (1.4)

PH
141.9
54.2 (15.9)
40.1 (3.8) — 27.9 (−1.7)
40.1 (3.8) — 27.9 (−1.7)
30.9 (1.3) — 40.1 (3.8)
30.9 (1.3)

COOMe
173.5
53.7 (15.4)
39.2 (2.9) — 28.1 (−1.5)
39.2 (2.9) — 28.1 (−1.5)
29.9 (.3) — 39.2 (2.9)
29.9 (.3)

SnMe₃
38.8 (.5)
40.3 (4.0) — 30.1 (.5)
40.3 (4.0) — 30.1 (.5)
31.4 (1.8) — 40.3 (4.0)
31.4 (1.8)

CH₃
12.7
44.3 (5.6)
41.0 (4.2) — 27.2 (−2.9)
41.0 (4.2) — 27.2 (−2.9)
31.0 (.9) — 41.0 (4.2)
31.0 (.9)

CH₂TMS
14.9
45.5 (7.2)
41.2 (4.9) — 27.1 (−2.5)
41.2 (4.9) — 27.1 (−2.5)
30.3 (.7) — 41.2 (4.9)
30.3 (.7)

COOH
180.2
53.8 (15.5)
39.2 (2.9) — 28.0 (−1.6)
39.2 (2.9) — 28.0 (−1.6)
29.9 (.3) — 39.2 (2.9)
29.9 (.3)

)₂
47.5 (8.8)
38.7 (1.9) — 27.7 (−2.4)
38.7 (1.9) — 27.7 (−2.4)
30.6 (.5) — 38.7 (1.9)
30.6 (.5)

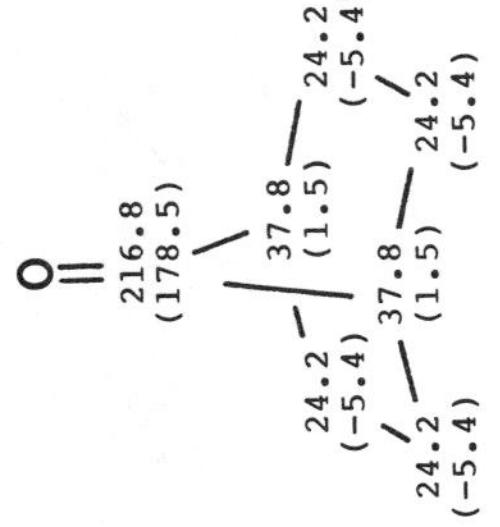

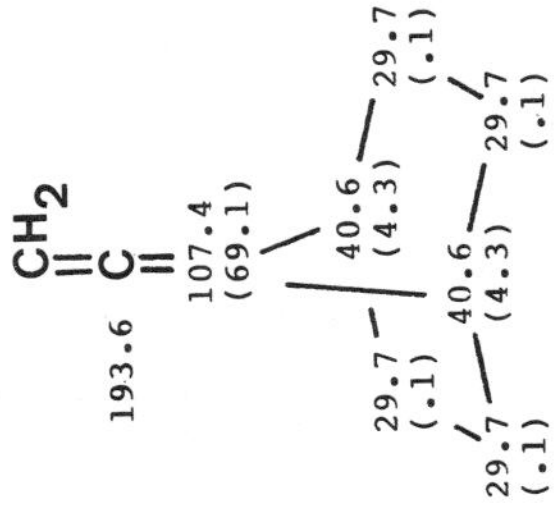

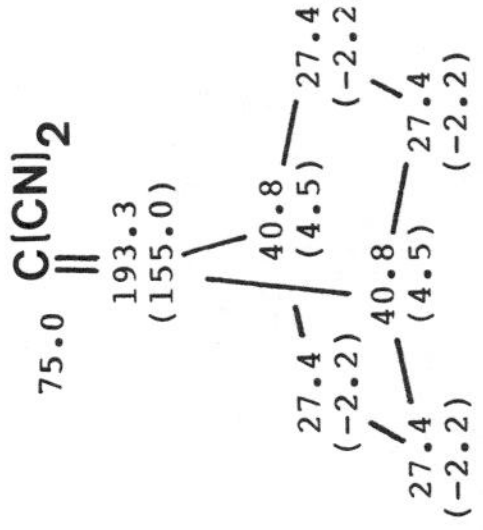

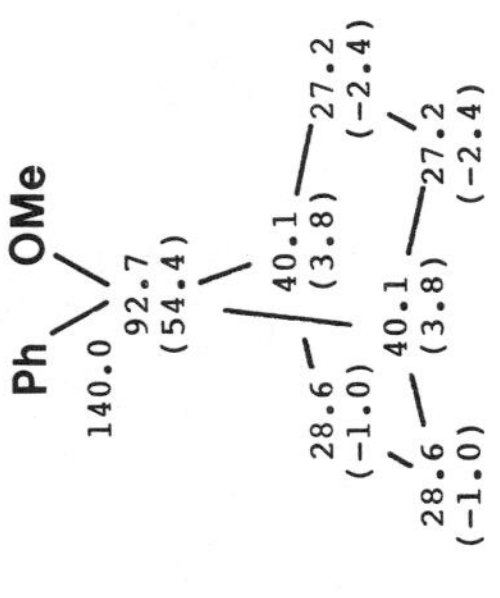

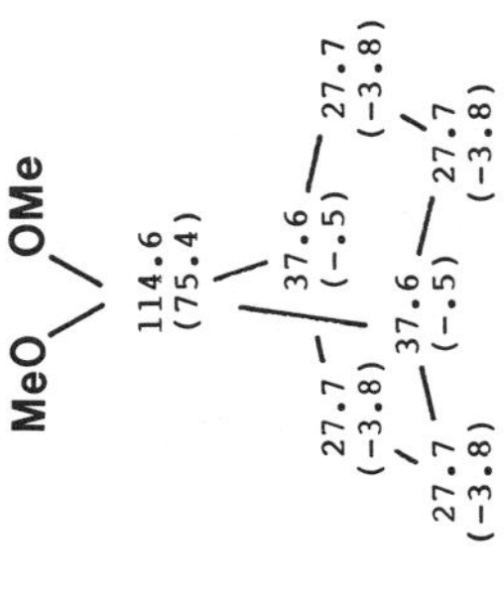

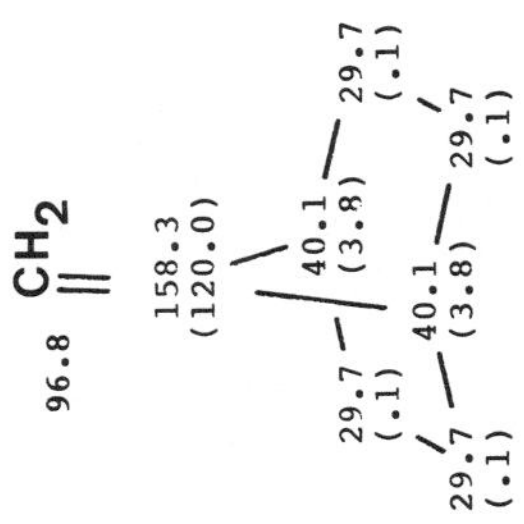

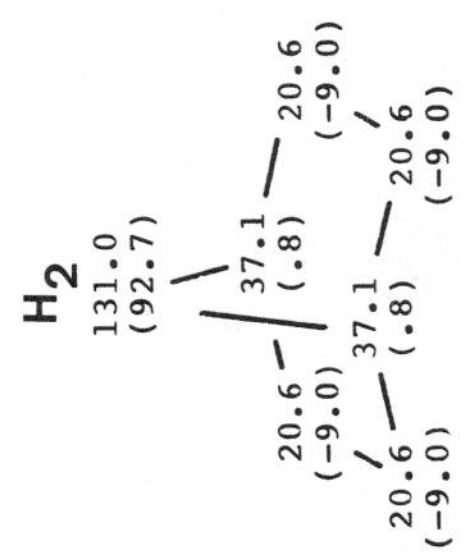

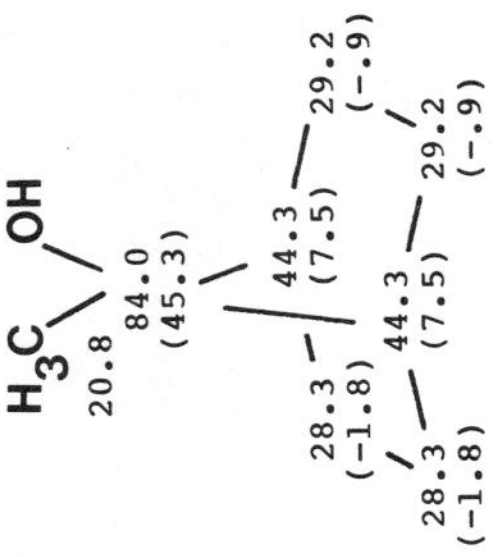

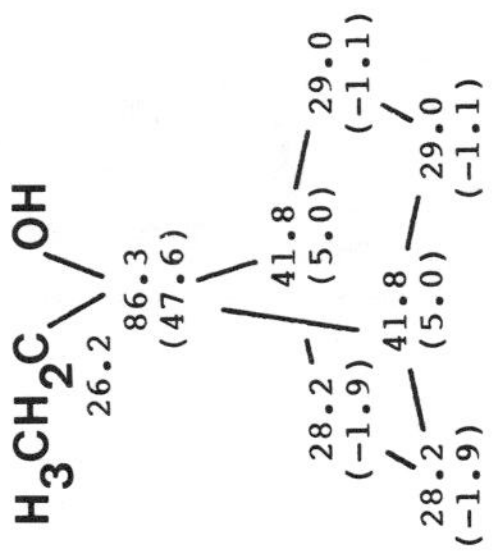

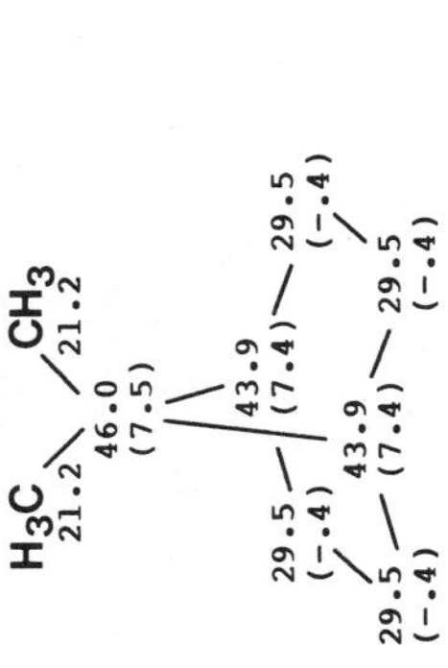

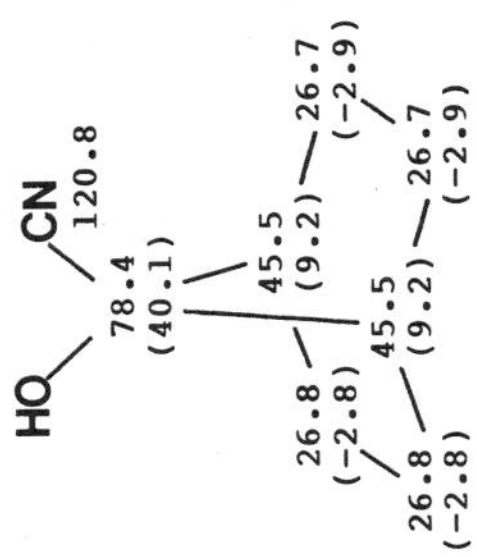

O
216.8
(178.5)
37.8 (1.5) — 24.2 (-5.4)
24.2 (-5.4)
37.8 (1.5) — 24.2 (-5.4)
24.2 (-5.4)

CH₂=C (CH_2=C)
193.6
107.4 (69.1)
40.6 (4.3) — 29.7 (.1)
29.7 (.1)
40.6 (4.3) — 29.7 (.1)
29.7 (.1)

C(CN)₂ ($C(CN)_2$)
75.0
193.3 (155.0)
40.8 (4.5) — 27.4 (-2.2)
27.4 (-2.2)
40.8 (4.5) — 27.4 (-2.2)
27.4 (-2.2)

Ph OMe
140.0 92.7 (54.4)
40.1 (3.8) — 27.2 (-2.4)
28.6 (-1.0) — 27.2 (-2.4)
40.1 (3.8)
28.6 (-1.0)

MeO OMe
114.6 (75.4)
37.6 (-.5) — 27.7 (-3.8)
27.7 (-3.8)
37.6 (-.5) — 27.7 (-3.8)
27.7 (-3.8)

CH₂ (CH_2)
96.8
158.3 (120.0)
40.1 (3.8) — 29.7 (.1)
29.7 (.1)
40.1 (3.8) — 29.7 (.1)
29.7 (.1)

H₂ (H_2)
131.0 (92.7)
37.1 (.8) — 20.6 (-9.0)
20.6 (-9.0)
37.1 (.8) — 20.6 (-9.0)
20.6 (-9.0)

H₃C OH (H_3C OH)
20.8 84.0 (45.3)
44.3 (7.5) — 29.2 (-.9)
28.3 (-1.8) — 29.2 (-.9)
44.3 (7.5)
28.3 (-1.8)

H₃CH₂C OH (H_3CH_2C OH)
26.2 86.3 (47.6)
41.8 (5.0) — 29.0 (-1.1)
28.2 (-1.9) — 29.0 (-1.1)
41.8 (5.0)
28.2 (-1.9)

H₃C CH₃ (H_3C CH_3)
21.2 21.2 46.0 (7.5)
43.9 (7.4) — 29.5 (-.4)
29.5 (-.4) — 29.5 (-.4)
43.9 (7.4)
29.5 (-.4)

HO CN
120.8 78.4 (40.1)
45.5 (9.2) — 26.7 (-2.9)
26.8 (-2.8) — 26.7 (-2.9)
45.5 (9.2)
26.8 (-2.8)

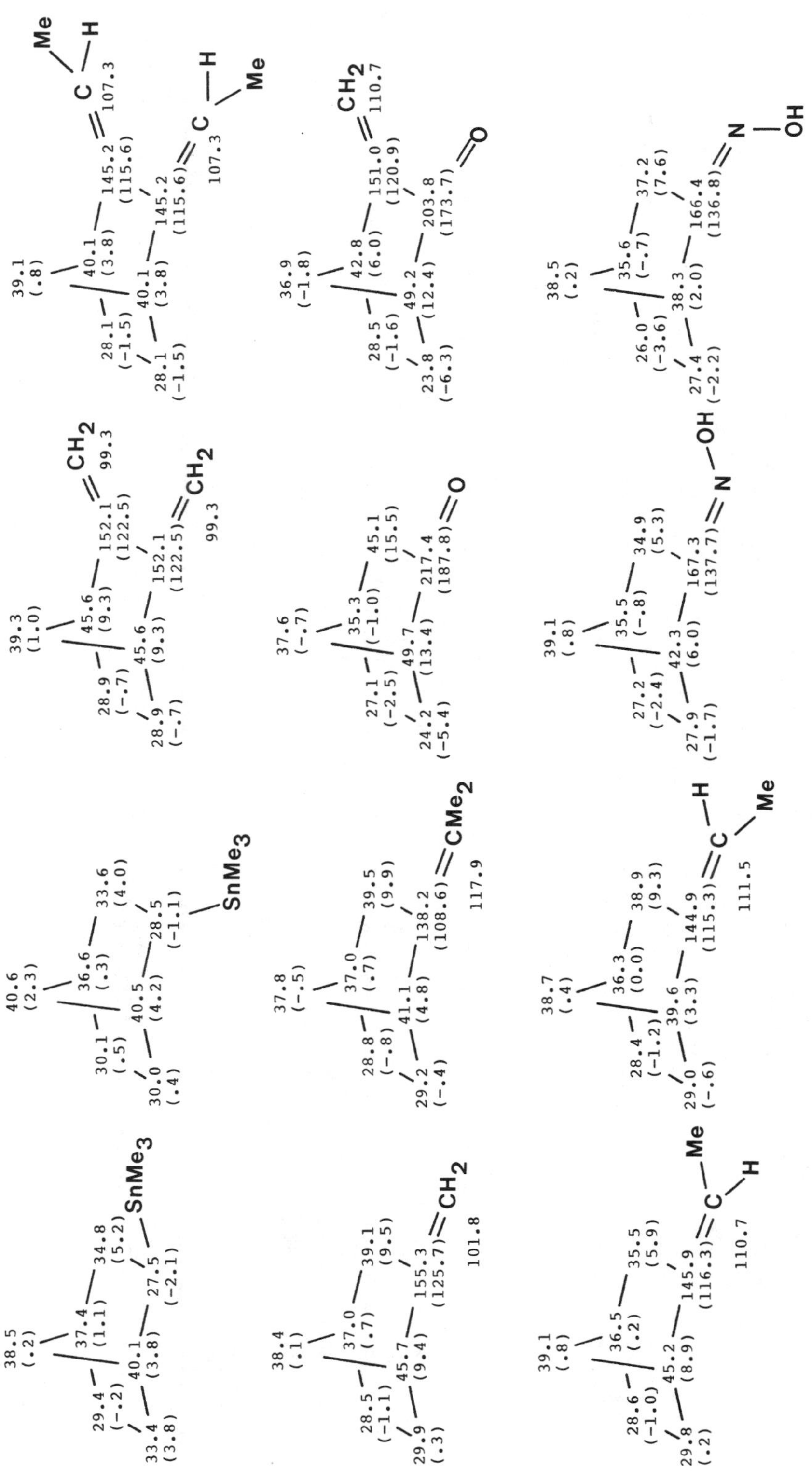

Structure 1 (Me/H isopropylidene, top)

39.1 (.8); 40.1 (3.8); 145.2 (115.6); 28.1 (-1.5); 28.1 (-1.5); 40.1 (3.8); 145.2 (115.6); C 107.3 — Me, H; C 107.3 — H, Me

Structure 2 (=CH₂ / =O, top middle)

36.9 (-1.8); 42.8 (6.0); 151.0 (120.9) =CH₂ 110.7; 28.5 (-1.6); 23.8 (-6.3); 49.2 (12.4); 203.8 (173.7) =O

Structure 3 (=N-OH, top right)

38.5 (.2); 35.6 (-.7); 37.2 (7.6); 26.0 (-3.6); 27.4 (-2.2); 38.3 (2.0); 166.4 (136.8) =N-OH

Structure 4 (=CH₂, middle left)

39.3 (1.0); 45.6 (9.3); 152.1 (122.5) =CH₂ 99.3; 28.9 (-.7); 28.9 (-.7); 45.6 (9.3); 152.1 (122.5) =CH₂ 99.3

Structure 5 (=O, middle)

37.6 (-.7); 35.3 (-1.0); 45.1 (15.5); 27.1 (-2.5); 24.2 (-5.4); 49.7 (13.4); 217.4 (187.8) =O

Structure 6 (=N-OH, middle right)

39.1 (.8); 35.5 (-.8); 34.9 (5.3); 27.2 (-2.4); 27.9 (-1.7); 42.3 (6.0); 167.3 (137.7) =N-OH

Structure 7 (SnMe₃, lower left)

40.6 (2.3); 36.6 (.3); 33.6 (4.0); 30.1 (.5); 30.0 (.4); 40.5 (4.2); 28.5 (-1.1) SnMe₃

Structure 8 (=CMe₂, lower middle)

37.8 (-.5); 37.0 (.7); 39.5 (9.9); 28.8 (-.8); 29.2 (-.4); 41.1 (4.8); 138.2 (108.6) =CMe₂ 117.9

Structure 9 (=C H/Me, lower right)

38.7 (.4); 36.3 (0.0); 38.9 (9.3); 28.4 (-1.2); 29.0 (-.6); 39.6 (3.3); 144.9 (115.3) =C — H, Me 111.5

Structure 10 (SnMe₃, bottom left)

38.5 (.2); 37.4 (1.1); 34.8 (5.2); 29.4 (-.2); 33.4 (3.8); 40.1 (3.8); 27.5 (-2.1) SnMe₃

Structure 11 (=CH₂, bottom middle)

38.4 (.1); 37.0 (.7); 39.1 (9.5); 28.5 (-1.1); 29.9 (.3); 45.7 (9.4); 155.3 (125.7) =CH₂ 101.8

Structure 12 (=C Me/H, bottom right)

39.1 (.8); 36.5 (.2); 35.5 (5.9); 28.6 (-1.0); 29.8 (.2); 45.2 (8.9); 145.9 (116.3) =C — Me, H 110.7

sat. BICYCLO[2.2.1]HEPTANES

Str	Substituents	C1	C2	C3	C4	C5	C6	C7	Substituent Shifts	References
1	PARENT	36.3	29.6	29.6	36.3	29.6	29.6	38.3		7,1-4,9,16,20,25,32,38,42,47, 51,104,108-9,234,245,268
2	1-Br	62.2	40.0	31.6	34.8	31.6	40.0	48.2		4
3	1-Br/2-=O	66.4	207.5	43.6	33.6	29.4	34.4	46.7		191
4	1-Br/2-=O/A7-Me S7-Me	77.2	208.7	42.3	40.5	28.0	32.7	48.9	19.4,19.9	161
5	1-Br/2-=O/X3-Me N3-Me	66.3	212.8	46.9	44.6	25.7	35.0	44.5	24.0,22.0	69
6	1-Br/4-F	55.6	40.4	34.6	98.4	34.6	40.0	50.0		264
7	1-CH2CH2CH2-2N	55.9	53.6	29.9	43.5	32.3	27.5	40.4	27.7,27.7	64,66
8	1-CH2CH2CH2-2X	56.5	48.1	40.0	39.3	29.6	33.7	41.1	28.8,33.7	64,66
9	1-CH2COOEt/2-=O/X3-Me N3-Me	55.3	220.9	47.6	45.3	30.5	30.1	38.9	34.1,23.2,21.5	69
10	1-CH2OH	50.1	31.8	32.3	37.3	30.5	31.8	40.6	66.9	146
11	1-CH2OH/2-=CH2/X3-Me N3-Me	56.2	165.2	43.3	47.0	24.6	31.0	39.6	98.3,29.2,25.9	50
12	1-CH2OH/2-=O/X3-Me N3-Me	60.2	223.8	48.0	45.2	24.2	27.1	37.4	62.2,23.1,21.4	69
13	1-CH2OMe/2-=O	51.1	215.6	45.5	33.8	27.7	26.4	39.8	70.3	191
14	1-CH2SO3H/2-=O/A7-Me S7-Me	58.0	208.1	42.5	42.5	24.4	26.7	47.4	47.4,19.6,19.6	195
15	1-Cl	69.9	38.4	30.9	34.8	30.9	38.4	46.8		4
16	1-Cl/2-=CH2/X3-Me N3-Me	73.4	163.0	42.7	44.9	25.7	37.9	46.3	101.3,29.8,26.3	50
17	1-Cl/2-=O	73.9	207.9	45.6	32.5	29.9	33.0	44.1		191
18	1-Cl/2-=O/X3-Me N3-Me	73.8	213.0	47.2	43.5	25.2	33.2	43.6	23.9,21.9	69
19	1-Cl/4-F	64.5	39.0	34.1	98.2	34.1	39.0	49.1		264
20	1-Cl/X2-Cl	72.6	64.9	43.9	34.3	30.4	36.6	42.2		237
21	1-Cl/X3-Cl	66.9	51.1	60.9	44.4	27.5	36.7	43.1		237
22	1-CN/4-F	33.1	33.7	31.4	100.6	31.4	33.7	45.1	121.7	264
23	1-COMe/4-F	54.6	31.8	32.3	101.6	32.3	31.8	43.4	210.2	264
24	1-CONH2/2-=CH2/X3-Me N3-Me	60.7	163.9	43.0	47.7	24.6	31.4	41.6	0,101.0,29.6,26.0	50
25	1-CONH2/4-F		32.8	32.3	101.4	32.3	32.8	43.6		264
26	1-COOH	52.1	32.9	30.0	37.8	30.0	32.9	42.3	183.5	4,146
27	1-COOH/2-=CH2/X3-Me N3-Me	59.5	162.4	42.9	47.4	24.5	31.9	40.7	180.7,101.0,29.6,26.0	50
28	1-COOH/2-=O/X3-Me N3-Me	62.8	216.3	48.0	45.2	24.0	28.2	38.5	175.2,23.4,21.6	69
29	1-COOH/4-F	47.2	32.1	32.1	101.6	32.1	31.4	44.0	181.4	264
30	1-COOH/4-OH	50.1	33.7	35.2	82.2	35.2	33.7	46.4	181.6	264
31	1-COOMe/4-OMe	48.2	35.5	31.2	87.0	31.2	35.5	43.0	181.2	264
32	1-COOMe	52.2	33.1	30.0	35.6	30.0	31.9	42.3	176.6	4
33	1-COOMe/2-=CH2/X3-Me N3-Me	59.7	163.0	42.8	47.3	24.5	31.9	40.6	174.1,100.6,29.6,26.0	50
34	1-COOMe/2-=O/X3-Me N3-Me	63.9	215.3	47.7	45.3	24.1	27.1	38.5	171.0,23.6,21.7	69
35	1-COOMe/4-F	47.4	32.3	32.2	101.6	32.2	33.7	44.0	175.1	264
36	1-COOMe/4-OH	49.0	33.0	35.2	81.7	35.2	33.0	46.7	176.2	264
37	1-Et	48.4	34.2	31.2	37.5	31.2	34.2	43.2	28.5	3
38	1-Et/2-=O	57.9	214.6	45.8	34.1	29.4	28.7	40.7	21.6	3,95
39	1-Et/N2-OH	51.9	74.9	40.5	36.5	31.0	25.0	40.9	24.6	3,95
40	1-Et/X2-OAc	50.9	76.9	41.4	35.7	29.3	29.3	40.2	23.4	3
41	1-Et/X2-OH	51.7	74.8	43.1	35.5	30.3	29.3	38.9	23.1	3,95
42	1-Et/X3-OAc	47.9	43.7	77.9	42.3	25.7	32.6	39.7	27.6	3
43	1-F	103.9	32.0	29.6	33.0	29.6	32.0	41.0		264
44	1-F/4-F	96.5	33.1	33.1	96.5	33.1	33.1	44.4		264,143
45	1-F/4-I	98.3	35.2	43.6	30.0	43.6	35.2	52.2		264
46	1-F/4-Me	102.4	33.8	36.5	39.9	36.5	33.8	47.4	22.1	264
47	1-F/4-NO2	97.4	32.5	33.5	85.3	33.5	32.5	44.8		264
48	1-F/4-Ph	102.2	33.3	36.6	46.7	36.6	33.3	45.0	145.4	264
49	1-I	37.7	43.0	32.1	34.6	32.1	43.0	50.8		4
50	1-I/2-=O	44.9	208.8	42.3	35.4	29.9	37.4	49.2		190

Str	Substituents	C1	C2	C3	C4	C5	C6	C7	Substituent Shifts	References
51	1-I/N2-I	43.0	43.3	44.5	35.6	31.9	41.8	47.6		190
52	1-I/X2-I	45.3	41.0	44.9	36.6	31.8	40.6	47.6		190
53	1-1Pr/2-=0/4-Me	63.2	215.7	52.7	40.8	35.3	28.3	44.9	27.3,21.1	87,90
54	1-1Pr/4-Me	52.5	32.7	37.6	43.7	37.6	32.7	48.3	32.6,21.4	90
55	1-Me	43.7	36.7	31.2	37.8	31.2	36.7	45.2	20.9	4,1,2,25,173,268
56	1-Me/2-=CH2	48.7	158.3	39.6	35.9	30.2	36.7	46.0	17.5,99.8	33,2,25
57	1-Me/2-=CH2/A7-Me S7-Me	51.7	159.3	37.2	45.2	28.3	35.4	47.5	12.6,101.6,19.0,19.7	33,241
58	1-Me/2-=CH2/X3-Me N3-Me	50.2	168.8	43.1	47.8	25.8	35.8	44.5	18.4,97.6,29.6,26.2	33,30,50
59	1-Me/2-=NMe/A7-Me S7-Me	53.6	183.5	35.2	43.9	27.5	32.1	47.1	11.3,19.1,19.6	26
60	1-Me/2-=NNO2/A7-Me S7-Me	53.4	189.4	43.5	35.2	26.8	31.7	48.9	10.4,18.7,19.4	68
61	1-Me/2-=NOH[ANTI]/A7-Me S7-Me	51.6	169.0	33.1	43.8	27.3	32.6	48.1	11.0,18.5,19.5	195
62	1-Me/2-=0	53.6	218.4	45.5	34.3	28.9	31.5	44.1	13.8	7,1,19
63	1-Me/2-=0 3-=N2/4-Me/A7-Me S7-Me	59.4	201.9	64.4	51.6	34.4	30.9	51.6	10.2,11.9,15.7,18.1	68
64	1-Me/2-=0 3-=N2/A7-Me S7-Me	58.2	202.2	61.2	49.7	34.7	30.8	48.7	9.5,18.7,20.3	68
65	1-Me/2-=0 3-=0/4-Br/A7-Me S7-Me	58.6	199.4	195.2	73.8	32.5	29.7	47.4	9.9,15.9,19.1	145
66	1-Me/2-=0 3-=0/4-C1/A7-Me S7-Me	58.5	200.1	195.6	79.1	31.3	29.0	47.0	9.8,14.9,18.4	145
67	1-Me/2-=0 3-=0/4-COOH/A7-Me S7-Me	59.1	201.8	205.0	67.9	26.1	28.7	46.6	9.1,173.4,15.8,19.9	145
68	1-Me/2-=0 3-=0/4-Me/A7-Me S7-Me	58.7	205.0	205.0	58.7	29.6	29.6	44.4	9.3,9.3,14.8,19.0	145
69	1-Me/2-=0 3-=0/4-NO2/A7-Me S7-Me	60.0	198.3	189.1	99.8	26.5	27.2	47.5	9.4,14.8,18.8	145
70	1-Me/2-=0 3-=0/A7-Me S7-Me	58.7	204.7	202.7	58.2	22.3	30.1	42.6	8.8,17.5,21.0	145,171,195
71	1-Me/2-=0/4-Br/A7-Me S7-Me	56.8	213.3	52.1	64.1	37.5	29.2	51.0	10.4,16.7,17.9	68
72	1-Me/2-=0/4-C CH/A7-Me S7-Me	58.1	215.3	48.1	43.5	33.8	29.1	50.3	10.0,83.6,16.7,18.0	177
73	1-Me/2-=0/4-C1/A7-Me S7-Me	57.9	212.8	50.8	70.1	36.3	28.6	50.6	10.3,15.8,17.2	68
74	1-Me/2-=0/4-COMe/A7-Me S7-Me	60.5	215.7	45.4	60.9	30.6	29.3	50.5	9.3,208.5,17.3,18.3	68
75	1-Me/2-=0/4-CONH2/A7-Me S7-Me	60.3	215.7	45.6	54.7	30.7	29.2	50.4	9.6,174.8,17.3,18.3	68
76	1-Me/2-=0/4-COOH/A7-Me S7-Me	60.3	215.4	45.3	54.8	30.4	29.1	50.8	9.4,178.6,17.2,18.2	68
77	1-Me/2-=0/4-COOMe/A7-Me S7-Me	60.0	218.7	45.4	45.4	30.4	29.1	50.6	9.5,172.6,17.2,18.3	68
78	1-Me/2-=0/4-Me/A7-Me S7-Me	59.6	218.7	49.0	45.4	34.3	29.6	48.1	10.0,15.4,15.8,17.4	68
79	1-Me/2-=0/4-NH2/A7-Me S7-Me	59.0	215.9	49.0	61.9	34.6	28.5	48.2	10.4,15.0,16.9	68
80	1-Me/2-=0/4-NO2/A7-Me S7-Me	60.3	210.0	46.0	91.1	31.1	27.7	51.8	9.6,16.3,17.5	68
81	1-Me/2-=0/4-OH/A7-Me S7-Me	58.5	215.7	48.1	79.7	33.2	27.9	48.1	10.0,15.1,17.0	68
82	1-Me/2-=0/A7-Me	55.6	218.5	46.1	39.0	26.4	27.8	48.1	11.5,10.5	7
83	1-Me/2-=0/A7-Me S7-Me	57.5	218.7	43.2	43.1	27.1	27.9	46.6	9.2,19.1,19.8	7,1,2,19,28,54,57,68,125,195
84	1-Me/2-=0/N3-Br/A7-Me S7-Me	57.4	209.6	54.2	50.1	23.2	31.0	46.6	10.3,20.5,20.5	19,2,195
85	1-Me/2-=0/N3-COOH/A7-Me S7-Me	57.9	211.0	55.4	46.6	22.2	31.0	46.1	9.4,170.9,18.6,19.2	195
86	1-Me/2-=0/N3-Me/A7-Me S7-Me	58.3	221.4	43.9	48.2	19.9	29.1	45.3	9.6,11.7,19.3,19.4	7
87	1-Me/2-=0/S7-Me	56.0	218.7	41.0	38.7	28.2	31.0	45.8	11.8,10.9	7
88	1-Me/2-=0/X3-Me	53.1	220.9	48.7	40.7	29.7	31.1	47.7	14.0,14.3	160
89	1-Me/2-=0/X3-Me N3-Me	53.8	222.6	45.6	45.3	24.9	31.8	41.0	14.6,23.3,21.6	7,2,19,28,30,54,86,89,148,195
90	1-Me/2-=0/X3-Me N3-Me/5-=0	52.8	220.0	45.6	60.0	213.3	45.1	41.5	14.4,21.6,21.6	30
91	1-Me/2-=0/X5-Me N5-Me	56.1	218.8	41.9	47.4	37.6	34.9	39.1	14.5,31.2,26.6	153
92	1-Me/2-=S/A7-Me S7-Me	69.7	269.3	56.4	46.7	28.6	36.0	45.3	14.5,21.1,21.1	149,184,226
93	1-Me/2-=S/X3-Me N3-Me	66.7	280.4	58.1	47.5	29.0	32.9	49.8	19.6,25.5,26.8	233,148
94	1-Me/2-=Se/X3-Me N3-Me	72.6	215.8	63.3	47.3	27.6	42.6	44.2	20.8,26.5,25.0	176,233
95	1-Me/4-Me	45.3	38.7	38.7	45.3	38.7	38.7	52.4	21.9,21.9	2,25
96	1-Me/4-Me/7-=0	41.9	31.3	31.3	41.9	31.3	31.3	219.6	14.8,14.8	197
97	1-Me/7-=0	40.5	31.1	24.1	38.9	24.1	31.1	218.6	14.1	197
98	1-Me/7-Me	44.7	33.8	28.9	42.5	38.6	36.6	48.2	18.6,10.7	2,25
99	1-Me/A7-Me S7-Me	46.8	36.6	28.5	46.0	28.5	36.6	49.7	15.7,19.1	175,1,2,25
100	1-Me/N2-CH2OH/A7-Me S7-Me	47.6	46.6	46.0	46.0	27.4	29.5	49.7	15.3,65.1,19.3,18.9	241

sat. BICYCLO[2.2.1]HEPTANES (cont.)

Str	Substituents	C1	C2	C3	C4	C5	C6	C7	Substituent Shifts	References
101	1-Me/N2-CHO/A7-Me S7-Me	49.9	57.1	28.0	45.5	28.8	30.6	49.5	14.9,202.5,18.7,18.3	53
102	1-Me/N2-C1/A7-Me S7-Me	50.6	67.1	40.2	45.3	28.3	28.3	47.7	13.2,20.5,18.4	2
103	1-Me/N2-COOH/A7-Me S7-Me	49.9	50.2	32.0	46.0	28.4	31.1	50.3	14.6,177.0,19.5,19.1	241
104	1-Me/N2-Me	46.6	39.9	40.4	37.5	32.0	29.3	47.7	19.5,15.2	2,25
105	1-Me/N2-NHMe/A7-Me S7-Me	48.6	65.6	37.6	45.0	27.5	28.4	48.4	14.3,19.8,18.7	26
106	1-Me/N2-NO2/A7-Me S7-Me	49.7	92.3	32.0	43.9	26.9	28.2	51.9	13.8,19.4,18.6	141
107	1-Me/N2-OAc/A7-Me S7-Me	48.7	79.9	36.8	45.0	28.1	27.1	47.8	13.5,19.7,18.9	54,195,125
108	1-Me/N2-OAc/X3-Me N3-Me	48.2	86.1	39.5	48.5	25.9	26.6	41.5	19.4,29.7,20.1	54
109	1-Me/N2-OH	47.8	77.6	40.6	36.8	31.3	26.9	44.4	18.4	34,1,272
110	1-Me/N2-OH X2-Me/X3-Me N3-Me	52.0	79.8	42.9	49.2	25.5	29.4	41.0	16.0,22.1,26.6,26.3	78,153
111	1-Me/N2-OH X2-Ph	52.4	81.7	49.0	36.4	31.2	27.0	43.9	17.3,147.6	161
112	1-Me/N2-OH X2-Ph/A7-Me	53.7	82.2	47.6	41.0	28.5	27.0	45.0	15.0,147.4,10.8	160
113	1-Me/N2-OH X2-Ph/X3-Me N3-Me	52.9	84.1	45.7	49.1	24.1	33.7	41.9	17.4,145.2,30.2,21.3	153
114	1-Me/N2-OH X2-Ph/X5-Me N5-Me	54.0	80.7	42.8	47.6	37.3	46.6	42.3	18.2,147.8,32.0,25.9	153
115	1-Me/N2-OH/A7-Me	49.4	78.5	40.2	41.6	28.5	24.0	46.6	16.1,10.5	34
116	1-Me/N2-OH/A7-Me S7-Me	49.5	77.2	39.0	45.2	28.3	26.0	48.0	13.3,20.2,18.7	34,2,23,29,54,125,195
117	1-Me/N2-OH/S7-Me	48.4	76.1	38.3	41.2	29.7	28.6	49.5	16.2,10.1	34
118	1-Me/N2-OH/X3-Me N3-Me	49.3	84.8	39.3	48.2	25.3	26.3	41.3	19.6,30.9,20.5	28,30,34,54,153
119	1-Me/N2-OH/X5-Me N5-Me	49.6	76.2	35.3	48.2	37.6	43.7	44.0	19.4,31.8,26.1	28
120	1-Me/N2-OTMS/A7-Me S7-Me	49.0	76.3	38.2	45.0	28.1	25.9	48.1	0	220
121	1-Me/N2-OTs/X3-Me N3-Me	49.0	94.0	39.4	48.0	25.7	26.0	41.1	18.8,29.3,21.0	153
122	1-Me/N3-Me	45.3	45.8	35.9	43.5	24.2	37.6	47.5	21.9,18.2	2,25
123	1-Me/X2-CH2OH/A7-Me S7-Me	47.7	51.3	35.1	46.0	28.0	40.6	47.4	13.0,66.4,20.9,20.9	241
124	1-Me/X2-C1/A7-Me S7-Me	49.8	67.6	42.8	46.4	27.3	36.6	47.5	13.6,20.4,20.4	2
125	1-Me/X2-COOH/A7-Me S7-Me	50.1	51.7	33.1	46.1	28.2	40.2	48.4	15.3,176.7,20.6,19.9	241
126	1-Me/X2-F N2-F	49.0	130.5	43.6	35.6	29.9	29.1	43.3	12.5	1
127	1-Me/X2-F	45.7	41.0	40.4	39.2	31.1	37.5	41.6	17.8,19.9	2,25
128	1-Me/X2-NHMe/A7-Me S7-Me	48.3	69.2	38.3	45.2	27.4	37.0	46.6	12.0,20.6,20.5	26
129	1-Me/X2-NHPh/A7-Me S7-Me	48.4	61.6	40.4	44.9	27.1	36.5	46.8	11.9,20.1,20.1,149.0	103
130	1-Me/X2-OAc/A7-Me S7-Me	48.7	80.7	38.8	45.3	27.2	33.9	47.0	11.5,20.3,20.0	195,54
131	1-Me/X2-OCHO/X5-Me N5-Me	48.3	78.4	36.3	47.3	37.3	50.1	40.9	16.8,30.9,26.6	153
132	1-Me/X2-OH	47.4	77.3	43.1	35.9	30.3	33.1	40.1	16.1	34,1
133	1-Me/X2-OH N2-Me/A7-Me S7-Me	52.2	78.6	48.2	46.1	31.4	31.6	49.3	10.3,27.1,21.4,21.9	241
134	1-Me/X2-OH N2-Ph/A7-Me S7-Me	53.5	83.6	45.5	45.7	26.6	31.3	50.5	9.8,146.2,21.7,21.7	158
135	1-Me/X2-OH N2-Ph/S7-Me	52.2	82.4	46.1	41.6	28.0	33.4	52.2	12.7,145.4,12.5	160
136	1-Me/X2-OH N2-Ph/X3-Me	52.7	83.4	47.8	43.9	31.1	31.3	42.7	15.0,14.5,143.7	160
137	1-Me/X2-OH/A7-Me	48.6	78.3	42.2	40.4	30.4	27.4	42.3	13.8,10.0	34
138	1-Me/X2-OH/A7-Me S7-Me	49.0	79.9	40.5	45.1	27.2	34.0	46.4	11.3,20.5,20.2	158,2,28,29,34,44,195
139	1-Me/X2-OH/S7-Me	48.1	78.3	40.6	41.1	28.6	35.3	47.3	14.1,11.4	34
140	1-Me/X2-OH/X3-Me N3-Me	49.1	86.2	43.5	48.3	25.6	33.8	40.9	17.1,23.1,26.3	34,54
141	1-Me/X2-OH/X5-Me N5-Me	49.5	75.9	38.2	47.2	37.4	50.6	39.9	16.8,31.0,26.7	28
142	1-Me/X2-OTMS/A7-Me S7-Me	48.2	78.4	38.7	45.5	30.3	33.7	45.7	0	220
143	1-Me/X2-SMe N2-PO(OMe)2/A7-Me S7-Me	54.6	55.3	34.7	46.1	26.2	34.2	50.6	12.7,21.0,22.2	198
144	1-Me/X3-Me	44.9	47.5	38.3	45.3	32.0	36.2	42.0	21.4,22.8	2,25
145	1-N=)2[Z]	81.4	32.3	30.8	34.3	30.8	32.3	47.8		181
146	1-N=)2[E]	82.6	32.3	30.1	36.0	30.1	32.3	42.0		181
147	1-NH2/2-=CH2/X3-Me N3-Me	66.5	168.2	42.4	45.5	25.3	35.3	45.5	97.2,29.6,26.3	50
148	1-NH2/2-=O/X3-Me N3-Me	69.0	221.7	46.6	42.0	24.9	35.3	43.4	23.5,21.7	69
149	1-NH2/4-F	57.3	37.0	33.4	99.9	33.4	37.0	48.7		264
150	1-NHAc	62.6	33.7	29.9	35.3	29.9	33.7	41.7		4

Str	Substituents	C1	C2	C3	C4	C5	C6	C7	Substituent Shifts	References
151	1-NO$_2$/2-=CH$_2$/X3-Me N3-Me	96.2	158.8	43.0	44.7	24.7	32.2	41.1	101.0,29.6,26.1	50
152	1-NO$_2$/2-=O/X3-Me N3-Me	97.8	207.9	48.0	42.9	24.0	27.3	39.0	24.0,21.8	69
153	1-OAc/4-F	81.8	32.8	32.9	97.6	32.9	32.8	44.4		264
154	1-OEt	87.1	31.7	30.0	33.8	30.0	31.7	40.7		4
155	1-OEt/2-=CH$_2$/X3-Me N3-Me			38.6	44.6	24.6	33.3	44.6	98.9,29.5,27.0	222
156	1-OH	82.8	35.4	30.3	34.8	30.3	35.4	43.9		4
157	1-OH/2-=CH$_2$/X3-Me N3-Me	84.7	165.9	42.0	44.1	24.4	33.6	43.4	97.2,29.4,26.3	50
158	1-OH/2-=O/X3-Me N3-Me	85.6	220.7	46.2	42.0	24.5	29.6	40.3	23.7,21.7	69
159	1-OH/4-F	76.6	35.9	33.4	98.3	33.4	35.9	46.8		264
160	1-OMe	87.7	31.0	29.9	33.9	29.9	31.0	40.0		4
161	1-OSO$_2$CF$_3$/2-=CH$_2$/X3-Me N3-Me	100.9	159.8	40.4	43.6	24.1	33.3	42.6	101.6,0,0	222
162	1-OSO$_2$CF$_3$/X2-Me N2-Me/3-=CH$_2$	104.4	39.7	160.3	41.8	28.0	29.1	46.0	26.1,23.1,102.8	222
163	1-Ph	51.3	37.2	30.9	37.3	30.9	37.2	42.8	146.3	4
164	1-Ph-Me	52.7	34.6	31.0	35.7	31.0	34.6	43.9	147.0	4
165	1-Ph-Me/2-=CH$_2$/A7-Me S7-Me	60.3	158.8	38.1	46.3	27.4	33.0	50.0	136.5,105.2,19.2,21.7	153
166	1-Ph/2-=CH$_2$/S7-Me	58.9	156.3	36.8	42.1	28.6	35.8	50.0	141.3,105.8,12.8	160
167	1-Ph/2-=CH$_2$/X3-Me N3-Me	58.7	169.3	42.8	47.0	25.1	33.4	42.3	143.5,100.6,29.9,26.1	158
168	1-Ph/X2-OAc	54.9	78.3	38.6	35.5	30.5	32.8	40.9	141.8	195
169	1-SnMe$_3$	33.5	34.1	30.5	36.1	30.5	34.1	42.0		155,156
170	1-SnMe$_3$/4-F	31.3	34.3	33.2	104.7	33.2	34.3	44.5		264
171	1-t-Bu/4-I	52.2	32.4	44.2	38.3	44.2	32.4	50.7	31.3	111
172	N2-)2	39.9	46.1	30.4	37.3	30.4	23.2	43.1	46.1	275
173	2-=CH$_2$	45.7	155.3	39.1	37.0	28.5	29.9	38.4	101.8	33,2,25,92
174	2-=CH$_2$ 3-=CH$_2$	45.6	152.1	152.1	45.6	28.9	28.9	39.3	99.3,99.3	24,107
175	2-=CH$_2$/5-=CH$_2$ 6-=CH$_2$	54.5	149.0	149.0	54.5	149.0	149.0	40.4	4*101.4	127
176	2-=CH$_2$ 3-=CH$_2$/6-=CH$_2$/7-=CMe$_2$	56.3	148.6	148.6	56.3	148.6	148.6	137.5	4*100.7,116.9	126
177	2-=CH$_2$ 3-=CH$_2$/5-O-6	45.5	147.0	147.0	45.5	50.9	50.9	26.2	103.3,103.3	24
178	2-=CH$_2$ 3-=CH$_2$/X5-O-6X	45.5	147.0	147.0	45.5	50.9	50.9	26.2	103.3,103.3	266
179	2-=CH$_2$ 3-=CH$_2$/X5-OH/X6-OH	52.6	147.2	147.2	52.6	73.1	73.1	32.3	103.0,103.0	24
180	2-=CH$_2$/7-=O	48.9	144.7	35.6	40.9	23.6	24.6	48.9	106.9	70
181	2-=CH$_2$/A7-Me S7-Me	54.0	156.2	37.7	45.2	29.0	28.7	46.2	103.4,20.8,21.8	33
182	2-=CHCH=CH$_2$/X3-Me N3-Me	41.9	160.5		47.9	28.1	23.8	37.3	116.6,28.8,25.6	75
183	2-=CHCOMe/X3-Me N3-Me	43.7	178.2	44.3	46.8	27.6	23.3	37.5	115.9,28.3,25.4	75
184	Z2-=CHMe	39.6	144.9	38.9	36.3	28.4	23.3	38.7	111.5	185,96
185	E2-=CHMe	45.2	145.9	35.5	36.5	28.6	29.8	39.1	110.7	185,96
186	E2-=CHMe E3-=CHMe	40.1	145.2	145.2	40.1	28.1	28.1	39.1	107.3	178
187	Z2-=CHMe/A7-OAc	42.5	140.6	36.1	38.9	26.5	26.5	80.5	115.1	96
188	E2-=CHMe/A7-OAc	47.7	141.6	33.3	39.0	26.7	27.3	80.4	114.4	96
189	E2-=CHMe/S7-OAc	42.2	142.8	35.7	39.0	25.9	25.9	81.3	114.7	96
190	E2-=CHMe/S7-OAc	47.5	143.8	32.7	39.1	25.9	26.8	81.2	113.8	96
191	2-=CMe$_2$	41.1	138.2	39.5	37.0	28.8	29.2	37.8	117.9	59
192	2-=NOH[ANTI]	42.3	167.3	34.9	35.5	27.2	27.9	39.1		163,57,195
193	2-=NOH[SYN]	38.3	166.4	37.2	35.6	26.0	27.4	38.5		163,57
194	2-=O	49.7	217.4	45.1	35.3	27.1	24.2	37.6		7,1,2,19,28,30-32,57,68,92,95
195	2-=O 3-=CH$_2$	49.2	203.8	151.0	42.8	28.5	23.8	36.9	110.7	1,195
196	2-=O/4-Br	49.5	210.8	54.0	53.9	37.5	25.5	46.8		191
197	2-=O/4-CH$_2$OMe	50.5	216.1		47.7	29.7	24.6	39.5	75.5	191
198	2-=O/4-Cl	49.8	208.5	53.1	64.1	36.2	24.6	45.7		191
199	2-=O/4-Et	50.6	213.6	48.5	47.6	31.6	25.1	41.4	28.1	101
200	2-=O/4-I	49.1	211.7	56.4	27.1	40.8	26.1	49.2		190

sat. BICYCLO[2.2.1]HEPTANES (cont.)

Str	Substituents	C1	C2	C3	C4	C5	C6	C7	Substituent Shifts	References
201	2==O/4-Me	51.6	214.6	51.0	42.8	34.2	25.7	43.6	20.8	67,101
202	2==O/4-Me/5-==CH_2/X6-Me N6-Me	48.6	219.4	42.4	62.1	148.8	50.1	42.4	18.2,101.9,28.9,27.1	30
203	2==O/4-Me/N5-OH/X6-Me N6-Me	63.1	219.4	38.7	47.7	83.1	40.6	42.6	18.1,21.1,20.7	30
204	2==O/5-==CH_2 6-==CH_2	60.3	210.0	43.8	52.9	148.8	142.4	38.6	102.5,106.1	24
205	2==O/5-==CH_2/A7-Me S7-Me	59.3	217.0	42.9	52.9	151.0	30.7	45.1	105.9,21.1,21.0	274
206	2==O/5-==CH_2/X6-Me N6-Me	62.0	215.1	42.3	43.4	162.0	39.1	33.8	100.0,24.8,23.7	30
207	2==O/5-==O	47.4	213.7	36.4	47.4	213.7	36.4	33.8		30
208	2==O/5-==O/A7-Me S7-Me	57.5	213.1	37.2	57.7	213.1	37.2	44.4	21.2,21.2	274
209	2==O/6-==CH_2	60.4	211.9	43.4	34.6	36.1	143.4	38.3	109.5	230
210	2==O/7-==CH_2	54.4	212.7	48.2	40.6	26.9	23.9	153.0	101.4	70
211	2==O/A7-Me	54.3	215.1	47.3	40.2	24.9	21.3	42.8	12.1	1
212	2==O/A7-Me S7-Me	58.1	217.5	44.1	43.3	27.6	22.7	45.2	21.6,20.7	7
213	2==O/A7-NO_2	53.2	208.4	46.1	39.4	25.0	21.5	87.8		215
214	2==O/E6-==CHMe	59.6	210.6	43.5	40.0	33.5	149.9	38.3	118.8	98
215	2==O/N3-Br	48.5	208.1	56.2	42.9	25.7	23.4	35.6		19
216	2==O/N3-Me	50.2	219.8	48.2	40.4	20.9	25.4	37.2	10.7	7,1,19
217	2==O/N3-OSO_2CF_3	48.1	205.2	87.6	40.4	19.4	26.3	31.3		256
218	2==O/N5-Me	51.1	217.6	39.0	40.3	32.4	33.1	38.7	17.1	7,1
219	2==O/N5-Me N6-Me	57.4	216.9	39.4	41.2	34.4	33.9	37.9	11.9,13.6	7
220	2==O/N5-Me/X6-Me	57.7	217.2	38.1	41.2	42.9	40.2	36.0	15.9,19.7	7
221	2==O/N5-OOH X5-OMe/A7-Me S7-Me	59.1	216.6	38.5	51.9	114.0	32.3	38.4	21.8,23.3	274
222	2==O/N6-Et	53.6	213.1	45.7	39.7	28.0	40.2	38.7	27.3	98
223	2==O/N6-Me	55.8	216.7	45.8	44.2	35.8	32.3	38.7	18.7	7,1
224	2==O/S7-Me	55.2	215.9	40.3	40.0	29.1	24.5	43.9	12.9	1
225	2==O/S7-NO_2	51.9	209.9	40.6	40.5	26.3	20.9	89.9		215
226	2==O/X3-Br	49.5	208.6	50.4	45.1	26.9	24.4	35.6		19
227	2==O/X3-CH_2COMe	49.3	217.2	48.8	39.7	28.0	23.3	34.4	42.5	260
228	2==O/X3-Cl	48.6	209.8	59.4	44.2	25.6	23.7	34.2		195
229	2==O/X3-Me	49.4	220.0	48.2	41.5	28.0	23.8	34.9	14.1	7,1,160
230	2==O/X3-Me N3-Me	50.0	222.0	46.8	46.1	23.3	24.5	34.9	23.2,21.4	7,1,28,30
231	2==O/X3-Me N3-Me/5-==O	48.5	217.9	44.8	60.2	213.6	38.6	32.6	21.6,21.6	30
232	2==O/X3-Me N3-Me/N6-Me	56.3	219.3	47.7	46.6	30.6	33.4	31.7	23.9,20.0,18.1	7,89
233	2==O/X3-Me N3-Me/X6-Me	57.0	220.0	46.0	46.9	31.5	33.5	33.8	23.2,21.3,21.0	7,89
234	2==O/X3-OSO_2CF_3	47.6	205.7	85.0	41.1	23.5	23.0	33.8		256
235	2==O/X3-OTMS	47.6	215.3	76.0	42.9	24.1	23.6	33.8		256
236	2==O/X5-CH_2OH N5-OH/A7-Me A S-Me	59.7	219.3	38.5	41.8	75.6	37.7	47.0	69.2,23.2,24.3	274
237	2==O/X5-Me	50.5	216.1	45.5	41.8	34.5	33.3	34.2	22.0	7,1
238	2==O/X5-Me N5-Me	52.3	217.9	41.5	46.6	36.2	41.3	37.4	31.2,26.6	27,28
239	2==O/X5-Me/N6-Me	56.8	216.8	46.1	42.0	43.4	41.8	35.6	21.0,17.2	7
240	2==O/X5-Me/X6-Me	57.9	217.6	45.5	43.2	38.1	34.9	31.6	15.7,14.9	7
241	2==O/X5-OOH N5-OMe/A7-Me S7-Me	59.2	216.6	48.2	39.7	113.8	34.9	46.0	22.1,23.6	274
242	2==O/X5-$SnMe_3$	49.6	213.7	48.2	39.7	24.1	28.1	37.0	48.6	88
243	2==O/X6-CH_2COMe	54.5	214.3	44.0	35.7	35.8	31.8	34.4		260
244	2==O/X6-Et	54.2	214.0	44.0	36.0	35.7	33.0	34.3	28.2	3,95,98
245	2==O/X6-Me	56.6	216.4	44.2	35.9	37.3	31.0	34.2	20.8	7,1
246	2==O/X6-Me N6-Me	61.7	216.6	44.1	31.1	35.8	43.8	36.8	29.7,28.2	7,27,30
247	2==O/X6-$SnMe_3$	50.5	211.3	43.3	34.4	35.0	18.9	36.4		88
248	2==O/Z6-==CHMe	54.7	209.7	43.2	47.9	36.5	149.9	37.8	120.2	98
249	2==S/X3-Me N3-Me	65.0	277.8	58.4	47.9	26.0	28.4	37.2	23.3,18.0	184
250	X2-Br	46.4	53.1	43.8	37.0	28.0	27.4	35.5		36,108,113,195

sat. BICYCLO[2.2.1]HEPTANES (cont.)

Str	Substituents	C1	C2	C3	C4	C5	C6	C7	Substituent Shifts	References
251	N2-Br	43.8	53.3	41.4	36.9	29.4	24.4	39.2		36,113
252	X2-Br/A7-Me S7-Me	52.1	52.3	43.1	45.4	29.2	26.8	46.4	22.1,22.0	108
253	N2-CX	36.0	37.1	38.4	39.9	22.6	30.6	40.8	47.1	275
254	X2-CH2CH2Br	40.8	35.5	40.0	36.5	28.7	30.0	37.6	32.0	221
255	N2-CH2CH2CH2-5N	42.2	34.1	34.1	42.2	33.3	34.1	43.7	27.0,27.0	63
256	N2-CH2CH2CH2-6N	41.9	33.9	33.7	37.8	33.7	33.9	41.9	27.0,27.0	211,64
257	N2-CH2CH2CH=[E]CMeCH2OAc X2-Me/3-=CH2	45.4	44.8	166.2	46.8	29.1	23.8	37.0	38.5,25.2,99.6	244
258	N2-CH2CH2CH=[E]CMeCH2OH X2-Me/3-=CH2	45.4	45.0	166.6	46.9	29.8	23.8	37.2	38.9,25.3,99.4	244
259	X2-CH2CH2CH=[Z]CMeCH2OAc N2-Me/3-=CH2	44.9	44.9	165.8	46.9	29.8	23.8	37.1	41.3,22.6,100.0	135
260	N2-CH2CH2CH=[Z]CMeCH2OAc X2-Me/3-=CH2	45.4	45.0	166.5	46.9	29.2	23.9	37.2	39.2,25.3,99.6	135,244
261	N2-CH2CH2CH=[Z]CMeCH2OH X2-Me/3-=CH2	45.4	45.0	166.6	46.8	29.1	23.7	37.0	39.5,25.3,99.4	244
262	N2-CH2CH2CHO X2-Me/3-=CH2	45.2	44.5	165.5	46.8	29.0	23.7	37.0	40.4,25.3,100.2	244
263	X2-CH2CH2OH	41.3	38.3	38.6	36.6	29.0	30.2	35.4	39.9	195
264	X2-CH2CH2Cl/X3-=CH2Cl/X5-CH2Cl/X6-CH2Cl/7-=CMe2	45.6	47.4	47.4	45.6	47.4	47.4	134.2	4*43.6,125.5	126
265	N2-CH2CN X2-Me/N3-CH2CN X3-Me	49.3	43.8	43.8	49.3	23.3	23.3	34.5	24.7,24.3,24.7,24.3	74
266	X2-CH2COCH=CMe2 N2-Me/3-=CH2	44.8	45.0	165.5	46.9	29.6	23.7	37.2	54.1,23.2,100.6	273
267	X2-CH2COMe	41.3	36.8	38.2	37.6	28.7	30.0	35.4	51.1	259,260
268	X2-CH2COO-3X	41.8	41.0	86.4	42.0	23.3	28.0	33.3	31.4	218
269	N2-CH2COO-6N/X5-I	39.1	30.2	35.8	46.9	33.2	91.4	38.3	33.1	217
270	X2-CH2COOH	41.2	37.9	38.4	36.9	28.7	29.9	35.2	41.2	195
271	N2-CH2NH3Cl	36.6	39.6	34.5	37.8	29.6	22.4	38.1	41.5	195
272	N2-CH2OH	38.3	45.0	34.2	36.2	29.1	30.0	35.3	66.9	142,1,2,195
273	N2-CH2OH	37.9	42.5	33.7	36.7	30.1	28.0	35.8	64.9	142,1,2,195
274	X2-CH2OH N2-CH2OH	40.3	47.2	38.2	37.2	29.1	24.8	37.8	67.5,66.2	1
275	X2-CH2OH X2-Me/N3-CH2OH X3-Me	49.7	45.9	45.9	49.7	22.6	22.6	35.5	68.3,22.9,68.3,22.9	74
276	X2-CH2OH/N3-CH2OH	39.4	50.7	49.1	39.2	22.6	30.4	37.7	66.0,64.2	142
277	N2-CH2OH/N3-CH2OH	40.9	43.3	40.9	40.9	22.8	22.8	40.2	61.8,61.8	142
278	N2-CH2OH/N3-CH2OH/N5-CH2OH/N6-CH2OH	44.7	44.3	44.3	44.7	44.3	44.3	42.2	4*67.4	213
279	N2-CH2OH/N3-CH2OH/N5-OH/N6-OH	43.4	46.2	46.2	41.7	70.8	70.8	32.9	62.0,62.0	213
280	X2-CH2OH/N3-Me	39.4	53.5	38.7	41.7	21.7	30.1	37.3	66.3,16.5	142
281	N2-CH2OH/N3-Me	39.5	43.1	34.3	43.0	21.8	29.3	39.7	61.3,11.2	142
282	X2-CH2OH/X3-CH2OH	40.5	48.6	48.6	40.5	29.7	29.7	34.7	63.4,63.4	142
283	X2-CH2OH/X3-CH2OH/X5-CH2OH/X6-CH2OH/7-=CMe2	42.8	46.8	46.8	42.8	46.8	46.8	138.1	119.0	126
284	X2-CH2OH/X3-Me	39.5	48.6	39.4	44.5	21.7	29.6	32.6	63.2,15.2	142
285	N2-CH2OH/X3-Me	38.5	52.5	40.8	43.4	30.0	21.7	36.5	64.0,21.7	142
286	N2-CH2OMs X2-Me/N3-CH2OMs X3-Me	47.8	45.4	45.4	47.8	23.1	23.1	35.3	73.6,21.9,73.6,21.9	74
287	X2-CH2TMS	41.6	44.8	38.3	37.1	28.7	30.1	34.9	26.0	229
288	X2-CH=CH2	42.7	46.4	37.6	37.0	29.4	31.4	35.8	144.3	99,5
289	N2-CH=CH2	42.5	44.7	35.6	37.8	30.5	23.3	40.3	141.9	99,5
290	X2-CHO/N3-CHO	38.6	53.3	53.3	37.7	24.5	28.8	37.3	200.6,201.8	128
291	X2-Cl	45.9	61.7	43.4	36.3	28.0	26.5	35.0		36,108,113
292	N2-Cl	43.5	60.6	40.8	36.9	29.4	22.2	37.9		36,113
293	X2-Cl N2-Me	55.4	94.2	56.0	37.7	27.7	25.5	37.9		237
294	X2-Cl N2-Me	51.5	76.3	50.5	37.7	27.7	24.2	38.6	29.2	92
295	X2-Cl/A7-Cl	51.7	59.0	41.7	42.4	25.3	24.3	63.4		237
296	X2-Cl/A7-Me S7-Me	52.0	62.4	42.9	45.1	28.2	26.9	46.3	22.1,22.0	108
297	X2-Cl/N3-Cl	46.8	69.6	70.3	44.2	21.2	27.1	35.3		36,113
298	X2-Cl/N3-Cl	43.9	61.6	61.6	43.9	21.7	21.7	35.8		237
299	N2-Cl/N5-Cl	46.3	60.7	36.2	43.6	58.4	38.4	34.6		237
300	X2-Cl/S7-Cl	50.2	59.1	41.2	43.6	26.8	25.1	63.8		237

Str	Substituents	C1	C2	C3	C4	C5	C6	C7	Substituent Shifts	References
301	X2-Cl/X5-Cl	45.8	59.9	40.2	45.8	59.9	40.2	31.8	123.4	237
302	X2-CN	42.3	31.1	36.4	36.5	28.5	28.6	37.4	123.6	1,257,258
303	N2-CN	40.2	30.2	35.6	37.0	29.4	25.2	38.7	122.6	1
304	X2-CN N2-Me	46.5		44.5	37.1	28.2	22.1	39.4	126.8,22.5	49
305	X2-CN/X5-O-6X	42.5	27.0	32.2	36.7	49.9	49.2	25.2	122.0	123
306	N2-CN/X5-O-6X	39.9	27.8	31.5	37.0	50.2	48.4	26.4	121.1	123
307	X2-CN/X5-OH	41.7	29.9	31.8	44.4	73.3	41.0	34.2	124.8	112
308	N2-CN/X5-OH	39.9	29.5	31.2	44.7	73.8	38.1	35.4	123.7	112
309	N2-CN/X5-TMS	41.1	30.0	38.9	38.9	29.8	28.8	40.7	122.4	144
310	X2-CN/X6-OTs	47.0	26.5	34.8	35.3	38.5	81.7	34.1	121.8	60
311	N2-CN/X6-TMS	42.3	34.2	35.2	37.8	32.8	24.6	38.8	122.3	144
312	N2-CO3C-3N	40.2	50.0	50.0	40.2	25.0	25.0	42.3	172.6,172.6	142
313	X2-CO3C-3X	41.0	49.1	49.1	41.0	27.4	27.4	34.4	173.5,173.5	142
314	X2-COMe/X3-Me N3-Me	40.5	66.1	42.1	48.9	24.0	28.6	37.5	198.0,25.2,27.8	235
315	N2-COMe/X3-Me N3-Me	41.4	63.3	38.1	49.7			37.3	197.9	235
316	X2-CONH2 N2-Me	45.2	47.8	41.8	37.4	28.7	23.5	38.8	181.7,23.9	49
317	X2-CONMe2/X5-O-6X	40.1	41.1	30.6	36.7	50.4	51.5	23.9	173.1	123
318	N2-COO-6N	47.6	38.9	35.5	39.2	37.6	82.2	40.2	183.3	219,104
319	N2-COO-6N/X5-I	46.5	37.4	34.6	46.8	29.8	89.9	37.4	179.1	202,183,217
320	X2-COO-7S/N3-COO-6N	47.7	40.7	40.7	51.3	80.4	30.9	85.4	175.3,175.5	212
321	N2-COOCH2-3N	41.4	47.8	42.9	37.1	22.5	26.4	41.4	180.7,69.7	219
322	X2-COOEt N2-COOEt	43.9	61.4	39.0	36.8	28.1	25.3	39.8	171.9,170.8	1
323	N2-COOEt/N3-CH2OH	40.9	47.6	44.9		23.3	25.0	42.5	176.1,62.8	219
324	X2-COOEt/N6-OEt	44.3	45.3	32.4	38.0	28.7	82.3	39.9	176.6,	219
325	X2-COOH	41.0	46.5	34.2	36.1	28.7	29.5	36.6	182.6	142,1,48,195,257
326	N2-COOH	40.6	46.1	31.8	37.1	29.1	24.9	40.3	181.8	142,1,195
327	X2-COOH/N3-Me	41.6	54.8	39.9	41.6	21.4	29.8	38.7	182.7,16.6	142
328	N2-COOH/N3-Me	41.0	47.5	35.0	40.2	21.8	23.3	40.2	180.7,13.6	142,48
329	N2-COOH/N5-O-6N	40.8	44.6	29.0	37.3	62.2	62.7	50.3	180.9	123
330	X2-COOH/X3-CH2Cl	41.4	52.3	47.9	41.2	29.7	24.7	37.7	175.5,48.3	47
331	N2-COOH/X3-CH2Cl/N5-O-6N	42.1	49.9	45.2	40.9	61.7	61.4	47.5	174.1,47.4	47
332	N2-COOH/X3-Me	40.7	51.5	40.0	43.1	29.3	28.7	34.5	180.8,16.6	142
333	N2-COOH/X3-Me	41.0	55.6	38.8	43.7	29.5	24.0	36.9	181.5,21.5	142
334	X2-COOH/X3-Me N3-Me	40.9	59.0	42.5	48.1	24.0	21.5	37.5	180.7,25.4,27.5	188
335	N2-COOH/X3-Me N3-Me	40.9	56.2	38.5	49.1	24.5	28.5	37.6	180.6,31.9,22.8	188,187
336	X2-COOH/X5-O-6X	41.6	42.4	29.9	36.7	51.1	51.8	24.2	179.8	123
337	X2-COOH/X6-OH	48.5	40.9	33.9	34.9	42.1	73.2	32.5	177.3	64,66
338	N2-COOH/X6-OH	48.6	42.5	31.4	36.9	43.5	70.1	36.7	176.5	59
339	N2-COOMe	41.1	46.5	34.3	36.2	28.8	25.0	36.5	176.2	195,1,225
340	N2-COOMe	40.6	46.0	32.1	37.2	29.2	25.0	40.3	175.1	195,1,47,225
341	N2-COOMe X2-Me/N3-COOMe X3-Me	49.1	55.2	55.2	49.1	25.2	25.2	35.8	176.0,22.4,176.0,22.4	74
342	N2-COOMe/N3-COOMe/X6-COOMe	42.9	45.7	45.7	42.9	44.4	44.4	36.8	171.7,171.7,173.0,173.0	127
343	N2-COOMe/N5-O-6N	41.1	44.7	29.2	37.6	62.6	62.6	50.4	174.9	123,47
344	N2-COOMe/X3-CH2Cl	41.4	52.5	48.0	41.1	29.6	24.5	37.6	173.8,51.6	47
345	N2-COOMe/X3-CH2Cl/N5-O-6N	42.1	50.1	45.2	40.9	61.8	61.4	47.5	173.2,47.5	47
346	N2-COOMe/X3-CH2Cl/X5-O-6X	41.6	51.7	45.2	41.4	50.8	49.6	24.9	173.1,47.3	47
347	X2-COOMe/X3-COOMe	40.2	51.3	51.3	40.2	28.9	28.9	36.0	173.1,173.1	56
348	N2-COOMe/X3-COOMe/X6-COOMe/7-=CMe2	42.7	49.2	49.2	42.7	49.2	49.2	133.1	4*171.7,124.5	126
349	X2-COOMe/X5-O-6X	41.6	42.4	30.0	36.7	50.8	51.8	24.1	175.1	123,195
350	N2-COOMe/X5-O-6X	40.3	44.9	28.4	37.2	50.8	48.9	27.4	174.1	123,47,195

sat. BICYCLO[2.2.1]HEPTANES (cont.)

Str	Substituents	C1	C2	C3	C4	C5	C6	C7	Substituent Shifts	References
351	X2-COOMe/X6-OH	48.9	42.3	33.4	35.4	41.2	73.9	33.0	176.0	60
352	N2-COOMe/X6-OH	47.8	41.8	30.9	36.3	42.7	70.2	36.2	175.0	59
353	X2-Et	41.0	44.6	38.2	36.8	29.1	30.4	35.4	29.8	3+5
354	N2-Et	39.8	42.3	37.3	37.3	30.4	22.4	40.1	25.9	3+5
355	X2-F	41.8	95.8	39.8	34.4	27.9	22.2	34.7		108,1
356	X2-F N2-F	45.2	131.4	43.0	36.5	27.7	21.1	37.2		1
357	X2-F N2-F/A7-Me	49.4	130.9	44.4	40.4	24.8	18.2	42.4	11.8	1
358	X2-F N2-F/N3-Me	44.3	129.6	45.9	41.5	20.4	21.2	36.3	9.3	1
359	X2-F N2-F/N5-Me	46.3	131.3	35.7	41.6	32.5	29.5	38.8	16.4	1
360	X2-F N2-F/N6-Me	49.8	131.8	43.8	36.4	36.6	32.6	39.0	18.1	1
361	X2-F N2-F/S7-Me	49.1	131.7	40.6	41.4	28.0	22.3	45.5	12.5	1
362	X2-F N2-F/X3-Me	45.2	131.2	47.1	43.9	28.9	21.3	34.5	12.2	1
363	X2-F N2-F/X5-Me	45.9	131.3	43.8	42.8	34.9	30.9	33.7	21.8	1
364	X2-F N2-F/X6-Me	51.7	131.3	42.2	36.8	38.1	27.4	33.7	20.8	1
365	X2-I	47.9	29.9	45.1	37.9	28.7	28.4	36.2		1
366	N2-I	45.1	32.4	43.7	37.3	29.9	28.7	36.3		108,41
367	X2-I/A7-Me S7-Me	52.6	25.4	44.0	45.5	45.5	26.8	46.8	22.4,22.1	108
368	X2-Me	43.0	36.4	39.8	36.9	28.7	30.0	34.8	22.3	34+7,1,2,25,34
369	N2-Me	41.7	34.0	38.4	37.7	30.2	22.1	40.2	17.4	34+7,1,2,25,34
370	X2-Me N2-CO3C-3N X3-Me	48.7	57.0	57.0	48.7	25.5	25.5	37.7	17.9,175.8,175.8,17.9	74
371	X2-Me N2-Me	48.1	36.9	47.2	38.8	28.7	25.0	38.6	31.6,27.2	1,2,25
372	X2-Me N2-Me/3=CH2	48.1	41.7	165.9	46.9	28.9	23.8	37.4	29.4,25.8,99.2	33,1,30,50,54,125,133-4,195,2
373	X2-Me N2-Me/5=CH2	48.2	36.4	46.9	47.8	154.3	34.1	39.0	31.4,26.9,102.1	54
374	X2-Me N2-Me N3-CH2TMS	47.6	38.4	49.5	43.8	20.8	25.6	37.9	33.0,23.0,14.1	144
375	X2-Me/3=CH2	42.7	43.2	161.8	45.9	28.9	29.0	35.4	20.0,100,100.4	33,2,25
376	N2-Me/3=CH2	41.3	42.3	161.3	46.5	30.7	21.4	39.4	15.1,100.4	33,2,25
377	X2-Me/N3-Me	44.4	45.2	44.0	42.4	21.2	30.3	36.9	21.1,15.9	7,2,25
378	N2-Me/N3-Me	43.1	34.8	34.8	43.1	21.8	21.8	39.7	11.8,11.8	7
379	N2-Me/N3-Me/A7-OMe S7-Ph	46.6	31.2	31.2	46.6	20.3	20.3	92.0	11.3,11.3,139.0	194
380	N2-Me/N3-Me/S7-OMe A7-Ph	46.3	32.3	32.3	46.3	19.1	19.1	92.2	11.6,11.6,139.5	194
381	X2-Me/X3-Me	44.5	39.9	39.9	44.5	29.7	29.7	31.9	15.7,15.7	7
382	N2-Me/X5-TMS	42.9	34.4	40.7	43.6	31.0	26.1	40.0	18.8	144
383	N2-Me/X6-TMS	44.1	37.7	40.5	38.8	34.0	20.1	38.3	17.5	144
384	N2-N(OSO2Ph)-3N/7=CH2	40.7	46.7	46.7	40.7	25.2	25.2	162.5	97.2	250
385	N2-N(OSO2Ph)-3N/7=O	36.7	39.4	39.4	36.7	19.4	19.4	203.7		250
386	N2-N(OSO2Ph)-3N/A7-CH2OH	38.4	49.8	38.4	38.4	23.9	23.9	64.6	59.9	250
387	X2-N(OSO2Ph)-3X/7=CH2	41.0	41.8	41.8	41.0	24.8	24.8	147.1	103.6	250
388	X2-N(OSO2Ph)-3X/7=O	41.3	41.6	41.6	41.3	19.5	19.5	204.2		250
389	N2-N(Ph)-3N/A7-OH	41.4	42.5	42.5	41.4	23.8	23.8	85.3		248
390	N2-N(Ph)-3N/A7-OH S7-Me	45.2	44.5	44.5	45.2	25.4	25.4	93.1	20.3	248
391	X2-NH2	45.7	55.4	42.5	36.4	28.9	27.0	34.3		1
392	N2-NH2	43.6	53.4	40.6	38.0	30.7	20.6	39.0		1
393	N2-NH2Cl	39.5	51.9	34.2	36.5	21.5	21.5	38.4		195
394	N2-O-3N	37.7	62.0	62.0	37.7	25.5	25.5	50.4		40,94,123
395	N2-O-3N/7=CH2	41.4	58.1	58.1	41.4	24.4	24.4	163.4	95.4	250
396	N2-O-3N/A7-OH	41.5	55.1	55.1	41.5	22.4	22.4	84.0		247
397	N2-O-3N/A7-OTs	40.3	53.0	53.0	40.3	22.4	22.4	86.8		247
398	N2-O-3N/A7-OTs S7-Me	45.2	55.1	55.1	45.2	24.0	24.0		18.1	247
399	N2-O-3N/S7-Br	44.4	61.7	61.7	44.4	24.5	24.5	69.0		94
400	N2-O-3N/S7-CH2Br A7-Br	39.4	56.4	56.4	39.4	26.0	26.0	82.7	47.5	250

sat. BICYCLO[2.2.1]HEPTANES (cont.)

Str	Substituents	C1	C2	C3	C4	C5	C6	C7	Substituent Shifts	References
401	X2-O-3X	36.8	51.0	51.0	36.8	25.3	25.3	26.3		40,47,94,123
402	X2-O-3X N2-COEt	37.8	63.2	58.8	37.8	25.9	23.7	30.3		276
403	X2-O-3X/N5-OCOO-6N	40.0	46.4	46.4	40.0	78.9	78.9	20.7		93
404	X2-O-3X/S7-Br	42.3	51.4	51.4	42.3	24.3	24.3	43.0		94
405	X2-OAc	41.6	77.5	39.7	35.5	28.3	24.4	35.3		35,3,104
406	N2-OAc	40.7	75.6	37.3	37.0	29.7	21.3	37.6		3
407	X2-OAc/5=CH2	42.2	76.1	39.4	44.6			35.7	103.1	100
408	X2-OAc/5=CH2 6=CH2	50.8	75.9	38.6	44.5	150.6	146.5	35.8	100.2,104.0	24
409	N2-OAc/5=CH2 6=CH2	49.2	74.1	36.1	44.7	151.0	145.7	37.8	100.0,104.0	24
410	X2-OAc/6=CH2	50.9	76.1	38.8	36.2	37.4	148.6	36.1	107.0	100
411	X2-OAc/6=CMe2	46.9	76.2	38.8	36.5	36.8	132.2	36.3	123.8	98
412	X2-OAc/A7-Me	45.6	78.2	40.7	39.5	25.3	22.0	40.4	11.7	34,35
413	X2-OAc/A7-NO2	45.2	74.2	37.5	39.2	25.6	22.0	89.4		215
414	X2-OAc/E5=CHMe	42.2	76.7	39.9	44.6	144.0	30.8	36.3	112.2	96
415	X2-OAc/E6=CHMe	50.9	76.4	39.2	36.3	34.9	140.5	36.3	116.2	96
416	X2-OAc/N5-CH=CH2	42.7	76.9	33.0	42.4	40.9	30.4	36.6	140.3	97
417	X2-OAc/N5-Et	42.3	76.8	32.6	38.7	40.7	32.1	36.8	25.2	3
418	X2-OAc/N5-Me	42.9	77.0	32.3	40.9	32.8	33.5	37.1	17.0	97
419	X2-OAc/N6-Me	46.9	72.7	40.5	37.4	36.9	31.8	37.1	17.0	97
420	N2-OAc/N6-OTs	43.2	74.1	35.8	35.7	36.3	81.5	35.5		65
421	X2-OAc/S7-Me	45.5	78.5	38.3	41.3	28.6	26.2	43.9	13.0	34,35
422	X2-OAc/S7-NO2	45.9	76.3	36.4	37.9	25.5	22.5	89.7		215
423	X2-OAc/X3-Me	42.1	79.3	42.8	42.8	29.2	24.2	32.7	13.7	34,35
424	X2-OAc/X5-CH=CH2	41.8	76.8	39.6	41.3	44.4	32.3	31.7	143.1	97
425	X2-OAc/X5-Et	42.0	76.8	40.4	40.2	43.5	32.7	32.2	29.5	3
426	X2-OAc/X5-Me	42.3	76.6	40.1	42.2	35.3	34.3	31.7	22.0	97
427	X2-OAc/X5-Me N5-Me/E6=CHMe	46.9	75.0	33.6	47.5	41.1	149.7	33.9	29.0,25.8,114.5	98
428	X2-OAc/X5-Me N5-Me/Z6=CHMe	53.8	76.1	34.2	50.1	41.1	149.7	34.2	26.4,24.5,116.2	98
429	X2-OAc/X6-CH=CH2	47.3	76.9	39.3	35.8	35.9	40.8	32.8	142.3	97
430	X2-OAc/X6-Et	46.3	77.2	39.3	35.7	36.8	39.1	32.6	28.9	3
431	X2-OAc/X6-Me	48.4	77.0	38.9	36.1	38.5	31.4	32.1	21.5	97
432	N2-OAc/X6-OTs	46.3	72.3	35.6	36.0	40.0	79.6	34.4		65
433	X2-OAc/Z5=CHMe	42.1	76.7	39.2	39.0	143.2	33.8	36.1	113.1	96
434	X2-OAc/Z6=CHMe	45.9	75.5	38.4	35.9	37.9	139.3	35.8	117.1	96
435	N2-OBS/5-=CH2 6-=CH2	50.1	80.6	36.4	44.5	150.1	143.8	37.6	100.7,106.5	24
436	N2-OCH2-6N/X3-I	44.2	89.6	37.8	46.8	36.9	37.3	37.7	74.1	202
437	X2-OCHO	41.9	77.0	39.7	35.8	28.3	24.0	35.4		81
438	X2-OCH[R]MeCH[R]MeO-2N [1R]	45.8	115.6	45.5	35.6	28.1	22.0	37.6		114
439	X2-OCH[R]MeCH[R]MeO-2N [1S]	44.9	115.5	45.5	35.8	28.5	22.0	37.6		114
440	X2-OCOCH2-7S	41.7	80.9	37.8	40.5	29.1	22.2	41.1	31.7	218
441	X2-OCOCH2CH2-2N	46.4	93.6	29.7	45.7	28.0	22.2	30.6		130
442	X2-OCOCH2CH2-2N/A7-I	53.8	87.7	43.2	44.6	29.1	21.3	29.3	26.8	
443	X2-OCOCH2CH2-2N	46.8	91.3	27.8	46.7	23.0	17.3	29.3		130
444	X2-OCOCH2CH2CH2-2N/A7-I	54.3	88.3	43.9	44.7	26.9	21.9	30.1	29.1	
445	X2-OH	44.2	80.9	37.8	35.4	28.1	24.4	34.4		34,1-3,81,195
446	N2-OH	42.5	72.9	39.4	37.2	29.9	20.0	37.6		34,1-3,195
447	X2-OH N2-Et	46.5	80.2	47.9	36.7	28.5	23.6	37.2	31.4	97
448	X2-OH N2-iPr	47.4	82.4	46.8	37.6	28.4	22.8	37.2	33.4	59
449	X2-OH N2-Me	49.3	77.9	48.7	37.0	27.9	24.0	37.4	25.8	34,2,77,272
450	X2-OH N2-Ph/A7-Me S7-Me	55.2	83.9	43.7	45.5	26.5	24.3	48.2	146.2,23.2,23.7	161

74

sat. BICYCLO[2.2.1]HEPTANES (cont.)

Str	Substituents	C1	C2	C3	C4	C5	C6	C7	Substituent Shifts	References
451	X2-OH N2-Ph/X3-Me	48.9	83.4	45.1	44.3	29.7	23.4	34.8	145.2,15.0	160
452	N2-OH X2-CH2CH=CH2/X3-Me N3-Me	50.1	79.3	43.0	47.4	24.0	21.3	34.5	42.2,26.6,22.0	75
453	N2-OH X2-Et	46.9	79.4	44.6	37.2	28.7	22.4	38.8	34.8	3
454	N2-OH X2-iPr	45.3	81.5	44.5	37.5	28.3	22.8	38.6	36.1	59,221
455	N2-OH X2-Me	48.4	77.2	47.0	37.3	28.3	22.0	38.7	30.4	34,2,3,72,78,92,150,162
456	N2-OH X2-Me/X3-Me	50.0	79.9	49.6	45.0	29.8	21.8	35.9	24.7,16.1	150
457	N2-OH X2-Me/X3-Me N3-Me	51.1	78.8	42.1	49.9	24.1	21.2	34.8	26.4,27.1,21.8	150
458	N2-OH X2-Me/X5-Me N5-Me	54.1	79.1	44.7	52.5	38.9	41.4	41.7	34.6,35.3,29.4	28
459	N2-OH X2-Ph	47.1	80.7	46.6	37.6	29.1	22.3	38.8	149.1	157
460	N2-OH X2-Ph/X3-Me	48.3	83.4	51.5	44.7	29.3	22.8	38.1	145.9,19.1	160
461	N2-OH X2-Ph/X3-Me N3-Me	48.8	81.5	44.8	49.6	23.6	22.3	36.4	147.1,30.0,21.4	154
462	X2-OH/4-Et	44.8	75.1	46.0	47.7	32.9	25.8	38.7	27.7	101
463	N2-OH/4-Et	43.1	72.5	43.2	48.9	34.1	21.5	41.7	28.7	101
464	X2-OH/4-Me	45.7	75.9	49.0	44.0	35.5	26.2	41.2	20.4	101
465	N2-OH/4-Me	43.9	73.2	46.0	44.6	37.0	21.9	44.1	21.5	101
466	N2-OH/4-Me/5=CH2/X6-Me N6-Me	52.6	75.4	44.5	50.1		43.4	43.1	18.7,0,31.8,27.5	30
467	X2-OH/5=CH2 6-=CH2	54.2	73.6	41.3	44.9	151.2	147.8	35.2	100.1,102.9	24
468	N2-OH/5=CH2 6-=CH2	52.0	70.6	39.8	45.3	151.7	145.3	37.8	100.4,105.8	24
469	X2-OH/A7-Me	48.5	75.4	43.4	39.5	25.3	25.2	39.6	11.9	34
470	X2-OH/A7-Me S7-Me	51.4	77.6	41.9	44.4	27.5	25.7	45.7	22.2,22.4	34,161
471	N2-OH/A7-Me S7-Me	44.9	72.4	39.9	44.5	29.4	18.9	47.3	22.0,21.0	161
472	X2-OH/A7-NO2	47.9	72.1	39.6	39.1	25.7	21.9	89.4		215
473	X2-OH/E6=CHMe	53.5	73.8	41.2	35.8	34.7	141.7	35.3	114.6	98
474	N2-OH/E6=CHMe	51.7	70.8	39.1	36.4	36.1	138.9	37.8	117.9	98
475	X2-OH/N3-Me	45.4	82.5	46.6	40.9	21.0	25.0	36.5	15.0	34,1
476	N2-OH/N3-Me	43.5	72.2	36.5	42.2	21.8	19.7	36.7	10.2	34,1
477	X2-OH/N3-NMe2	39.3	79.5	80.0	45.4	20.5	21.9	34.7		151
478	N2-OH/N3-NMe2	39.8	67.8	69.0	43.1	20.4	21.5	34.7		151
479	N2-OH/N3-OH/N5-Me/N6-Me	44.4			44.4	43.3	43.3	35.4		137
480	N2-OH/N3-OH/X5-Me/X6-Me	45.3			45.3	35.5	35.5	44.4		137
481	X2-OH/N5-CH=CH2	44.8	74.1	35.2	42.6	40.8	30.2	35.8	140.8	99
482	X2-OH/N5-Me	45.5	74.6	34.8	40.7	32.4	33.3	36.2	16.9	34
483	X2-OH/N5-Me	43.7	73.3	31.5	42.7	34.3	27.6	39.5	16.3	34
484	X2-OH/N5-Me/N6-Me	51.2	69.0	35.2	42.4	33.3	31.5	35.9	11.8,11.3	34
485	N2-OH/N5-Me/N6-Me	48.2	76.2	31.1	43.8	35.1	35.5	38.8	11.6,14.2	34
486	X2-OH/N5-Me/X6-Me	52.4	74.8	34.3	41.5	42.7	40.7	33.4	15.5,20.2	34
487	X2-OH/N5-Me/X6-Me	50.2	73.2	30.9	43.5	44.6	33.4	36.4	14.8,20.6	34
488	X2-OH/N6-Br	50.9	70.7	41.9	36.2	40.3	49.0	34.7		65
489	N2-OH/N6-Br	49.0	76.1	37.7	39.8	42.3	46.7	37.5		65
490	X2-OH/N6-CH2Br	47.8	68.5	42.7	36.8	35.4	40.7	35.3	36.1	65
491	X2-OH/N6-Et	47.2	68.3	42.7	36.4	36.3	39.8	35.8	25.2	62
492	X2-OH/N6-Me	49.4	69.0	42.9	36.8	36.5	31.5	36.5	16.9	96
493	N2-OH/N6-Me	47.0	76.4	40.0	39.1	37.9	35.4	39.4	19.4	34
494	X2-OH/N6-NMe2	48.0	67.9	43.6	36.6	35.4	67.1	35.0		34,65
495	N2-OH/N6-OAc	44.2	75.1	39.9	36.7	38.2	77.4	36.3		65
496	X2-OH/S7-Me	48.2	76.3	41.0	41.0	26.5	26.5	43.7	13.5	65
497	X2-OH/S7-NO2	48.4	74.4	39.3	38.6	25.6	22.9	89.9		215
498	N2-OH/X3-CH2NMe2	41.0	78.1	48.7	43.2	30.4	19.8	35.0	64.8	195
499	X2-OH/X3-Me	44.8	76.5	43.2	42.8	29.2	24.6	31.9	13.4	34,1
500	N2-OH/X3-Me	43.2	81.8	45.8	43.8	30.1	19.4	34.4	20.1	34,1

sat. BICYCLO[2.2.1]HEPTANES (cont.)

Str	Substituents	C1	C2	C3	C4	C5	C6	C7	Substituent Shifts	References
501	X2-OH/X3-Me N3-Me	46.3	83.9	42.8	48.0	25.1	23.9	35.2	23.2,26.2	34
502	N2-OH/X3-Me N3-Me	44.1	80.5	38.0	48.4	24.7	18.3	33.9	30.6,20.2	34
503	N2-OH/X3-Me N3-Me/N6-Me	49.5	83.9	37.4	48.6	31.8	33.2	35.5	31.2,20.1,18.9	89,34
504	N2-OH/X3-Me N3-Me/X6-Me	51.1	80.6	37.4	49.3	34.8	24.1	29.7	30.6,19.7,21.7	34,89
505	N2-OH/X3-NMe$_2$	42.5	78.2	76.6	39.0	28.4	19.4	34.3		152
506	X2-OH/X3-OH/N5-CH=CH$_2$	43.5	74.0	69.2	48.2	40.6	29.3	32.7	138.9	263
507	X2-OH/X3-OH/N5-Me N6-Me	47.4	74.0	47.4	47.8	40.3	42.6	35.0		137
508	X2-OH/X3-OH/X5-CH=CH$_2$	42.6	69.2	73.5	47.8	40.3	31.6	28.7	141.8	263
509	X2-OH/X3-OH/X5-Me/X6-Me	46.4	74.0	41.8	46.4	41.8	41.8	26.1		137
510	X2-OH/X5-CH=CH$_2$	44.2	73.8	41.8	41.3	44.6	32.0	31.5	143.5	99
511	X2-OH/X5-Me	44.8	74.1	42.5	42.0	34.9	34.3	30.7	21.9	34
512	N2-OH/X5-Me	43.0	72.1	40.3	43.5	36.3	30.2	34.1	22.4	34
513	X2-OH/X5-Me N5-Me/E6-=CHMe	50.0	72.1	36.5	47.8	40.8	151.5	33.0	29.4,25.7,112.5	98
514	X2-OH/X5-Me N5-Me/Z6-=CHMe	57.2	73.4	36.5	50.4	40.4	151.5	32.9	26.4,24.4,114.4	98
515	X2-OH/X5-Me/N6-Me	50.5	69.0	43.4	43.4	43.9	41.7	33.2	20.9,15.5	34
516	N2-OH/X5-Me/N6-Me	47.9	75.8	40.8	45.2	45.8	44.3	36.2	21.2,17.7	34
517	N2-OH/X5-Me/X6-Me	52.9	74.8	42.5	43.5	38.5	34.5	28.4	15.4,15.3	34
518	N2-OH/X5-Me/X6-Me	50.7	72.7	40.2	45.2	40.4	28.9	31.6	15.8,15.2	34
519	X2-OH/X6-CH=CH$_2$	49.8	74.0	41.3	35.6	36.0	41.0	32.0	142.8	99
520	X2-OH/X6-Et	48.8	74.2	41.2	35.6	36.9	39.5	31.7	28.9	96
521	X2-OH/X6-Me	51.1	74.6	41.3	35.7	38.1	31.3	31.2	21.4	34
522	N2-OH/X6-Me	49.2	72.8	38.5	37.9	40.0	25.3	34.3	21.8	34
523	N2-OH/X6-Me N6-Me	52.7	76.1	37.1	38.3	47.8	33.2	37.2	33.2,28.8	34
524	N2-OH/Z6-=CHMe	46.5	71.8	39.1	36.2	39.1	138.2	37.5	118.8	98
525	X2-OMe	39.7	83.8	39.2	36.4	28.5	24.3	35.1		36
526	N2-OMe	39.2	81.5	37.0	36.0	29.7	20.0	36.9		36
527	X2-OMe/E6-=CHMe	49.7	84.0	39.1	36.0	35.3	142.1	35.8	115.0	98
528	X2-OMe/N3-NMe$_2$	38.8	89.2	77.8	39.2	20.9	25.5	35.0		151
529	N2-OMe/N3-NMe$_2$	40.0	79.1	69.8	39.8	20.5	21.0	34.0		151
530	N2-OMe/N6-OTs	42.7	82.2	36.3	36.0	37.6	82.4	35.6		65
531	N2-OMe/X6-OTs	45.6	79.3	35.9	36.0	40.0	80.2	34.3		65
532	N2-OMs	41.5	82.6	37.0	36.5	29.3	20.8	37.2		196
533	X2-OOtBu/N3-Br	40.7	94.1	56.3	43.0	24.6	23.9	34.6		172
534	X2-OOtBu/X3-Br	40.0	84.3	54.9	46.7	27.8	24.4	33.0		172
535	X2-OTMS	43.8	73.4	41.2	35.5	28.6	24.7	35.5		220
536	N2-OTMS	42.1	71.2	38.4	37.7	29.9	20.1	40.0		220
537	N2-OTMS X2-Me	47.0	72.8	45.9	36.8	27.9	21.5	39.0		220
538	X2-OTs	42.1	85.3	39.7	35.4	27.9	23.9	35.0		60
539	X2-OTs/N6-Br	49.0	82.1	39.8	36.2	40.1	46.9	35.2		65
540	X2-OTs/X6-Br	45.8	81.2	38.6	36.1	42.4	52.5	32.7		60
541	N2-OTs/X6-Br	45.6	80.8	35.4	37.2	34.8	51.1	43.5		61
542	N2-OTs/X6-NMe$_2$·HBr	42.5	79.6	39.1	35.8		62.7			61
543	X2-OTs/X6-NO$_2$	48.9	80.4	38.5	44.8	36.1	83.3	33.0		60
544	X2-OTs/X6-OAc	48.0	81.0	38.7	34.6	38.6	72.7	32.5	169.9	60
545	X2-OTs/X6-OMe	46.8	82.0	34.6	34.3	38.1	79.6	32.0		60
546	X2-OTs/X6-SMe	47.4	83.4	39.1	35.3	37.2	43.4	32.4		60
547	N2-P(OMe)$_2$		44.3	39.1	35.3	29.5	31.5	40.7		189
548	N2-P(OMe)$_2$		42.7	30.4	29.6		25.8	40.7		189
549	X2-PCl$_2$	38.5	54.5	33.1	37.0	28.5	31.2	37.0		189
550	N2-PCl$_2$	39.4	53.4	33.3	35.3	29.5	24.8	40.5		189

sat. BICYCLO[2.2.1]HEPTANES (cont.)

Str	Substituents	C1	C2	C3	C4	C5	C6	C7	Substituent Shifts	References
551	N2-Ph-Cl	45.5	42.5	40.6	37.6	30.1	22.8	34.4	142.3	157
552	X2-PMe2	38.2	44.2	33.7	35.7	28.2	30.7	36.0		189
553	N2-PMe2	38.0	43.3	33.8		28.9	23.9	39.5		189
554	X2-PMe3 I	38.1	35.8	32.1	36.4	27.4	31.9	37.7		189
555	N2-PMe3 I			33.6		31.5	28.4	42.9		189
556	N2-PO(OEt)2/N3-PO(OEt)2/X5-O-6X	41.4	38.5	38.5	41.4	49.4	49.4	28.2	202	
557	X2-PO(OMe)2	38.1	37.8	31.9	36.3	28.5	31.8	37.1		189,8
558	N2-PO(OMe)2	38.9		30.9	37.2	29.2	25.4	40.7		189
559	X2-PO(OPh)2	38.1	38.3	31.9	36.0	28.2	31.5	37.0		55
560	X2-POCl2	36.3	55.0	32.6	36.8	27.9	31.0	39.0		55
561	X2-POMe2	37.7	42.8	31.3	36.3	28.4	32.1	37.3		189
562	N2-POMe2	39.1	41.8	29.7		29.3	25.1	41.4		189
563	X2-PSMe2	38.4	44.5	32.4	36.4	28.4	32.4	36.9		189
564	N2-PSMe2	40.0	42.9	30.5	37.5	28.9	24.3	41.9		189
565	X2-SePh/N3-Cl	44.5	54.2	68.0	44.6	21.6	29.6	36.3		115,116
566	X2-SePh/N3-Cl/5-==CH2	40.9	53.2	66.9	48.5	134.7	31.0	36.9	104.8	116
567	X2-SePh/N3-OMe	40.3	50.5	88.7	43.7	19.5	29.8	35.2		116
568	X2-SePh/X3-OMe	41.6	49.9	84.3	44.6	29.4	29.7	37.2		116
569	X2-SnMe3	40.1	27.5	34.8	37.4	29.4	33.4	38.5		140,42,43
570	N2-SnMe3	40.5	28.5	33.6	36.6	30.1	30.0	40.6		140,42,43
571	7-)2	38.7	27.7	27.7	38.7	30.6	30.6	47.5		3
572	7-==)2	37.1	20.6	20.6	37.1	20.6	20.6	131.0		106
573	7-==C(CN)2	40.8	27.4	27.4	40.8	27.4	27.4	193.3	75.0	106
574	7-==C=CH2	40.6	29.7	29.7	40.6	29.7	29.7	107.4	193.6	214
575	7-==CH2	40.1	29.7	29.7	40.1	29.7	29.7	158.3	96.8	45,46
576	7-=O	37.8	24.2	24.2	37.8	24.2	24.2	216.8		32,17,95
577	7-Br	42.7	27.3	27.3	42.7	27.3	27.3	58.3		108,110
578	7-CH2OH	37.7	27.8	27.8	37.7	31.0	27.3	57.0	61.1	147
579	7-CH2TMS	41.2	27.1	27.1	41.2	30.3	30.3	45.5	14.9	229
580	7-Cl	42.3	26.7	26.7	42.3	27.4	27.4	66.3		108
581	S7-CN A7-OH	45.5	26.7	26.7	45.5	26.8	26.8	78.4	120.8	249
582	7-COOH	39.2	28.0	28.0	39.2	29.9	26.8	53.8	180.2	194,48
583	7-COOMe	39.2	28.1	28.1	39.2	29.9	29.9	53.7	173.5	225
584	7-F	38.4	26.5	26.5	38.4	25.2	25.2	99.2		108
585	7-I	43.9	28.3	28.3	43.9	26.9	26.9	35.6		108
586	7-Me	41.0	27.2	27.2	41.0	31.0	31.0	44.3	12.7	1,2,25
587	A7-Me S7-Me	43.9	29.5	29.5	43.9	29.5	29.5	46.0	21.2,21.2	108
588	7-OH	40.4	26.8	26.8	40.4	27.0	27.0	79.7		6,3,16
589	S7-OH A7-Et	41.8	29.0	29.0	41.8	28.2	28.2	86.3	26.2	3
590	S7-OH A7-Me	44.3	29.2	29.2	44.3	28.3	28.3	84.0	20.8	3
591	7-OMe 7-Ph	40.1	27.2	27.2	40.1	28.6	28.6	92.7	140.0	194
592	A7-OMe S7-OMe	37.6	27.7	27.7	37.6	27.7	27.7	114.6		16
593	7-OTMS	39.7	26.6	26.6	39.7	27.0	27.0	78.2		220
594	7-PCl2	42.2	27.5	27.5	42.2	31.5	31.5	65.4		110
595	7-Ph	40.1	27.9	27.9	40.1	30.9	30.9	54.2	141.9	194
596	7-PMe2	39.4	27.6	27.6	39.4	31.7	31.7	54.9		110
597	S7-SePh/X2-Cl	42.6	69.2	26.8	35.9	24.1	24.5	56.7		116
598	S7-SePh/X2-OMe	46.8	90.6	35.3	35.3	24.6	24.6	55.8		116
599	7-SnMe3	40.3	30.1	30.1	40.3	31.4	31.4	38.8		156
600	X2-OBS/5-=CH2 6-==CH2	51.5	83.8	38.8	44.4	149.6	145.2	35.6	101.0,105.3	24

5.2 Unsaturated bicyclo [2.2.1] heptanes

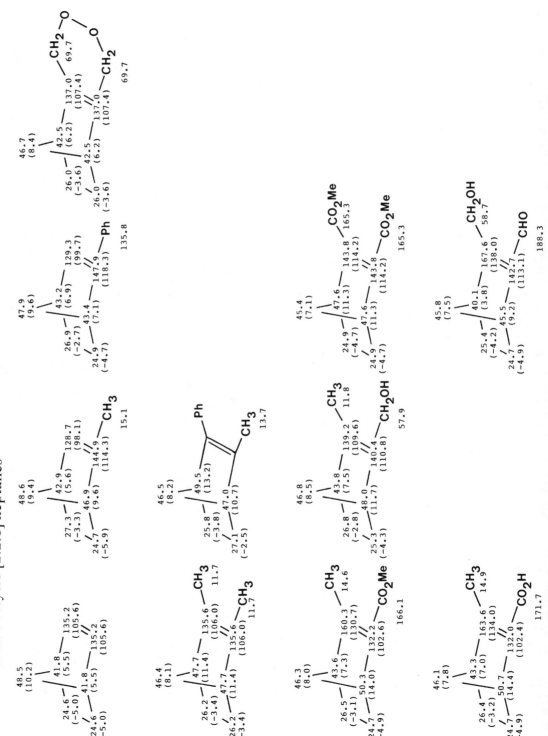

Str	Substituents	C1	C2	C3	C4	C5	C6	C7	Substituent Shifts	References
1	2=	41.8	135.2	135.2	41.8	24.6	24.6	48.5		7,1,2,21,25,32,123,164,174,195
2	2=-CH2OOCH2-3	42.5	137.0	137.0	42.5	26.0	26.0	46.7	69.7,69.7	265,267
3	2=/5-=CH2	41.9	134.1	136.2	50.8	150.7	33.5	50.0	103.2	33,119,216,223
4	2=/5-=CH2 6-=CH2	50.7	136.1	136.1	50.7	148.4	148.4	51.4	101.1,101.1	24,107
5	2=/5-=CMe2	42.3	134.2	134.2	46.5	134.5	32.8	50.3	119.9	98
6	2=/7-=C(CN)2	45.0	134.5	134.5	45.0	23.0	23.0	192.8	69.6	106
7	2=/7-=CH2	45.2	134.9	134.9	45.2	24.8	24.8	162.9	90.1	46
8	2=/7-=CHCH2CH2Me	40.4	136.2	135.7	45.3	24.6	25.0	155.1	105.8	194
9	2=/7-=CHOMe	39.8	136.2	135.7	42.4	24.7	25.7	126.5	134.6	194
10	2=/7-=O	45.3	133.2	133.2	45.3	21.0	21.0	205.3		32,17,194
11	2=/A7-COOH	43.1	135.6	135.6	43.1	22.5	22.5	60.9	177.9	194
12	2=/A7-COOMe	43.2	135.7	135.7	43.2	22.6	22.6	61.1	172.2	194
13	2=/A7-Me	46.0	137.8	137.8	46.0	21.8	21.8	53.3	14.4	1
14	2=/A7-Me S7-Me	50.6	135.1	135.1	50.6	23.8	23.8	55.3	21.7,22.0	179
15	2=/A7-OH	45.6	134.3	134.1	45.6	21.4	21.4	82.0		6,16
16	2=/A7-OH S7-Me	50.2	135.3	135.3	50.2	23.3	23.3	89.7	20.2	162
17	2=/A7-OH S7-Ph	48.7	135.0	135.0	48.7	23.5	23.5	93.0	143.5	194
18	2=/A7-OMe S7-OMe	44.6	134.2	134.2	44.6	23.7	23.7	119.5		16,194
19	2=/A7-OMe S7-Ph	46.8	134.4	134.4	46.8	23.6	23.6	97.7	139.8	194
20	2=/A7-PCl2	44.0	136.6	136.6	44.0	22.1	22.1	71.2		110
21	2=/A7-PMe2	44.1	137.3	137.3	44.1	22.8	22.8	64.5		110
22	2=/A7-PMe3I	43.5	137.7	137.7	43.5	23.8	23.8	52.5		110
23	2=/A7-PSMe2	43.7	138.5	138.5	43.7	22.8	22.8	58.6		110
24	2=/A7-SnMe3	45.4	137.8	137.8	45.4	24.9	24.9	51.2		156
25	2=/E5-=CBrPh/X6-=Me N6-Me	56.9	133.3	138.5	55.3	151.1	46.8	47.1	112.4,28.5,28.3	129
26	2=/E5-=CClPh/X6-=Me N6-Me	56.4	133.4	138.4	52.1	148.4	45.5	47.1	122.2,28.6,28.3	129
27	2=/E5-=CHMe	41.8	134.4	135.9	50.5	142.0	31.1	50.3	112.7	119,2,25,195
28	2=/E5-=CHMe E6-=CHMe	45.3	136.2	136.2	45.3	141.3	141.3	50.9	109.7	178
29	2=/E5-=CHMe/X6-Me N6-Me	53.2	133.7	137.1	46.2	151.0	42.3	47.9	115.5,29.8,28.8	98
30	2=/N5-)2	45.4	45.1	31.1	42.6	136.7	132.7	49.2	45.1	275
31	2=/N5-Br/X6-Br	44.5	137.2	137.2	45.6	52.8	52.6	30.2	72.1	120
32	2=/N5-CH(OH)Me{R,R}	42.4	137.6	131.7	44.8	49.2	30.2	49.8	72.1	98
33	2=/N5-CH(OH)Me{R,S}	42.4	136.9	133.0	44.0	47.9	29.7	49.2	71.1	98
34	2=/N5-CH(OMe)Me	42.7	137.2	133.5	44.5	44.2	29.9	49.4	80.8	98
35	2=/N5-CH=CH2	42.9	137.3	133.2	48.1	43.2	33.5	49.7	143.2	5,10,99,102
36	2=/N5-CH=CH2/N6-Me	48.0	135.2	135.5	48.5	49.5	37.9	49.4	141.0,15.6	224
37	2=/N5-CH=CH2/X6-Me	48.4	138.1	133.3	47.9	53.3	40.9	46.1	142.8,20.1	224
38	2=/N5-CH=[E]CHMe	42.9	137.0	132.9	48.2	42.2	33.1	49.6	136.8	224
39	2=/N5-CH=[Z]CHMe	43.0	136.8	132.8	48.5	36.1	34.1	49.8	136.4	224
40	2=/N5-CMe=CH2	42.9	136.7	132.2	45.9	45.1	29.8	49.7	147.1	10,98
41	2=/N5-COMe/X6-CH2CH(OMe)2	47.2	137.9	131.6	45.9	59.3	36.6	46.4	207.5,38.2	201
42	2=/N5-COMe/X6-Me N6-Me	54.7	135.7	135.1	46.7	64.7	43.5	46.9	209.7,31.5,24.3	235,98
43	2=/N5-Me	43.1	136.7	132.1	47.2	32.5	33.7	50.1	19.3	7,1,2,25
44	2=/N5-Me/N6-Me	49.1	135.0	135.0	49.1	35.7	35.7	49.3	14.4,14.4	7
45	2=/N5-Me/N6-Me/A7-OMe S7-Ph	53.8	133.9	133.9	53.8	33.9	33.9	98.0	13.3,13.3,140.5	194
46	2=/N5-Me/N6-Me/S7-OMe A7-Ph	51.5	130.3	130.3	51.5	30.9	30.9	97.0	11.8,11.8	194
47	2=/N5-Me/X6-Me	49.4	137.9	133.0	48.1	42.5	41.5	46.6	18.6,20.2	7,2,25
48	2=/S7-Br	44.3	132.8	132.8	44.3	22.7	22.7	66.0		110
49	2=/S7-CH2CH2OH	45.9	132.7	132.7	45.9	25.8	25.8	57.1	30.8	81
50	2=/S7-CN A7-OH	50.0	134.7	134.7	50.0	21.5	21.5	82.7	122.2	249

unsat. BICYCLO[2.2.1]HEPTANES (cont.)

Str	Substituents	C1	C2	C3	C4	C5	C6	C7	Substituent Shifts	References
51	2=/S7-CN A7-OTs	50.2	133.5	133.5	50.2	21.3	21.3	86.5	127.8	249
52	2=/S7-COOH	44.0	133.4	133.4	44.0	24.7	24.7	62.4	178.6	194
53	2=/S7-COOMe	44.2	133.4	133.4	44.2	24.7	24.7	62.7	173.0	194
54	2=/S7-I	50.9	134.4	134.4	50.9	21.8	21.8	45.6		225
55	2=/S7-Me	47.8	132.4	132.4	47.8	25.9	25.9	54.7	12.5	1
56	2=/S7-OH	47.4	131.9	131.9	47.4	22.2	22.2	86.9		6
57	2=/S7-OH A7-Ph	49.3	134.3	134.3	49.3	22.4	22.4	93.1	140.4	194
58	2=/S7-OMe A7-Ph	46.1	132.4	132.4	46.1	22.5	22.5	98.8	139.0	194
59	2=/S7-PCl2		134.8	134.8		25.5	25.5	76.7		110
60	2=/S7-PMe2		133.9	133.9		24.3	24.3	66.8		110
61	2=/S7-SnMe3	46.1	136.0	136.0	46.1	26.3	26.3	50.4		156
62	2=/X5-Br/A7-Br	44.0	135.8	137.8	48.7	54.1	29.1	56.4		120
63	2=/X5-Br/S7-Br	44.4	133.0	134.5	51.4	56.0	33.0	59.2		120
64	2=/X5-Br/X6-Br	44.5	137.4	137.4	44.5	52.4	52.4	30.2		120
65	2=/X5-Br/X6-OMe	46.1	139.5	133.0	45.7	54.0	82.8	49.4		120
66	2=/X5-CH=CH2	42.2	136.8	137.3	48.1	42.6	33.5	45.9	143.6	5,10,99,102
67	2=/X5-CH=CH2/N6-Me	47.7	134.1	137.4	48.9	51.8	40.8	47.2	142.8,18.3	224
68	2=/X5-CH=CH2/X6-Me	48.8	136.8	137.9	45.8	49.0	37.9	42.4	141.3,17.9	224
69	2=/X5-CH=[E]CHMe	41.5	136.6	137.1	48.5	41.7	33.1	45.5	133.6	224
70	2=/X5-CH=[Z]CHMe	42.2	136.0	137.1	47.4	35.7	34.1	45.7	136.4	224
71	2=/X5-CMe=CH2	42.3	135.9	136.7	45.5	44.7	31.5	45.3	149.5	10,98
72	2=/X5-COMe/X6-Me N6-Me	54.4	136.6	137.6	46.1	61.2	42.0	47.0	211.6,26.0,29.3	235,98
73	2=/X5-Me	42.3	135.9	137.0	48.3	32.5	34.6	44.8	21.6	7,1,2,25,183
74	2=/X5-Me N5-Me	43.7	136.0	135.7	53.2	37.6	41.5	48.3	30.7,28.2	27,182
75	2=/X5-Me/X6-Me	49.5	136.8	136.8	49.5	34.5	34.5	41.8	16.7,16.7	7
76	2=/X5-OCOCF3 N5-Me	41.9	140.4	132.2	51.7	94.5	41.3	48.1	22,2	223
77	2=/X5-OEt/A7-Br	44.0	134.8	139.0	47.0	83.9	29.2	57.0		120
78	2=/X5-OMe/A7-Br	43.9	134.6	139.0	46.9	82.2	29.4	56.7		120
79	2=/Z5-CHMe	41.4	133.7	136.9	44.8	141.4	33.6	49.5	113.1	119,2,25,195
80	1-Me/2-Me 2=	50.9	146.2	129.5	41.8	29.6	31.8	54.7	17.7,12.4	2,25
81	1-Me/2-OEt 2=/A7-Me S7-Me	52.9	165.3	95.7	49.5	27.8	31.7	54.9	9.8,19.2,19.9	167
82	1-Me/2=	49.9	140.0	135.8	43.3	28.0	32.6	55.0	18.0	1
83	1-Me/2=/A7-Me S7-Me	52.9	139.8	134.3	52.7	24.8	31.9	56.6	13.4,19.7,19.9	1
84	1-Me/2=/N5-CH=CH2	50.6	141.3	137.2	49.0	45.8	39.9	55.6	18.8,143.1	102
85	1-Me/2=/N6-CH=CH2	53.5	137.1	137.2	42.8	35.6	47.7	55.6	17.1,142.2	102
86	1-Me/2=/X5-CH=CH2	50.0	141.4	136.7	49.2	45.2	40.1	51.6	18.7,143.7	102
87	1-Me/2=/X6-CH=CH2	52.3	141.9	137.1	42.4	35.7	47.6	50.9	16.6,142.2	102
88	2-CH2OH 2=/3-Me	48.0	140.4	139.2	43.8	26.8	25.3	46.8	57.9,11.8	273
89	2-CHO 2= 3-CH2OH	45.5	142.7	167.6	40.1	25.4	24.7	45.8	188.3,58.7	128,267
90	2-COOH 2= 3-Me	50.7	132.0	163.6	43.3	26.4	24.7	46.1	171.7,14.9	273
91	2-COOMe 2= 3-COOMe	47.6	143.8	143.8	49.5	24.9	24.9	45.4	165.3	242
92	2-COOMe 2= 3-COOMe/X5-I/X6-I	58.1	143.1	143.1	58.2	32.2	32.2	43.9	163.5	121
93	2-COOMe 2= 3-Me	50.3	132.2	160.3	43.6	26.5	24.7	46.3	166.1,14.6	273
94	2-Me 2=	46.9	144.9	128.7	42.9	27.3	24.7	48.6	15.1	2,25
95	2-Me 2= 3-Me	47.7	135.6	135.6	47.7	26.2	26.2	46.4	11.7,11.7	54
96	2-Me 2= 3-Ph	47.0			49.5	25.8	27.1	46.5	13.7,0	160
97	2-Me 2=/N5-CH=CH2	47.7	147.0	126.1	48.9	45.8	32.6	49.5	14.8,143.3	102
98	2-Me 2=/N6-CH=CH2	53.5	143.2	129.8	43.4	33.5	43.3	50.3	17.6,144.0	102
99	2-Me 2=/X5-CH=CH2	47.2	146.9	129.5	49.1	45.0	32.3	45.2	15.3,144.2	102
100	2-Me 2=/X6-CH=CH2	53.2	146.3	130.3	43.1	35.4	42.6	45.4	15.1,144.2	102

unsat. BICYCLO[2.2.1]HEPTANES (cont.)

Str	Substituents	C1	C2	C3	C4	C5	C6	C7	Substituent Shifts	References
101	2-Ph 2=	43.4	147.9	129.3	43.2	26.9	24.9	47.9	135.8	195
102	2-Ph-Me 2=/N6-Me/A7-Me S7-Me	57.2	135.4	128.3	52.0	33.6	31.0	57.9	19.2,22.2,22.5	153
103	2-PO(OEt)$_2$ 2= 3-PO(OEt)$_2$/N5-Br X6-Br	41.6	150.4	148.3	41.7	59.9	54.6	33.3		202
104	2-PO(OEt)$_2$ 2= 3-PO(OEt)$_2$/X5-O-6X	50.1	154.8	154.8	50.1	57.8	57.8	39.7		202
105	1-Me/2-=2/5=	59.3	213.5	37.8	38.6	143.0	135.6	56.1	12.0	2
106	1-Me/N2-OH/5=	54.3	77.1	40.7	43.3	139.2	136.8	54.3	17.0	2
107	1-Me/X2-OH/5=	53.8	74.6	38.9	41.1	140.2	139.4	50.9	14.7	2
108	2-=0/5-Me 5=	56.2	214.4	36.7	44.4	153.5	121.9	49.4	16.2	7
109	2-=0/5=	55.8	214.7	37.1	40.0	142.7	130.4	50.8		7,2,32
110	2-=0/5- 6-Me	60.6	215.3	38.1	39.9	135.3	141.2	50.5	15.7	7
111	2-=0/N3-Me/5=	56.3	215.8	42.5	45.0	140.7	130.1	49.8	16.5	7
112	2-=0/X3-Me N3-Me/5=	56.4	219.1	42.9	50.6	143.1	130.3	47.9	24.4,26.8	7
113	2-=0/X3-Me/5=	55.6	217.8	40.0	46.0	143.7	131.4	46.9	15.7	7
114	X2-CH$_2$OBs/5=	43.4	38.3	29.4	41.6	137.1	135.9	44.8	74.9	183
115	X2-CH$_2$OH N2-CH$_2$OH/5=	44.7	48.7	39.2	42.1	137.1	135.2	47.0	70.6,69.7	195
116	X2-CH$_2$OH N2-Me/5=	47.9	43.6	37.4	43.1	136.8	135.5	47.6	72.0,22.9	182
117	N2-CH$_2$OH X2-Me/5=	49.7	43.7	37.1	42.6	136.3	135.4	47.6	70.9,24.9	182
118	X2-CH$_2$OH/5=	43.3	41.6	29.6	41.8	136.5	136.7	45.0	67.3	142,2,183,195
119	N2-CH$_2$OH/5=	43.6	41.7	28.9	42.2	137.2	132.2	49.5	66.3	142,2,195
120	X2-CH$_2$OH/N3-CH$_2$OH/5=	44.6	46.9	47.9	44.8	133.5	138.0	47.1	66.4,65.9	142
121	N2-CH$_2$OH/N3-CH$_2$OH/5=	46.5	45.1	45.1	46.5	134.4	134.4	49.9	63.2	142
122	X2-CH$_2$OH/N3-Me/5=	44.6	50.6	37.7	47.3	134.2	137.5	46.8	66.7,19.1	142
123	X2-CH$_2$OH/N3-Me/5=	45.5	44.5	35.4	49.0	135.7	134.8	49.3	63.3,13.9	142,147
124	X2-CH$_2$OH/X3-CH$_2$OH/5=	45.8	43.5	33.5	45.8	137.5	137.5	43.9	64.7	142
125	X2-CH$_2$OH/X3-Me/5=	43.5	44.6	34.3	49.5	137.4	136.9	42.1	64.7,16.1	142
126	N2-CH$_2$OH/X3-Me/5=	44.3	51.4	36.9	48.8	138.4	132.9	46.2	66.3,21.0	142
127	X2-CH$_2$SePh/N2-C1/5=	55.6	78.8	43.3	42.9	133.2	139.8	45.1	49.3	116
128	X2-CHO/5=	44.2	51.7	27.0	41.8	138.4	135.4	45.8	202.9	195
129	N2-CHO/5=	44.8	52.1	27.5	42.7	137.8	132.0	49.6	203.6	195
130	X2-CN/5=	47.6	27.4	32.3	42.0	138.1	134.4	47.4	122.2	195
131	N2-CN/5=	45.8	27.1	32.5	42.5	138.7	133.0	48.6	122.1	165,112,123,195,253,257,258
132	X2-COC1/N3-COC1/5=	47.3	59.3	59.9	46.7	135.2	137.3	48.3	174.3,172.8	195
133	X2-CONMe$_2$/5=	45.6	41.4	30.6	40.7	138.1	135.9	46.6	173.6	123
134	X2-CONMe$_2$/5=	45.2	42.0	30.8	42.4	135.4	133.5	49.1	172.4	123
135	X2-COOH/5=	50.6	49.6	37.6	43.0	138.8	133.6	49.1	185.9,24.2	182
136	N2-COOH X2-Me/5=	51.0	50.1	37.8	42.7	137.9	135.4	47.0	184.4,26.5	182
137	X2-COOH/5=	46.8	43.3	30.4	41.7	138.2	135.8	46.5	183.1	142,123,183
138	N2-COOH/5=	45.7	43.4	29.2	42.6	137.9	132.5	49.7	181.3	142,123,183
139	X2-COOH/N3-Me/5=	47.4	51.4	39.4	47.4	135.4	136.6	48.2	183.1,19.0	142
140	N2-COOH/N3-Me/5=	45.9	48.6	38.3	48.4	136.5	133.9	49.1	180.7,15.6	142
141	X2-COOH/N3-Ph/5=	48.3	50.7	48.7	48.3	136.1	136.6	48.3	181.8,142.8	18
142	N2-COOH/N3-Ph/5=	49.7	51.9	50.4	46.0	134.5	137.3	49.7	179.2,141.1	18
143	X2-COOH/X3-Me/5=	47.1	45.5	36.9	48.5	139.2	136.3	44.1	182.4,17.2	142
144	N2-COOH/X3-Me/5=	45.9	52.6	38.0	48.9	138.8	133.4	46.1	181.5,20.9	142
145	X2-COOH/X3-Ph/5=	48.5	48.9	46.2	45.6	140.4	137.4	45.9	180.0,140.5	18
146	X2-COOH/X3-Ph/5=	48.1	52.2	47.6	46.3	139.2	134.7	47.3	180.6,144.0	18
147	X2-COOMe/5=	46.6	42.8	30.4	41.8	138.1	135.9	46.5	175.2	165,123,195,253
148	N2-COOMe/5=	45.7	43.2	29.4	42.7	137.5	132.7	49.7	173.6	165,123,195,253
149	N2-COOOC-3N/5=/7-=CMe$_2$	46.6		46.6	46.6	135.7	135.7	146.0	170.7,170.7,112.5	231,232
150	N2-O-3N/5=	43.0	51.8	51.8	43.0	130.7	130.7	61.8		40

81

unsat. BICYCLO[2.2.1]HEPTANES (cont.)

Str	Substituents	C1	C2	C3	C4	C5	C6	C7	Substituent Shifts	References
151	X2-O-3X/5=	43.4	59.0	59.0	43.4	141.0	141.0	40.4		40
152	X2-OAc/5=	47.7	75.2	35.0	41.1	141.4	133.1	46.6		104,168,195
153	N2-OAc/5=	45.3	74.5	34.1	41.8	138.6	131.7	47.2		168,195
154	X2-OCH[R]MeCH[R]MeO-2N/5= [1R]	51.3	117.6	41.9	40.7	139.5	133.2	49.2		114
155	X2-OCH[R]MeCH[R]MeO-2N/5= [1S]	51.3	117.6	41.9	41.0	139.9	133.2	48.8		114
156	X2-OH N2-CH=CH2/5=	54.9	81.1	42.0	42.9	139.0	134.0	48.1	145.0	196
157	X2-OH N2-Me/5=	54.2	78.6	43.1	42.0	138.1	134.3	48.2	27.5	77
158	N2-OH X2-CH=CH2/5=	53.4	80.4	43.1	42.9	139.8	133.6	48.6	144.7	196
159	N2-OH X2-Me/5=	54.2	78.0	44.5	43.2	138.1	134.5	49.3	29.5	2,77
160	N2-OH X2-Ph/5=	53.3	82.0	44.7	43.3	140.8	133.7	49.1	147.4	240
161	N2-OH X2-tBu/5=	48.6	85.3	40.2	42.5	140.6	136.7	50.3	37.4	59
162	X2-OH/3-=CMe2/5=	50.1	72.7	137.9	44.5	139.2	133.9	46.4	124.8	129
163	X2-OH/5=	50.1	72.3	36.9	40.7	140.2	133.5	45.6		34,2
164	N2-OH/5=	48.2	72.3	37.6	42.9	140.0	131.1	48.2		34,2
165	X2-OH/N3-Me/5=	51.4	80.2	44.6	46.4	137.4	134.1	47.2	18.1	34
166	N2-OH/N3-Me/5=	49.4	73.6	38.7	47.0	138.6	132.9	48.3	13.1	34
167	X2-OH/X3-Me N3-Me/5=	51.8	80.1	40.6	52.9	138.9	133.0	45.8	24.2,28.3	34
168	N2-OH/X3-Me N3-Me/5=	49.7	81.3	41.9	54.1	141.1	131.7	44.4	29.9,22.2	34
169	X2-OH/X3-Me/5=	50.6	72.6	36.8	47.9	141.1	133.6	42.6	14.8	34
170	N2-OH/X3-Me/5=	48.7	81.1	44.9	49.3	141.1	131.5	44.9	19.4	34
171	X2-OMe N2-CH=CH2/5=	51.5	87.2	38.1	42.0	139.5	133.9	47.8	140.8	193
172	N2-OMe X2-CH=CH2/5=	49.6	86.4	39.5	42.3	138.2	133.9	48.1	142.4	196
173	X2-PO(OiPr)2/N3-COMe/5=/5=	46.7	37.4	54.3	45.2	133.0	138.7	48.1	204.3	186
174	X2-PO(OiPr)2/N3-COOMe/5=	45.7	39.7	46.9	45.1	133.8	139.0	47.8	172.9	186
175	N2-PO(OiPr)2/X3-COOMe/5=	46.9	43.1	47.7	46.5	141.1	134.6	49.3	172.3	186
176	N2-PO(OiPr)2/X3-COMe/5=	45.0	39.4	54.5	47.6	135.8	136.8	47.4	206.4	186
177	X2-PO(OiPr)2/X3-COOMe/5=	44.6	42.1	45.1	46.9	137.3	139.0	46.1	174.0	186
178	N2-PO(OiPr)2/X3-COOMe/5=	45.1	41.2	46.6	48.8	135.8	136.4	47.9	174.6	186
179	X2-PO(Ph)2/X3-PO(Ph)2/5=	46.5	41.6	41.6	46.5	138.4	138.4	29.7		202
180	X2-POPh2/N3-POPh2/5=	47.0	38.6	39.6	47.2			48.2		202
181	X2-SePh/N3-Cl/5=	49.5	50.6	64.9	49.6	134.9	137.0	46.3	0	115,116
182	N2-SePh/X3-Cl/5=	52.2	53.0	65.8	47.7	137.6	131.8	46.7	0	115,116
183	X2-SPh/N3-Cl/5=	48.9	58.5	64.7	49.4	134.4	136.8	46.0	0	117
184	N2-SPh/X3-Cl/5=	50.2	56.1	65.1	47.4	137.4	132.8	46.6	0	117
185	2=/5=	50.2	143.1	143.1	50.2	143.1	143.1	75.1		20,1,2,16,25,51,58,195,234,271
186	2=/5=/7-=2	49.9	143.3	143.3	49.9	143.3	143.3	139.0		106
187	2=/5=/7-=C(CN)2	50.7	140.0	140.0	50.7	140.0	140.0	191.3	59.3	106
188	2=/5=/7-=C=CH2	52.2	142.2	142.2	52.2	142.2	142.2			214
189	2=/5=/7-=CH2	53.5	141.9	141.9	53.5	141.9	141.9	177.1	78.5	46,45,106
190	2=/5=/7-=CMe2		142.7	142.7		142.7	142.7	165.7	93.7	106
191	2=/5=/7-CH2OAc	51.4	140.0	140.0	51.4	144.4	144.4	82.6	64.8	52
192	2=/5=/7-CH2OBs	51.9	140.8	140.8	51.9	145.0	145.0	82.4	72.4	52
193	2=/5=/7-CH2OH	51.2	140.0	140.0	51.2	144.6	144.6	87.0	62.3	85
194	2=/5=/7-CHO	50.1	141.3	141.3	50.1	143.0	143.0	92.0	202.6	85
195	2=/5=/7-COOH	51.0	141.2	141.2	51.0	142.7	142.7	84.2	176.8	85
196	2=/5=/7-COOMe	50.6	140.4	140.4	50.6	142.1	142.1	84.0	169.6	85
197	2=/5=/7-OH	56.1	139.1	139.1	56.1	138.0	138.0	102.4		16
198	2=/5=/7-OtBu	55.5	139.8	139.8	55.5	137.2	137.2	104.3		195
199	2=/5=/A7-Me S7-Me	60.0	142.4	142.4		142.4	142.4	84.8	24.3,24.3	58
200	2=/5=/A7-OMe S7-OMe	53.7	138.9	138.9	53.7	138.9	138.9	132.8		16

unsat. BICYCLO[2.2.1]HEPTANES (cont.)

Str	Substituents	C1	C2	C3	C4	C5	C6	C7	Substituent Shifts	References
201	1-CF3/2-CF3 2=/5=	52.5			52.5	142.2	142.2	73.4	0	82
202	2-CF3 2= 3-CF3/5=/7-=CH2	56.0			56.0	139.5	139.5		122.3,122.3,85.3	106
203	2-CF3 2= 3-CF3/5=/7-=CMe2	52.3			52.3	142.3	142.3	161.2	121.9,121.9,100.8	106
204	2-CN 2= 3-CN/5=/7-=CMe2	55.2	140.7	140.7	55.2	141.6	141.6	161.2	113.2,113.2,103.7	106
205	2-COOH 2= 3-Me/5=	58.6	138.1	173.4	50.9	144.3	140.2	71.0	171.7,17.4	273
206	2-COOMe 2= 3-COOMe/5-COOMe 5= 6=COOMe/7-=CMe2	56.3	150.6	150.6	56.3	150.6	150.6	158.8	4*163.9,105.6	106
207	2-COOMe 2= 3-COOMe/5=	51.9	152.4	152.4	51.9	142.4	142.4	73.0	165.4,165.4	105
208	2-COOMe 2= 3-COOMe/5=/7-=CHMe	51.9	150.6	151.8	56.5	142.2	141.5	167.1	165.4,164.6,92.9	106
209	2-COOMe 2= 3-COOMe/5=/7-=CMe2	53.5	151.8	151.8	53.5	142.3	142.3	162.1	164.9,164.6,100.0	106
210	2-COOMe 2= 3-COOMe/5=/7-=CPh2	54.8	151.2	151.2	54.8	142.3	142.3	164.0	164.6,164.6,112.9	106
211	2-COOMe 2= 3-Me/5=	58.3	138.4	169.8	50.8	144.1	140.5	71.0	166.3,17.1	273
212	2-COOMe 2=/5=/7-=CMe2	51.8	148.9	154.4	50.1	143.4	141.6	164.1	164.8,97.9	106
213	2-OTMS 2=/5=	53.6	173.0	109.1	48.1	145.4	141.2	72.5		243
214	2-PO(OEt)2 2= 3-PO(OEt)2/5=	57.2	156.7	156.7	57.2	142.1	142.1	73.2		202
215	2= 2-CH2OOCH2-3/5=	49.5	144.8	144.8	49.5	142.2	142.2	71.3	71.7,71.7	128

References to Bicyclo[2.2.1]heptanes and related* systems

1. Grutzner, J. B., Jautelat, M., Dence, J. B., Smith, R. A. and Roberts, J. D., *J. Am. Chem. Soc.* **92**, 7107 (1970).

2. Lippmaa, E., Pehk, T., Paasivirta, J., Belikova, N. and Plate, A., *Org. Magn. Reson.* **2**, 581 (1970).

3. Lippmaa, E., Pehk, T., Belikova, N. A., Bobyleva, A. A., Kalinichenko, A. N., Ordubadi, M. D. and Plate, A. F., *Org. Magn. Reson.* **8**, 74 (1976).

4. Poindexter, G. S. and Kropp, P. J., *J. Org. Chem.* **41**, 1215 (1976).

5. Nakai, N., Iwasa, S., Ishii, Y. and Ogawa, M., *Shikizai Kyokaishi* **51**, 132 (1978).

6. Stothers, J. B. and Tan, C. T., *Can. J. Chem.* **55**, 841 (1977).

7. Stothers, J. B., Tan, C. T. and Teo, K. C., *Can. J. Chem.* **51**, 2893 (1973).

8. Buchanan, G. W. and Benezra, C., *Can. J. Chem.* **54**, 231 (1976).

9. Olah, G. A. and White, A. M., *J. Am. Chem. Soc.* **91**, 3954 (1969).

10. Iwase, S., Nakata, M., Hamanaka, S. and Ogawa, M., *Bull. Chem. Soc. Japan* **49**, 2017 (1976).

*11. Nakagawa, K., Iwase, S., Ishii, Y., Hamanaka, S. and Ogawa, M., *Bull. Chem. Soc. Japan* **50**, 2391 (1977).

*12. Murakhovskaya, A. S., Stepanyants, A. U., Zimina, K. I., Aref'ev, O. A. and Epishev, V. I., *Bull. Acad. Sci. USSR, Div. Chem. Sci.* 847 (1975).

*13. Kazimirchik, I. V., Lukin, K. A., Borisenko, A. A., Bebikh, G. F., Yarovoi, S. S. and Zefirov, N. S., *J. Org. Chem. USSR* **18**, 507 (1982).

*14. Cheng, A. K. and Stothers, J. B., *Org. Magn. Reson.* **9**, 355 (1977).

*15. Waddington, M. D. and Jennings, P. W., *Organometallics* **1**, 385 (1982).

16. Bicker, R., Kessler, H. and Zimmerman, G., *Chem. Ber.* **111**, 3200 (1978).

17. Bicker, R., Kessler, H., Steigel, A. and Zimmerman, G., *Chem. Ber.* **111**, 3215 (1978).

18. Nicolas, L., Beugelmans-Verrier, M. and Guilhem. J., *Tetrahedron* **37**, 3847 (1981).

19. Wenkert, E., Clouse, A. O., Cochran, D. W. and Doddrell, D., *J. Chem. Soc. Chem. Commun.* 1433 (1969).

20. Tori, K., Tsushima, T., Tanida, H., Kushida, K. and Satoh, S., *Org. Magn. Reson.* **6**, 324 (1974).

21. Tori, K., Ueyama, M., Tsuji, T., Matsumura, H., Tanida, H., Iwamura, H., Kushida, K., Nishida, T. and Satoh, S., *Tetrahedron Lett.* 327 (1974).

*22. Lippmaa, E., Pehk, T. and Paasivirta, J., *Org. Magn. Reson.* **5**, 277 (1973).

23. Briggs, J., Hart, F. A., Moss, G. P. and Randall, E. W., *J. Chem. Soc. Chem. Commun.* 364 (1971).

24. Quarroz, D., Sonney, J.-M., Chollet, A., Florey, A. and Vogel, P., *Org. Magn. Reson.* **9**, 611 (1977).

25. Pehk, T. I., Lippmaa, E. T., Belikova, N. A. and Plate, A. F., *Doklady Physical Chemistry* **195**, 930 (1970).

26. Kiyooka, S. and Suzuki, K., *Bull. Chem. Soc. Japan* **47**, 2081 (1974).

27. Yates, P. and Cong, D. D., *Org. Magn. Reson.* **20**, 199 (1982).

28. Schneider, H.-J. and Weigand, E. F., *Tetrahedron* **31**, 2125 (1975).

29. Levy, G. C. and Komoroski, R. A., *J. Am. Chem. Soc.* **96**, 678 (1974).

30. Werstiuk, N. H., Taillefer, R., Bell, R. A. and Sayer, B. G., *Can. J. Chem.* **50**, 2146 (1972).

31. Werstiuk, N. H., Taillefer, R., Bell, R. A. and Sayer, B., *Can. J. Chem.* **51**, 3010 (1973).

32. Stothers, J. B., Swenson, J. R. and Tan, C. T., *Can. J. Chem.* **53**, 581 (1975).

33. Grover, S. H. and Stothers, J. B., *Can. J. Chem.* **53**, 589 (1975).

34. Stothers, J. B., Tan, C. T. and Teo, K. C., *Can. J. Chem.* **54**, 1211 (1976).
35. Stothers, J. B. and Teo, K. C., *Can. J. Chem.* **54**, 1222 (1976).
36. Wilson, N. K. and Stothers, J. B., *Topics in Stereochemistry*, vol. 8, Eliel, E. L and Allinger, N. L., Eds., Wiley-Interscience, New York, 1974, p. 43.
37. Kelly, D. P. and Brown, H. C., *J. Am. Chem. Soc.* **97**, 3897 (1975).
38. Olah, G. A., White, A. M., DeMember, J. R., Commeyras, A. and Lui, C. Y., *J. Am. Chem. Soc.* **92**, 4627 (1970).
*39. Olah, G. A. and Liang, G., *J. Am. Chem. Soc.* **95**, 3792 (1973).
40. Davies, S. G. and Whitham, G. H., *J. Chem. Soc. Perkin Trans. II* 861 (1975).
41. Smith, C. A. and Grutzner, J. B., *J. Org. Chem.* **41**, 367 (1976).
42. Kitching, W., Marriott, M., Adcock, W. and Doddrell, D., *J. Org. Chem.* **41**, 1671 (1976).
43. Doddrell, D., Burfitt, I., Kitching, W., Bullpitt, M., Lee, C.-H., Mynott, R. J., Considine, J. L., Kuivila, H. G. and Sarma, R. H., *J. Am. Chem. Soc.* **96**, 1640 (1974).
44. Gansow, O. A., Willcott, M. R. and Lenkinski, R. E., *J. Am. Chem. Soc.* **93**, 4295 (1971).
45. Hoffmann, R. W., Schuttler, R., Schafer, W. and Schweig, A. *Angew. Chem., Int. Ed. Engl.* **11**, 512 (1972).
46. Hoffmann, R. W., Kurz, H., Reetz, M. T. and Schuttler, R., *Chem. Ber.* **108**, 109 (1975).
47. Zimmermann, D., Reisse, J., Coste, J., Plenat, F. and Christol, H., *Org. Magn. Reson.* **6**, 492 (1974).
48. Marshall, J. L. and Miiller, D. E., *J. Am. Chem. Soc.* **95**, 8305 (1973).
49. Reetz, M. T., Chatziiosifidis, I., Kunzer, H. and Muller-Starke, K., *Tetrahedron* **39**, 961 (1983).
50. Morris, D. G. and Murray, A. M., *J. Chem. Soc. Perkin Trans. II* 539 (1975).
51. Lippmaa, E. and Pehk, T., *Kemian Teollisuus* **24**, 1001 (1967).
52. Bly, R. S., Bly, R. K. and Shibata, T., *J. Org. Chem.* **48**, 101 (1983).
53. Himmele, W. and Siegel, H., *Tetrahedron Lett.* 907 (1976).
54. Bohlmann, F., Zeisberg, R. and Klein, E., *Org. Magn. Reson.* **7**, 426 (1975).
55. Hanstock, C. C., Tebby, J. C. and Coates, H., *J. Chem. Res., Synop.* 110 (1982).
56. James, D. E. and Stille, J. K., *J. Org. Chem.* **41**, 1504 (1976).
57. Hawkes, G. E., Herwig, K. and Roberts, J. D., *J. Org. Chem.* **39**, 1017 (1974).
58. Jefford, C. W., Wallace, T. W., Can, N.-T. and Rimbault, C. G., *J. Org. Chem.* **44**, 689 (1979).
59. Grob, C. A., Gunther, B. and Waldner, A., *Helv. Chim. Acta* **64**, 2709 (1981).
60. Fischer, W., Grob, C. A. and von Sprecher, G., *Helv. Chim. Acta* **63**, 806 (1980).
61. Fischer, W., Grob, C. A., von Sprecher, G. and Waldner, A., *Helv. Chim. Acta* **63**, 816 (1980).
62. Fischer, W., Grob, C. A., von Sprecher, G. and Waldner, A., *Helv. Chim. Acta* **63**, 928 (1980).
63. Kanel, H.-R., Capraro, H.-G. and Ganter, C., *Helv. Chim. Acta* **65**, 1032 (1982).
64. Klester, A. M. and Ganter, C., *Helv. Chim. Acta* **66**, 1200 (1983).
65. Grob, C. A., Gunther, B. and Hanreich, R., *Helv. Chim. Acta* **65**, 2288 (1982).
66. Jaggi, F. J. and Ganter, C., *Helv. Chim. Acta* **63**, 866 (1980).
67. Cargill, R. L., Bushey, D. F., Ellis, P. D., Wolff, S. and Agosta, W. C., *J. Org. Chem.* **39**, 573 (1974).
68. Morris, D. G. and Murray, A. M., *J. Chem. Soc. Perkin Trans. II* 1579 (1976).

69. Brown, F. C. and Morris, D. G., *J. Chem. Soc. Perkin Trans. II* 125 (1977).
70. Schippers, P. H., van der Ploeg, J. P. M. and Dekkers, H. P. J. M., *J. Am. Chem. Soc.* **105**, 84 (1983).
*71. Subramanyam, R., Bartlett, P. D., Moltrasio Iglesias, G. Y., Watson, W. H. and Galloy, J., *J. Org. Chem.* **47**, 4491 (1982).
72. Bartlett, P. D. and Wu, C., *J. Am. Chem. Soc.* **105**, 100 (1983).
*73. Watson, W. H., Galloy, J., Bartlett, P. D. and Roof, A. A. M., *J. Am. Chem. Soc.* **103**, 2022 (1981).
74. Kreiser, W., Janitschke, L., Voss, W., Ernst, L. and Sheldrick, W. S., *Chem. Ber.* **112**, 397 (1979).
75. Kreiser, W. and Janitschke, L., *Chem. Ber.* **112**, 408 (1979).
*76. Kreiser, W., Janitschke, L. and Ernst, L., *Tetrahedron* **34**, 131 (1978).
77. Paasivirta, J. and Laihia, K., *Org. Magn. Reson.* **7**, 596 (1975).
78. Barfield, M., Burfitt, I. and Doddrell, D., *J. Am. Chem. Soc.* **97**, 2631 (1975).
*79. Belikova, N. A., Arbuzov, V. A., Pekhk, T. I., Bobyleva, A. A. and Plate, A. F., *J. Org. Chem. USSR* **17**, 2195 (1981).
*80. Bobyleva, A. A., Belikova, N. A., Baranova, S. V., Arbuzov, V. A., Pekhk, T. I. and Plate, A. F., *J. Org. Chem. USSR* **17**, 1183 (1981).
81. Bly, R. S., Bly, R. K., Hamilton, J. B., Hsu, J. N. C. and Lillis, P. K., *J. Am. Chem. Soc.* **99**, 216 (1977).
82. Brinker, V. H. and Fleischbauer, I., *Tetrahedron* **37**, 4495 (1981).
*83. Schleyer, P. v. R., Grubmuller, P., Maier, W. F., Vostrowsky, O., Skattebol, L. and Holm, K. H., *Tetrahedron Lett.* **21**, 921 (1980).
*84. Jaggi, F. G. and Ganter, C., *Helv. Chim. Acta* **63**, 214 (1980).
85. Stapersma, J. and Klumpp, G. W., *Tetrahedron* **37**, 187 (1981).
86. Bose, A. K., Sugiura, M. and Srinivasan, P. R., *Tetrahedron Lett.* 1251 (1975).
87. Agosta, W. C. and Wolff, S., *J. Org. Chem.* **41**, 2605 (1976).
88. Kuivila, H. G., Maxfield, P. L., Tsai, K.-H. and Dixon, J. E., *J. Am. Chem. Soc.* **98**, 104 (1976).
89. Johnson, A. L., Stothers, J. B. and Tan, C. T., *Can. J. Chem.* **53**, 212 (1975).
90. Berger, S., *Org. Magn. Reson.* **14**, 65 (1980).
*91. Takaishi, N., Inamoto, Y., Tsuchihashi, K., Yashima, K. and Aigami, K., *J. Org. Chem.* **40**, 2929 (1975).
92. Kelly, D. P., Underwood, G. R. and Barron, P. F., *J. Am. Chem. Soc.* **98**, 3106 (1976).
93. Saksena, A. K., Mangiaracina, P., Brambilla, R., McPhail, A. T. and Onan, K. D., *Tetrahedron Lett.* 1729 (1978).
94. Zefirov, N. S., Kasyan, L. I., Gnedenkov, L. Y., Shashkov, A. S. and Cherepanova, E. G., *Tetrahedron Lett.* 949 (1979).
95. Belikova, N. A., Bobyleva, A. A., Kalinichenko, A. N., Plate, A. F., Pekhk, T. I. and Lippmaa, E. T., *J. Org. Chem. USSR* **10**, 241 (1974).
96. Bobyleva, A. A., Dubitskaya, N. F., Belikova, N. A., Pekhk, T. I., Lippmaa, E. T. and Plate, A. F., *J. Org. Chem. USSR* **13**, 1939 (1977).
97. Belikova, N. A., Lermontov, S. A., Pekhk, T. I., Lippmaa, E. T. and Plate, A. F., *J. Org. Chem. USSR* **14**, 2101 (1978).
98. Belikova, N. A., Lermontov, S. A., Skornyakova, T. G., Pekhk, T. I., Lippmaa, E. T. and Plate, A. F., *J. Org. Chem. USSR* **15**, 436 (1979).

99. Bobyleva, A. A., Belikova, N. A., Kalinichenko, A. N., Baryshnikov, A. T., Dubitskaya, N. F., Pekhk, T. I., Lippmaa, E. T. and Plate, A. F., *J. Org. Chem. USSR* **16**, 1397 (1980).

100. Lermontov, S. A., Belikova, N. A., Skornyakova, T. G., Pekhk, T. I., Lippmaa, E. T. and Plate, A. F., *J. Org. Chem. USSR* **16**, 1982 (1980).

101. Belikova, N. A., Ordubadi, M. D., Bobyleva, A. A., Dubitskaya, N. F., Loshkareva, L. N., Pekhk, T. I., Lippmaa, E. T. and Plate, A. F., *J. Org. Chem. USSR* **15**, 277 (1979).

102. Titova, L. F., Bazhanov, Y. V., Belikova, N. A., Lapuka, L. F., Pekhk, T. I. and Lippmaa, E. T., *J. Org. Chem. USSR* **18**, 721 (1982).

103. Lysenkov, V. I., Pekhk, T. I., Klyuev, A. Y. and Zheleznyak, T. L., *J. Org. Chem. USSR* **18**, 2309 (1982).

104. Barron, P. F., Doddrell, D. and Kitching, W., *J. Organomet. Chem.* **132**, 351 (1977).

105. Hughes, R. P., Krishnamachari, N., Locke, C. J. L., Powell, J. and Turner, G., *Inorg. Chem.* **16**, 314 (1977).

106. Knothe, L., Werp, J., Babsch, H., Prinzback, H. and Fritz, H., *Liebigs Ann. Chem.* 709 (1977).

107. Pfeffer, H. U. and Klessinger, M., *Org. Magn. Reson.* **9**, 121 (1977).

108. Wiberg, K. B., Pratt, W. E. and Bailey, W. F., *J. Org. Chem.* **45**, 4936 (1980).

109. Della, E. W., Hine, P. T. and Patney, H. K., *J. Org. Chem.* **42**, 2940 (1977).

110. Quin, L. D. and Littlefield, B., *J. Org. Chem.* **43**, 3508 (1978).

111. Wiberg, K. B., Pratt, W. E. and Bailey, W. F., *J. Am. Chem. Soc.* **99**, 2297 (1977).

112. Iyoda, J. and Ishikawa, Y., *Osaka Kogyo Gijutsu Shikensho Kiho* **27**, 1 (1976).

113. Bach, R. D., Holubka, J. W. and Taaffee, T. H., *J. Org. Chem.* **44**, 35 (1979).

114. Bordeaux, D. and Gagnaire, G., *Tetrahedron Lett.* **23**, 3353 (1982).

115. Beaulieu, P. L., Morisset, V. M. and Garratt, D. G., *Can. J. Chem.* **58**, 1005 (1980).

116. Garratt, D. G. and Kabo, A., *Can. J. Chem.* **58**, 1030 (1980).

117. Garratt, D. G. and Beaulieu, P. L., *J. Org. Chem.* **44**, 3555 (1979).

*118. Beaulieu, P. L., Kabo, A. and Garratt, D. G., *Can. J. Chem.* **58**, 1014 (1980).

119. Garratt, D. G., Beaulieu, P. L. and Morisset, V. M., *Can. J. Chem.* **58**, 1021 (1980).

120. Garratt, D. G., *Can. J. Chem.* **58**, 1327 (1980).

121. McCulloch, A. W., McInnes, A. G., Smith, D. G. and Walter, J. A., *Can. J. Chem.* **54**, 2013 (1976).

*122. Asmus, P. and Klessinger, M., *Liebigs Ann. Chem.* 2169 (1975).

123. Shashkov, A. S., Cherepanova, E. G., Kas'yan, L. I., Gnedenkov, L. Y. and Bombushkar, M. F., *Bull. Acad. Sci. USSR, Div. Chem. Sci.* **29**, 382 (1980).

*124. Baldwin, J. E. and Barden, T. C., *J. Am. Chem. Soc.* **105**, 6656 (1983).

125. Burgar, M. I., Karba, D. and Kikelj, D., *Farm. Vestnik* **30**, 253 (1979).

126. de Picciotto, L., Carrupt, P.-A. and Vogel, P., *J. Org. Chem.* **47**, 3796 (1982).

127. Pilet, O. and Vogel, P., *Helv. Chim. Acta* **64**, 2563 (1981).

128. Hagenbuck, J.-P., Birbaum, J.-L., Metral, J.-L. and Vogel, P., *Helv. Chim. Acta* **65**, 887 (1982).

129. Mayr, H. and Halberstadt-Kausch, I. K., *Chem. Ber.* **115**, 3479 (1982).

130. bin Sadikun, A. and Davies, D. I., *J. Chem. Soc. Perkin Trans. I* 2461 (1982).

131. Kelly, D. P., Farquharson, G. J., Giansiracusa, J. J., Jensen, W. A., Hugel, H. M., Porter, A. P., Rainbow, I. J. and Timewell, P. H., *J. Am. Chem. Soc.* **103**, 3539 (1981).

*132. Reitz, T. J. and Grunewald, G. L., *Org. Magn. Reson.* **21**, 596 (1983).

133. Offermann, W., *Org. Magn. Reson.* **20**, 203 (1982).

134. Chatterjee, N., *J. Mag. Reson.* **33**, 241 (1979).

135. Brunke, E.-J., Hammerschmidt, F.-J. and Struwe, H., *Tetrahedron Lett.* 21, 2405 (1980).
*136. Dawson, B. A. and Stothers, J. B., *Org. Magn. Reson.* 21, 217 (1983).
137. Kleinpeter, E., Kuhn, H. and Muhlstadt, M., *Org. Magn. Reson.* 8, 279 (1976).
*138. Kleinpeter, E. and Borsdorf, R., *J. Prakt. Chem.* 319, 458 (1977).
*139. Matoba, Y., Kagayama, T., Ishii, Y. and Ogawa, M., *Org. Magn. Reson.* 17, 144 (1981).
140. Kitching, W., *Org. Magn. Reson.* 20, 123 (1982).
141. Bowman, W. R., Golding, B. T. and Watson, W. P., *J. Chem. Soc. Perkin Trans. II* 731 (1980).
142. Brouwer, H., Stothers, J. B. and Tan, C. T., *Org. Magn. Reson.* 9, 360 (1977).
143. Della, E. W., Cotsaris, E. and Hine, P. T., *J. Am. Chem. Soc.* 103, 4131 (1981).
144. Yuryev, V. P., Salimgareyeva, I. M., Kaverin, V. V., Khalilov, K. M. and Panasenko, A. A., *J. Organomet. Chem.* 171, 167 (1979).
145. Brown, F. C., Morris, D. G. and Murray, A. M., *Tetrahedron* 34, 1845 (1978).
146. Barfield, M., Brown, S. E., Canada, Jr., E. D., Ledford, N. D., Marshall, J. L., Walter, S. R. and Yakali, E., *J. Am. Chem. Soc.* 102, 3355 (1980).
147. Marshall, J. L., Conn, S. A. and Barfield, M., *Org. Magn. Reson.* 9, 404 (1977).
148. Andrieu, C. G., Debruyne, D. and Paquer, D., *Org. Magn. Reson.* 11, 528 (1978).
149. Demarco, P. V., Doddrell, D. and Wenkert, E., *J. Chem. Soc. Chem. Commun.* 1418 (1969).
150. Stothers, J. B. and Teo, K. C., *Org. Magn. Reson.* 9, 712 (1977).
151. Kunz, H., Lindig, M., Bicker, R. and Bock, H., *Chem. Ber.* 111, 2282 (1978).
152. Madgzinski, L. J., Somasekharen Pillay, K., Richard, H. and Chow, Y. L., *Can. J. Chem.* 56, 1657 (1978).
153. Coxon, J. M. and Steel, P. J., *Aust. J. Chem.* 32, 2441 (1979).
154. Coxon, J. M., Pojer, P. M., Steel, P. J., Rae, I. D. and Jones, A. J., *Aust. J. Chem.* 31, 1747 (1978).
155. Della, E. W. and Patney, H. K., *Aust. J. Chem.* 32, 2243 (1979).
156. Kuivila, H. G., Considine, J. L., Sarma, R. H. and Mynott, R. J., *J. Organomet. Chem.* 111, 179 (1976).
157. Coxon, J. M., Robinson, W. T. and Steel, P. J., *Aust. J. Chem.* 32, 167 (1979).
158. Coxon, J. M., Steel, P. J., Coddington, J. M., Rae, I. D. and Jones, A. J., *Aust. J. Chem.* 31, 1223 (1978).
*159. Butler, O., Coxon, J. M. and Steel, P. J., *Aust. J. Chem.* 36, 955 (1983).
160. Coxon, J. M. and Steel, P. J., *Aust. J. Chem.* 33, 2455 (1980).
161. Brown, R. F. C., Coddington, J. M., Coxon, J. M., Jones, A. J., Rae, I. D. and Steel, P. J., *Aust. J. Chem.* 31, 2727 (1978).
162. Servis, K. L. and Shue, F.-F., *J. Am. Chem. Soc.* 102, 7233 (1980).
163. Geneste, P., Durand, R., Kamenka, J.-M., Beierbeck, H., Martino, R. and Saunders, J. K., *Can. J. Chem.* 56, 1940 (1978).
164. Hoffmann, E. G., Jolly, P. W., Kusters, A., Mynott, R. and Wilke, G., *Z. Naturforsch. B* 31, 1712 (1976).
165. Sera, A., Takagi, K., Nakamura, M. and Seguchi, K., *Bull. Chem. Soc. Japan* 54, 1271 (1981).
*166. Hoffmann, R. W., Kurz, H. R., Becherer, J. and Reetz, M. T., *Chem. Ber.* 111, 1264 (1978).
167. Strobel, M. P., Andrieu, C. G., Paquer, D., Vazeux, M. and Pham, C. C., *Nouv. J. Chim.* 4, 101 (1980).
168. Quarroz, D. and Vogel, P., *Helv. Chim. Acta* 62, 335 (1979).

*169. Reetz, M. T. and Sauerwald, M., *Chem. Ber.* **114**, 2355 (1981).
*170. Vogel, P., Delseth, R. and Quarroz, D., *Helv. Chim. Acta* **58**, 508 (1975).
171. Rodina, L. L., Kuruts, I. and Korobitsyna, I. K., *J. Org. Chem. USSR* **17**, 1711 (1981).
172. Bloodworth, A. J. and Courtneidge, J. L., *J. Chem. Soc. Perkin Trans. I* 3258 (1981).
173. Della, E. W. and Pigou, P. E., *J. Am. Chem. Soc.* **104**, 862 (1982).
174. Smith, W. B., *Org. Magn. Reson.* **17**, 124 (1981).
175. Sugiura, M., Takao, N. and Ueji, S., *Org. Magn. Reson.* **18**, 128 (1982).
176. Cullen, E. R., Guziec, Jr., F. S., Murphy, C. J., Wong, T. C. and Andersen, K. K., *J. Chem. Soc. Perkin Trans. II* 473 (1982).
177. Morris, D. G., Shepherd, A. G., Walker, M. F. and Jemison, R. W., *Aust. J. Chem.* **35**, 1061 (1982).
178. Paquette, L. A., Schaefer, A. G. and Blount, J. F., *J. Am. Chem. Soc.* **105**, 3642 (1983).
179. Paquette, L. A., Doecke, C. W., Kearney, F. R., Drake, A. F. and Mason, S. F., *J. Am. Chem. Soc.* **102**, 7228 (1980).
*180. Pilet, O., Chollet, A. and Vogel, P., *Helv. Chim. Acta* **62**, 2341 (1979).
181. Schmittel, M., Schulz, A., Ruchardt, C. and Hadicke, E., *Chem. Ber.* **114**, 3533 (1981).
182. Ho, H. T., Ivin, K. J. and Rooney, J. J., *Makromol. Chem.* **183**, 1629 (1982).
183. Ivin, K. J., Lapienis, G. and Rooney, J. J., *Polymer* **21**, 436 (1980).
184. Ramnath, N., Ramesh, V. and Ramamurthy, V., *J. Org. Chem.* **48**, 214 (1983).
185. van der Velden, G., *Macromolecules* **16**, 85 (1983).
186. Haslinger, E., Ohler, E. and Robien, W., *Monatsh. Chem.* **113**, 1321 (1982).
187. Buchbauer, G., Vitek, R., Hirsch, M. C., Kurz, C., Cech, B. and Vas, E. M., *Monatsh. Chem.* **113**, 1433 (1982).
188. Buchbauer, G., Haslinger, E., Robien, W. and Vitek, R., *Monatsh. Chem.* **114**, 113 (1983).
189. Quin, L. D., Gallagher, M. J., Cunkle, G. T. and Chesnut, D. B., *J. Am. Chem. Soc.* **102**, 3136 (1980).
190. Camenzind, H., Krebs, E.-P. and Keese, R., *Helv. Chim. Acta* **65**, 2042 (1982).
191. Luef, W., Vogeli, U.-C. and Keese, R., *Helv. Chim. Acta* **66**, 2729 (1983).
*192. Risch, N., Emig, P., Scheffler, G., Pohle, H. and Henkel, G., *Arzneim.-Forsch.* **32**, 1409 (1982).
193. Buchbauer, G. and Schmidmayer, I., *Monatsh. Chem.* **109**, 751 (1978).
194. Peoples, III., P. R., Ph.D. Thesis, Purdue University, (1979).
195. *Sadtler Standard Carbon-13 NMR Spectra*, Sadtler Research Laboratories, Inc., Philadelphia, Pa.
196. Whitesell, J. K. and Minton, M. A., The University of Texas at Austin, unpublished results.
197. Kirmse, W. and Streu, J., *Synthesis* 994 (1983).
198. Thiem, J. and Meyer, B., *Org. Magn. Reson.* **11**, 50 (1978).
*199. Casanova, J., Waegell, B., Koukoua, G. and Toure, V., *J. Org. Chem.* **44**, 3976 (1979).
*200. Hamlin, J. E. and Toyne, K. J., *J. Chem. Soc. Perkin Trans. I* 2731 (1981).
201. Yates, P. and Douglas, S. P., *Can. J. Chem.* **60**, 2760 (1982).
202. Kyba, E. P. and Whitesell, M. A., The University of Texas at Austin, unpublished results.
*203. Schmitz, L. R. and Sorensen, T. S., *J. Am. Chem. Soc.* **104**, 2600 (1982).
*204. Luh, T.-Y. and Lei, K. L., *J. Org. Chem.* **46**, 5328 (1981).
*205. Luh, T.-Y. and Lei, K. L., *J. Chem. Soc. Chem. Commun.* 214 (1981).
*206. Baird, M. S., Sadler, P., Hatem, J., Zahra, J.-P. and Waegell, B., *J. Chem. Soc. Chem. Commun.* 452 (1979).

*207. Tobe, Y., Hayauchi, Y., Sakai, Y. and Odaira, Y., *J. Org. Chem.* **45**, 637 (1980).

*208. Tobe, Y., Terashima, K., Sakai, Y. and Odaira, Y., *J. Am. Chem. Soc.* **103**, 2307 (1981).

*209. Kleinpeter, E., Kuhn, H., Muhlstadt, M., Jancke, H. and Zeigan, D., *J. Prakt. Chem.* **324**, 609 (1982).

*210. Fujikura, Y., Ohsugi, M., Inamoto, Y., Takaishi, N. and Aigami, K., *J. Org. Chem.* **43**, 2608 (1978).

211. Takaishi, N., Fujikura, Y., Inamoto, Y. and Aigami, K., *J. Org. Chem.* **42**, 1737 (1977).

212. Taylor, E. C., Jagdmann, Jr., G. E. and McKillop, A., *J. Org. Chem.* **45**, 3373 (1980).

213. Butler, D. N. and Munshaw, T. J., *Can. J. Chem.* **59**, 3365 (1981).

214. Butler, D. N. and Gupta, I., *Can. J. Chem.* **60**, 415 (1982).

215. Shitkin, V. M., Druzhkova, T. N., Zlotin, S. G., Krayushkin, M. M. and Sevost'yanova, V. V., *Bull. Acad. Sci. USSR, Div. Chem. Sci.* 1183 (1979).

216. Weissberger, E. and Page, G., *J. Am. Chem. Soc.* **99**, 147 (1977).

217. Davies, D. I. and Dowle, M. D., *J. Chem. Soc. Perkin Trans. I* 2267 (1976).

218. Davies, D. I. and Dowle, M. D., *J. Chem. Soc. Perkin Trans. I* 227 (1978).

219. Storm, D. R. and Koshland, Jr., D. E., *J. Am. Chem. Soc.* **94**, 5805 (1972).

220. Schneider, H.-J., *J. Am. Chem. Soc.* **94**, 3636 (1972).

221. Okazawa, N. E. and Sorensen, T. S., *Can. J. Chem.* **60**, 2180 (1982).

222. Garcia Martinez, A. and Gomez Marin, M., *An. Quim.* **74**, 339 (1978).

223. Brigodiot, M. and Marechal, E., *Polymer Bulletin* **4**, 45 (1981).

224. Ishii, Y., Nakagawa, K., Yuki, H., Iwase, S., Hamanaka S. and Ogawa, M., *J. Japan Petrol. Inst. (Sekiyu Gakkaishi)* **25**, 58 (1982).

225. Camenzind, H., Vogeli, V. C. and Keese, R., *Helv. Chim. Acta* **66**, 168 (1983).

226. Wilson, S. R., Caldera, P. and Jester, M. A., *J. Org. Chem.* **47**, 3319 (1982).

*227. Chollet, A. and Vogel, P., *Helv. Chim. Acta* **61**, 732 (1978).

*228. Berman, S. S., Denisov, Y. V., Murakhovskaya, A. S., Stepanyants, G. U. and Petrov, A. A., *Neftekhimiya* **14**, 3 (1974).

229. Daniels, R. G. and Paquette, L. A., *J. Org. Chem.* **46**, 2901 (1981).

230. Vinkovic, V. and Majerski, Z., *J. Am. Chem. Soc.* **104**, 4027 (1982).

231. Schonholzer, S., Slongo, M., Rentsch, C. and Neuenschwander, M., *Makromol. Chem.* **181**, 37 (1980).

232. Uebersax, B., Neuenschwander, M. and Kellerhals, H.-P., *Helv. Chim. Acta* **65**, 74 (1982).

233. Wijekoon, W. M. D., Bunnenberg, E., Records, R. and Lightner, D. A., *J. Phys. Chem.* **87**, 3034 (1983).

234. Lippmaa, E., Pehk, T. and Past, J., *Eesti NSV Tead. Akad. Toim. Fus., Mat.* **16**, 345 (1967).

235. Buchbauer, G., *Sci. Pharm.* **45**, 196 (1977).

*236. Ohsugi, M., Inamoto, Y., Takaishi, N., Fujikura, Y. and Aigami, K., *Synthesis* 632 (1977).

237. Laichia, K., Paasivirta, J., Pikkarainen, H. and Aho-Pulliainen, S., *Org. Magn. Reson.* **22**, 117 (1984).

*238. Farcasiu, D., Slutsky, J., Schleyer, P. v. R., Overton, K. H., Luk, K. and Stothers, J. B., *Tetrahedron* **33**, 3269 (1977).

*239. Jagt, J. C. and vanLeusen, A. M., *Recl. Trav. Chim. Pays-Bas* **96**, 145 (1977).

240. Snowden, R. L., *Helv. Chim. Acta* **66**, 1031 (1983).

241. Walter, S. R., Marshall, J. L., McDaniel, Jr., C. R., Canada, Jr., E. D. and Barfield, M., *J. Am. Chem. Soc.* **105**, 4185 (1983).

242. Tabor, D. C., White, F. H., Collier, IV, L. W. and Evans, Jr., S. A., *J. Org. Chem.* **48**, 1638 (1983).

243. Ragaushkas, A. J. and Stothers, J. B., *Can. J. Chem.* **61**, 2254 (1983).

244. Snowden, R. L., Sonnay, P. and Ohloff, G., *Helv. Chim. Acta* **64**, 25 (1981).

245. Kitching, W. and Drew, G. M., *J. Org. Chem.* **46**, 2695 (1981).

*246. Andersen, N. H., Bissonette, P., Liu, C.-B., Shunk, B., Ohta, Y., Tseng, C.-L. W., Moore, A. and Huneck, S., *Phytochem.* **16**, 1731 (1977).

247. Gassman, P. G., Schaffhausen, J. G. and Raynolds, P. W., *J. Am. Chem. Soc.* **104**, 6408 (1982).

248. Gassman, P. G., Schaffhausen, J. G., Starkey, F. D. and Raynolds, P. W., *J. Am. Chem. Soc.* **104**, 6411 (1982).

249. Gassman, P. G. and Talley, J. J., *J. Am. Chem. Soc.* **102**, 4138 (1980).

250. Hoffmann, R. W. and Havel, N., *Chem. Ber.* **116**, 389 (1983).

*251. Ishii, Y., Kawahara, M., Noda, T., Ishigaki, H. and Ogawa, M., *Bull. Chem. Soc. Japan* **56**, 2181 (1983).

*252. Nakagawa, K., Ishii, Y. and Ogawa, M., *Chem. Lett.*, 21 (1977).

253. Nakagawa, K., Ishii, Y. and Ogawa, M., *Tetrahedron* **32**, 1429 (1976).

*254. Roth, K., *Z. Naturforsch. B* **32**, 76 (1977).

*255. Harnisch, J., Baumgartel, O., Szeimies, G., Van Meerssche, M., Germain, G. and Declercq, J.-P., *J. Am. Chem. Soc.* **101**, 3370 (1979).

256. Creary, X. and Geiger, C. C., *J. Am. Chem. Soc.* **105**, 7123 (1983).

257. Elmes, P. S. and Jackson, W. R., *Aust. J. Chem.* **35**, 2041 (1982).

258. Jackson, W. R. and Lovel C. G., *Aust. J. Chem.* **35**, 2053 (1982).

259. Jaouhari, R., Filliatre, C., Maillard, B. and Villenave, J. J., *Tetrahedron* **38**, 3137 (1982).

260. Villenave, J. J., Jaouhari, R., Baratchart, M. and Filliarte, C., *Bull. Soc. Chim. Belg.* **92**, 167 (1983).

*261. Hatem, J., Zahra, J. P. and Waegell, B., *Tetrahedron* **39**, 2175 (1983).

*262. Formacek, V. and Kubeczka, K.-H., *Vorkommen und Analytik Atherischer Ole*, Kubeczka, K. -H., Ed., Thieme, Stuttgart, 1979, p130ff.

263. Matoba, Y., Ohnishi, M., Kagohashi, M., Ishii, U. and Ogawa, M., *J. Japan Petrol. Inst. (Sekiyu Gakkaishi)* **26**, 15 (1983).

264. Adcock, W., Abeywickrema, A. N. and Kok, G. B., *J. Org. Chem.* **49**, 1387 (1984).

265. Hagenbuch, J.-P., Vogel, P., Pinkerton, A. A. and Schwarzenbach, D., *Helv. Chim. Acta* **64**, 1818 (1981).

266. Gergely, V., Akhavin, Z. and Vogel, P., *Helv. Chim. Acta* **58**, 871 (1975).

267. Hagenbuch, J. P. and Vogel, P., *Tetrahedron Lett.* 561 (1979).

268. Della, E. W. and Pigou, P. E., *J. Am. Chem. Soc.* **106**, 1085 (1984).

*269. Schnurpfeil, D., *J. Prakt. Chem.* **325**, 481 (1983).

*270. Henseling, K.-O. and Weyerstahl, P., *Chem. Ber.* **108**, 2803 (1975).

271. Johnson, L. F. and Jankowski, W. C., *Carbon-13 NMR Spectra*, Wiley-Interscience, New York, 1972.

272. Banert, K., Kirmse, W. and Wroblowsky, H.-J., *Chem. Ber.* **116**, 3591 (1983).

273. Simmroso, F.-M. and Wyerstahl, P., *Liebigs Ann. Chem.* 1089 (1981).

274. Sternback, D. D., Duke University, private communication.

275. Nakai, N., Iwase, S., Ishii, Y. and Ogawa, M., *J. Japan Petrol. Inst. (Sekiyu Gakkaishi)* **21**, 415 (1978).

276. Freerksen, R.W., Selikson, S.J., Wroble, R.R., Kyler, K.S. and Watt, *J. Org. Chem.* **48**, 4087 (1983).

Chapter 6

Bicyclo[2.2.2]octanes

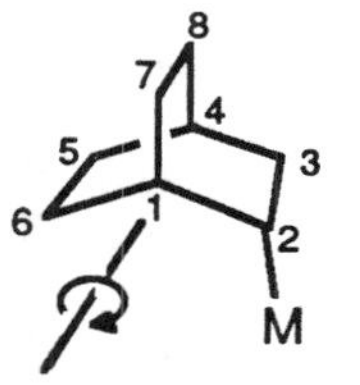

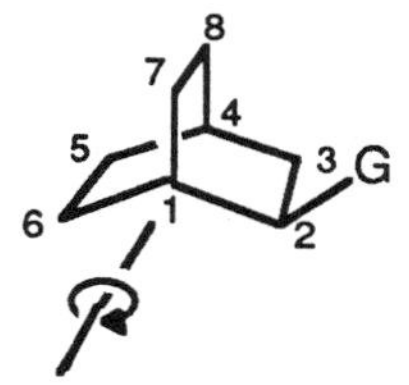

Designating stereochemical relationships in bicyclo[2.2.2]octanes is complicated by the absence of a bridge that is unique. We have developed a straightforward system, published elsewhere, that is depicted schematically above. Briefly, a sense of rotation (clockwise or counterclockwise) is defined for an axis that passes through both bridgehead atoms based upon the ordering of assigned numbers for the atoms adjacent to atom 1. Substituents that are oriented with this sense of rotation are defined as M (mit) while those opposed are G (gegen). For monosubstituted compounds the numbering system is defined so that the substituent is M. Relatively unusual dihedral angles are found about all of the bonds. As with the bicyclo[2.2.1]heptanes, caution should be exercised in applying $\Delta\delta$ effects observed here to other systems.

Whitesell, J.K. and Minton, M.A. *J. Org. Chem.,* **50**, 509 (1985).

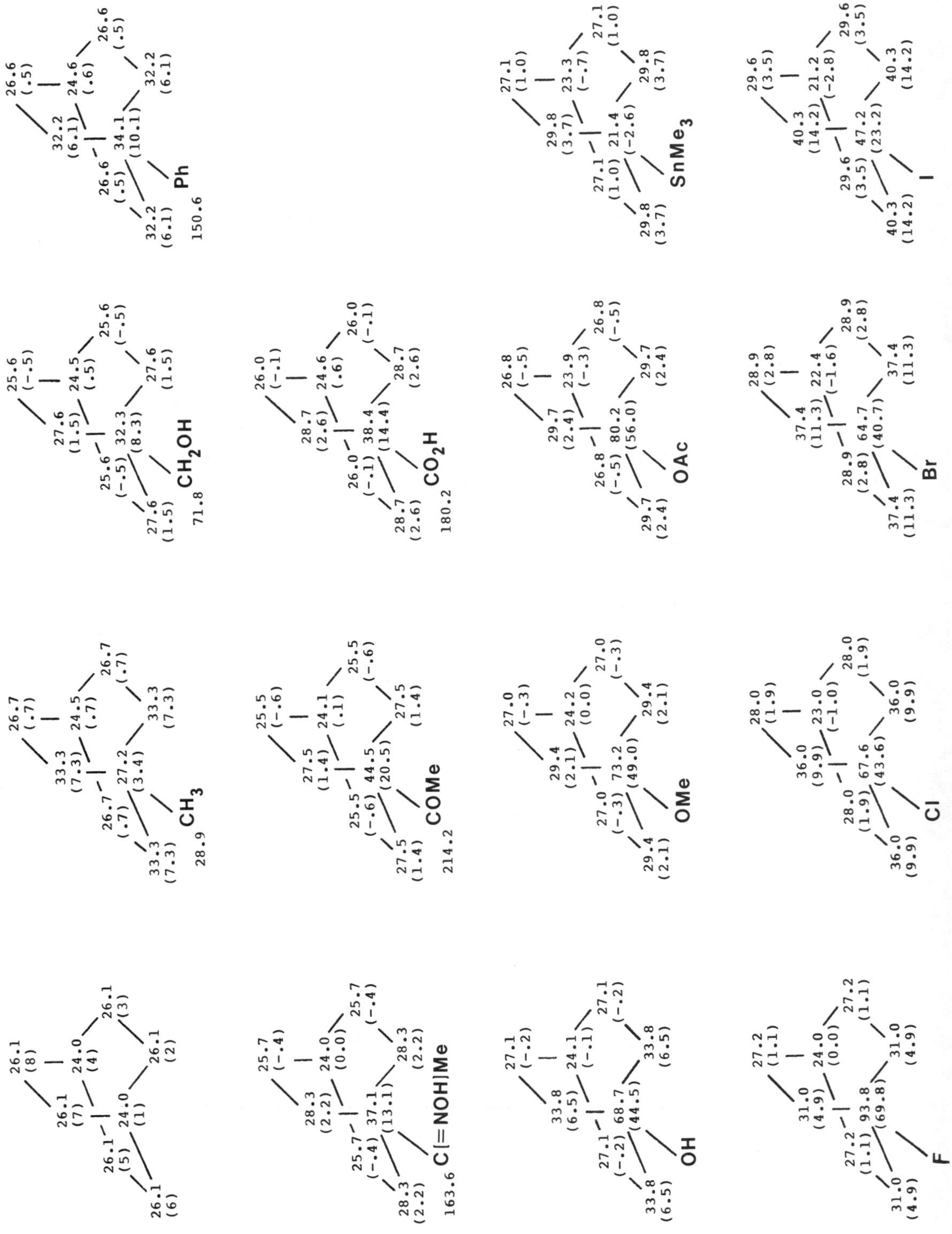

Ph (150.6)

26.6 (.5)　24.6 (.6)　26.6 (.5)
32.2 (6.1)　34.1 (10.1)　32.2 (6.1)
26.6 (.5)
32.2 (6.1)

SnMe₃

27.1 (1.0)　23.3 (−.7)　27.1 (1.0)
29.8 (3.7)　21.4 (−2.6)　29.8 (3.7)
27.1 (1.0)
29.8 (3.7)

I

29.6 (3.5)　21.2 (−2.8)　29.6 (3.5)
40.3 (14.2)　47.2 (23.2)　40.3 (14.2)
29.6 (3.5)
40.3 (14.2)

CH₂OH (71.8)

25.6 (−.5)　24.5 (.5)　25.6 (−.5)
27.6 (1.5)　32.3 (8.3)　27.6 (1.5)
25.6 (−.5)
27.6 (1.5)

CO₂H (180.2)

26.0 (−.1)　24.6 (.6)　26.0 (−.1)
28.7 (2.6)　38.4 (14.4)　28.7 (2.6)
26.0 (−.1)
28.7 (2.6)

OAc

26.8 (−.5)　23.9 (−.3)　26.8 (−.5)
29.7 (2.4)　80.2 (56.0)　29.7 (2.4)
26.8 (−.5)
29.7 (2.4)

Br

28.9 (2.8)　22.4 (−1.6)　28.9 (2.8)
37.4 (11.3)　64.7 (40.7)　37.4 (11.3)
28.9 (2.8)
37.4 (11.3)

CH₃ (28.9)

26.7 (.7)　24.5 (.7)　26.7 (.7)
33.3 (7.3)　27.2 (3.4)　33.3 (7.3)
26.7 (.7)
33.3 (7.3)

COMe (214.2)

25.5 (−.6)　24.1 (.1)　25.5 (−.6)
27.5 (1.4)　44.5 (20.5)　27.5 (1.4)
25.5 (−.6)
27.5 (1.4)

OMe

27.0 (−.3)　24.2 (0.0)　27.0 (−.3)
29.4 (2.1)　73.2 (49.0)　29.4 (2.1)
27.0 (−.3)
29.4 (2.1)

Cl

28.0 (1.9)　23.0 (−1.0)　28.0 (1.9)
36.0 (9.9)　67.6 (43.6)　36.0 (9.9)
28.0 (1.9)
36.0 (9.9)

C(=NOH)Me (163.6)

26.1 (8)　24.0 (4)　26.1 (3)
26.1 (7)　24.0 (1)　26.1 (2)
26.1 (5)
26.1 (6)

25.7 (−.4)　24.0 (0.0)　25.7 (−.4)
28.3 (2.2)　37.1 (13.1)　28.3 (2.2)
25.7 (−.4)
28.3 (2.2)

OH

27.1 (−.2)　24.1 (−.1)　27.1 (−.2)
33.8 (6.5)　68.7 (44.5)　33.8 (6.5)
27.1 (−.2)
33.8 (6.5)

F

27.2 (1.1)　24.0 (0.0)　27.2 (1.1)
31.0 (4.9)　93.8 (69.8)　31.0 (4.9)
27.2 (1.1)
31.0 (4.9)

CH₃ substituent (30.9 / 30.9):
25.1 (−1.0), 26.5 (2.5), 42.7 (16.6), 23.4 (−2.7), 35.1 (11.1), 30.6 (4.5), 25.1 (−1.0), 23.4 (−2.7); CH_3 30.9

CH₃ / CO₂H (26.6 / 186.2):
25.3 (−.9), 25.2 (1.1), 36.6 (10.4), 21.6 (−4.6), 32.1 (8.0), 44.0 (17.8), 24.3 (−1.9), 24.3 (−1.9); CH_3 26.6, CO_2H 186.2

CO₂Et (175.8):
24.9 (−1.2), 23.5 (−.5), 27.9 (1.8), 26.0 (−.1), 27.3 (3.3), 41.6 (15.5), 25.0 (−1.1), 21.6 (−4.5); CO_2Et 175.8

CH₂OH (66.0):
25.6 (−.5), 24.4 (.4), 30.2 (4.1), 27.0 (.9), 24.8 (.8), 38.4 (12.3), 26.1 (0.0), 20.7 (−5.4); CH_2OH 66.0

CH=CH₂ (143.6):
25.6 (−.4), 24.8 (1.0), 32.5 (6.5), 27.1 (1.1), 30.1 (6.3), 40.4 (14.4), 26.3 (.3), 21.1 (−4.9); $CH{=}CH_2$ 143.6

CO₂H (183.0):
25.1 (−1.0), 23.7 (−.3), 28.0 (1.9), 26.3 (.2), 27.5 (3.5), 41.9 (15.8), 25.3 (−.8), 21.9 (−4.2); CO_2H 183.0

CH₂CH₃ (29.1):
25.8 (−.2), 25.3 (1.5), 34.6 (8.6), 27.9 (1.9), 28.5 (4.7), 38.4 (12.4), 26.6 (.6), 21.1 (−4.9); CH_2CH_3 29.1

CH(OH)Me (71.2):
24.1 (−2.0), 24.0 (0.0), 30.6 (4.5), 27.1 (1.0), 25.8 (1.8), 43.8 (17.7), 25.6 (−.5), 20.6 (−5.5); CH(OH)Me 71.2

COMe (209.4):
24.5 (−1.6), 23.4 (−.6), 25.9 (−.2), 25.9 (−.2), 26.3 (2.3), 49.5 (23.4), 24.6 (−1.5), 21.0 (−5.1); COMe 209.4

OH / CH₃ (30.3):
24.4 (−1.7), 26.2 (2.2), 44.1 (18), 23.3 (−2.8), 36.6 (12.6), 71.7 (45.6), 24.6 (−1.5), 21.1 (−5.0); CH_3 30.3, OH

CH₃ (21.2):
25.2 (−.9), 25.2 (1.2), 35.8 (9.7), 27.5 (1.4), 30.4 (6.4), 30.4 (4.3), 26.2 (.1), 20.4 (−5.7); CH_3 21.2

CH(OH)Me (69.2):
24.8 (−1.3), 23.7 (−.3), 30.4 (4.3), 26.4 (.3), 24.2 (.2), 43.8 (17.7), 25.7 (−.4), 20.3 (−5.8); CH(OH)Me 69.2

CHO (204.1):
25.0 (−1.1), 23.5 (−.5), 24.7 (−1.4), 26.0 (−.1), 25.2 (1.2), 49.2 (23.1), 25.0 (−1.1), 21.6 (−4.5); CHO 204.1

OH:
24.5 (−1.6), 24.8 (.8), 37.5 (11.4), 23.8 (−2.3), 31.7 (7.7), 69.5 (43.4), 25.6 (−.5), 18.6 (−7.5); OH

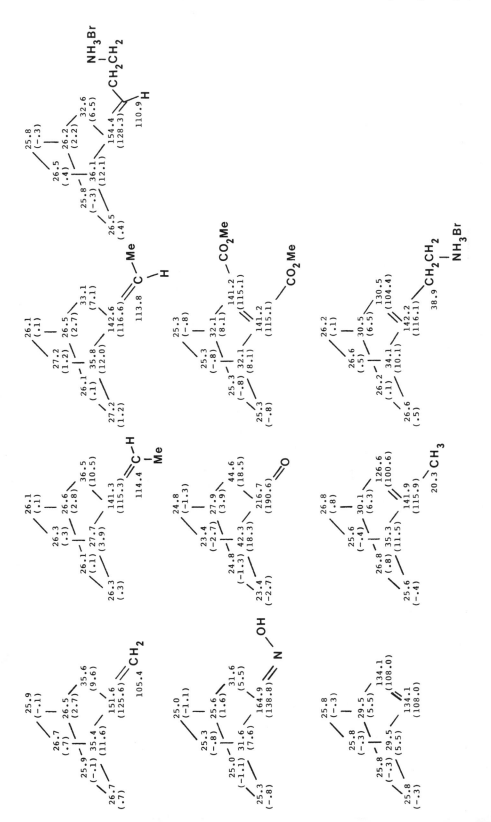

BICYCLO[2.2.2]OCTANES

Str	Substituents	C1	C2	C3	C4	C5	C6	C7	C8	Substituent Shifts	References
1	PARENT	24.0	26.1	26.1	24.0	26.1	26.1	26.1	26.1		9,2-4,6,8,10,11,42,2,66
2	1-Br	64.7	37.4	28.9	22.4	28.9	37.4	37.4	28.9		22,66,91
3	1-Br/2-=0/4-OMe	66.2	202.5	47.8	72.5	31.1	33.9	33.9	31.1		99
4	1-Br/4-Br	61.5	36.5	36.5	61.5	36.5	36.5	36.5	36.5		14
5	1-Br/4-Ph	64.0	38.0	35.0	32.6	35.0	38.0	36.5	35.0	148.0	5,52
6	1-C(anti=NOH)Me	37.1	28.3	25.7	24.0	25.7	28.3	28.3	25.7	163.6	17
7	1-CH2COOH/4-Ph	31.1	31.5	32.3	34.4	32.3	31.5	31.5	32.3	45.8,149.6	56
8	1-CH2OH	32.3	27.6	25.6	24.5	25.6	27.6	27.6	25.6	71.8	28,15
9	1-CH2OH/4-Ph	33.4	28.5	32.0	25.2	32.0	28.5	28.5	32.0	71.6,149.9	56
10	1-Cl	67.6	36.0	28.0	23.0	28.0	36.0	36.0	28.0		42,2,66
11	1-Cl/4-Ph	67.4	36.6	34.2	33.4	34.2	36.6	36.6	34.2	147.9	5,1,52
12	1-CN/4-Ph	27.5	30.1	30.8	33.4	30.8	30.0	30.0	30.8	0,148.0	5
13	1-COMe	44.5	27.5	25.5	24.1	25.5	27.5	27.5	25.5	214.2	17
14	1-COMe/4-Ph	45.1	28.1	31.7	34.8	31.7	28.1	28.1	31.7	0,149.1	5,52
15	1-COOEt/2-/4-Me	44.6	131.6	138.6	34.2	33.5	30.8	30.8	33.5	176.4,25.2	49
16	1-COOEt/4-Me	38.7	28.8	32.8	27.4	32.8	28.9	28.9	32.8	178.0,27.8	49
17	1-COOEt/4-Ph	60.1	28.8	31.8	34.7	31.8	28.8	28.8	31.8	0,149.2	5
18	1-COOH	38.4	28.7	26.0	24.6	26.0	28.7	28.7	26.0	180.2	28,15
19	1-Et/4-Ph	30.8	31.0	32.6	34.9	32.6	31.0	31.1	32.6		5,53
20	1-F	93.8	31.0	27.2	24.0	27.2	31.0	31.1	27.2	0,150.3	42,2,55
21	1-F/4-Br	90.8	33.7	38.6	59.4	38.6	33.7	33.7	38.6		55
22	1-F/4-CHO	93.9	30.1	26.9	42.7	26.9	30.1	30.1	26.9		55
23	1-F/4-Cl	91.5	32.7	37.0	64.1	37.0	32.7	32.7	37.0	0	55
24	1-F/4-CN	91.9	30.0	31.1	27.2	31.1	30.4	30.4	31.1	0	55
25	1-F/4-COOH	93.9	30.4	29.5	38.1	29.5	30.4	30.4	29.5		55
26	1-F/4-Et	94.9	31.3	31.6	30.4	31.6	31.3	31.3	31.6	32.6	55
27	1-F/4-F	91.6	31.5	31.5	91.6	31.5	31.5	31.5	31.5		55
28	1-F/4-I	89.6	34.8	41.9	82.2	41.9	34.8	34.8	41.9		55
29	1-F/4-iPr	94.8	31.2	29.1	39.8	29.1	31.2	31.2	29.1	34.8	55
30	1-F/4-Me	94.9	31.4	34.4	32.7	34.4	31.2	31.4	34.4	26.9	55
31	1-F/4-NH2	93.5	31.5	35.8	27.7	35.8	31.5	31.5	35.8		55
32	1-F/4-NO2	92.0	30.7	31.0	46.5	35.8	31.0	30.7	31.0		55
33	1-F/4-OH	92.6	31.6	34.4	67.9	34.4	31.6	31.6	34.4		55
34	1-F/4-Ph	94.5	31.5	33.5	34.3	33.5	31.6	31.5	33.5	147.8	5,1,52,54,55
35	1-F/4-t-Bu	94.7	31.2	26.9	35.0	26.9	31.2	31.2	26.9	33.9	55
36	1-I	47.2	40.3	29.6	21.2	29.6	40.3	40.3	29.6		42,66
37	1-I/4-Ph	45.9	41.0	35.8	31.5	35.8	41.0	41.0	35.8	148.5	5
38	1-iPr/4-Ph	33.4	28.4	32.6	34.7	32.6	28.4	28.4	32.6	0,150.4	5,53
39	1-Me	27.2	33.3	26.7	24.5	26.7	33.3	30.3	26.7	28.9	10,27
40	1-Me/2-=0	42.4	217.9	44.5	28.1	25.5	30.9	30.9	25.5	20.2	74
41	1-Me/2-=0/4-Me	42.2	216.9	50.4	32.5	32.7	31.4	31.4	32.7	19.9,27.0	73
42	1-Me/2-=0/5=/G8-Me M8-Me	49.8	213.1	36.4	49.8	147.1	124.1	46.8	35.0	17.7,21.9,31.1,28.6	6
43	1-Me/2-=0/5=	49.0	213.5	40.5	32.3	136.7	133.9	30.3	30.3	17.6	74
44	1-Me/2-=0/5=/G8-Me M8-Me	50.6	213.5	36.8	44.3	138.3	132.3	47.2	35.4	17.6,31.7,28.8	74
45	1-Me/2-=0/5=/G3-OAc M3-Me/5=/G7-COOMe/G8-COOMe	50.3	201.4	79.4	42.6	134.9	131.5	49.5	44.5	15.6,21.7,171.2,171.7	74,78
46	1-Me/2-=0/G3-OAc M3-Me/5=/G7-COOOC-8G	50.4	201.4	77.0	43.2	134.9	134.0	47.0	42.5	14.7,23.4,170.0,171.4	74,78
47	1-Me/2-=0/G3-OH M3-Me/5=/G7-COOOC-8G	49.9	208.7	70.7	44.8	134.5	133.7	46.5	42.5	14.5,25.4,169.3,172.0	74,78
48	1-Me/2-=0/G5-COOOC-6G	44.2	211.0	40.9	30.8	44.2	47.5	30.4	24.3	17.5,173.4,171.7	74
49	1-Me/2-=0/G5-COOOC-6G/G8-Me M8-Me	47.4	209.9	37.8	41.0	42.2	46.6	47.7	31.4	17.5,173.1,170.3,29.0,29.8	74
50	1-Me/2-=0/G5-Me M5-Me	44.3	217.8	41.9	39.1	31.3	47.6	29.3	22.8	20.3,30.4,30.0	74

BICYCLO[2.2.2]OCTANES (cont.)

Str	Substituents	C1	C2	C3	C4	C5	C6	C7	C8	Substituent Shifts	References
51	1-Me/2-=O/G5-Me M5-Me/G8-Me	45.0	217.7	35.8	45.4	32.3	47.3	39.3	25.4	20.1,29.5,29.5,20.6	6
52	1-Me/2-=O/G5-Me M5-Me/M8-Me	44.8	217.2	44.8	45.3	32.5	47.3	38.6	34.4	20.1,32.3,32.8,22.6	6
53	1-Me/2-=O/M3-OAc G3-Me	42.3	212.6	83.9	35.2	21.8	31.0	30.1	19.0	20.3,22.3	74
54	1-Me/2-=O/M3-OAc G3-Me/5=	48.3	207.8	79.7	39.4	134.9	133.3	29.0	19.9	17.8,22.3	74
55	1-Me/2-=O/M3-OAc G3-Me/5=/7=	55.9	199.6	75.4	45.1	134.4	135.7	135.3	133.1	15.2,22.5	74
56	1-Me/2-=O/M3-OAc G3-Me/M5-COOC-6M	44.9	206.6	80.2	39.0	41.8	46.2	39.3	18.7	17.2,21.8,172.3,170.3	74,78
57	1-Me/2-=O/M3-OH G3-Me/5= 5-COOMe/7=	57.2	204.5	67.8	48.4	138.8	142.7	135.5	134.5	14.9,26.8,164.8	74
58	1-Me/2-=O/M3-OH G3-Me/5= 6-COOMe/7=	56.7	204.8	68.4	51.7	146.3	136.6	135.1	134.3	14.2,27.1,164.7	74
59	1-Me/2-=O/M3-OH G3-Me/G6-COOH	44.5	216.3	74.9	37.9	25.8	48.6	30.4	22.0	18.0,23.1,180.1	78
60	1-Me/2-=O/M3-OH G3-Me/M5-COOC-6M	43.9	214.3	73.7	40.0	40.8	46.0	25.1	17.9	16.9,23.1,172.5,170.1	74,78
61	1-Me/4-Ph	28.0	33.8	32.8	34.6	32.8	33.8	33.8	32.8	28.0,150.2	5,52,53
62	1-Me/G2-OH//5-Me 5=/G8-Me M8-Me	40.3	73.3	32.2	48.2	145.3	127.6	41.2	34.5	21.8,21.9,31.5,28.3	9
63	1-Me/M2-COOOC-3M/5= 6-OMe	39.3	50.1	46.4	32.7	93.0	160.6	32.9	25.8	17.8,172.2,173.1	75
64	1-Me/M2-COOOC-3M/5= 6-OMe/G8-Me M8-Me	41.3	48.8	43.9	43.7	93.7	159.2	49.5	34.6	17.8,171.1,173.8,30.5,29.3	75
65	1-Me/M2-Me	29.9	35.9	36.9	26.7	27.0	27.5	35.4	25.9	25.7,17.9	10
66	1-Me/M2-OH/5-Me 5=/G8-Me M8-Me	41.0	74.6	35.7	48.3	145.6	124.6	47.9	34.3	21.6,21.9,30.9,29.3	9
67	1-Me/M3-Me	28.8	43.3	31.2	31.4	21.3	33.7	32.6	28.3	28.9,21.2	10
68	1-N3/4-Me	58.9	31.1	33.7	27.4	33.7	31.1	33.1	33.7	27.3	59
69	1-NH2/4-Ph	47.5	33.1	35.2	34.3	35.2	33.1	33.1	35.2	149.1	5
70	1-NHAc/4-Ph	51.6	31.1	32.5	34.3	32.5	31.1	31.1	32.5	149.0	5
71	1-NO2/4-Ph	84.8	30.5	32.2	34.7	32.2	30.5	30.5	32.2	147.1	5
72	1-OAc	80.2	29.7	26.8	23.9	26.8	29.7	29.7	26.8		66
73	1-OAc/4-Ph	80.6	30.2	33.1	34.2	33.1	30.2	30.2	33.1	148.4	5,1,52
74	1-OH	68.7	33.8	27.1	24.1	27.1	33.8	33.8	27.1		66,15
75	1-OH/4-Ph	69.6	34.3	33.5	34.3	33.5	34.3	34.3	33.5	148.6	5,1
76	1-OMe	73.2	29.4	27.0	24.2	27.0	29.4	29.4	27.0		66,2
77	1-OMe/4-Ph	73.6	29.7	33.1	34.1	33.1	29.7	29.7	33.1	148.7	5,1,52
78	1-OMe/M2-COOOC-3M/5=/G7-Me	80.3	46.3	44.2	31.2	130.3	134.5	31.8	33.9	169.0,172.3,17.4	75
79	1-Ph	34.1	32.2	32.2	24.6	32.2	32.2	30.2	26.6	150.6	5,1,52,53,56
80	1-Pr/2-=O/3-=O/4-Pr/5-COOMe 5= 6-COOMe/7=	58.3	179.8	179.8	58.3	140.1	140.1	134.1	134.1	28.1,28.1,179.7,179.7	89
81	1-Pr/2-=O/3-=O/4-Pr/5-COOMe 5=/7=	57.8	181.2	181.0	58.3	136.4	143.3	134.7	134.0	29.8,28.0,163.7	89
82	1-Pr/2-=O/3-=O/4-Pr/5-Ph 5= 6-Ph/7=	60.0	182.6	182.6	60.0	144.6	144.6	133.8	133.8	29.7,29.7,135.1,135.1	89
83	1-SiMe3/4-Ph	16.5	26.8	32.1	34.4	32.1	26.8	26.8	27.1	150.7	53
84	1-SnMe3	21.4	29.8	27.1	23.3	27.1	29.8	29.8	27.1		24
85	1-SnMe3/4-Ph	22.1	30.8	33.2	33.7	33.2	30.8	30.8	33.2	150.6	5,53
86	1-tBu/4-Ph	35.6	26.0	32.6	34.2	32.6	26.0	26.0	32.6	34.2,150.4	5,52,53
87	E2-=CHCH2NH3Br	36.1	154.4	32.6	26.2	25.8	26.5	26.5	25.8	110.9	68
88	E2-=CHMe	35.8	142.6	33.1	26.5	26.1	27.2	27.2	26.1	113.8	10
89	E2-=NOH	31.6	164.9	31.6	25.6	25.0	25.3	25.3	25.0		18
90	2-=CH2	35.4	151.6	35.4	26.5	25.9	26.7	26.7	25.9	105.4	10
91	2-=CH2 3-=CH2/5-=CH2 6-=CH2/7-=CH2 8-=CH2	59.2	144.4	144.4	59.2	144.4	144.4	144.4	144.4	6*104.8	103
92	2-=CH2/3-=CH2	37.0	150.0	150.0	37.0	26.7	26.7	26.7	26.7	103.5,103.5	43
93	2-=O	42.3	216.7	44.6	27.9	24.8	23.4	23.4	24.8		9,4,6,8,12,50,
94	2-=O/3-Me	42.7	221.9	45.7	38.5	22.4	23.5	23.4	22.4	23.7,23.7	9,6,13 66,74
95	2-=O/3-Me/4-Me	42.3	222.2	48.7	36.7	30.0	23.4	23.4	30.0	20.3,20.3,21.3	6
96	2-=O/4-Br	40.6	210.2	54.9	56.8	36.1	24.7	24.7	36.1		66
97	2-=O/4-Cl	40.8	210.4	53.6	63.3	34.9	23.7	23.7	34.9		66
98	2-=O/4-Me	42.1	216.6	50.5	32.6	31.9	23.7	23.7	31.9	27.0	6
99	2-=O/4-Me/5=	48.7	212.1	47.0	37.4	141.5	127.7	23.8	32.1	24.1	6
100	2-=O/4-Me/5=/G7-Me M7-Me	61.6	211.6	45.4	37.9	140.6	127.8	35.7	48.8	24.0,29.4,30.8	6

98

Str	Substituents	C1	C2	C3	C4	C5	C6	C7	C8	Substituent Shifts	References
101	2-=O/4-Me/G6-Me M6-Me	54.1	216.3	49.1	33.7	48.8	31.2	19.7	30.8	26.9,31.9,28.8	6
102	2-=O/4-Me/G6-OH M6-Me	55.6	212.0	48.8	33.5	49.5	72.1	20.5	31.5	26.5,28.6	95
103	2-=O/4-OAc	41.6	211.9	48.6	79.1	29.0	22.2	22.2	29.0		66
104	2-=O/4-OH	41.5	213.3	51.7	70.0	32.6	22.5	22.5	32.6		66
105	2-=O/4-OMe	41.5	212.4	47.6	74.2	28.4	22.3	22.3	28.4		66
106	2-=O/5-=CH2 6-=CH2/7-=CH2 8-=CH2	67.1	205.8	44.1	48.4	143.8	139.3	139.3	143.8	2*106.2,2*108.9	97
107	2-=O/5-=CH2/6-=CH2	54.8	211.3	44.2	38.2	145.9	141.4	23.9	24.9	105.2,108.1	60
108	2-=O/5-=O	45.1	211.4	40.3	45.1	211.4	40.3	22.2	22.2		4
109	2-=O/5=	48.6	212.4	40.4	32.4	136.8	128.3	22.5	22.2		8,6,74
110	2-=O/6-=CH2	54.4	212.1	44.1	28.2	34.3	143.4	24.3	24.6	110.8	79
111	2-=O/6-=O	63.9	206.6	44.0	28.1	44.0	206.6	22.2	23.8		65
112	2-=O/E5-=CHCl/E6-=CHCl	47.9	208.8	42.8	31.9	139.1	134.6	22.4	23.4	110.4,112.7	61
113	2-=O/G3-Me M3-Me/5-COOMe 5= 6-COOMe	49.5	213.5	42.9	46.1	145.7	132.1	22.3	20.7	23.8,26.4,164.7,166.6	23
114	2-=O/G3-Me M3-Me/5-COOMe 5= 6-COOMe/7=	57.9	206.0	39.6	51.9	143.8	135.7	128.3	136.5	26.9,28.3,164.7,166.1	23
115	2-=O/G3-Me M3-Me/5=	48.8	216.7	43.8	44.1	138.6	126.1	21.8	20.7	24.3,27.5	6
116	2-=O/G3-Me M3-Me/G5-Me/G6-Me	51.1	221.1	46.1	46.1	36.1	35.5	23.0	26.0	25.6,27.3,13.2,16.8	9
117	2-=O/G3-Me/5=	48.4	215.1	42.5	38.2	136.0	128.1	23.1	18.6	14.4	6
118	2-=O/G3-Me/G5-Me/G6-Me	51.0	219.6	42.0	41.2	33.5	34.1	24.6	21.8	13.1,14.9,16.8	9
119	2-=O/G5-Me/G6-Me	50.7	216.8	39.8	35.5	32.7	34.6	23.5	26.3	14.9,16.4	9
120	2-=O/G6-Cl	50.5	211.4	43.9	28.4	37.7	55.2	22.7	23.3		9
121	2-=O/M3-Me	42.3	220.1	47.2	33.9	20.2	24.2	22.7	26.1	13.5	50
122	2-=O/M3-Me/4-Me	42.3	219.9	48.7	36.7	26.8	24.2	22.8	34.1	10.4,24.7	9,6
123	2-=O/M3-Me/5=	48.4	214.1	44.4	39.1	136.0	127.2	21.9	24.8	17.5	6
124	2-=O/M3-Me/G5-Me/G6-Me	50.2	220.3	47.6	41.7	34.5	36.3	22.6	29.2	15.4,16.3,16.9	6
125	2-=O/M3-Me/5-OAc	42.0	212.4	75.8	33.5	19.8	25.6	23.0	21.3		9
126	2-=O/M3-OH	41.3	219.6	76.1	35.0	19.4	26.8	23.1	21.2		74
127	2-=O/M3-OH G3-Me	41.8	220.7	75.6	38.2	21.9	25.4	22.1	20.6	23.3	74,75
128	2-=O/M6-Cl	50.9	212.9	43.3	27.9	38.0	54.4	17.3	24.2		50
129	2-=O/Z6-=CHMe	47.4	212.7	44.4	28.3	35.0	133.2	23.8	24.9	120.7	79
130	2-=O/Z6-=CHOMe	48.0	212.8	44.3	27.9	29.3	111.8	24.6	24.9	142.3	79
131	M2-Br/G3-Br	35.0	61.4	61.4	35.0	25.6	19.3	19.3	25.6		46
132	M2-CH(OH)Me	24.2	43.8	30.4	23.7	25.7	20.3	26.4	24.8	69.2	80
133	M2-CH(OH)Me	25.8	43.8	30.6	24.0	25.6	20.6	27.1	24.1	71.2	80
134	2-CH2CH2NH3Br 2=	34.1	142.2	130.5	30.5	26.2	26.7	26.6	26.2	38.9	68
135	M2-CH2OH	24.8	38.4	30.2	24.4	26.1	20.7	27.0	25.6	66.0	16,10
136	M2-CH2OH/5=	31.3	40.5	30.3	29.8	134.9	131.9	26.0	24.8	67.0	16,10
137	G2-CH2OH/5=	30.4	38.9	30.4	29.7	134.1	135.8	19.3	26.0	65.7	16,10
138	M2-CH2OH/G3-CH2OH	26.7	45.0	30.6	26.7	25.6	21.2	27.2	21.2	66.6,66.6	16
139	M2-CH2OH/G3-CH2OH/5=	32.3	46.9	44.6	32.0	135.7	132.0	25.9	19.3	67.5,66.0	16
140	M2-CH2OH/G3-CH2OH/G5-CH2OH/M6-CH2OH/M7-CH2OH/G8-CH2OH	29.8	37.1	37.1	29.8	37.1	37.1	37.1	37.1	6*62.6	103
141	M2-CH2OH/G3-Me	25.6	48.2	34.5	30.9	27.3	20.4	26.9	20.2	64.8,20.0	16
142	M2-CH2OH/G3-Me/5=	32.1	49.6	35.1	36.1	136.6	131.2	26.2	18.4	67.0,19.4	16
143	M2-CH2OH/M3-CH2OH	28.2	41.2	41.2	28.2	21.2	21.2	26.8	26.8	64.3,64.3	16
144	M2-CH2OH/M3-CH2OH/5=	34.5	45.6	45.6	34.5	133.0	133.0	25.5	25.5	65.0,65.0	16
145	G2-CH2OH/M3-Me/5=	30.8	48.7	37.0	37.0	132.6	135.5	18.1	26.5	64.6,22.7	16
146	M2-CH=CH2	30.1	40.4	32.5	24.8	26.3	21.1	27.1	25.6	143.6	10
147	M2-CHO	25.2	49.2	24.7	23.5	25.0	21.6	26.0	25.0	204.1	80
148	M2-CHO/5=	30.4	50.5	26.4	29.0	135.5	130.3	24.8	24.5	202.9	80
149	G2-CHO/5=	29.9	49.8	25.0	29.0	133.0	135.0	20.8	24.5	203.3	80
150	M2-COCl/G5-COCl	27.8	53.3	24.8	27.8	53.3	24.8	25.2	25.2	175.7,175.7	82

BICYCLO[2.2.2]OCTANES (cont.)

Str	Substituents	C1	C2	C3	C4	C5	C6	C7	C8	Substituent Shifts	References
151	M2-COMe	26.3	49.5	25.9	23.4	24.6	21.0	25.9	24.5	209.4	80
152	M2-COOEt	27.3	41.6	27.9	23.5	25.0	21.6	26.0	24.9	175.8	80
153	M2-COOEt G2-Me/3-=CH2/4-Me/5-=O	36.9	49.9	150.2	50.5	211.6	42.3	21.2	31.5	175.1,26.1,111.8,16.5	79
154	M2-COOH	27.5	41.9	28.0	23.7	25.3	21.9	26.3	25.1	183.0	16,4
155	M2-COOH G2-Me	32.1	44.0	36.6	25.2	24.3	24.3	21.6	25.3	26.6,186.2	4,94
156	M2-COOH G2-Me/5-=O	35.8	43.2	33.6	42.7	214.3	42.9	20.8	22.2	182.3,26.1	4
157	M2-COOH G2-Me/G5-OH	33.4	43.4	30.0	32.5	68.3	35.8	20.1	22.8	182.8,26.6	4,7,44
158	M2-COOH G2-Me/G5-OH M5-C≡CH	33.3	42.4	30.9	37.5	67.7	42.9	19.7	21.8	0,26.6,89.0	41
159	M2-COOH G2-Me/G5-OH M5-CH=CH2	33.9	42.5	32.0	36.7	72.2	39.1	19.8	21.1	0,26.0,143.8	41
160	M2-COOH G2-Me/G5-OH M5-Et	34.7	43.0	32.6	35.0	72.5	41.6	20.2	21.8	0,26.3,33.9	41
161	M2-COOH G2-Me/G5-OH M5-Me	35.0	42.5	32.6	37.7	70.8	42.5	20.2	22.4	0,26.3,28.8	41,94
162	M2-COOH G2-Me/M5-OH	33.1	42.8	34.3	32.8	68.0	35.5	20.9	17.6	181.8,26.3	7,4
163	M2-COOH/5-=O	31.2	40.9	25.8	41.9	216.4	40.2	25.3	22.6	180.1	4
164	M2-COOH/5=	32.4	42.8	29.6	29.4	135.3	131.5	24.4	25.4	182.2	16
165	G2-COOH/5=	32.4	42.7	28.1	29.4	133.8	135.2	21.1	25.1	182.2	16
166	M2-COOH/G3-Me	28.6	51.0	32.8	30.3	26.6	21.4	26.4	19.8	182.6,20.2	16
167	M2-COOH/G3-Me/5=	32.8	51.4	35.6	35.8	136.9	131.0	25.8	17.7	182.2,19.7	16
168	M2-COOH/G5-OH	28.6	41.5	21.3	31.2	68.5	33.7	25.0	22.8	180.0	4,7
169	G2-COOH/M3-Me/5=	33.0	51.7	34.9	36.5	133.3	133.3	19.8	25.7	182.0,22.6	16
170	2-COOMe 2= 3-COOMe	32.1	141.2	141.2	32.1	25.3	25.3	25.3	25.3	0,0	48
171	2-COOMe 2= 3-COOMe/G5-O-6G	35.0	135.6	135.6	35.0	47.1	47.1	21.4	21.4		48
172	2-COOMe 2= 3-COOMe/M5-O-6M	36.9	143.8	143.8	36.9	58.3	58.3	22.5	22.5		48
173	2-COOMe 2=/G5-COOMe/G6-COOMe/M7-COOMe/M8-COOMe	35.6	132.6	140.7	35.0	45.7	45.7	45.7	45.7	164.3,4*171.3	103
174	M2-COOMe/G3-CH2Cl	28.4	47.6	40.7	26.3	26.0	21.3	26.0	22.8	175.0,47.6	48
175	M2-COOMe/G3-CH2Cl/G5-O-6G	30.5	46.5	38.0	28.7	52.3	51.1	21.5	16.2	173.9,46.2	48
176	M2-COOMe/G3-CH2Cl/M5-O-6M	31.5	48.1	40.0	30.0	53.5	50.8	23.6	16.7	174.6,46.6	48
177	M2-COOMe/G3-COOMe	27.5	44.1	44.1	27.5	25.5	21.9	25.5	21.5	174.1,174.1	48
178	M2-COOMe/G3-COOMe/G5-COOMe/G6-COOMe/M7-COOMe/G8-COOMe	33.8	40.1	40.1	33.8	40.1	40.1	40.1	40.1	6*172.5	103
179	M2-COOMe/G3-COOMe/G5-O-6G	30.2	42.7	42.5	30.8	51.6	51.0	21.1	18.2	174.0,175.1	48
180	M2-COOMe/G3-COOMe/M5-O-6M	31.7	44.0	42.8	31.0	52.9	51.0	22.9	18.7	174.0,174.6	48
181	M2-COOMe/G5-O-6G	30.0	41.1	26.7	27.3	53.0	51.7	22.0	21.8	175.3	48
182	G2-COOMe/M3-CH2Cl/G5-O-6G	31.2	44.0	42.4	29.6	51.8	51.9	18.2	23.1	174.9,47.5	48
183	G2-COOMe/M3-CH2Cl/M5-O-6M	32.4	45.2	42.3	30.9	51.0	52.9	18.9	23.9	174.0,48.3	48
184	M2-COOMe/M3-COOMe	26.8	44.1	44.1	26.8	21.4	21.4	25.8	25.8	174.1,174.1	48
185	M2-COOMe/M3-COOMe/5=	32.1	47.1	47.1	32.1	131.9	131.9	24.1	24.1	172.8,172.8	37,25,46
186	M2-COOMe/M3-COOMe/M5-Br/G6-Br	32.7	46.3	51.9	32.8	67.9	68.2	24.3	20.8	171.8,171.8	46
187	M2-COOMe/M3-COOMe/M5-Br/M6-Br	33.6	52.0	52.0	33.6	66.9	66.9	20.9	20.9	171.6,171.6	46
188	M2-COOMe/M3-COOMe/M5-C1/G6-C1	35.8	47.2	52.3	35.9	68.5	68.6	25.3	20.2	172.1,172.3	46
189	M2-COOMe/M3-COOMe/M5-C1/M6-C1	36.6	51.1	51.1	36.6	67.3	67.3	20.1	20.1	172.3,172.3	46
190	M2-COOMe/M3-COOMe/M5-O-6M	30.5	44.9	44.9	30.5	51.1	51.1	23.3	23.3	173.5,173.5	48
191	M2-COOMe/M3-COOMe/M5-SePh/G6-C1	37.0	40.5	44.1	31.2	50.8	69.3	22.9	15.1	172.3,172.3	46
192	M2-COOMe/M3-COOMe/M5-SePh/M6-C1	36.0	43.5	44.8	32.3	59.8	61.6	20.0	18.5	171.6,171.6	46
193	M2-COOMe/M5-O-6M	30.8	41.3	26.6	27.5	53.0	50.6	23.8	22.6	176.0	48
194	M2-COOOC-3M	26.1	44.3	44.3	26.1	21.5	21.5	24.2	24.2	174.1,174.1	16,102
195	M2-COOOC-3M/5=	31.7	44.8	44.8	31.7	133.1	133.1	23.0	23.0	172.9,172.9	16,37,46,102
196	M2-COOOC-3M/M5-Br/G6-Br	35.3	40.8	44.6	35.5	54.2	54.2	26.3	18.6	171.4,171.4	46
197	M2-COOOC-3M/M5-Br/M6-Br	36.5	44.2	44.2	36.5	54.4	54.4	17.8	17.8	171.4,171.4	46
198	M2-COOOC-3M/M5-SePh/G6-C1	37.1	40.3	42.2	31.2	51.3	67.7	20.2	15.1	172.8,172.8	46
199	M2-Et	28.5	38.4	34.6	25.3	26.6	21.1	27.9	25.8	29.1	10
200	M2-Me	30.4	30.4	35.8	25.2	26.2	20.4	27.5	25.2	21.2	9,4,10

BICYCLO[2.2.2]OCTANES (cont.)

Str	Substituents	C1	C2	C3	C4	C5	C6	C7	C8	Substituent Shifts	References
201	2-Me 2-Me	35.1	30.6	42.7	26.5	25.1	23.4	23.4	25.1	30.9,30.9	9,10
202	2-Me 2=	35.3	141.9	126.6	30.1	26.8	25.6	25.6	26.8	20.3	10
203	M2-Me/G3-Me	31.5	40.1	40.1	31.5	27.5	20.2	27.5	20.2	19.2,19.2	9,10
204	M2-Me/M3-Me	31.8	33.3	33.3	31.8	20.3	20.3	27.4	27.4	15.2,15.2	9,10
205	G2-Me/M5-Me	32.0	29.8	37.3	31.8	29.8	20.3	27.3	20.8	20.8,20.8	10
206	G2-Me/M6-Me	37.0	32.7	34.8	26.6	34.8	32.7	15.1	26.6	21.3,21.3	10
207	M2-NCO/G5-NCO	31.8	52.0	30.6	31.8	52.0	30.6	22.6	22.6	0,0	82
208	M2-NH_3Cl/G5-NH_3Cl	28.9	49.7	25.7	28.9	49.7	25.7	23.7	23.7		82
209	G2-O-3G/5-=CH_2 6-=CH_2	40.2	53.4	53.4	40.2	145.6	145.6	23.9	23.7	106.6,106.6	60
210	G2-O-3G/E5-=CHCl E6-=CHCl	33.3	51.7	51.7	33.3	138.7	138.7	21.9	21.9	110.9,110.9	61,100
211	M2-O-3M	27.6	53.2	53.2	27.6	23.7	23.7	22.9	22.9		48
212	M2-O-3M/5-Me 5= 6-Me/7-=CH_2 8-=CH_2	50.5	48.8	48.8	50.5	126.9	126.9	142.9	142.9	15.9,15.9,104.9,104.9	81
213	M2-O-3M/E5-=CHCl E6-=CHCl	32.3	50.2	50.2	32.3	137.0	137.0	21.4	21.4	109.9,109.9	61
214	G2-OAc/G5-Me	29.4	72.4	29.6	31.0	29.2	33.2	18.6	27.0	21.2	71
215	G2-OAc/G6-Me	35.2	69.5	35.1	25.8	34.8	29.9	20.9	24.7	20.5	71
216	M2-OH	31.7	69.5	37.5	24.8	25.6	18.6	23.8	24.5		9,4,50
217	M2-OH G2-C(Me)=CH_2/6-=CH_2	44.9	74.7	39.8	27.0	34.5	147.9	22.0	24.1	147.8,110.7	79
218	M2-OH G2-CH_2Ph/5=	39.9	75.6	45.6	30.9	135.9	132.2	21.2	24.2	45.6	96
219	M2-OH G2-CH=CH_2/6-=CH_2	47.9	72.6	41.5	27.0	34.3	147.6	22.0	24.4	143.5,110.3	79
220	M2-OH G2-CH=CH_2/Z6-=CHMe	40.4	73.2	41.9	27.2	35.1	137.0	21.4	24.5	143.6,120.2	79
221	M2-OH G2-Et/5=	40.3	76.1	45.3	31.1	136.1	132.4	21.0	23.9	32.8	96
222	M2-OH G2-Me	36.6	71.7	44.1	26.2	24.6	21.1	23.3	24.4	30.3	50,94
223	M2-OH G2-Me/G3-OH M3-Me	38.8	77.4	73.4	38.8	22.8	20.7	20.4	20.7	24.7,24.7	75
224	M2-OH G2-Ph/5=	43.8	77.4	43.8	31.2	136.7	132.1	20.4	24.3	145.2	96
225	G2-OH M2-C(Me)=CH_2/6-=CH_2	45.1	75.4	40.0	27.3	33.8	148.9	21.5	24.7	149.9,108.4	79
226	G2-OH M2-CH_2Ph/5=	40.0	75.5	42.2	31.1	133.9	133.3	19.5	24.6	49.1	96
227	G2-OH M2-CH=CH_2/6-=CH_2	47.7	73.2	42.1	27.0	34.0	148.4	20.9	24.4	146.3,108.8	79
228	G2-OH M2-CH=CH_2/Z6-=CHMe	40.0	73.8	42.4	27.2	34.7	138.2	20.2	24.6	146.4,117.7	79
229	G2-OH M2-CH=CH_2/Z6-=CHOMe	41.9	73.6	42.6	26.4	28.9	116.5	21.6	24.6	146.6,140.7	79
230	G2-OH M2-CMe=CH_2/5=	41.2	75.2	42.2	31.2	133.7	133.4	19.8	24.5	142.9	96
231	G2-OH M2-Et/5=	31.7	76.0	42.2	41.4	134.0	133.8	20.2	24.9	36.4	72,96
232	G2-OH M2-Ph/5=	42.7	76.7	44.5	31.3	134.0	133.4	20.6	24.3	150.0	96
233	G2-OH/4-Me	31.7	69.8	44.5	28.6	32.8	19.4	24.2	32.0	28.0	9
234	G2-OH/4-Me/5=/G7-Me M7-Me	49.9	72.9	40.5	35.4	136.8	133.4	34.7	51.5	25.0,33.3,31.8	9
235	M2-OH/4-Me/G6-Me M6-Me	43.0	72.8	43.4	31.0	50.1	30.7	22.5	30.6	27.9,32.3,32.5	9
236	M2-OH/5-=CH_2 6-=CH_2/7-=CH_2 8-=CH_2	56.5	68.5	39.5	48.1	145.9	141.3	143.3	145.4	104.8,109.1,107.1,104.6	97
237	M2-OH/5-=CH_2/6-=CH_2	44.4	68.5	38.9	36.2	148.4	143.9	22.9	24.8	103.6,108.1	60
238	G2-OH/5-=CH_2/6-=CH_2	44.2	68.6	37.2	36.8	147.9	146.5	18.2	25.8	103.4,105.5	60
239	M2-OH/5=	37.6	70.2	38.9	30.1	135.9	129.9	22.1	24.6		9
240	G2-OH/5=	37.6	68.8	35.4	30.1	135.2	132.0	17.3	24.1		9
241	M2-OH/E5-=CHCl/E6-=CHCl	37.8	68.7	36.8	29.7	141.1	137.6	21.4	23.1	108.7,112.4	61
242	G2-OH/E5-=CHCl/E6-=CHCl	37.5	67.0	35.1	30.4	141.1	139.6	16.8	24.1	108.9,110.2	61
243	M2-OH/G3-Me	32.9	78.1	41.8	31.3	27.1	18.4	24.1	19.7	18.6	9
244	M2-OH/G3-Me M3-Me	33.7	77.9	35.9	36.5	22.7	17.7	24.2	22.5	29.9,22.4	9,13
245	G2-OH/G3-Me M3-Me/5=	39.9	79.9	40.1	43.3	138.7	127.9	20.9	21.7	29.7,24.5	9
246	G2-OH/G3-Me M3-Me/5=	39.2	76.7	35.1	42.6	136.4	130.1	14.9	21.9	22.7,32.5	9
247	M2-OH/G3-Me M3-Me/G5-Me/G6-Me	40.6	80.2	37.1	43.9	36.2	32.8	21.4	25.8	32.8,25.0,18.6,17.5	9
248	M2-OH/G3-Me M3-Me/G5-Me/G6-Me	40.2	79.6	34.9	39.7	34.0	33.5	26.8	21.0	18.3,14.6,17.6	9
249	M2-OH/G5-Me G6-Me	39.1	71.4	31.6	32.7	33.1	33.3	25.8	26.4	14.8,17.7	9
250	G2-OH/G5-Me/G6-Me	40.1	65.0	32.3	33.1	33.1	32.9	19.9	27.1	15.4,14.6	9

BICYCLO[2.2.2]OCTANES (cont.)

Str	Substituents	C1	C2	C3	C4	C5	C6	C7	C8	Substituent Shifts	References
251	M2-OH/G5-OH	32.5	69.1	30.4	32.5	69.1	30.4	22.8	22.8		4
252	M2-OH/M3-Me	31.8	70.9	35.8	31.3	20.2	18.6	23.7	26.5	12.8	9
253	M2-OH/M5-OH	32.5	68.6	30.5	32.4	68.2	35.6	17.3	23.4		4
254	G2-OH/M5-OH	32.7	68.4	34.6	32.7	68.4	34.6	18.4	18.4		4
255	M2-PO(OiPr)$_2$/G3-COMe/5=	30.2	36.8	53.2	32.5	132.6	134.6	27.0	19.5	205.9	58
256	G2-PO(OiPr)$_2$/M3-COMe/5=	30.4	35.6	52.6	33.6	130.2	137.0	21.3	25.4	204.8	58
257	M2-SePh/G3-Cl	31.4	53.8	66.6	34.0	24.9	20.2	24.9	18.5		36,46
258	M2-SePh/G3-OMe	30.2	50.3	85.3	28.4	26.6	20.9	23.0	17.9		36
259	M2-SPh/G3-Cl	30.5	58.0	65.7	33.9	24.9	19.1	25.8	18.7		46
260	2=	29.5	134.1	134.1	29.5	25.8	25.8	25.8	25.8		9,6,8,10,37,46
261	2=/5-=CH$_2$	31.5	132.9	133.5	41.1	149.4	34.9	25.1	26.6	104.1	69
262	2=/5-=CH$_2$ 6-=CH$_2$/7-=CH$_2$ 8-=CH$_2$	53.2	132.2	132.2	53.2	144.3	144.3	144.3	144.3	4*103.9	97,101
263	2=/5-=CH$_2$/6-=CH$_2$	42.3	133.2	133.2	42.3	147.4	147.4	26.2	26.2	102.6,102.6	43
264	2=/5=	36.9	134.1	134.1	36.9	134.1	134.1	24.7	24.7		70
265	2=/5=/G7-Me M7-Me/8-=CH$_2$	50.4	134.6	131.7	51.7	131.7	134.6	42.9	155.9	31.7,31.7,103.3	62
266	2=/E5-=C(C1)Ph/G6-Me M6-Me	46.8	129.5	136.8	39.0	146.7	41.3	20.6	23.3	123.2,30.0,28.3	19
267	2=/E5-=CHC1/E6-=CHC1	34.9	133.0	133.0	34.9	140.2	140.2	23.7	23.7	107.1,107.1	61,100
268	2=/E5-=CHMe	31.9	133.8	134.1	41.6	141.8	32.9	25.7	27.2	113.3	72
269	2=/G5-CH$_2$Cl/G6-CH$_2$Cl/M7-CH$_2$Cl/M8-CH$_2$Cl	37.6	132.8	132.8	37.6	44.5	44.5	44.5	44.5	4*44.6	97
270	2=/G5-CH=CH$_2$	30.0	134.7	132.1	35.9	42.7	33.9	24.6	26.3	145.1	72
271	2=/G5-COMe/M6-CH$_2$CH(OMe)$_2$	34.5	136.6	129.9	33.2	58.8	34.5	18.4	26.4	209.3,37.7	26
272	2=/G5-Me	30.5	134.1	132.0	36.8	32.2	36.0	24.3	26.6	23.0	9,10
273	2=/G5-Me M5-Me	31.5	136.0	131.9	41.7	33.3	42.8	23.7	21.8	32.1,30.1	9
274	2=/G5-Me/G6-Me	38.5	132.9	132.9	38.5	36.6	36.6	26.0	26.0	17.1,17.1	9
275	2=/G5-Me/M6-Me	36.6	136.3	131.5	37.4	41.5	40.8	17.7	27.0	22.2,18.6	9
276	2=/M5-CH=CH$_2$	30.1	133.9	135.7	35.7	40.9	31.8	26.4	19.6	143.0	72
277	2=/M5-Me	30.8	132.9	136.4	36.0	30.4	34.9	26.4	18.9	20.5	9,10
278	2=/Z5-=CHMe	31.9	133.1	135.0	33.7	140.4	36.2	25.7	26.3	113.3	72
279	Z2-=CHMe	27.7	141.3	36.5	26.6	26.1	26.3	26.3	26.1	114.4	10

102

References to Bicyclo[2.2.2]octanes and related* systems

1. Kitching, W., Adcock, W., Khor, T. C. and Doddrell, D., *J. Org. Chem.* **41**, 2055 (1976).
2. Maciel, G. E. and Dorn, H. C., *J. Am. Chem. Soc.* **93**, 1268 (1971).
3. Della, E. W., Hine, P. T. and Patney, H. K., *J. Org. Chem.* **42**, 2940 (1977).
4. Garratt, P. J. and Riguera, R., *J. Org. Chem.* **41**, 465 (1976).
5. Adcock, W. and Khor, T.-C., *J. Am. Chem. Soc.* **100**, 7799 (1978).
6. Stephens, K. R., Stothers, J. B. and Tan, C. T., *Mass Spectrometry and NMR Spectroscopy in Pesticide Chemistry*, R. Haque and F. J. Giros, Eds. 179 (1974).
7. Davalian, D., Garratt, P. J. and Riguera, R., *J. Org. Chem.* **42**, 368 (1977).
8. Stothers, J. B., Swenson, J. R. and Tan, C. T., *Can. J. Chem.* **53**, 581 (1975).
9. Stothers, J. B. and Tan, C. T., *Can. J. Chem.* **54**, 917 (1976).
10. Pekhk, T. T., Lippmaa, E. T., Sokolova, I. M., Vorob'eva, N. S., Gervits, E. S., Bobyleva, A. A., Kalinichenko, A. N. and Belikova, N. A., *J. Org. Chem. USSR* **12**, 1207 (1976).
11. Boaz, H., *Tetrahedron Lett.* 55 (1973).
12. Van Binst, G. and Tourwe, D., *Org. Magn. Reson.* **4**, 625 (1972).
13. Hudyma, D. M., Stothers, J. B. and Tan, C. T., *Org. Magn. Reson.* **6**, 614 (1974).
14. Dannenberg, J. J., Prociv, T. M. and Hutt, C., *J. Am. Chem. Soc.* **96**, 913 (1974).
15. Morris, D. G. and Murray, A. M., *J. Chem. Soc., Perkin Trans. II* 734 (1975).
16. Brouwer, H., Stothers, J. B. and Tan, C. T., *Org. Magn. Reson.* **9**, 360 (1977).
17. Hawkes, G. E., Herwig, K. and Roberts, J. D., *J. Org. Chem.* **39**, 1017 (1974).
18. Geneste, P., Durand, R., Kamenko, J.-M., Beierbeck, H., Martino, R. and Saunders, J. K., *Can. J. Chem.* **56**, 1940 (1976).
19. Mayr, H., Schutz, F. and Halberstadt-Kausch, I. K., *Chem. Ber.* **115**, 3516 (1982).
*20. Beierbeck, H. and Saunders, J. K., *Can. J. Chem.* **55**, 3161 (1977).
*21. de Meijere, A., Schallner, O., Weitemeyer, C. and Spielmann, W., *Chem. Ber.* **112**, 908 (1979).
22. Gunther, H., Herrig, W., Seel, H., Tobias, S., de Meijere, A. and Schrader, B., *J. Org. Chem.* **45**, 4329 (1980).
23. Yates, P. and Stevens, K. E., *Tetrahedron* **37**, 4401 (1981).
24. Della, E. W. and Patney, H. K., *Aust. J. Chem.* **32**, 2243 (1979).
25. Williamson, K. L., Ul Hasan, M. and Clutter, D. R., *J. Magn. Reson.* **30**, 367 (1978).
26. Yates, P. and Douglas, S. P., *Can. J. Chem.* **60**, 2760 (1982).
27. Della, E. W. and Pigou, P. E., *J. Am. Chem. Soc.* **104**, 862 (1982).
28. Barfield, M., Brown, S. E., Canada, E. D., Jr., Ledford, N. D., Marshall, J. L., Walter, S. R. and Yakali, E., *J. Am. Chem. Soc.* **102**, 3355 (1980).
*29. Murray, R. K., Jr. and Goff, D. L., *J. Org. Chem.* **43**, 3844 (1978).
*30. Taka-Ishi, N., Inamoto, Y. and Aigami, K., *Chem. Lett.* 1185 (1973).
*31. Majerski, K. M. and Majerski, Z., *Tetrahedron Lett.* 4915 (1973).
*32. Takaishi, N., Inamoto, Y. and Aigami, K., *J. Org. Chem.* **40**, 276 (1975).
*33. Takaishi, N., Inamoto, Y. and Aigami, K., *Synth. Commun.* **4**, 225 (1974).
34. Takaishi, N., Inamoto Y, Tsuchihashi, K., Yahima, K. and Aigami K, *J. Org. Chem.* **40**, 2929 (1975).
*35. Takaishi, N., Fujikura, Y., Inamoto, Y., Ikeda, H. and Aigami, K., *J. Chem. Soc., Chem. Commun.* 371 (1975).
36. Garratt, D. G. and Kabo, A., *Can. J. Chem.* **58**, 1030 (1980).
37. Garratt, D. G., Ryan, M. D. and Kabo, A., *Can. J. Chem.* **58**, 2329 (1980).

*38. Aigami, K., Inamoto, Y., Takaishi, N., Fujikura, Y., Takatsuki, A. and Tamura, G., *J. Med. Chem.* **19**, 536 (1976).

*39. Aigami, K., Inamoto, Y., Takaishi, N. and Fujikura, Y., *Phytochem.* **16**, 41 (1977).

*40. Asmus, P. and Klessinger, M., *Liebigs Ann. Chem.* 2169 (1975).

41. Riguera, R., *Tetrahedron* **34**, 2039 (1978).

42. Wiberg, K. B., Pratt, W. E. and Bailey, W. F., *J. Org. Chem.* **45**, 4936 (1980).

43. Pfeffer, H. U. and Klessinger, M., *Org. Magn. Reson.* **9**, 121 (1977).

*44. Alberts, V., Brecknell, D. J., Carman, R. M. and Smith, S. S., *Aust. J. Chem.* **34**, 1719 (1981).

*45. Carman, R. M. and Smith, S. S., *Aust. J. Chem.* **34**, 1285 (1981).

46. Garratt, D. G., Ryan, M. D. and Beaulieu, P. L., *J. Org. Chem.* **45**, 839 (1980).

*47. Damin, B., Forestiere, A., Garapon, J. and Sillion, B., *J. Org. Chem.* **46**, 3552 (1981).

48. Christol, H., Laffite, C., Plenat, F. and Renard, G., *Org. Magn. Reson.* **17**, 110 (1981).

49. Berger, S., *Org. Magn. Reson.* **14**, 65 (1980).

50. Berger, S., *J. Org. Chem.* **43**, 209 (1978).

*51. Mehta, G., Singh, V., Pandey, P. N., Chaudhury, B. and Duddeck, H., *Chem. Lett.* 59 (1980).

52. Adcock, W. and Aldous, G. L., *J. Organomet. Chem.* **201**, 411 (1980).

53. Adcock, W. and Aldous, G. L., *J. Organomet. Chem.* **202**, 385 (1980).

54. Adcock, W. and Abeywickrema, A. N., *J. Org. Chem.* **47**, 2945 (1982).

55. Adcock, W. and Abeywickrema, A. N., *J. Org. Chem.* **47**, 2951 (1982).

56. Adcock, W. and Cox, D. P., *J. Org. Chem.* **44**, 3004 (1979).

*57. Murakhovskaya, A. S., Stepanyants, A. U., Zimina, K. I., Aref'ev, O. A. and Epishev, V. I., *Bull. Acad. Sci. USSR , Div. Chem. Sci.* 847 (1975).

58. Haslinger, E., Ohler, E. and Robien, W., *Monatsh. Chem.* **113**, 1321 (1982).

59. Quast, H. and Seiferling, B., *Liebigs Ann. Chem.* 1553 (1982).

60. Avenati, M., Pilet, O., Carrupt, P.-A. and Vogel, P., *Helv. Chim. Acta* **65**, 178 (1982).

61. Avenati, M. and Vogel, P., *Helv. Chim. Acta* **65**, 204 (1982).

62. Hoffmann, R. W., Kurz, H., Reetz, M. T. and Schuttler, R., *Chem. Ber.* **108**, 109 (1975).

*63. Capraro, H.-G. and Ganter, C., *Helv. Chim. Acta* **59**, 97 (1976).

*64. Bohm, M. C., Carr, R. V. C., Gleiter, R. and Paquette, L. A., *J. Am. Chem. Soc.* **102**, 7218 (1980).

65. Whitesell, J. K. and Minton, M. A., The University of Texas at Austin, unpublished results.

66. Duddeck, H. and Wolff, P., *Org. Magn. Reson.* **9**, 528 (1977).

*67. Gabioud, R. and Vogel, P., *Tetrahedron* **36**, 149 (1980).

68. Roche, V. F., Roche, E. B. and Nagel, D. L., *J. Org. Chem.* **47**, 1368 (1982).

69. Belikova, N. A., Bobyleva, A. A., Dzhigirkhanova, A. V., Pekhk, T. I., Lippmaa, E. T. and Plate, A. F., *J. Org. Chem. USSR* **13**, 1759 (1977).

70. Bobyleva, A. A., Dzhigirkhanova, A. V., Belikova, N. A., Pekhk, T. I., Lippmaa, E. T., Plate, A. F. and Gubarevich, T. M., *J. Org. Chem. USSR* **14**, 915 (1978).

71. Belikova, N. A., Lermontov, S. A., Pekhk, T. I., Lippmaa, E. T. and Plate, A. F., *J. Org. Chem. USSR* **14**, 2101 (1978).

72. Lermontov, S. A., Belikova, N. A., Skornyakova, T. G., Pekhk, T. I., Lippmaa, E. T. and Plate, A. F., *J. Org. Chem. USSR* **16**, 1982 (1980).

73. Jacquesy, J.-C., Jacquesy, R. and Patoiseau, J.-F., *Tetrahedron* **32**, 1699 (1976).

74. Langford, G. E., Auksi, H., Gosbee, J. A., MacLachlan, F. N. and Yates, P., *Tetrahedron* **37**, 1091 (1981).

75. Yates, P. and Langford, G. E., *Can. J. Chem.* **59**, 344 (1981).

*76. Fujikura, Y., Aigami, K., Takaishi, N., Ikeda, H. and Inamoto, Y., *Chem. Lett.* 507 (1976).

*77. Farina, M., Morandi, C., Mantica, E. and Botta, D., *J. Org. Chem.* **42**, 2399 (1977).

78. Auksi, H. and Yates, P., *Can. J. Chem.* **59**, 2510 (1981).

79. White, J. B., Ph.D. Dissertation, University of Texas at Austin, 1983,

80. Wenkert, E., Cochran, D. W., Gottlieb, H. E., Hagaman, E. W., Braz Filho, R., Matos, F. G. A. and Madruga, M. I. L. M., *Helv. Chim. Acta* **59**, 2437 (1976).

81. Gabioud, R., Chapuis, G. and Vogel, P., *J. Org. Chem.* **47**, 3316 (1982).

82. Askani, R., Eichenauer, G. and Kohler, J., *Chem. Ber.* **115**, 748 (1982).

*83. Seto, H., Hirokawa, S., Fujimoto, Y. and Tatsuno, T., *Chem. Lett.* 989 (1983).

*84. Inamoto, Y., Takaishi, N., Fujikura, Y., Aigami, K. and Tsuchihashi, K., *Chem. Lett.* 631 (1976).

*85. Wiseman, J. R., Vanderbilt, J. J. and Butler, W. M., *J. Org. Chem.* **45**, 667 (1980).

*86. Wiseman, J. R. and Vanderbilt, J. J., *J. Am. Chem. Soc.* **100**, 7730 (1978).

*87. Adam, W., De Lucchi, O., Peters, K., Peters, E.-M. and von Schnering, H. G., *J. Am. Chem. Soc.* **104**, 161 (1982).

*88. Askani, R. and Schwertfeger, W., *Chem. Ber.* **110**, 3046 (1977).

89. Liao, C. C. and Lin, H. S., *J. Chin. Chem. Soc.* [II] **30**, 69 (1983).

*90. Takaishi, N., Fujikura, Y. and Inamoto, Y., *Synthesis* 293 (1983).

91. Wiberg, K. B., Pratt, W. E. and Bailey, W. F., *J. Org. Chem.* **45**, 4936 (1980).

92. Vorob'eva, N. S., Aref'ev, O. A., Pekhk, T. I., Denisov, Y. V. and Petrov, A. A., *Neftekhimiya* **15**, 659 (1975).

*93. Momose, T., Masuda, K., Furusawa, S., Muraoka, O. and Itooka, T., *Synth. Commun.* **12**, 1039 (1982).

94. Riguera, R. and Garratt, P. J., *An. Quim.* **74**, 216 (1978).

95. Duc, D. K. M., Fetizon, M., Hanna, I. and Olesker, A., *Tetrahedron Lett.* **22**, 3847 (1981).

96. Snowden, R. L. and Schulte-Elte, K. H., *Helv. Chim. Acta* **64**, 2193 (1981).

97. Gabioud, R. and Vogel, P., *Helv. Chim. Acta* **66**, 1134 (1983).

*98. Macas, T. S. and Yates, P., *Tetrahedron Lett.* **24**, 147 (1983).

99. Adcock, W., Abeywickrema, A. N. and Kok, G. B., *J. Org. Chem.* **49**, 1387 (1984).

100. Avenati, M., Hagenbuch, J.-P., Mahaim, C. and Vogel, P., *Tetrahedron Lett.* **21**, 3167 (1980).

101. Narbel, P., Boschi, T., Roulet, R., Vogel, P., Pinkerton, A. A. and Schwarzenbach, D., *Inorg. Chim. Acta* **36**, 161 (1979).

102. *Sadtler Standard Carbon-13 NMR Spectra*, Sadtler Research Laboratories, Inc., Philadelphia, Pa.

103. Pilet, O., Birbaum, J.-L. and Vogel, P., *Helv. Chim. Acta* **66**, 19 (1983).

104. Della, E. W. and Pigou, P. E., *J. Am. Chem. Soc.* **106**, 1085 (1984).

*105. Inamoto, Y., Aigami, K., Takaishi, N., Fujikura, Y., Tsuchihashi, K. and Ikeda, H., *J. Org. Chem.* **42**, 3833 (1977).

106. Capraro, H.-G. and Ganter, C., *Helv. Chim. Acta* **63**, 1347 (1980).

Chapter 7

Bicyclo[3.1.0]hexanes

The terms exo and endo can be readily used for the cis-fused bicyclo[3.1.0]hexanes.

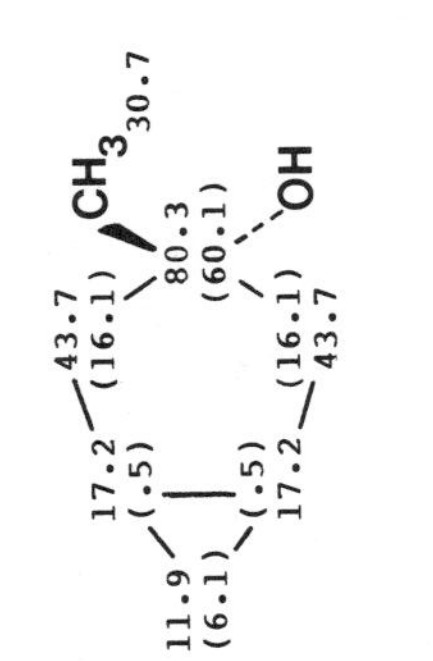

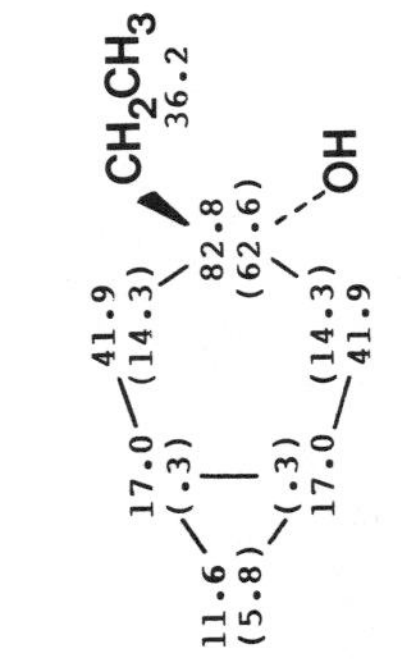

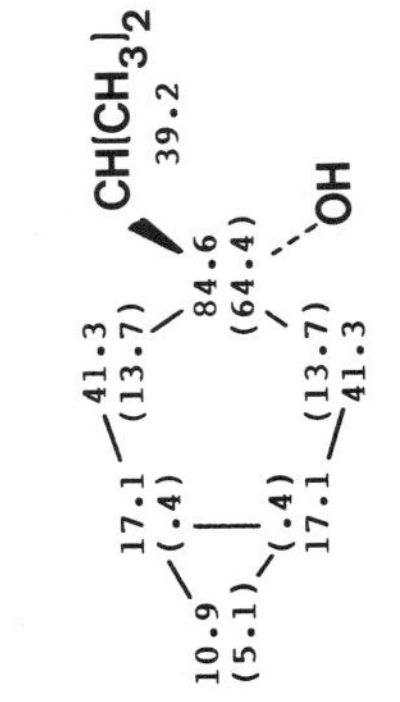

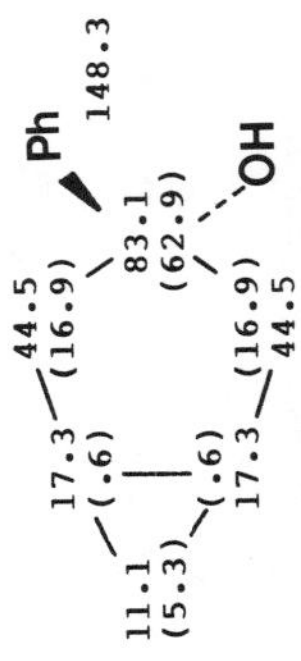

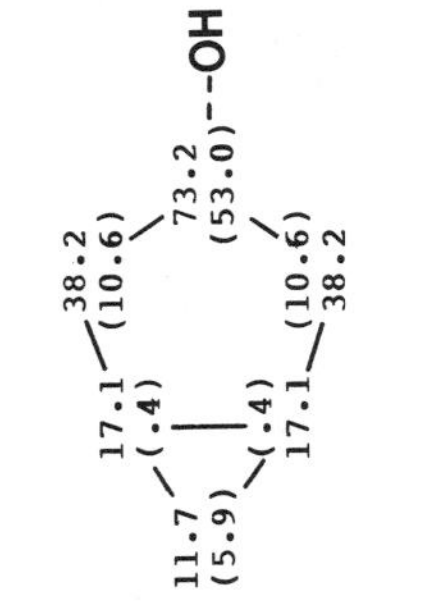

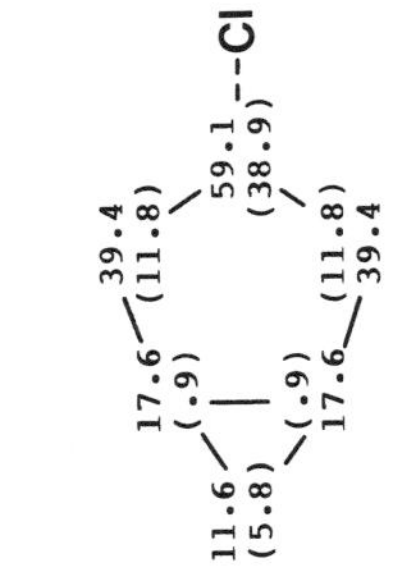

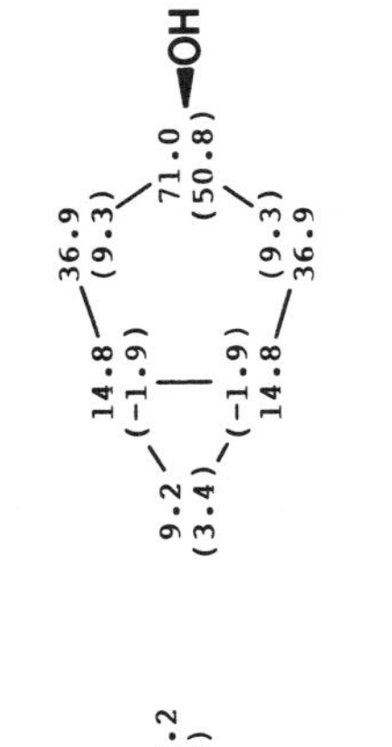

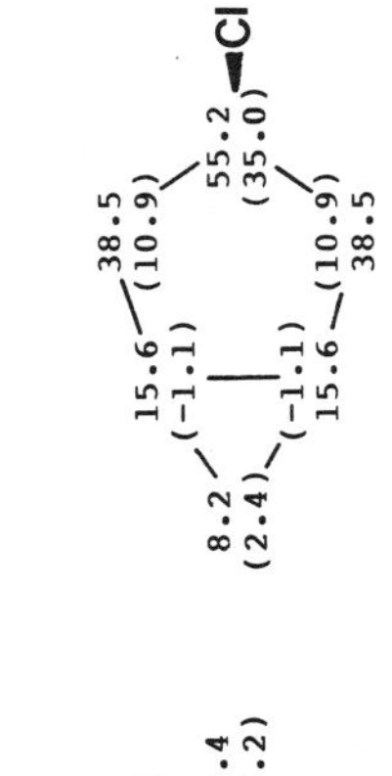

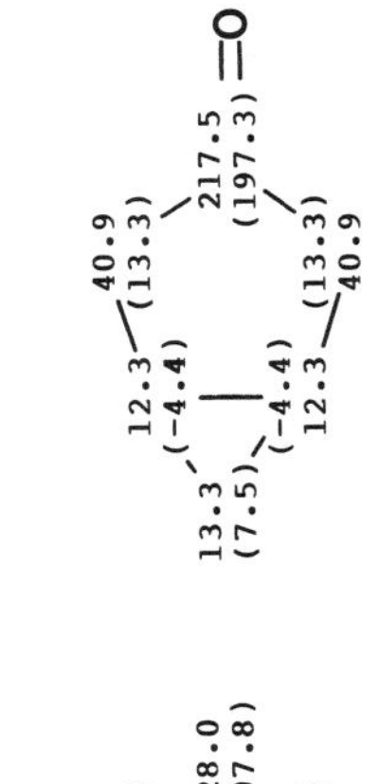

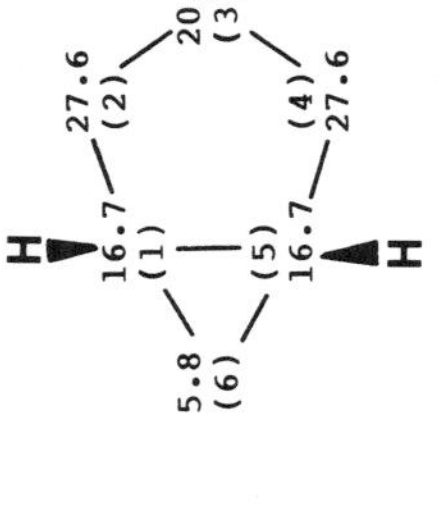

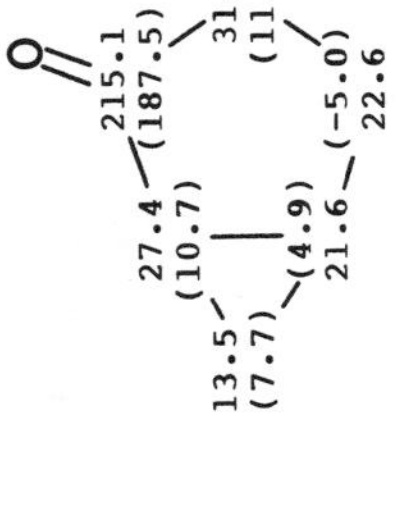

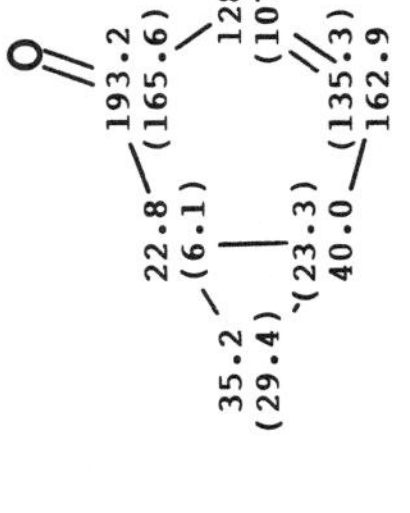

Structure 1 (CH₃ / OH)

CH$_3$ 30.7
80.3 (60.1) OH
43.7 (16.1) / (16.1) 43.7
17.2 (.5) / (.5) 17.2
11.9 (6.1)

Structure 2 (CH₂CH₃ / OH)

CH$_2$CH$_3$ 36.2
82.8 (62.6) OH
41.9 (14.3) / (14.3) 41.9
17.0 (.3) / (.3) 17.0
11.6 (5.8)

Structure 3 (CHlCH₃)₂ / OH)

CHlCH$_3$)$_2$ 39.2
84.6 (64.4) OH
41.3 (13.7) / (13.7) 41.3
17.1 (.4) / (.4) 17.1
10.9 (5.1)

Structure 4 (Ph / OH)

Ph 148.3
83.1 (62.9) OH
44.5 (16.9) / (16.9) 44.5
17.3 (.6) / (.6) 17.3
11.1 (5.3)

Structure 5 (OH)

73.2 (53.0)--OH
38.2 (10.6) / (10.6) 38.2
17.1 (.4) / (.4) 17.1
11.7 (5.9)

Structure 6 (Cl)

59.1--Cl
39.4 (11.8) / (11.8) 39.4
17.6 (.9) / (.9) 17.6
11.6 (5.8)

Structure 7 (OH)

71.0 (50.8) OH
36.9 (9.3) / (9.3) 36.9
14.8 (-1.9) / (-1.9) 14.8
9.2 (3.4)

Structure 8 (Cl)

55.2 (35.0) Cl
38.5 (10.9) / (10.9) 38.5
15.6 (-1.1) / (-1.1) 15.6
8.2 (2.4)

Structure 9 (=O)

217.5 (197.3) =O
40.9 (13.3) / (13.3) 40.9
12.3 (-4.4) / (-4.4) 12.3
13.3 (7.5)

Structure 10 (H, H)

20.2 (3)
27.6 (2) / (4) 27.6
16.7 (1) / (5) 16.7
H H
5.8 (6)

Structure 11 (O)

O
215.1 (187.5) 31.4 (11.2)
27.4 (10.7) / (-5.0) 22.6
13.5 (7.7) / (4.9) 21.6

Structure 12 (O)

O
193.2 (165.6) 128.0 (107.8)
22.8 (6.1) / (135.3) 162.9
35.2 (29.4) / (23.3) 40.0

BICYCLO[3.1.0]HEXANES

Str	Substituents	C1	C2	C3	C4	C5	C6	Substituent Shifts	References
1	PARENT	16.7	27.6	20.2	27.6	16.7	5.8		12,1
2	1-Et/X2-OH/3-Et 3=/4-Ph/5-Ph	31.2	78.3	134.7	142.2	35.3	12.9		13
3	1-iPr/2-=0 3=/4-Me	40.5	193.1	123.7	176.8	20.9	37.8	26.5,18.5	6,7
4	1-iPr/2-=0/N4-Me	43.5	186.1	40.8	25.9	31.7	13.4	28.5,18.1	6
5	1-iPr/3-=0/N4-Me	27.4	41.6	217.2	45.3	25.5	14.7	32.7,12.5	12,23
6	1-iPr/3-=0/X4-Me	29.6	39.6	220.2	47.2	25.5	18.7	32.9,19.7,19.9	12,6,8,23
7	1-iPr/4-==CH2	37.6	27.5	29.0	154.0	30.2	16.1	32.7,101.8	8
8	1-iPr/N2-OH/3=/4-Me	35.3	78.5	133.3	147.4	31.6	21.4	31.0,16.2	7
9	1-iPr/N2-OH/N4-Me	38.0	74.6	39.0	20.4	28.4	6.2	31.0,17.7	7
10	1-iPr/N3-OAc/4-==CH2	37.1	35.9	76.1	152.2	29.4	18.6	32.4,109.7	6
11	1-iPr/N3-OAc/N4-Me	33.1	36.9	76.5	41.4	28.9	12.5	32.7,12.5	6
12	1-iPr/N3-OH X3-Me/N4-Me	31.3	44.0	78.5	44.5	29.3	12.7	32.9,29.8,11.4	12
13	1-iPr/N3-OH X3-Me/X4-Me	33.8	41.9	82.0	47.3	30.5	17.2	33.3,26.7,18.8	12
14	1-iPr/N3-OH X3-Ph/N4-Me	31.8	46.8	82.4	46.1	29.1	13.3	33.1,148.2,11.0	12
15	1-iPr/N3-OH/4-==CH2	37.5	37.2	74.7	156.5	28.9	20.0	32.5,106.3	6
16	1-iPr/N3-OH/4N-Me	33.1	38.6	74.4	40.4	28.8	13.3	32.8,12.1	6
17	1-iPr/N4-Me	34.1	28.3	29.7	34.7	28.6	8.4	32.8,18.1	12
18	1-iPr/N4-Me/X3-C1	33.1	40.1	63.9	40.1	29.1	12.1	32.8,15.2	12
19	1-iPr/X2-OH/3=/4-Me	41.2	78.0	124.0	150.0	24.0	22.5	30.1,16.2	7
20	1-iPr/X2-OH/N4-Me	38.7	75.8	39.7	27.1	27.1	5.0	31.6,17.7	7
21	1-iPr/X3-C1/X4-Me	35.1	38.0	65.4	47.0	29.7	18.9	33.0,19.7	12
22	1-iPr/X3-OAc/N4-Me	30.8	33.9	79.0	39.8	26.2	11.2	33.1,15.9	6
23	1-iPr/X3-OH/N4-Me	30.1	37.0	77.0	42.7	26.7	11.0	33.1,15.9	6,8
24	1-iPr/X3-OH/X4-Me	31.2	33.2	72.3	37.5	28.4	14.4	33.4,14.6	8
25	1-iPr/X4-Me	33.9	24.8	28.8	34.6	30.1	13.1	32.9,19.8	12
26	1-Me/2-=0	32.6	215.8	32.0	21.7	28.7	20.5	13.9	2
27	1-Me/2-=0/5-Me	37.2	215.9	32.5	28.8	33.4	26.8	10.5,18.5	2
28	1-Me/2-=0/5-Me/X6-Me N6-Me	43.5	218.5	38.2	27.5	39.5	29.3	8.2,18.5,19.7,15.8	9
29	1-Me/2-=0/N4-CMe=CH2	33.4	213.3	36.4	40.3	31.7	18.4	13.7,145.9	19
30	1-Me/2-=0/X6-CH=[E]CHMe	38.5	215.3	32.4	22.0	34.2	32.6	9.5,128.2	15
31	1-Me/2-=0/X6-CH=CH2	38.6	215.0	32.2	21.8	34.2	33.2	9.5,134.3	15
32	1-Me/2-=0/X6-Me N6-Me	41.2	218.3	38.3	18.6	40.2	26.9	11.5,24.0,17.1	9
33	1-Me/2-Me 2=/3-==CH2/3-Me/5-Me N6-Me	33.0	141.6	130.7	156.7	39.1	37.6		5
34	1-Me/2-Me 2=/3-==CH2/3-Me/5-Me/X6-Me	30.2	148.1	128.1	161.9	36.5	38.2		5
35	1-Me/4-CMe=CH2	23.1	34.0	26.8	41.7	27.3	11.4	22.1,148.3	18
36	1-Me/4-CMe2	26.7	33.2	26.8	137.4	28.9	18.1	21.0,118.8	18
37	1-Me/N2-OH/N4-CMe=CH2	27.4	77.8	34.5	43.0	27.1	9.6	19.2,147.0	19
38	1-Me/N3-C1	25.1	45.1	59.1	39.8	24.7	17.8		12
39	1-Me/N3-OH	24.5	44.1	73.4	38.7	24.4	18.4		12
40	1-NO2/5-Ph/N6-Ph	79.8	35.1	21.1	28.4	53.3	38.6		10
41	1-NO2/5-Ph/X6-Ph	79.7	40.8	19.4	33.3	47.4	37.5		10
42	1-Ph/2-=0 3=/4-Ph/5-Ph	46.5	202.8	131.8	169.4	45.0	40.2		13
43	1-Ph/N6-Ph	38.8	26.4	23.3	31.8	31.3	33.7		10
44	1-Ph/X2-OH/3-Ph 3=/4-Ph/5-Ph	36.4	78.0	130.9	140.4	36.7	13.8		13
45	1-Ph/X6-Ph	42.4	28.5	21.8	36.8	31.2	30.0		10
46	2-=0	27.4	215.1	31.4	22.6	21.6	13.5		2
47	2-=0 3=	22.8	193.2	128.0	162.9	40.0	35.2		13
48	2-=0 3=/4-Ph/5-Ph/X6-Me	51.6	205.8	135.3	163.3	49.3	46.2		13
49	2-=0 3=/6-==CH2	27.5	204.0	128.9	160.2	30.0	145.3	99.6	20
50	2-=0/5-Me	35.0	214.5	33.6	29.1	30.0	20.9	21.2	2

BICYCLO[3.1.0]HEXANES (cont.)

Str	Substituents	C1	C2	C3	C4	C5	C6	Substituent Shifts	References
51	2=O/5-Me/X6-Me N6-Me	47.3	215.2	38.5	28.1	39.4	30.0	18.7,23.8,17.6	9
52	2=O/X6-CH=[E]CHMe	36.1	212.9	32.4	22.9	29.7	28.9	129.9	15
53	2=O/X6-CH=CH2	35.9	212.1	32.3	22.7	30.2	28.9	136.4	15
54	2=O/X6-Me N6-Me	41.3	214.9	37.9	19.7	35.6	26.1	27.3,16.0	9
55	2-Me 2=/5-iPr	31.5	145.1	121.0	36.7	34.1	21.5	16.3,33.0	6,8
56	N2-Me/N6-OH X6-CH(COOEt)2	34.6	36.5	32.5	26.3	29.5	63.1	17.9,59.6	24
57	X2-Me/N6-OH X6-CH(COOEt)2	33.4	36.8	34.0	24.2	29.4	61.8	22.3,59.4	24
58	N2-OH/3=	18.1	78.1	130.9	137.7	24.0	16.4		13
59	X2-OH/3=	22.5	77.0	130.3	139.6	25.7	21.4		13
60	X2-OH/N2-Me/5-iPr	34.4	80.6	36.7	26.0	34.7	13.3	25.0,32.2	8
61	3=O	12.3	40.9	217.5	40.9	12.3	13.3		12,2,3,4
62	N3-Cl	17.6	39.4	59.1	39.4	17.6	11.6		12
63	X3-Cl	15.6	38.5	55.2	38.5	15.6	8.2		12
64	X3-Me/N6-OH X6-CH(COOEt)2	29.5	34.6	34.1	34.6	29.5	62.0	21.9,59.3	24
65	X3-Me/N6-OH X6-CH(COOH)2	30.8	35.8	35.5	35.8	30.8	63.1	22.8,60.1	24
66	N3-OH	17.1	38.2	73.2	38.2	17.1	11.7		12,3
67	X3-OH	14.8	36.9	71.0	36.9	14.8	9.2		12,3
68	N3-OH X3-iPr	17.1	41.3	84.6	41.3	17.1	10.9	39.2	12
69	N3-OH X3-Me	17.2	43.7	80.3	43.7	17.2	11.9	30.7	12
70	N3-OH X3-Ph	17.3	44.5	83.1	44.5	17.3	11.1	148.3	12
71	N3-OH X3-Et	17.0	41.9	82.8	41.9	17.0	11.6	36.2	12
72	X6-Cl N6-Cl	38.3	27.8	25.4	27.8	38.3	68.3		25

References to Bicyclo[3.1.0]hexanes and related* systems

1. Christl, M., *Chem. Ber.* **108**, 2781 (1975).
2. Grover, S. H., Marr, D. H., Stothers, J. B. and Tan, C. T., *Can. J. Chem.* **53**, 1351 (1975).
3. Buttrick, P. A., Holden, C. M. Y. and Whittaker, D., *J. Chem. Soc. Chem. Commun.* 534 (1975).
4. Abraham, R. J., Chadwick, D. J. and Sancassan, F., *Tetrahedron Lett.* 265 (1979).
5. Mamatyuk, V. I., Rezvukhin, A. I., Isaev, I. S., Buraev, V. I. and Koptyug, V. A., *J. Org. Chem. USSR* **10**, 662 (1974).
6. Abraham, R. J., Holden, C. M., Loftus, P. and Whittaker, D., *Org. Magn. Reson.* **6**, 184 (1974).
7. Holden, C. M., Rees, J. C., Scott, S. P. and Whittaker, D., *J. Chem. Soc. Perkin Trans. II* 1342 (1976).
8. Bohlmann, F., Zeisberg, R. and Klein, E., *Org. Magn. Reson.* **7**, 426 (1975).
9. Schuster, D. I. and Rao, J. M., *J. Org. Chem.* **46**, 1515 (1981).
10. Zimmerman, H. E., Roberts, L. C. and Arnold, R., *J. Org. Chem.* **42**, 621 (1977).
*11. Gooding, K. R., Jackson, W. R., Pincombe, C. F. and Rash, D., *Tetrahedron Lett.* 263 (1979).
12. Rees, J. C. and Whittaker, D., *Org. Magn. Reson.* **15**, 363 (1981).
13. Durr, H. and Albert, K.-H., *Org. Magn. Reson.* **11**, 69 (1978).
*14. Izac, R. R., Fenical, W. and Wright, J. M., *Tetrahedron Lett.* **25**, 1325 (1984).
15. Hudlicky, T., Koszyk, F. J., Kutchan, T. M. and Sheth, J. P., *J. Org. Chem.* **45**, 5020 (1980).
*16. Gunther, H. and Jikeli, G., *Chem. Ber.* **106**, 1863 (1973).
*17. Gunther, H. and Keller, T., *Chem. Ber.* **103**, 3231 (1970).
18. Lehmkuhl, H., Naydowaski, C., Benn, R., Rufinska, A., Schroth, G., Mynott, R. and Kruger, C., *Chem. Ber.* **116**, 2447 (1983).
19. D'yakonova, R. R., Musina, A. A., Gainullina, R. G. and Chernov, P. P., *Bull. Acad. Sci. USSR, Div. Chem. Sci.* **31**, 1182 (1982).
20. Rule, M., Matlin, A. R., Hilinski, E. F., Dougherty, D. A. and Berson, J. A., *J. Am. Chem. Soc.* **101**, 5098 (1979).
*21. Smith, R. A. J. and Hannah, D. J., *Tetrahedron* **35**, 1183 (1979).
*22. Hamon, D. P. G. and Richards, K. R., *Aust. J. Chem.* **36**, 2243 (1983).
23. Burgar, M. I., Karba, D. and Kikelj, D., *Farm. Vestnik* **30**, 253 (1979).
24. Sakai, T., Tabata, H. and Takeda, A., *J. Org. Chem.* **48**, 4618 (1983).

Chapter 8

Bicyclo[3.1.1]heptanes

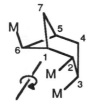 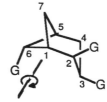

Using exo/endo and syn/anti to designate stereochemical relationships in bicyclo[3.1.1]heptanes creates problems because this system is based on the orientation of groups relative to the unique, smallest bridge. The M/G system defined in Chapter 6 for the bicyclo[2.2.2]octanes is equally applicable here. Briefly, a sense of rotation (clockwise or counterclockwise) is defined for an axis that passes through both bridgehead atoms based upon the ordering of assigned numbers for the atoms adjacent to atom 1. Substituents that are oriented with this sense of rotation are defined as M (mit) while those opposed are G (gegen). For monosubstituted compounds, the numbering system is defined so that the substituent is M.

Whitesell, J.K. and Minton, M.A. *J. Org. Chem.*, **50**, 509 (1985).

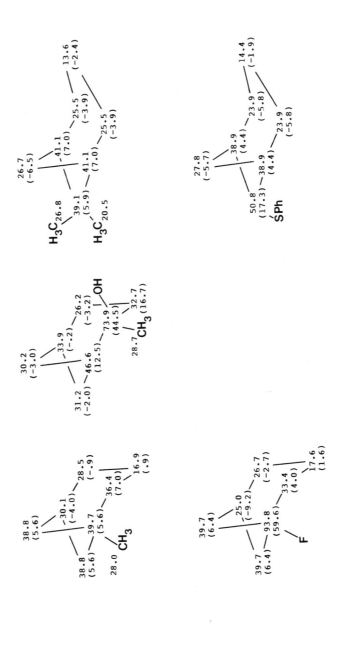

BICYCLO[3.1.1]HEPTANES

Str	Substituents	C1	C2	C3	C4	C5	C6	C7	Substituent Shifts	References
1	PARENT	34.1	29.4	16.0	29.4	34.1	33.2	33.2		29,7,16,18,27
2	1-Br/5-Br	51.7	37.8	20.3	37.8	51.7	55.5	55.5		27
3	1-Cl/2==O	71.0	204.5	33.2	24.4	30.0	41.1	41.4		17
4	1-COOH/5-COOH	41.5	29.1	15.5	29.1	41.5	37.4	37.4	176.2,176.2	27
5	1-COOMe/5-COOMe	42.1	29.4	15.8	29.4	42.1	37.8	37.8	175.3,175.3	27
6	1-F	93.8	33.4	17.6	26.7	25.0	39.7	39.7		18
7	1-F/2==O	94.0	207.4	33.0	25.1	24.7	37.2	37.2		18
8	1-Me	39.7	36.4	16.9	28.5	30.1	38.8	38.8	28.0	19,29
9	2==CH2/G3-OH/M6-Me G6-Me	50.5	155.3	66.7	34.4	39.8	40.3	31.5	110.9,26.0,21.8	11
10	2==CH2/M6-Me G6-Me	51.9	150.8	23.7	23.8	40.6	40.6	27.0	106.5,26.2,20.9	13,2,8-11,31,36
11	2==O 3= 4-Me/M6-Me G6-Me	57.4	202.6	121.0	169.6	49.6	53.5	40.5	23.3,26.5,22.0	8,9,11
12	2==O/G3-Me/M6-Me G6-Me	58.1	215.8	38.1	31.0	41.2	40.4	25.7	21.8,25.7,22.6	22
13	2==O/M3-Me/M6-Me G6-Me	57.8	216.3	37.1	31.0	41.2	43.0	25.6	14.1,26.3,22.0	22,15
14	2==O/M4-Me/M6-Me G6-Me	57.7	212.4	41.2	31.0	47.4	40.1	28.3	24.4,26.9,21.0	11+22
15	2==O/M6-Me G6-Me	58.0	211.6	32.7	21.5	40.7	41.1	25.2	25.9,22.0	13,11,15,22
16	G2-Br/3-==O/G4-Br/M6-Me G6-Me	46.7	51.8	200.0	51.8	46.7	42.2	28.2	26.9,19.5	12
17	G2-Br/3-==O/M6-Me G6-Me	47.3	54.5	204.0	43.8	41.7	41.1	30.2	26.5,19.9	12
18	2-CH(-SCH2CH2CH2S-) 2=/M6-Me G6-Me	45.1	148.0	121.4	31.3	40.6	38.2	31.9	52.3,26.1,21.5	25
19	2-CH2CH2OH 2=/M6-Me G6-Me	46.0	145.3	117.8	31.4	40.9	38.0	31.7	40.3,26.4,21.2	13,36
20	M2-CH2NH2/M6-Me G6-Me	44.1	45.7	20.4	26.2	41.6	38.7	33.5	48.4,28.1,23.3	36
21	2-CH2OH 2=/M6-Me G6-Me	43.4	147.8	117.7	31.2	41.0	38.0	31.6	65.8,26.3,21.1	8,11,36
22	M2-CH2OH/M6-Me G6-Me	43.1	44.4	18.9	26.1	41.6	38.7	33.3	67.5,28.1,23.2	13,8,11,36
23	G2-CH2OH/M6-Me G6-Me	42.4	37.5	18.5	23.4	41.2	39.2	24.4	66.0,26.9,20.2	13,8,11
24	2-CHO 2=/M6-Me G6-Me	38.2	151.6	147.5	31.1	40.7	37.6	33.0	191.0,25.7,20.9	8,11
25	G2-Cl M2-Me/M6-Me G6-Me	56.7	79.6	34.5	25.7	40.3	39.4	30.9	34.8,28.0,23.7	5
26	2-Et 2=/M6-Me G6-Me	46.0	150.0	114.6	31.3	41.2	38.0	31.7	29.8,26.7,21.2	6
27	2-Me 2=/6-==O/M7-Me G7-Me	67.9	138.7	118.4	33.0	62.6	205.8	30.1	23.0,14.7,27.3	35
28	2-Me 2=/G6-C(OTMS)=CH2 M6-Me	45.1	144.8	115.4	30.0	40.6	48.1	32.1	23.1,163.5,25.5	21
29	2-Me 2=/G6-CH2OH M6-Me	45.0	142.8	116.3	31.4	39.4	41.4	31.6	22.9,65.7,21.3	21
30	2-Me 2=/G6-CH=CH2 M6-Me	47.6	142.6	115.9	32.1	43.0	44.2	32.6	23.5,142.0,24.9	21
31	2-Me 2=/M6-Me G6-Me	43.4	142.8	116.6	30.0	41.4	52.3	31.6	23.0,26.5,20.9	21
32	2-Me 2=/M6-Me G6-Me	47.2	144.0	116.2	31.1	40.9	38.1	31.5	26.0,26.9,20.2	13,2-4,8-11,31,32,35,36
33	M2-Me G2-O-3G/M6-Me G6-Me	45.3	60.0	56.7	22.5	40.0	40.6	27.8		36
34	M2-Me/3-==CH2/M6-Me G6-Me	46.9	42.0	149.9	36.2	40.3	39.4	34.0	21.2,110.2,27.8,22.1	30
35	G2-Me/3-==O/M6-Me G6-Me	44.7	46.6	214.8	44.5	38.5	39.6	29.2	15.1,26.5,19.9	13
36	M2-Me/3-==O/M6-Me G6-Me	45.1	51.4	214.8	44.8	39.1	39.2	34.4	16.8,27.1,21.9	13
37	M2-Me/G3-CHO/M6-Me G6-Me	49.9	36.0	47.1	26.3	40.5	39.2	33.1	21.9,202.1,27.8,22.8	20
38	M2-Me/G3-OH/M6-Me G6-Me	48.0	47.1	70.8	38.8	41.9	38.3	34.1	20.8,27.8,23.7	5,8,11,13,23
39	G2-Me/G3-OH/M6-Me G6-Me	47.6	35.8	66.0	37.0	40.6	39.5	25.0	15.1,26.6,20.3	5
40	M2-Me/M3-OH/M6-Me G6-Me	48.0	40.5	64.0	37.4	40.7	38.9	30.2	15.1,27.7,22.3	5
41	G2-Me/M3-OH/M6-Me G6-Me	47.5	38.9	71.0	36.0	40.6	40.1	23.6	19.2,26.4,19.9	5
42	M2-Me/M6-Me G6-Me	48.4	36.2	24.0	26.7	41.6	38.9	34.1	23.0,28.4,23.3	5,4,11,13,15
43	G2-Me/M6-Me G6-Me	47.9	29.6	24.2	23.2	41.1	39.6	24.8	21.7,27.0,20.2	5,4,13
44	G2-O-3G M3-CH2CH2COOEt/M6-Me G6-Me	43.7	55.4	62.2	27.6	40.1	40.7	25.8	28.6,26.7,20.2	26
45	G2-O-3G M3-CH=[E]CHCOOEt/M6-Me G6-Me	42.3	58.6	60.8	27.8	40.1	40.5	25.8	148.3,26.7,20.4	13,5,14,34
46	G2-O-3G M3-CHO/M6-Me G6-Me	39.6	53.6	64.7	27.0	37.6	40.0	25.2	198.7,26.0,19.8	11
47	M2-OH G2-Me	46.6	73.9	32.7	26.2	33.9	30.2	31.2	28.7	33
48	M2-OH G2-Me/M6-Me G6-Me	53.6	76.0	31.8	26.4	40.9	38.5	28.5	31.8,27.7,23.7	5,11,14
49	G2-OH M2-Me/M6-Me G6-Me	54.6	75.0	31.9	25.0	40.8	38.3	27.4	31.4,27.8,23.5	5,4,13
50	G2-OH/3= 4-Me/M6-Me G6-Me	48.0	70.0	118.5	147.8	47.0	46.0	28.6	20.4,26.6,22.5	11

BICYCLO[3.1.1]HEPTANES (cont.)

Str	Substituents	C1	C2	C3	C4	C5	C6	C7	Substituent Shifts	References
51	M2-OH/M6-Me G6-Me	47.9	73.1	24.8	25.5	41.0	37.4	28.3	27.3,22.4	15,11
52	G2-OH/M6-Me G6-Me	48.0	69.4	23.3	22.7	40.5	39.4	25.7	26.7,20.0	15,11
53	2= 3-CH₂CH₂COOEt/M6-Me G6-Me	45.8	116.7	146.7	32.4	40.9	38.0	31.3	31.6,26.3,21.1	26
54	2= 3-CH=[E]CHCOOEt/M6-Me G6-Me	41.5	133.6	145.9	32.7	40.9	37.9	31.2	145.2,26.3,20.8	26
55	2=/M6-Me G6-Me	42.0	136.2	123.7	32.0	41.3	38.0	32.6	26.4,21.3	11
56	2-=O/M3-Me G3-Me/M6-Me G6-Me	58.4	218.2	42.9	37.6	42.0	40.8	25.9	33.8,27.5,26.4,22.6	22,15
57	3-=O/M6-Me G6-Me	38.2	44.8	211.2	44.8	38.2	39.0	32.6	26.3,20.3	12
58	G6-)2/M7-Me	39.6	20.1	14.4	20.1	39.6	37.5	34.9	37.5,11.2	28
59	M6-Me G6-Me	41.1	25.5	13.6	25.5	41.1	39.1	26.7	26.8,20.5	5,15
60	G6-OMe/M7-Me	40.3	19.3	14.3	19.3	40.3	76.2	26.8	4.5	28
61	G6-SPh	38.9	23.9	14.4	23.9	38.9	50.8	27.8		7
62	G6-SPh/M7-Ph	41.5	20.9	13.5	20.9	41.5	48.8	41.8	141.3	7

116

References to Bicyclo[3.1.1]heptanes and related* systems

*1. Andersen, N. H., Bissonette, P., Liu, C.-B., Shunk, B., Ohta, Y., Tseng, C.-L. W., Moore, A. and Huneck, S., *Phytochem.* **16**, 1731 (1977).

2. Hall, M. C., Kinns, M. and Wells, E. J., *Org. Magn. Reson.* **21**, 108 (1983).

3. Offermann, W., *Org. Magn. Reson.* **20**, 203 (1982).

4. Forsyth, D. A., Mahmoud, S. and Giessen, B. C., *Org. Magn. Reson.* **19**, 89 (1982).

5. Weigand, E. W. and Schneider, H.-J., *Org. Magn. Reson.* **12**, 637 (1979).

6. Brun, P., Casanova, J., Hatem, J., Zahra, J. P. and Waegell, B., *Org. Magn. Reson.* **12**, 537 (1979).

7. Szeimies, G., Schlosser, A., Philipp, F., Dietz, P. and Mickler, W., *Chem. Ber.* **111**, 1922 (1978).

8. Bohlmann, F., Zeisberg, R. and Klein, E., *Org. Magn. Reson.* **7**, 426 (1975).

9. Johnson, L. F. and Jankowski, W. C., "Carbon-13 NMR Spectra", Wiley-Interscience, New York (1972).

10. Jautelat, M., Grutzner, J. B. and Roberts, J. D., *Proc. Natl. Acad. Sci.* **65**, 288 (1970).

11. Holden, C. M. and Whittaker, D., *Org. Magn. Reson.* **7**, 125 (1975).

12. Reisse, J., Piccinni-Leopardi, C., Zahra, J. P., Waegell, B. and Fournier, *J., Org. Magn. Reson.* **9**, 512 (1977).

13. Lysenkov, V. I., Pekhk, T. I., Lippmaa, E. T. and Zheleznyak, T. L., *J. Org. Chem. USSR* **17**, 1436 (1981).

14. Kane, B. J., Marcelin, G. and Traynor, S. G., *J. Org. Chem.* **45**, 895 (1980).

15. Grover, S. H., Marr, D. H., Stothers, J. B. and Tan, C. T., *Can. J. Chem.* **53**, 1351 (1975).

16. Della, E. W., Cotsaris, E., Hine, P. T. and Pigou, P. E., *Aust. J. Chem.* **34**, 913 (1981).

17. Della, E. W. and Pigou, P. E., *Aust. J. Chem.* **36**, 2261 (1983).

18. Della, E. W., Cotsaris, E. and Hine, P. T., *J. Am. Chem. Soc.* **103**, 4131 (1981).

19. Della, E. W. and Pigou, P. E., *J. Am. Chem. Soc.* **104**, 862 (1982).

20. Himmele, W. and Siegel, H., *Tetrahedron Lett.* 907 (1976).

21. Bessiere, Y., Grison, C. and Boussac, G., *Tetrahedron* **34**, 1957 (1978).

22. Bessiere, Y., Barthelemy, M., Thomas, A. F., Pickenhagen, W. and Starkemann, C., *Nouv. J. Chim.* **2**, 365 (1978).

23. Ericsson, A., Kowalewski, J., Liljefors, T. and Stilbs, P., *J. Magn. Reson.* **38**, 9 (1980).

*24. Thomas, A. F., Thommen, W. and Becker, J., *Helv. Chim. Acta* **64**, 161 (1981).

25. Hoppmann, A., Weyerstahl, P. and Zummack, W., *Liebigs Ann. Chem.* 1547 (1977).

26. Marschall, H., Penninger, J. and Weyerstahl, P., *Liebigs Ann. Chem.* 49 (1982).

27. Gassman, P. G. and Proehl, G. S., *J. Am. Chem. Soc.* **102**, 6862 (1980).

28. Gassman, P. G. and Olson, K. D., *J. Am. Chem. Soc.* **104**, 3740 (1982).

29. Della, E. W. and Pigou, P. E., *J. Am. Chem. Soc.* **106**, 1085 (1984).

30. Brown, H. C. and Ford, T. M., *J. Org. Chem.* **46**, 647 (1981).

31. Burgar, M. I., Karba, D. and Kikelj, D., *Farm. Vestnik* **30**, 253 (1979).

32. Formacek, V. and Kubeczka, K.-H., "Vorkommen und Analytik Atherische Ole", Kubeczka, K.-H. Ed., Thieme, Stuttgart, 1979, p 130.

33. Banert, K., Kirmse, W. and Wroblowsky, H.-J., *Chem. Ber.* **116**, 3591 (1983).

34. Blunt, J. W. and Steel, P. J., *Aust. J. Chem.* **35**, 2561 (1982).

35. Uchio, Y., Matsuo, A., Kakayama, M. and Hayashi, S., *Tetrahedron Lett.* 2963 (1976).

36. *Sadtler Standard Carbon-13 NMR Spectra*, Sadtler Research Laboratories, Inc., Philadelphia, Pa.

Chapter 9

Bicyclo[3.2.0]heptanes

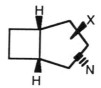

The terms exo and endo can be readily used for the cis-fused bicyclo[3.2.0]heptanes (no data for trans-fused compounds was found).

BICYCLO[3.2.0]HEPTANES

Str	Substituents	C1	C2	C3	C4	C5	C6	C7	Substituent Shifts	References
1	(5)1=/6-Me 6-Me/7-Me 7-Me	155.4	27.8	25.9	27.8	155.4	46.7	46.7	4*22.9	2
2	(5)1=/X6-(CH2)3-7X	150.5	29.2	26.0	29.2	150.5	45.9	45.9	2*25.8	2
3	(5)1=/X6-(CH2)4-7X	153.2	30.4	26.3	30.4	153.2	39.2	39.2	2*23.9	2
4	1-(CH2)3-5	55.3	39.2	31.0	39.2	55.3	27.7	27.7	2*39.2	14,17
5	X1-CH2OH	50.4	35.9	25.6	33.5	39.1	21.1	26.0	69.3	1
6	X1-CH2OH/N4-Me	50.6	35.6	33.3	37.7	43.7	14.5	26.3	69.4,13.7	6
7	1-CN/2=	44.0	135.2	128.8	39.9	42.2	24.6	31.7	121.8	21
8	1-COMe/X7-CH2TMS	69.0	39.5	36.9		47.6	32.9	28.0	211.2,24.7	19,20
9	1-Me/2-=0/3-Me 3= 4-CH2)2/5-Me/6-Me 6= 7-Me	59.0	207.8	133.8	171.8	57.1	145.1	139.8	8.6,8.1,26.4,10.6,13.2,11.6	13
10	1-Me/2-=0/3-Me 3= 4-Me/6-Me 6= 7-Me	55.7	205.6	134.0	165.7	58.3	141.1	141.6	9.1,7.9,16.1,13.7,12.6	12
11	1-Me/2-=0/3= 4-C=)2/5-Me/6-Me 6= 7-Me	61.4	208.1	132.1	172.1	57.2	144.2	140.9	8.9,131,10.5,14.1,11.9	13
12	1-Me/2-=0/3= 4-CH2)2/5-Me/6-Me 6= 7-Me	60.2	208.2	124.8	181.1	58.6	144.5	140.5	8.8,27.1,10.1,12.9,11.3	13
13	X1-OH/N2-Me/6= 7-COOEt	88.1	34.3	31.9	22.4	52.9	147.8	140.7	13.4,161.6	9
14	X1-OH/X2-Me/6= 7-COOEt	88.8	41.0	32.4	24.3	53.9	148.2	138.6	14.2,162.0	9
15	2-=0/5-nBu/6-=0	41.4	216.8	37.1	27.1	73.3	211.1	50.0	33.2	11
16	2-=0/N6-Me	42.3	221.7	38.2	20.9	37.3	29.3	31.6	16.6	1
17	2-=0/X3-Me N3-Me	44.2	173.6	48.4	43.9	31.0	25.6	22.7	26.4,24.1	1
18	2-=0/X6-Me	43.8	223.3	36.9	26.9	41.6	33.5	30.3	21.7	1
19	X2-CH=CHR/N3-OTBS/6-=0	33.8	55.9	80.9	38.9	63.1	212.3	52.5	133.8	18
20	2-COOH 2-/7-=0	72.2	132.2	145.1	40.4	26.5	53.6	204.0	164.4	10
21	2-COOH 2=/X6-C1 N6-C1/7-=0	68.9	132.2	146.4	36.8	50.4	90.9	191.3	163.5	10
22	X2-COOH/3=/X6-C1 N6-C1/7-=0	61.2	52.7	130.7	135.3	59.1	87.6	195.6	170.9	10
23	X2-COOH/7-=0	67.4	46.2	29.1	31.0	28.5	51.2	210.2	174.3	10
24	2-COSMe 2=/X6-C1 N6-C1/7-=0	67.9	137.9	142.7	36.6	49.6	90.0	189.7	186.6	10
25	X2-OH/X3-Me N3-Me	45.8	88.7	45.6	46.7	34.6	27.1	24.0	20.7,26.5	1
26	N2-OH/X3-Me N3-Me	42.8	80.9	45.8	45.5	36.0	26.0	16.3	27.8,22.5	1
27	2-/4-=CH2/N6-C(=CH2)-CH=CH-7N	48.6	140.5	135.1	157.9	52.7	52.7	48.6	102.9,157.9,140.5	15
28	2-/6-=0	37.4	133.6	132.3	35.0	62.2	210.8	54.4		7
29	2-/6-=0/7-=CMe2	47.0	131.5	130.7	33.9	57.3	202.7	139.8	146.2	4
30	2-/6-=0/N7-CH=CH2	43.1	129.9	134.5	34.2	59.6	212.2	68.0	129.9	16
31	2-/6-=0/X7-CH=CH2	43.0	132.5	132.1	34.9	59.8	212.5	71.3	132.4	16
32	2-/6-=0/X7-C1 N7-C1	58.8	137.0	128.5	35.3	59.6	197.2	88.3		6
33	2-/X6-CH2CHO/7-=0	72.1	133.6	125.8	40.3	33.7	58.6	208.5	43.1	3
34	2-/X6-CH2O-6N	40.4	133.4	131.9	33.2	43.5	61.1	37.9	51.6	6
35	2-/X6-CHBr2 N6-OH	36.5	133.2	132.5	32.9	44.8	77.6	40.3	58.3	8
36	2=/X6-Ph N6-OH/7-==CMe2	48.4	132.4	131.6	31.7	49.5	79.8	142.0	128.2	4
37	N6-(CH2)3-7N	38.1	28.5	26.6	28.5	38.1	38.1	38.1	28.5,28.5	5
38	X6-(CH2)3-7X	41.9	32.9	25.0	33.2	41.9	41.9	41.9	33.2,33.2	5
39	X6-CH2OAc	41.1	32.9	24.9	33.2	34.5	36.0	27.1	68.8	1
40	N6-CH2OAc	39.0	32.7	26.4	27.2	35.5	31.0	27.3	64.8	1
41	X6-CH2OH	40.9	33.0	25.0	33.2	34.6	39.4	26.9	67.6	1
42	N6-CH2OH	38.9	33.3	26.6	32.8	35.4	34.5	27.2	63.2	1

References to Bicyclo[3.2.0]heptanes and related* systems

1. Salomon, R. G., Coughlin, D. J., Ghosh, S. and Zagorski, M. G., *J. Am. Chem. Soc.* **104**, 998 (1982).
2. Fitjer, L., Kliebisch, U., Wehle, D. and Modaressi, S., *Tetrahedron Lett.* **23**, 1661 (1982).
3. Ayral-Kaloustian, S. and Agosta, W. C., *J. Org. Chem.* **48**, 1718 (1983).
4. Mayr, H. and Halberstadt-Kausch, I. K., *Chem. Ber.* **115**, 3479 (1982).
5. Salomon, R. G., Folting, K., Streib, W. E. and Kochi, J. K., *J. Am. Chem. Soc.* **96**, 1145 (1974).
6. Whitesell, J. K. and Minton, M. A., University of Texas at Austin, unpublished results.
7. Whitesell, J. K., Minton, M. A. and Flanagan, W. G., *Tetrahedron* **37**, 4451 (1981).
8. Whitesell, J. K. and Allen, D. E., University of Texas at Austin, unpublished results.
9. Clark, R. D. and Untch, K. G., *J. Org. Chem.* **44**, 248 (1979).
10. Gordon, E. M., Pluscec, J. and Ondetti, M. A., *Tetrahedron Lett.* **22**, 1871 (1981).
11. Berenjian, N., de Mayo, P., Sturgeon, M.-E., Sydnes, L. K. and Weedon, A. C., *Can. J. Chem.* **60**, 425 (1982).
12. Salakhutdinov, N. F., Detsina, A. N. and Koptyug, V. A., *J. Org. Chem. USSR* **17**, 1312 (1981).
13. Hogeveen, H. and van Kruchten, E. M. G. A., *J. Org. Chem.* **46**, 4734 (1981).
14. Tobe, Y., Terashima, K., Sakai, Y. and Odaira, Y., *J. Am. Chem. Soc.* **103**, 2307 (1981).
15. Uebersax, B., Neuenschwander, M. and Kellerhals, H.-P., *Helv. Chim. Acta* **65**, 74 (1982).
16. Danheiser, R. L. and Martinez-Davila, C., *Tetrahedron* **37**, 3943 (1981).
17. Bishop, R. and Landers, A. E., *Aust. J. Chem.* **32**, 2675 (1979).
18. Newton, R. F. and Wadsworth, A. H., *J. Chem. Soc. Perkin Trans. I* 823 (1982).
19. Pards, R., Zahra, J.-P. and Santelli, M., *Tetrahedron Lett.* 4557 (1979).
20. House, H. O., Gaa, P. C. and VanDerveer, D., *J. Org. Chem.* **48**, 1661 (1983).
21. Gassman, P. G. and Talley, J. J., *J. Am. Chem. Soc.* **102**, 4138 (1980).
*22. Browne, C. E., Ruehle, P. H., Dobbs, T. K. and Eisenbraun, E. J., *Org. Magn. Reson.* **12**, 553 (1979).

Chapter 10

Bicyclo[3.2.1]octanes

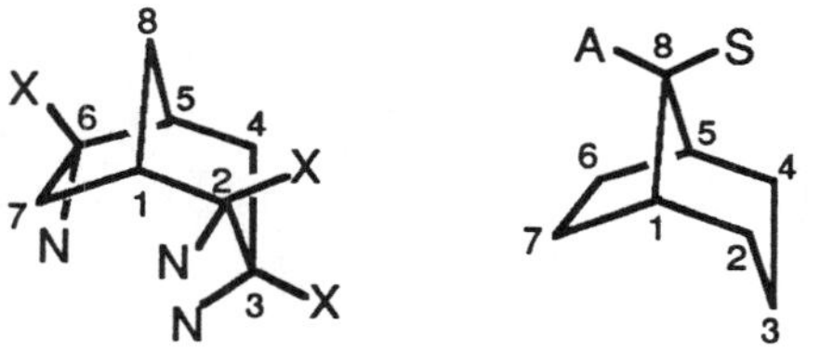

Exo/endo (relative to the one carbon bridge) and syn/anti are readily applied here. Syn always refers to an orientation of substituents on C-8 that is toward the three-carbon bridge. Conformational mobility is present in this system as the six-membered ring can be either a chair or rigid boat.

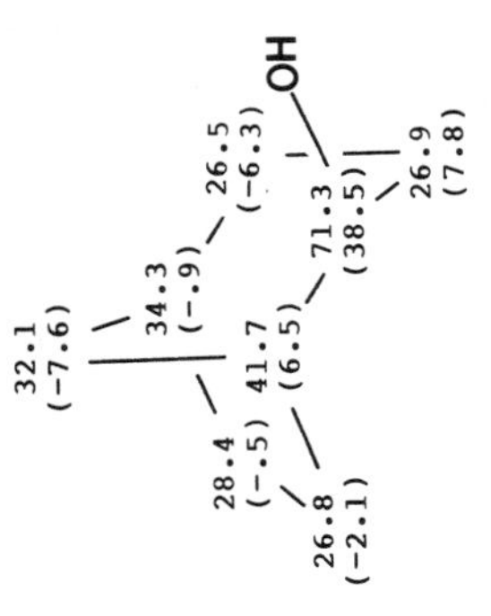

OH
32.1 (-7.6)
34.3 (-.9)
26.5 (-6.3)
71.3 (38.5)
26.9 (7.8)
28.4 (-.5)
41.7 (6.5)
26.8 (-2.1)

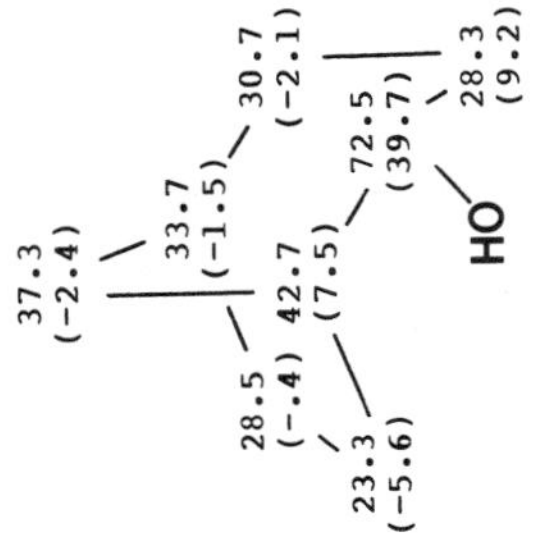

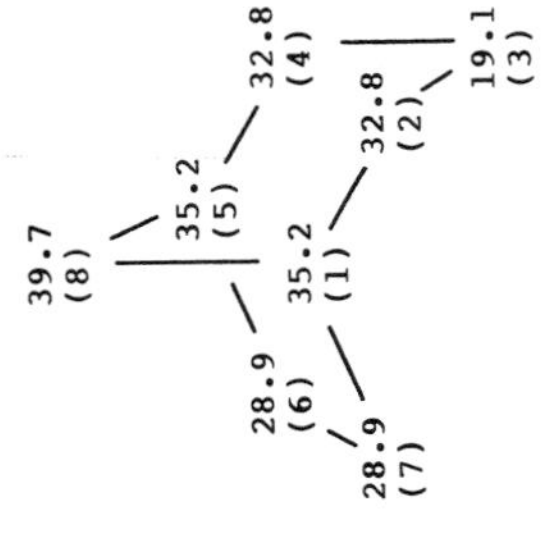

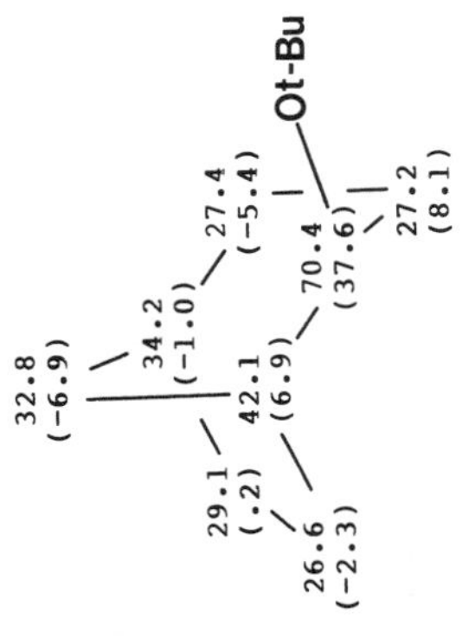

Ot-Bu
32.8 (-6.9)
34.2 (-1.0)
27.4 (-5.4)
70.4 (37.6)
27.2 (8.1)
29.1 (.2)
42.1 (6.9)
26.6 (-2.3)

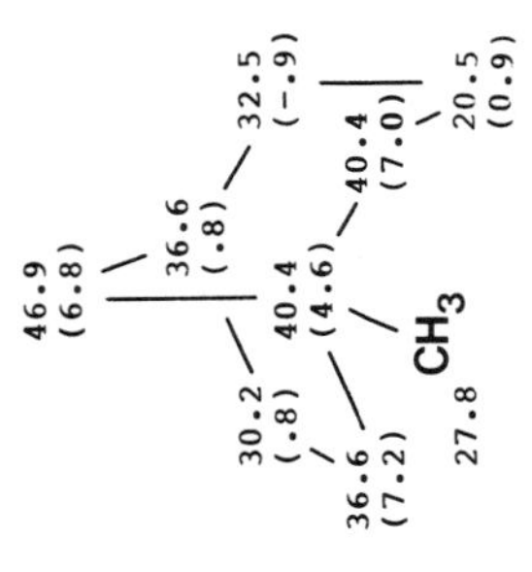

OAc
32.9 (-6.8)
34.1 (-1.1)
27.1 (-5.7)
74.0 (41.2)
23.8 (4.7)
28.9 (0.0)
38.7 (3.5)
26.8 (-2.1)

HO
37.3 (-2.4)
33.7 (-1.5)
30.7 (-2.1)
72.5 (39.7)
28.3 (9.2)
28.5 (-.4)
42.7 (7.5)
23.3 (-5.6)

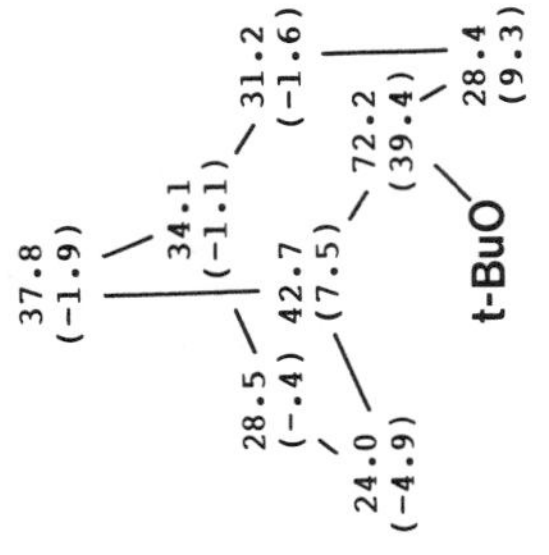

t-BuO
37.8 (-1.9)
34.1 (-1.1)
31.2 (-1.6)
72.2 (39.4)
28.4 (9.3)
28.5 (-.4)
42.7 (7.5)
24.0 (-4.9)

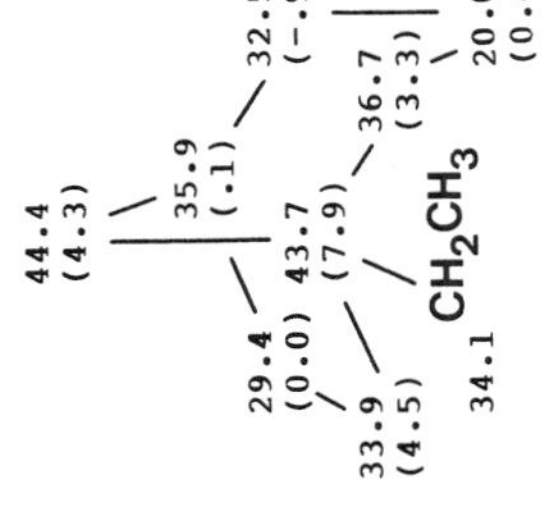

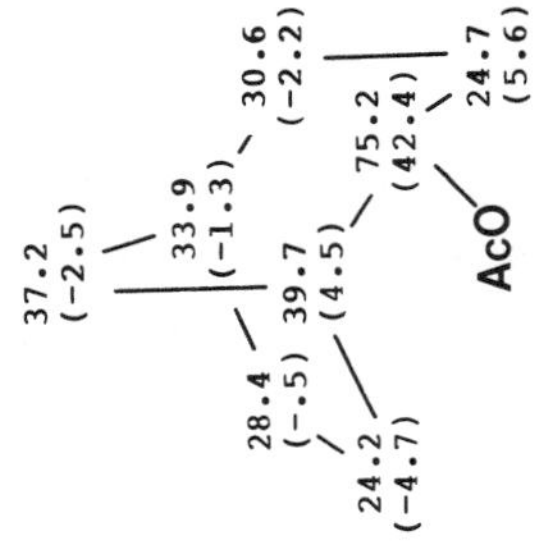

AcO
37.2 (-2.5)
33.9 (-1.3)
30.6 (-2.2)
75.2 (42.4)
24.7 (5.6)
28.4 (-.5)
39.7 (4.5)
24.2 (-4.7)

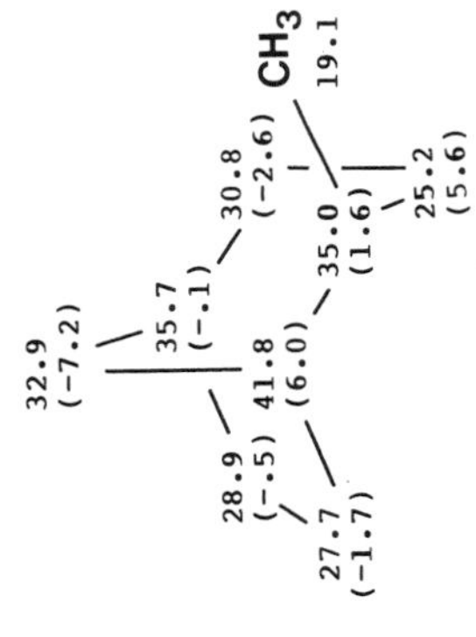

CH₃
46.9 (6.8)
36.6 (.8)
32.5 (-.9)
40.4 (7.0)
20.5 (0.9)
30.2 (.8)
40.4 (4.6)
36.6 (7.2)
27.8

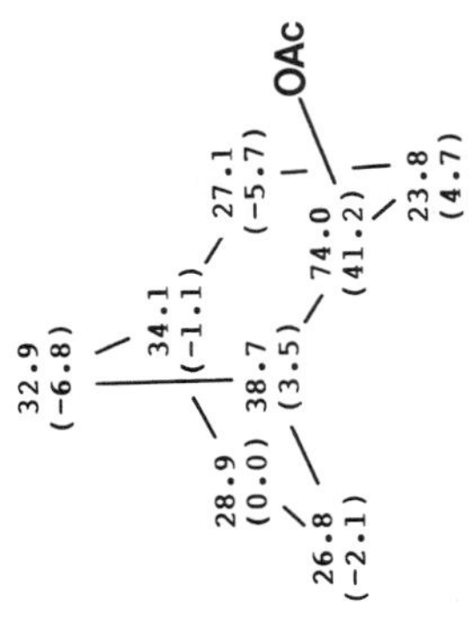

CH₂CH₃
44.4 (4.3)
35.9 (.1)
32.5 (-.9)
36.7 (3.3)
20.0 (0.4)
29.4 (0.0)
43.7 (7.9)
33.9 (4.5)
34.1

CH₃ 19.1
32.9 (-7.2)
35.7 (-.1)
30.8 (-2.6)
35.0 (1.6)
25.2 (5.6)
28.9 (-.5)
41.8 (6.0)
27.7 (-1.7)

39.7 (8)
35.2 (5)
32.8 (4)
35.2 (1)
32.8 (2)
19.1 (3)
28.9 (6)
28.9 (7)

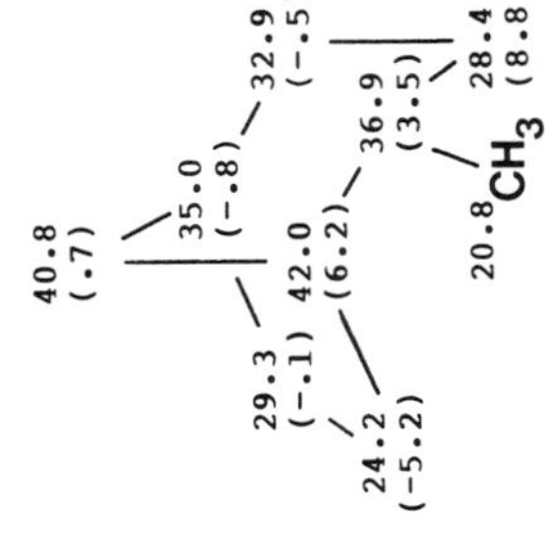

CH₃
40.8 (.7)
35.0 (-.8)
32.9 (-.5)
36.9 (3.5)
28.4 (8.8)
29.3 (-.1)
42.0 (6.2)
24.2 (-5.2)
20.8

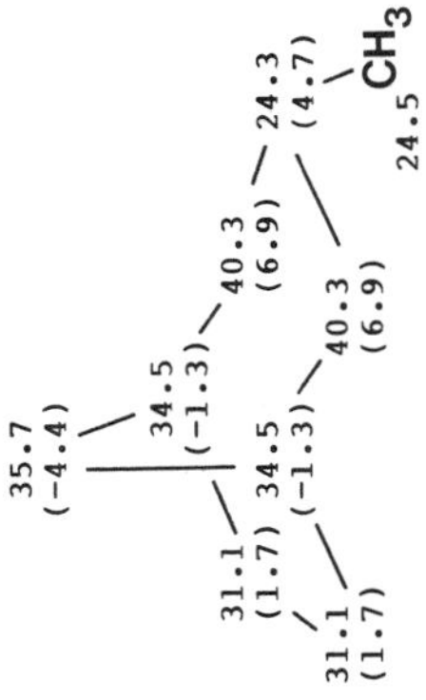

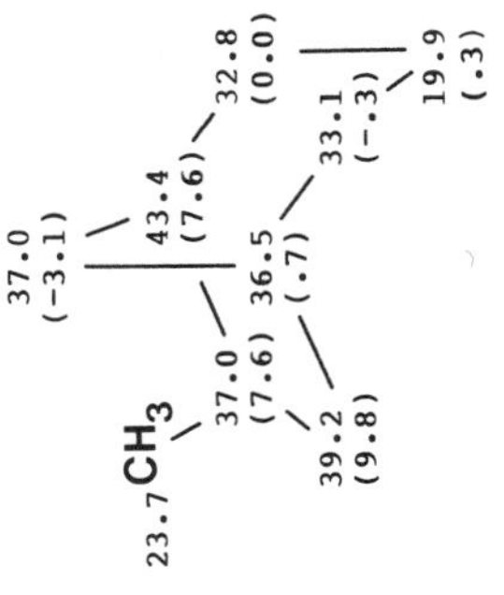

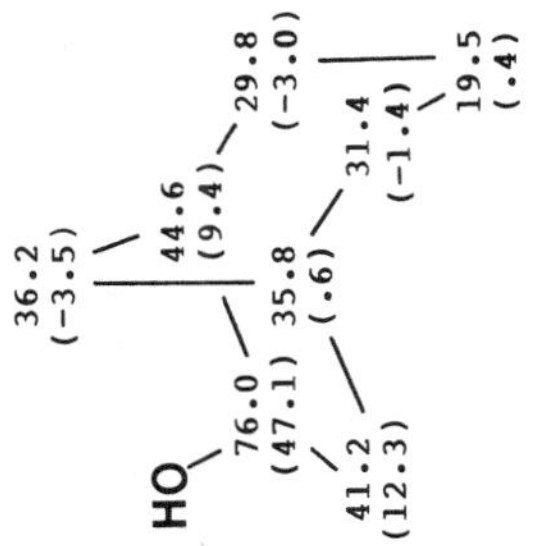

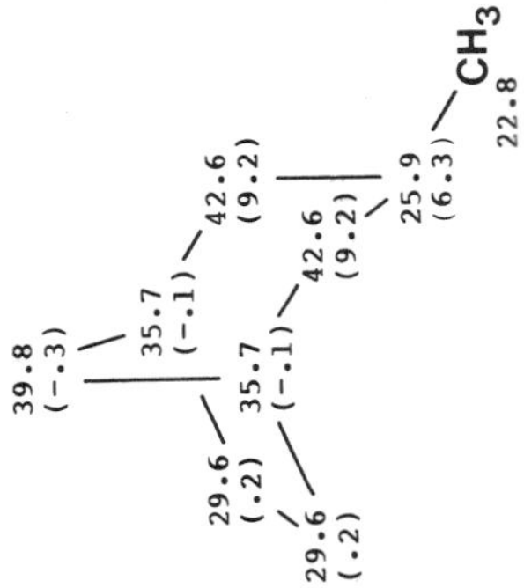

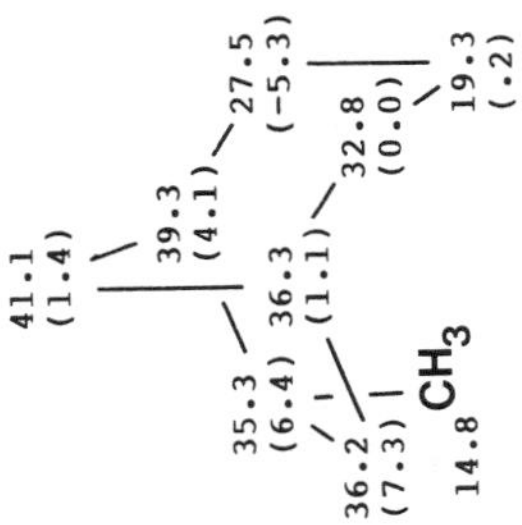

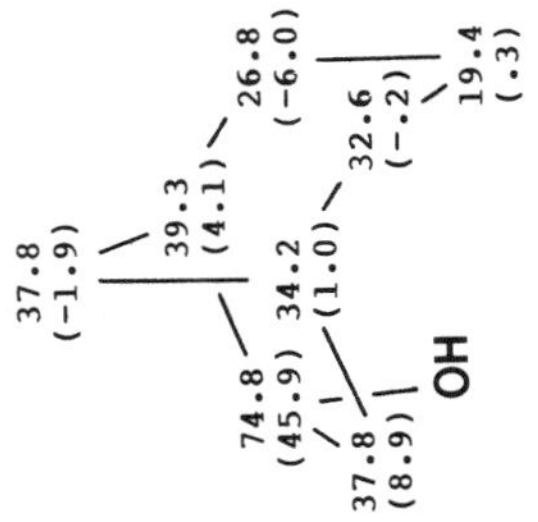

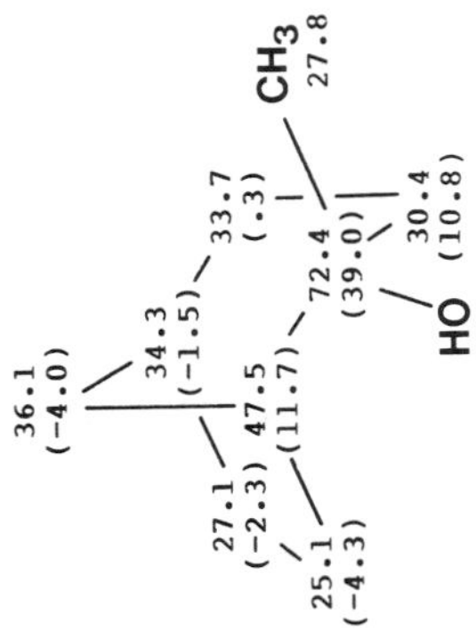

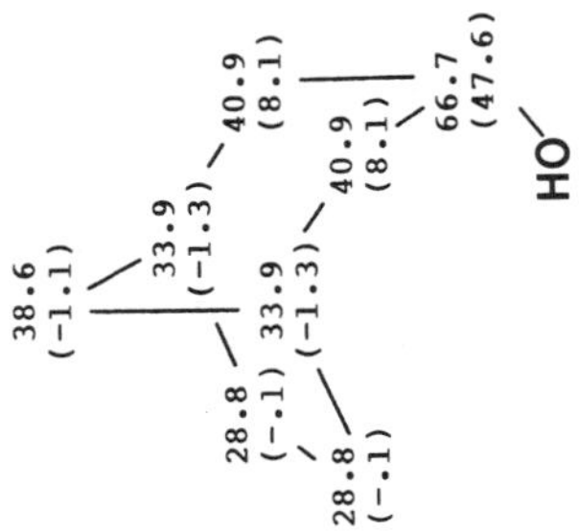

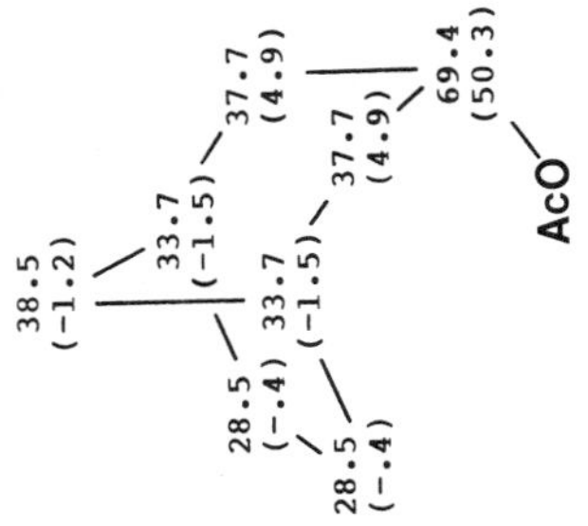

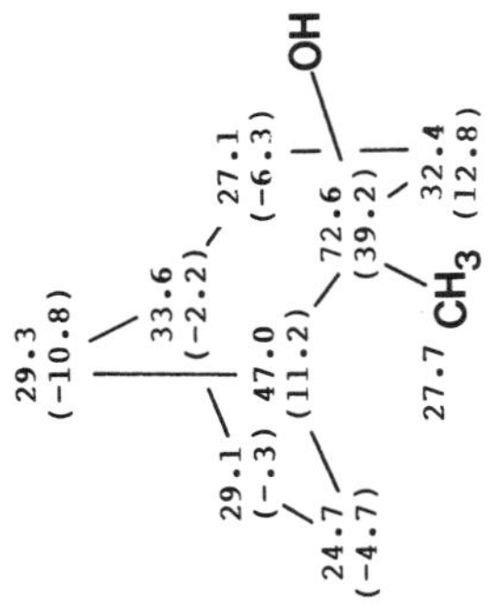

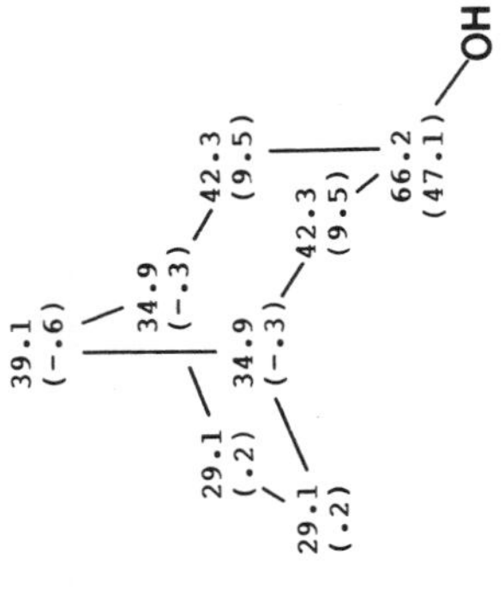

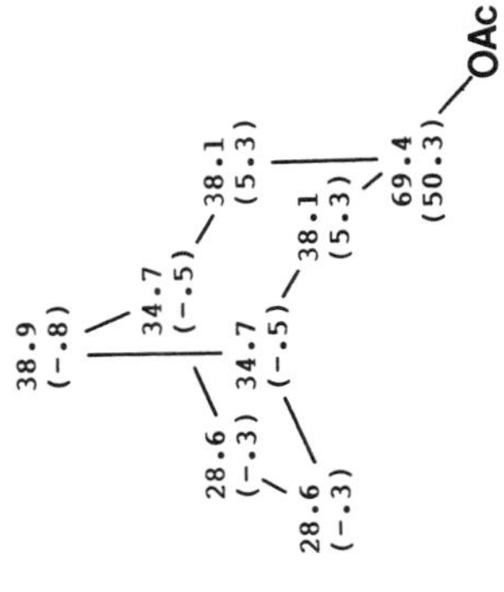

Row 1, structure 1:
35.7 (–4.4)
34.5 (–1.3)
40.3 (6.9)
24.3 (4.7)
CH₃ 24.5
31.1 (1.7)
34.5 (–1.3)
40.3 (6.9)
31.1 (1.7)

Row 1, structure 2:
37.0 (–3.1)
43.4 (7.6)
32.8 (0.0)
37.0 (7.6)
36.5 (–.7)
33.1 (–.3)
19.9 (.3)
23.7 CH₃
39.2 (9.8)

Row 1, structure 3:
36.2 (–3.5)
44.6 (9.4)
29.8 (–3.0)
35.8 (.6)
31.4 (–1.4)
19.5 (.4)
HO
76.0 (47.1)
41.2 (12.3)

Row 2, structure 1:
39.8 (–.3)
35.7 (–.1)
42.6 (9.2)
25.9 (6.3)
CH₃ 22.8
29.6 (.2)
35.7 (–.1)
42.6 (9.2)
29.6 (.2)

Row 2, structure 2:
41.1 (1.4)
39.3 (4.1)
27.5 (–5.3)
36.3 (1.1)
32.8 (0.0)
19.3 (.2)
35.3 (6.4)
36.2 (7.3)
14.8 CH₃

Row 2, structure 3:
37.8 (–1.9)
39.3 (4.1)
26.8 (–6.0)
34.2 (1.0)
32.6 (–.2)
19.4 (.3)
74.8 (45.9)
37.8 (8.9)
OH

Row 3, structure 1:
36.1 (–4.0)
34.3 (–1.5)
33.7 (.3)
CH₃ 27.8
72.4 (39.0)
30.4 (10.8)
27.1 (–2.3)
47.5 (11.7)
25.1 (–4.3)
HO

Row 3, structure 2:
38.6 (1.1)
33.9 (–1.3)
40.9 (8.1)
66.7 (47.6)
28.8 (–.1)
33.9 (–1.3)
40.9 (8.1)
28.8 (–.1)
HO

Row 3, structure 3:
38.5 (1.2)
33.7 (–1.5)
37.7 (4.9)
69.4 (50.3)
28.5 (–.4)
33.7 (–1.5)
37.7 (4.9)
28.5 (–.4)
AcO

Row 4, structure 1:
OH
29.3 (–10.8)
33.6 (–2.2)
27.1 (–6.3)
72.6 (39.2)
32.4 (12.8)
29.1 (–.3)
47.0 (11.2)
27.7 CH₃
24.7 (–4.7)

Row 4, structure 2:
39.1 (–.6)
34.9 (–.3)
42.3 (9.5)
66.2 (47.1)
29.1 (.2)
34.9 (–.3)
42.3 (9.5)
29.1 (.2)
OH

Row 4, structure 3:
38.9 (–.8)
34.7 (–.5)
38.1 (5.3)
69.4 (50.3)
28.6 (–.3)
34.7 (–.5)
38.1 (5.3)
28.6 (–.3)
OAc

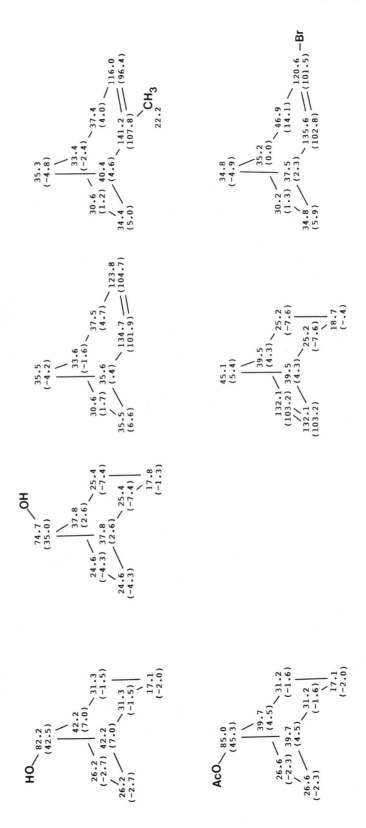

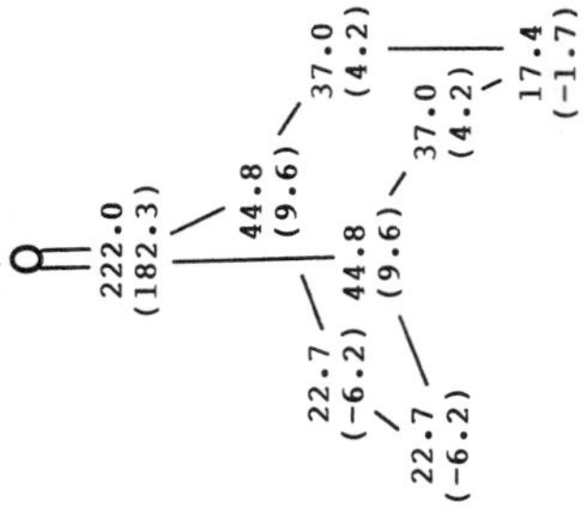

222.0
(182.3)
44.8
(9.6)
37.0
(4.2)
17.4
(−1.7)
37.0
(4.2)
44.8
(9.6)
22.7
(−6.2)
22.7
(−6.2)

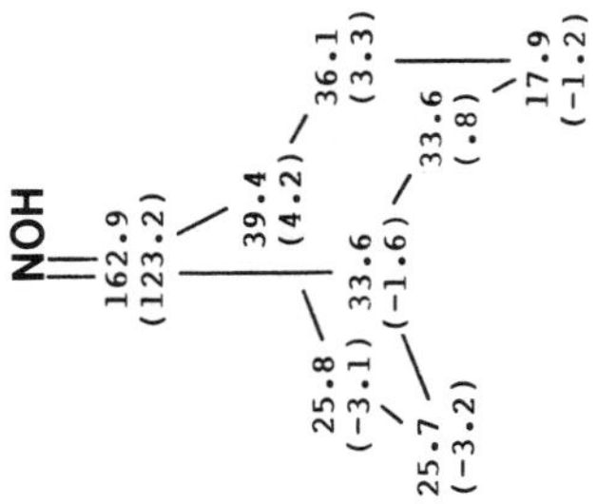

NOH
162.9
(123.2)
39.4
(4.2)
36.1
(3.3)
33.6
(.8)
17.9
(−1.2)
33.6
(−1.6)
25.8
(−3.1)
25.7
(−3.2)

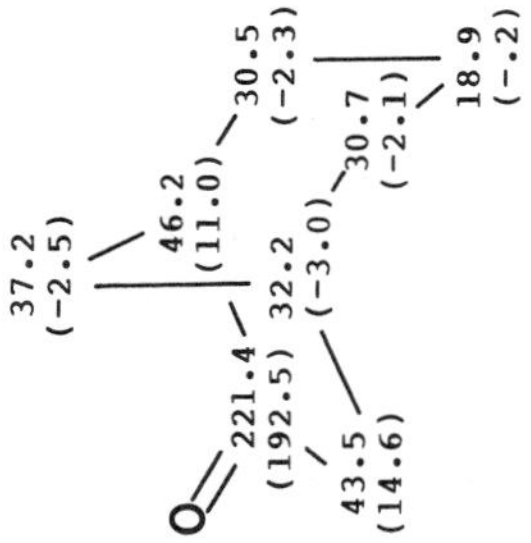

37.2
(−2.5)
46.2
(11.0)
30.5
(−2.3)
18.9
(−.2)
30.7
(−2.1)
32.2
(−3.0)
221.4
(192.5)
43.5
(14.6)

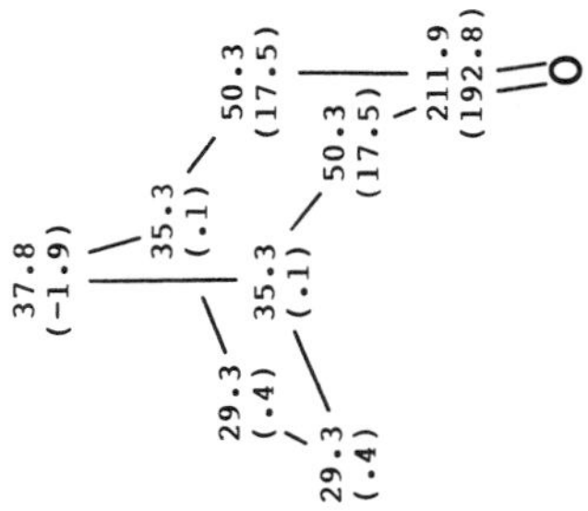

37.8
(−1.9)
35.3
(.1)
50.3
(17.5)
211.9
(192.8)
50.3
(17.5)
35.3
(.1)
29.3
(.4)
29.3
(.4)

NOH
39.3
(−.4)
34.9
(−.3)
38.6
(5.8)
159.0
(139.9)
33.1
(.3)
34.4
(−.8)
29.4
(.5)
28.5
(−.4)

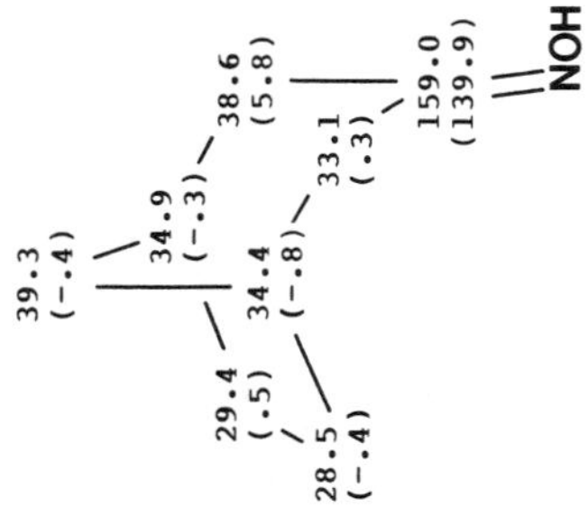

NNHTs
38.1
(−1.6)
34.9
(−.3)
42.6
(9.8)
160.4
(141.3)
35.2
(2.4)
34.4
(−.8)
29.4
(.5)
28.3
(−.6)

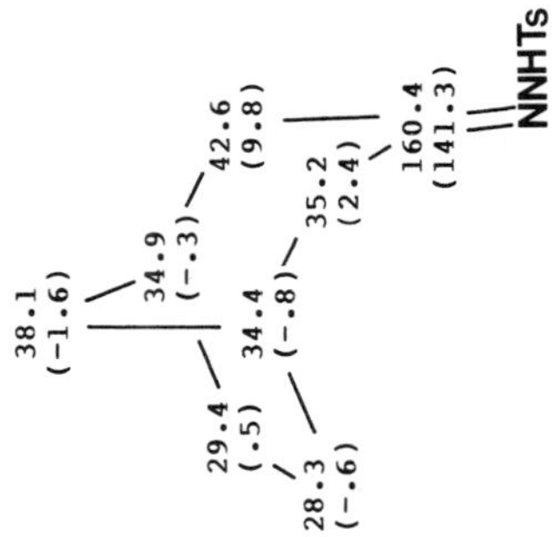

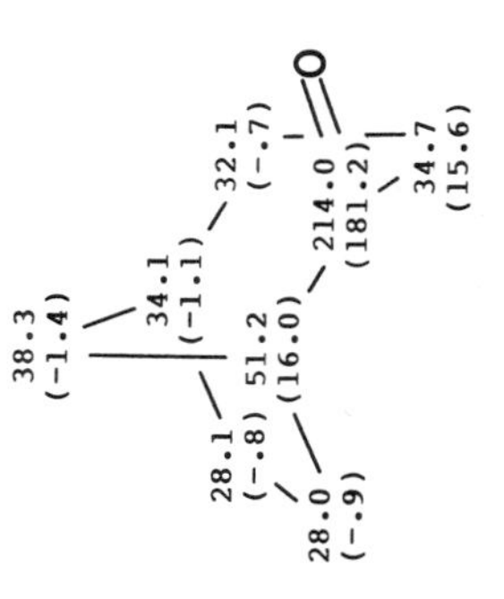

38.3
(−1.4)
34.1
(−1.1)
32.1
(−.7)
214.0
(181.2)
34.7
(15.6)
51.2
(16.0)
28.1
(−.8)
28.0
(−.9)

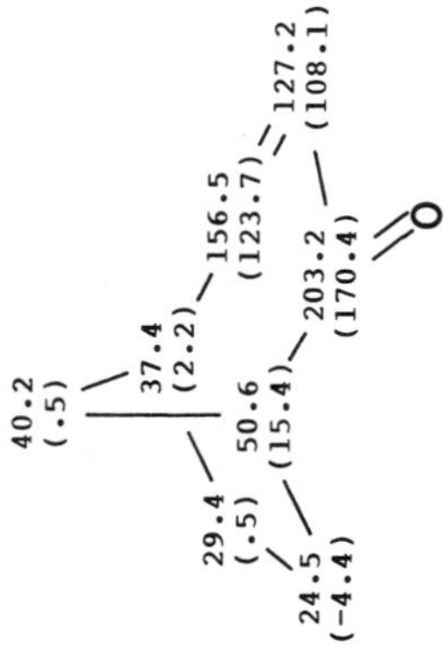

40.2
(.5)
37.4
(2.2)
156.5
(123.7)
127.2
(108.1)
203.2
(170.4)
50.6
(15.4)
29.4
(.5)
24.5
(−4.4)

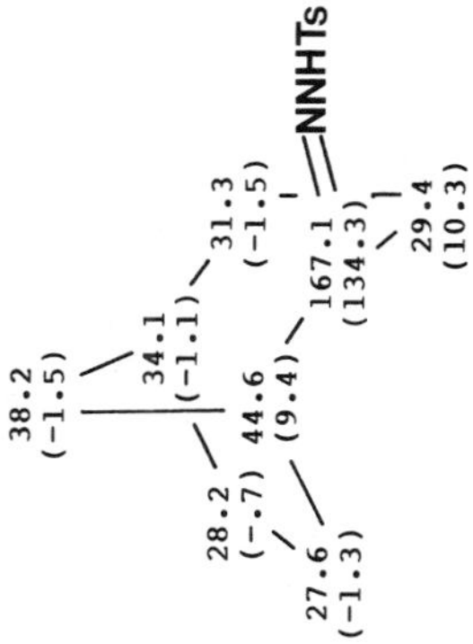

NNHTs
38.2
(−1.5)
34.1
(−1.1)
31.3
(−1.5)
167.1
(134.3)
29.4
(10.3)
44.6
(9.4)
28.2
(−.7)
27.6
(−1.3)

BICYCLO[3.2.1]OCTANES

Str	Substituents	C1	C2	C3	C4	C5	C6	C7	C8	Substituent Shifts	References
1	PARENT	35.2	32.8	19.1	32.8	35.2	28.9	28.9	39.7		1,3,4,15
2	1-Et	43.7	36.7	20.0	32.5	35.9	29.4	33.9	44.4	34.1	4,15
3	1-Et/2==0	57.0	210.4	35.3	32.7	35.4	29.2	33.4	43.3	27.1	4,15
4	1-Et/X2-OAc	47.1	73.6	25.0	28.3	35.4	29.0	32.1	39.3	29.5	4,15
5	1-Me	40.4	40.4	20.5	32.5	36.6	30.2	36.6	46.9	27.8	4,15
6	1-Me/2==0	53.2	211.3	35.2	33.0	35.9	29.6	36.1	45.8	20.0	4,15
7	1-Me/2==0 3=	52.7	202.2	126.8	155.3	38.4	30.2	32.1	47.2	20.0	16
8	1-Me/2=	40.4	139.4	123.3	37.3	34.8	31.8	43.4	43.1	24.4	16
9	1-Me/3==0/A8-Me S8-Me	44.5	53.9	212.6	47.0	45.8	27.4	36.2	43.1	18.8,23.6,20.8	34
10	1-Me/5-Me/S8-OH	41.4	31.0	19.9	31.0	41.4	32.4	32.4	83.1	25.5,25.5	41
11	1-Me/5-Me/S8-OTs	41.4	31.4	19.6	31.4	41.4	31.9	31.9	92.3	25.0,25.0	41
12	1-Me/X2-OH	44.6	75.0	27.7	27.4	36.4	28.4	31.9	38.8	24.3	4,15
13	1-Me/N2-OH	45.4	76.4	29.2	31.6	35.1	29.7	30.1	45.4	23.5	4,15
14	1-Me/X2-OH/3=	43.1	75.0	126.9	135.8	37.5	33.6	37.9	37.2	22.5	16
15	1-Me/N2-OH/3=	45.1	78.0	128.6	134.3	36.4	32.8	32.5	45.2	23.8	16
16	1-Me/N3-Me	39.8	47.1	25.4	39.2	36.2	31.5	37.5	44.3	28.6,24.7	26
17	1-Me/X7-Me	41.7	40.4	20.1	31.7	36.9	40.4	40.4	43.6	23.5,19.7	27
18	1-Me/N7-Me	41.7	34.2	19.6	32.1	37.2	42.0	34.6	48.3	26.3,12.5	27
19	1-Me/X2-OAc	43.6	77.0	25.0	28.7	35.9	28.5	35.5	40.1	24.0	4,15
20	1-Me/N2-OAc	44.1	78.0	26.0	31.3	35.3	29.7	32.5	45.4	23.1	4,15
21	1-OH/5-Me/6==0	73.4	37.8	20.6	36.7	52.5	218.3	50.8	50.8	19.7	56
22	1-OMe/5-Me/6==0	78.1	32.6	20.3	36.9	51.8	217.5	47.6	47.6	19.8	56
23	2-=NNHTs[anti]	44.6	167.1	29.4	31.3	34.1	28.2	27.6	38.2		5
24	2-=0	51.2	214.0	34.7	32.1	34.1	28.1	28.0	38.3		1,3,4,15,65
25	2-=0 3=	50.6	203.2	127.2	156.5	37.4	29.4	24.5	40.2		1
26	2-=0 3= 3-Br/6=	56.8	206.4	118.2	154.6	44.4	143.2	131.7	51.4		1
27	2-=0 3=/5-Me	50.6	200.7	126.1	159.8	43.1	37.0	25.8	46.2	23.5	16
28	2-=0 3=/6=	57.1	198.8	123.6	154.9	42.1	143.3	131.8	52.0		1
29	2-=0/5-Me	51.7	211.0	35.3	39.3	40.0	34.8	28.8	44.9	26.1	16
30	2-=0/6=	55.9	210.0	34.5	25.2	38.7	137.8	131.7	42.1		1
31	2-=0/N4-Me	50.3	213.8	43.4	37.2	40.4	22.4	28.1	38.6	19.4	57
32	2-=0/X3-Me	51.1	214.7	38.3	42.5	34.8	28.4	27.9	39.4	14.5	3,4
33	2-=0/X3-Me N3-Me	50.9	218.2	42.1	46.0	35.1	28.7	28.1	35.7	31.2,30.6	3
34	2-=0/X7-Me	59.2	210.3	35.7	31.8	35.2	38.4	35.7	35.7	22.8	4
35	X2-Br/3-==0	43.1	55.8	203.6	45.2	35.3	28.2	27.9	33.4		9
36	X2-Br/3-==0/N4-Br	43.2	54.3	195.8	58.6	46.3	24.3	27.9	35.3		9
37	X2-Br/3-==0/X4-Br	42.7	50.2	200.4	50.2	42.7	27.9	27.9	28.8		9
38	X2-Br/3-==0/N4-F	43.2	54.0	198.5	92.1	42.0	22.1	27.7	32.7		9
39	X2-Br/3-==0/X4-F	43.0	52.7	197.8	94.8	39.7	22.2	27.7	27.6		9
40	X2-Br/X3-OH/N4-Br	43.0	63.6	71.0	61.7	43.7	24.8	28.6	34.7		23
41	X2-Br/X3-OH/X4-Br	42.8	62.3	63.8	62.3	42.8					23
42	X2-CH2CN/3=/6= 6-Br/S8-OMs	43.8	32.8	128.7	126.3	47.1	129.0	128.0	76.2	20.2	28
43	X2-CH2CN/3=/6=/S8-OMs	42.3	33.5	129.5	127.5	40.3	139.0	128.1	78.4	21.2	28
44	X2-Cl N2-Me/5-Me/6-==0/8-==0	58.4	75.7	35.0	37.7	57.3	209.2	42.9	209.9	30.8,11.8	38
45	X2-Cl/8-==0	51.0	65.7	26.4	30.8	43.6	20.8	22.3	214.6		18
46	N2-Cl/8-==0	53.2	61.8	28.5	31.5	43.2	21.9	18.4	215.7		18
47	X2-Et N2-PhOMe/5-Me/6-==0/8-==0	54.3	52.4	26.8	39.0	58.9	211.4	42.4	215.1	30.8,136.4,11.8	38
48	X2-F/3-==0	40.5	95.2	205.9	47.9	35.5	27.9	22.4	31.8		9
49	N2-F/3-==0	43.2	98.6	206.2	48.6	36.8	28.8	22.2	36.5		9
50	X2-Me	41.8	35.0	25.2	30.8	35.7	28.9	27.7	32.9	19.1	4

BICYCLO[3.2.1]OCTANES (cont.)

Str	Substituents	C1	C2	C3	C4	C5	C6	C7	C8	Substituent Shifts	References
51	N2-Me	42.0	36.9	28.4	32.9	35.0	29.3	24.2	40.8	20.8	4
52	X2-Me N2-Me/3-=O	47.7	49.7		47.1	35.9	28.4	25.4	34.1	26.4,22.9	3
53	X2-Me N2-OMe/5-Me/6-=O/8-=O	54.2	80.9	31.7	35.8	58.3	210.8	40.6	213.4	20.6,11.6	38
54	X2-Me/3-=O	41.6	53.1	218.1	47.7	35.3	28.5	30.1	32.1	17.6	38
55	N2-Me/3-=O	42.7	51.4	213.0	49.8	36.6	29.2	24.3	39.8	12.8	3
56	X2-OAc	38.7	74.0	23.8	27.1	34.1	28.9	26.8	32.9		7,4,15
57	N2-OAc	39.7	75.2	24.7	30.6	33.9	28.4	24.2	37.2		7,4,15
58	X2-OAc/8-=O	47.4	79.5	22.6	32.0	44.0	21.3	20.1	216.8		18
59	N2-OAc/8-=O	49.7	74.8	23.3	28.7	43.5	22.3	18.2	215.7		18
60	N2-OAc/X3-Me	39.9	80.4	30.8	40.4	35.0	29.1	24.6	37.5		4,25
61	N2-OAc/X7-Me	47.3	75.2	25.0	30.1	35.0	38.4	31.0	34.4	18.7	25
62	X2-OH	41.7	71.3	26.9	26.5	34.3	28.4	26.8	32.1	23.1	7,4,15
63	N2-OH	42.7	72.5	28.3	30.7	33.7	28.5	23.3	37.3		7,4,15
64	X2-OH N2-Me	47.0	72.6	32.4	27.1	33.6	29.1	24.7	29.3	27.7	4
65	N2-OH X2-Me	47.5	72.4	30.4	33.7	34.3	27.1	25.1	36.1	27.8	4
66	X2-OH N2-Me/5-Me/6-=O/8-=O	57.5	80.4	32.2	38.5	58.1	211.0	42.7	213.9	28.3,11.8	38
67	X2-OH/3=	40.5	72.1	125.2	137.9	35.6	31.0	24.8	30.8		7,16,17
68	N2-OH/3=	41.8	73.7	127.1	135.9	35.2	32.5	21.3	37.5		7
69	X2-OH/3= 3-Br/6=	46.8	71.9	123.7	144.1	41.6	139.5	130.5	37.6		7
70	N2-OH/3= 3-Br/6=	47.3	71.9	124.8	143.6	41.6	138.7	130.5	44.3		1
71	X2-OH/3=/5-Me	41.1	70.9	125.3	141.1	40.5	38.9	26.3	37.9	23.9	16
72	N2-OH/3=/5-Me	42.1	72.5	127.2	139.0	40.3	40.4	28.3	44.3	28.3	16
73	X2-OH/3=/6=	46.0	66.2	126.3	144.5	39.5	138.2	130.3	37.9		1
74	N2-OH/3=/6=	47.2	67.1	127.1	144.1	38.9	137.1	129.8	44.7		1
75	X2-OH/5-Me	42.8	70.1	27.5	34.6	39.5	35.9	28.0	39.3	27.6	16
76	X2-OH/6=	46.4	66.1	27.4	21.9	39.2	134.8	132.9	37.3		7
77	N2-OH/6=	46.8	68.5	28.5	23.3	38.4	134.3	130.1	42.5		7
78	X2-OH/8-=O	51.4	78.4	24.9	32.4	44.5	21.5	20.8	219.3		18
79	N2-OH/8-=O	53.4	74.1	26.3	28.9	43.4	22.5	17.4	218.0		18
80	X2-OMe N2-Me/5-Me/6-=O/8-=O	54.2	84.2	29.7	37.8	57.9	210.9	42.3	212.0	21.7,11.8	38
81	X2-OtBu	42.1	70.4	27.2	27.4	34.2	29.1	26.6	32.8		8
82	N2-OtBu	42.7	72.2	28.4	31.2	34.1	28.5	24.0	37.8		8
83	X2-PhOMe N2-Et/5-Me/6-=O/8-=O	53.1	52.8	27.9	39.3	58.5	211.5	41.6	213.2	132.9,35.6,11.7	38
84	X2-PhOMe N2-Me/5-Me/6-=O/8-=O	54.7	49.2	29.6	39.1	57.9	211.4	41.9	213.0	135.9,31.2,11.7	38
85	N2-PhOMe X2-Me/5-Me/6-=O/8-=O	55.8	48.4	29.5	39.1	58.7	211.3	42.4	215.0	139.2,27.3,11.8	38
86	X2-PhOMe/5-Me/6-=O/8-=O	52.7	49.3	23.6	41.7	58.9	211.5	39.5	215.7	133.3,11.9	38
87	N2-PhOMe/5-Me/6-=O/8-=O	51.4	51.0	23.8	42.3	58.5	211.2	45.2	214.8	134.1,12.2	38
88	2=	35.6	134.7	123.8	37.5	33.6	30.6	35.5	35.5		38
89	2= 2-COOMe 3-OH/X4-Me N4-Me/A8-COOMe	48.5	103.4	177.5	42.6	36.8	23.0	31.8	49.5	172.1,27.9,23.3,174.0	1,4
90	2= 2-COOMe 3-OH/X4-Me N4-Me/S8-COOMe	48.6	102.1	175.7	40.5	34.4	26.1	32.3	49.3	172.4,27.4,25.3,173.4	24
91	2= 2-Et 3-Cl/X4-Me N4-Me/6=	44.6	141.4	132.9	40.3	51.9	131.9	139.9	39.5	26.9,28.4,23.1	24
92	2= 2-Me	40.4	141.2	116.0	37.4	33.4	30.6	34.4	35.3	22.2	37
93	2= 2-Me 3-Cl/X4-Me N4-Me/6=	46.5	136.0	133.3	40.6	51.9	132.2	139.4	39.3	19.9,28.3,23.2	4
94	2= 2-PhOMe/5-Me/6-=O/8-=O	48.5	144.6	119.2	45.9	57.1	211.2	50.1	213.3	130.7,12.4	37
95	2= 3-Br	37.5	135.6	120.6	46.9	35.2	30.2	34.8	34.8		38
96	2= 3-Br/6=	40.1	134.3	120.1	38.5	40.6	130.4	139.2	40.3		1
97	2= 3-Cl/X4-Me N4-Me/6=	40.1	129.9	139.7	41.2	51.5	131.3	140.7	39.8	27.9,22.4	37
98	2= 3-OEt/X4-Me N4-Me/6=	28.1	97.8	159.3	38.4	38.0	129.9	141.6	40.7	22.3,13.9	36
99	2=/4-=C=CMe2/6=	39.4	129.0	122.5	99.5	45.0	130.8	137.2	42.4	192.1	42
100	2=/5-Me	37.5	134.4	124.6	45.3	39.2	37.7	36.0	43.2	28.1	16

BICYCLO[3.2.1]OCTANES (cont.)

Str	Substituents	C1	C2	C3	C4	C5	C6	C7	C8	Substituent Shifts	References
101	2=/6==O/X7-Me N7-Me	43.2	133.9	124.6	33.2	45.2	225.5	54.1	30.4	24.3,20.9	2
102	2=/6=	38.7	134.1	123.8	28.7	38.3	130.2	139.7	40.7		1,43,44
103	2=/8==O	45.2	132.7	125.5	43.1	42.5	25.6	30.2	217.9		1
104	2=/A8-OH	42.3	132.6	124.2	37.5	40.7	28.3	32.6	78.1		7
105	2=/S8-OH	39.8	129.9	126.6	32.2	35.7	28.4	31.5	73.5		7
106	3==CH2/6=	39.4	37.1	146.0	37.1	39.4	133.6	133.6	43.6	112.3	31
107	3==NNHTs	34.4	35.2	160.4	42.6	34.9	29.4	28.3	38.1		5
108	3==NOH	34.4	33.1	159.0	38.6	34.9	29.4	28.5	39.3		6
109	3==NOH/6=	37.5	29.4	157.7	34.3	38.3	135.4	134.4	42.3		6
110	3==O	35.3	50.3	211.9	50.3	35.3	29.3	29.3	37.8		1,3,4,9
111	3=O/6=	38.3	46.4	210.0	46.4	38.3	135.3	135.3	41.8		1
112	3=O/6=/S8-CH2COOMe	41.3	40.3	209.3	40.3	41.3	137.1	137.1	43.4	33.0	47
113	X3-COPh/8==O	43.6	39.0	37.4	39.0	43.6	22.8	22.5	219.5	200.8	53
114	N3-COPh/8==O	43.0	37.8	37.4	37.8	43.0	22.5	22.5	221.3	202.6	53
115	X3-Me	35.7	42.6	25.9	42.6	35.7	29.6	29.6	39.8	22.8	4
116	N3-Me	34.5	40.3	24.3	40.3	34.5	31.1	31.1	35.7	24.5	4
117	X3-OAc	34.7	38.1	69.4	38.1	34.7	28.6	28.6	38.9		7,4
118	N3-OAc	33.7	37.7	69.4	37.7	33.7	28.5	28.5	38.5		7,4
119	N3-OCOPh	34.8	38.2	70.3	38.2	34.8	28.7	28.7	39.0	165.9	54
120	X3-OH	34.9	42.3	66.2	42.3	34.9	29.1	29.1	39.1		7,4,23
121	N3-OH	33.9	40.9	66.7	40.9	33.9	28.8	28.8	38.6		7,4,23
122	X3-OH/6=	39.1	35.2	67.1	35.2	39.1	132.6	132.6	44.9		7
123	N3-OH/6=	38.1	36.8	67.4	36.8	38.1	138.3	138.3	44.7		7
124	3=/6==O	30.2	31.9	128.0	126.0	46.7	212.4	43.7	34.3		1,2
125	3=/6=O/X7-Me N7-Me	41.8	29.9	128.4	126.9	47.3	215.9	48.2	30.0	26.3,23.0	2
126	3=/N6-OH	32.7	37.9	128.1	129.4	40.0	80.5	40.8	32.7		7
127	3=/X6-OH	33.3	36.4	125.9	130.2	44.0	78.0	42.7	31.4		7
128	X4-Cl/6==O	32.0	26.0	28.3	58.5	53.5	21.6	42.9	31.5		30
129	6==O	32.2	30.7	18.9	30.5	46.2	221.4	43.5	37.2		1,3,30
130	6==O/X7-Me	39.4	31.4	19.3	30.5	46.7	224.2	46.9	34.4	15.6	3
131	6==O/N7-Me	36.8	26.1	19.1	30.8	45.7	223.4	48.4	36.4	8.5	3
132	6==O/X7-Me N7-Me	43.1	28.1	19.2	31.2	45.8	225.4	48.1	34.4	25.2,17.9	2,3
133	X6-Me	36.5	33.1	19.9	32.8	43.4	225.4	37.0	37.0	23.7	4,19
134	N6-Me	36.3	32.8	19.3	27.5	39.3	35.3	36.2	41.1	14.8	19
135	X6-Me/X7-Me	45.4	32.8	20.1	32.8	45.4	39.2	39.2	34.4	17.2,17.2	14
136	N6-Me/X7-Me	46.6	27.2	13.4	19.9	38.5	41.0	43.4	32.7	17.1,22.4	14
137	X6-OH	35.8	31.4	19.5	29.8	44.6	76.0	41.2	36.2		7
138	N6-OH	34.2	32.6	19.4	26.8	39.3	74.8	37.8	37.8		7,30
139	6=	39.5	25.2	18.7	25.2	39.5	132.1	132.1	45.1		1
140	A8-OAc	39.7	31.2	17.1	31.2	39.7	26.6	26.6	85.0		7
141	A8-OH	42.2	31.3	17.1	31.3	42.2	26.2	26.2	82.2		7
142	8==NOH	33.6	33.6	17.9	36.1	39.4	25.8	25.7	162.9		6
143	8==O	44.8	37.0	17.4	37.0	44.8	22.7	22.7	222.0		1,18
144	S8-OH	37.8	25.4	17.8	25.4	37.8	24.6	24.6	74.7		7

References to Bicyclo[3.2.1]octanes and related* systems

1. Stothers, J. B., Swenson, J. R. and Tan, C. T., *Can. J. Chem.* **53**, 581 (1975).
2. Hudyma, D. M., Stothers, J. B. and Tan, C. T., *Org. Mag. Res.* **6**, 614 (1974).
3. Grover, S. H., Marr, D. H., Stothers, J. B. and Tan, C. T., *Can. J. Chem.* **53**, 1351, 2808 (1975).
4. Lippmaa, E., Pehk, T., Belikova, N. A., Bobyleva, A. A., Kalinichenko, A. N., Ordubadi, M. D. and Plate, A. F., *Org. Magn. Res.* **8**, 74 (1976).
5. Casanova, J. and Zahra, J.-P., *Tetrahedron Lett.* 1773 (1977).
6. Geneste, P., Durand, R., Kamenka, J.-M., Beierbeck, H., Martino, R. and Saunders, J. K., *Can. J. Chem.* **56**, 1940 (1978).
7. Stothers, J. B. and Tan, C. T., *Can. J. Chem.* **55**, 841 (1977).
8. Cheng, A. K. and Stothers, J. B., *Can. J. Chem.* **55**, 50 (1977).
9. Reisse, J., Piccinni-Leopardi, C., Zahra, J. P., Waegell, B. and Fournier, J., *Org. Magn. Reson.* **9**, 512 (1977).
*10. Stothers, J. B., Tan, C. T., Nickson, A., Huang, F., Sridhar, R. and Weglein, R., *J. Am. Chem. Soc.* **94**, 8581 (1972).
*11. Takaishi, N., Inamoto, Y. and Aigami, K., *J. Org. Chem.* **40**, 276 (1975).
*12. Takaishi, N., Inamoto, Y., Aigami, K. and Osawa, E., *J. Org. Chem.* **40**, 1483 (1975).
*13. Takaishi, N., Inamoto, Y. and Aigami, K., *J. Chem. Soc. Perkin Trans. I* 789 (1975).
14. Matveyeva, I. A., Sokolova, I. M., Pekhk, T. I. and Petrov, A. A., *Petroleum Chemistry* **15**, 160 (1975).
15. Kalinichenko, A. N., Bobyleva, A. A., Belikova, N. A., Plate, A. F., Pekhk, T. I. and Lippmaa, E. T., *J. Org. Chem. USSR* **10**, 1459 (1974).
16. Belikova, N. A., Bobyleva, A. A., Dzhigirkhanova, A. V., Pekhk, T. I., Lippmaa, E. T. and Plate, A. F., *J. Org. Chem. USSR* **13**, 1759 (1977).
17. Bobyleva, A. A., Dzhigirkhanova, A. V., Belikova, N. A., Pekhk, T. I., Lippmaa, E. T., Plate, A. F. and Gubarevich, T. M., *J. Org. Chem. USSR* **14**, 915 (1978).
18. Heumann, A. and Kolshorn, H., *J. Org. Chem.* **44**, 1575 (1979).
19. Jaggi, F. J., Buchs, P. and Ganter, C., *Helv. Chim. Acta* **63**, 872 (1980).
*20. Waddington, M. D. and Jennings, P. W., *Organometallics* **1**, 385 (1982).
*21. Johnson, T. H. and Cheng, S.-S., *Synth. Commun.* **10**, 381 (1980).
*22. Casanova, J., Waegell, B., Koukoua, G. and Toure, V., *J. Org. Chem.* **44**, 3976 (1979).
23. Zahra, J. P., Waegell, B., Faure, R. and Vincent, E. J., *Org. Magn. Reson.* **18**, 185 (1982).
24. Yates, P. and Stevens, K. E., *Tetrahedron* **37**, 4401 (1981).
25. Bobyleva, A. A., Belikova, N. A., Kalinichenko, A. N., Baryshnikov, A. T., Dubitskaya, N. F., Pekhk, T. I., Lippmaa, E. T. and Plate, A. F., *J. Org. Chem. USSR* **16**, 1397 (1980).
26. Matveeva, I. A., Sokolova, I. M., Pekhk, T. I. and Petrov, A. A., *Neftekhimiya* **15**, 17 (1975).
27. Matveeva, I. A., Sokolova, I. M., Pekhk, T. I. and Petrov, A. A., *Neftekhimiya* **15**, 646 (1975).
28. Mehta, G. and Srikrishna, A., *J. Org. Chem.* **46**, 1730 (1981).
29. Takaishi, N., Inamoto, Y., Tsuchihashi, K., Aigami, K. and Fujikura, Y., *J. Org. Chem.* **41**, 771 (1976).
30. Berger, S., *J. Org. Chem.* **43**, 209 (1978).
31. Garratt, P. J. and White, J. F., *J. Org. Chem.* **42**, 1733 (1977).
*32. Wiseman, J. R., Vanderbilt, J. J. and Butler, W. M., *J. Org. Chem.* **45**, 667 (1980).

*33. Tobe, Y., Terashima, K., Sakai, Y. and Odaira, Y., *J. Am. Chem. Soc.* **103**, 2307 (1981).

34. Crist, B. V., Rodgers, S. L. and Lightner, D. A., *J. Am. Chem. Soc.* **104**, 6040 (1982).

*35. Yates, P. and Langford, G. E., *Can. J. Chem.* **59**, 344 (1981).

36. Hoffmann, H. M. R. and Matthei, J., *Chem. Ber.* **113**, 3837 (1980).

37. Mayr, H. and Halberstadt-Kausch, *Chem. Ber.* **115**, 3479 (1982).

38. Kasturi, T. R., Reddy, S. M. and Murthy, P. S., *Org. Magn. Reson.* **20**, 42 (1982).

*39. Dawson, B. A. and Stothers, J. B., *Org. Magn. Reson.* **21**, 217 (1983).

*40. Jefford, C. W., Mareda, J., Gehret, J.-C. E., Kabengele, nT., Graham, W. D. and Burger, U., *J. Am. Chem. Soc.* **98**, 2585 (1976).

41. Whitesell, J. K., Matthews, R. S. and Solomon, P. A., *Tetrahedron Lett.* 1549 (1976).

42. Aue, D. H. and Meshishnek, M. J., *J. Am. Chem. Soc.* **99**, 223 (1977).

43. Kohler, F. H. and Hertkorn, N., *Chem. Ber.* **116**, 3274 (1983).

44. Christl, M., Leininger, H. and Bruckner, D., *J. Am. Chem. Soc.* **105**, 4843 (1983).

*45. Adam, W., De Lucchi, O., Peters, K., Peters, E.-M. and von Schnering, H. G., *J. Am. Chem. Soc.* **104**, 161 (1982).

*46. Bly, R. S., Bly, R. K. and Shibata, T., *J. Org. Chem.* **48**, 101 (1983).

47. Mander, L. N. and Wilshire, C., *Aust. J. Chem.* **32**, 1975 (1979).

*48. Inamoto, Y., Aigami, K., Fujikura, Y., Ohsugi, M., Takaishi, N., Tsuchihashi, K. and Ikeda, H., *Chem. Lett.* 25 (1978).

49. Heumann, A. and Kolshorn, H., *Tetrahedron* **31**, 1571 (1975).

*50. Paquettte, L. A., Jendralla, H., Jelich, K., Korp, J. D. and Bernal, I. *J. Am. Chem. Soc.* **106**, 433 (1984).

*51. Vorobeva, N. S., Arefev, O. A., Pekhk, T. I., Devisov, Y. V. and Petrov, A. A., *Neftekhimiya* **15**, 659 (1975).

*52. Takaishi, N., Inamoto, Y., Tsuchikashi, K., Yashima, K. and Aigami, K. *J. Org. Chem.* **40**, 2929 (1975).

53. Momose, T. and Muraoka, O., *Chem. Pharm. Bull.* **26**, 2217 (1978).

54. Momose, T. and Muraoka, O., *Chem. Pharm. Bull.* **26**, 2589 (1978).

*55. Cory, R. M., Burton, L. P. J. and Pecherle, R. G., *Synth. Commun.* **9**, 735 (1979)

56. House, H. O., Haack, J. L., McDaniel, W. C. and VanDerveer, D. *J. Org. Chem.* **48**, 1643 (1983).

57. Patel, V. and Stothers, J. B., *Can. J. Chem.* **58**, 2728 (1980).

*58. Ragauskas, A. J. and Stothers, J. B., *Can. J. Chem.* **61**, 2254 (1983).

*59. Trost, B. M. and Latimer, L. H., *J. Org. Chem.* **43**, 1031 (1978).

*60. McMurry, J. E. and Andrus, A., *Tetrahderon Lett.* **21**, 4687 (1980).

*61. Oppolzer, W. and Burford, S. C., *Helv. Chim. Acta* **63**, 788 (1980).

*62. Wenkert, E., Buckwalter, B. L., Burfitt, I. R., Gasic, M. J., Gottlieb, H. E., Hagaman, E. W., Schell, F. M. and Wovkulich, P. M., *Topics in Carbon-13 NMR Spectroscopy*, G.C. Levy, Ed., 1976, vol. 2, 81ff.

*63. Johnson, L. F. and Jankowski, W. C., *Carbon-13 NMR Spectra*, Wiley-Interscience, New York, 1972, spectrum 466.

*64. Thomas, A. F. and Ozainne, M. *Helv. Chim. Acta* **62**, 361 (1979).

65. *Sadtler Standard Carbon-13 NMR Spectra*, Sadtler Research Laboratories, Inc., Philadelphia, Pa.

Chapter 11

Bicyclo[3.2.2]nonanes

The problem of nomenclature is avoided here because, as a result of the scarcity of data, all data can be displayed on three-dimensional representations. Except in obvious cases, conformational biases should not be assumed from the drawings.

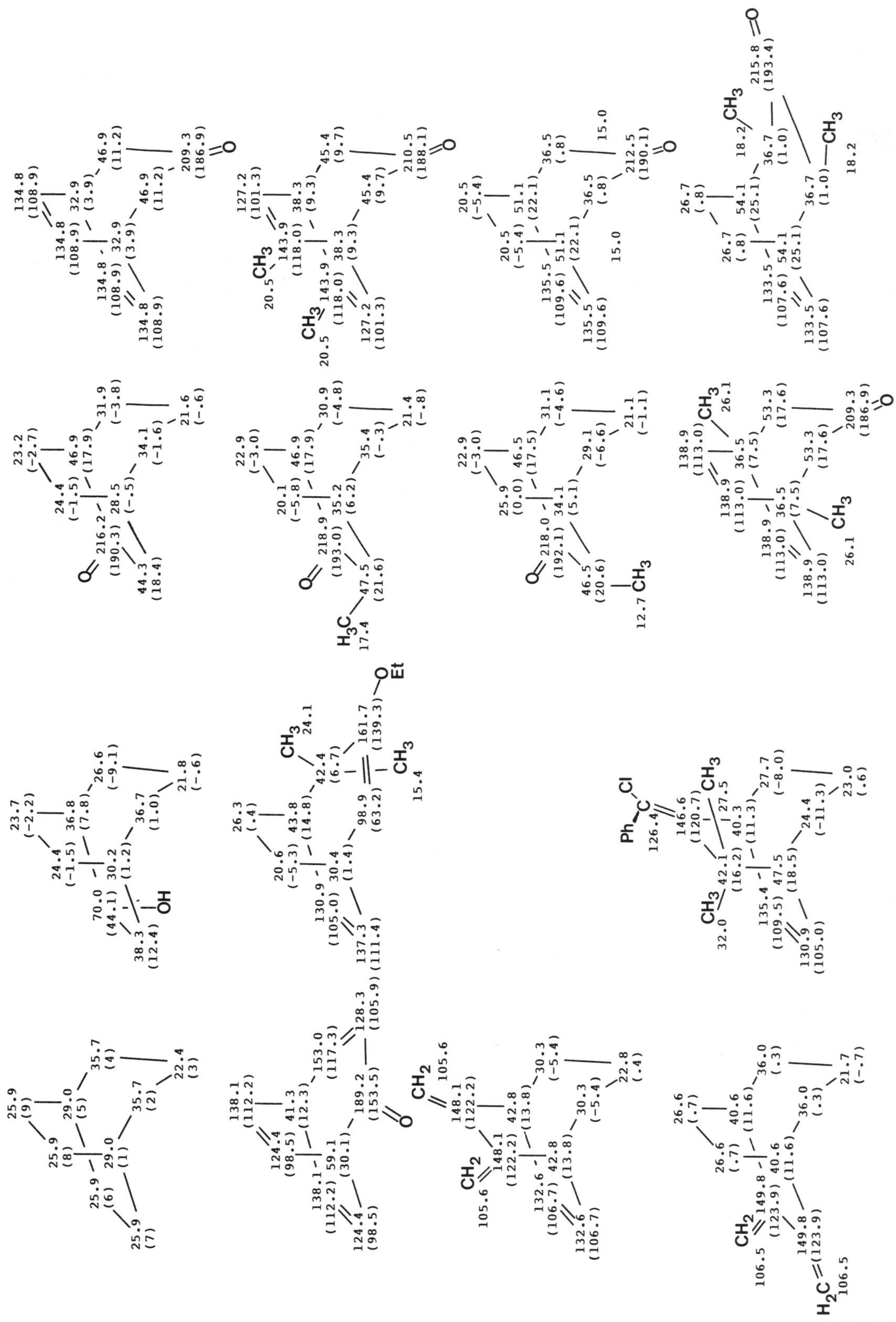

References

1. Grover, S. H., Marr, D. H., Stothers, J. B. and Tan, C. T., *Can. J. Chem.* **53**, 1351 (1975).
2. Stothers, J. B. and Tan, C. T., *Can. J. Chem.* **55**, 841 (1977).
3. Hoffmann, H. M. R. and Matthei, J., *Chem. Ber.* **113**, 3837 (1980).
4. Mayr, H., Schutz, F. and Halberstadt-Kausch, I. K., *Chem. Ber.* **115**, 3516 (1982).
5. Hill, A. E. and Hoffmann, H. M. R., *J. Am. Chem. Soc.* **96**, 4597 (1974).
6. Narula, A. S., *Tetrahedron Lett.* 1921 (1979).
7. Pfeffer, H. U. and Klessinger, M., *Org. Magn. Reson.* **9**, 121 (1977).
8. Engdahl, C., Jonsall, G. and Ahlberg, P., *J. Am. Chem. Soc.* **105**, 891 (1983).

Chapter 12

Bicyclo[3.3.0]octanes

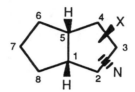

Stereochemistry for the cis-fused series in this system is readily specified by exo and endo. While there are two, relatively conformationally mobile rings present, this sytem is especially well behaved in terms of shift additivity. No significant quantity of data was found for trans-fused bicyclo[3.3.0]octanes.

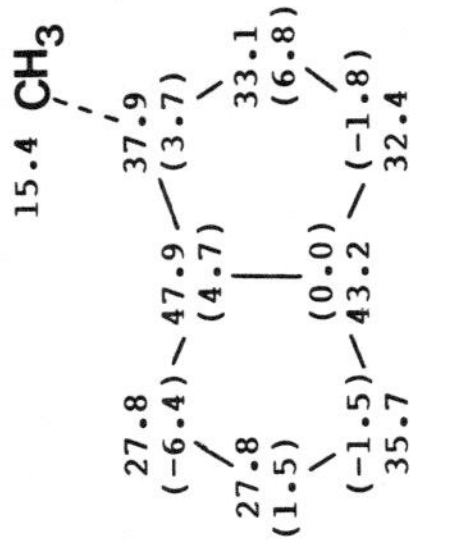
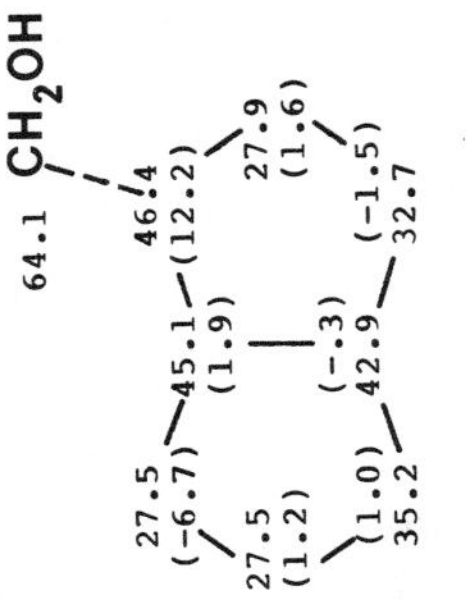
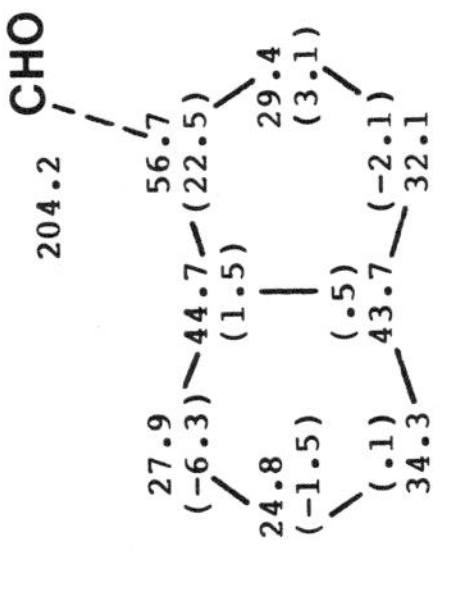
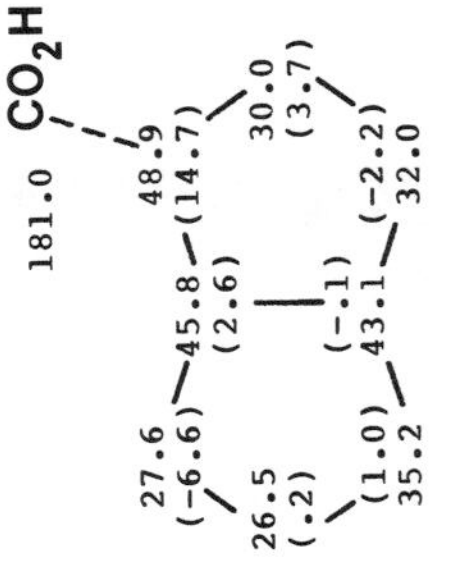
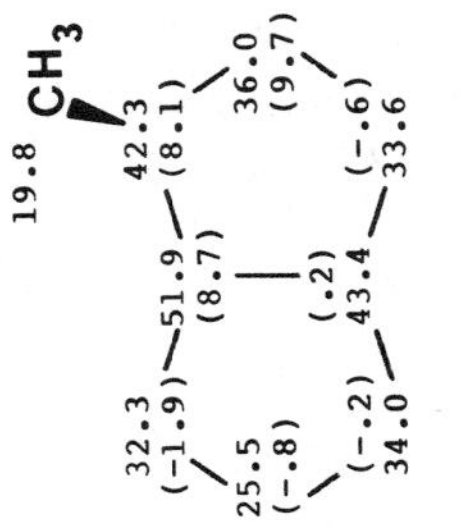
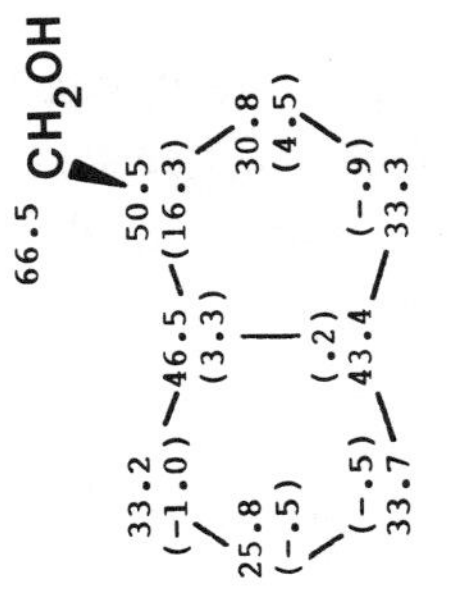
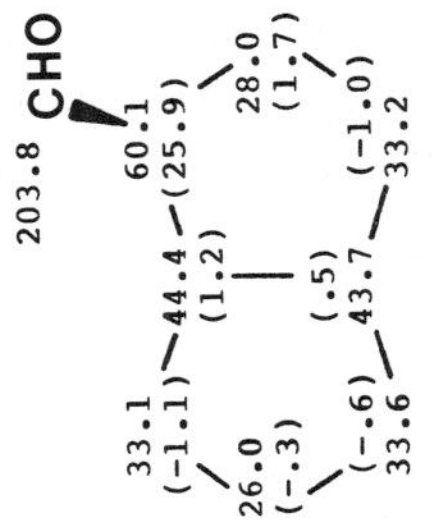
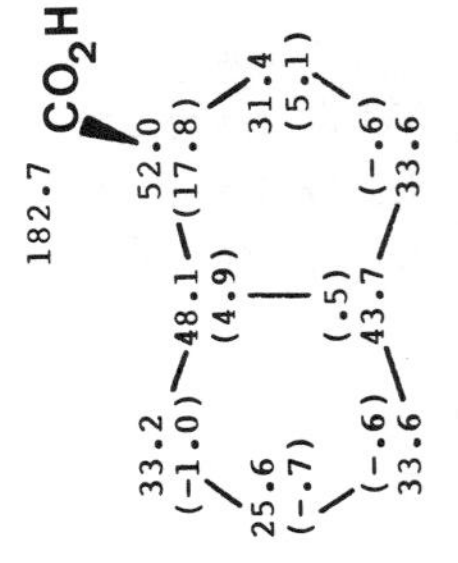
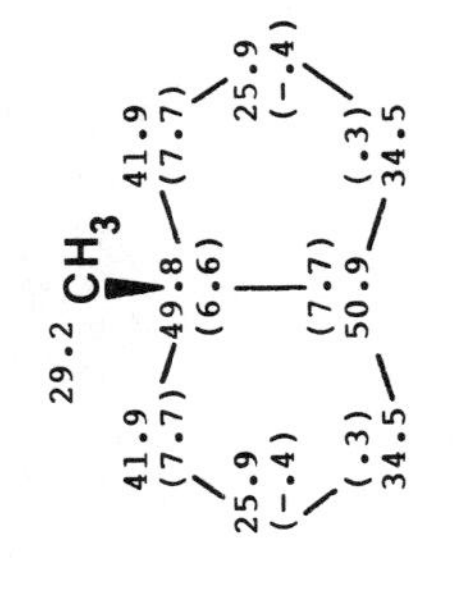
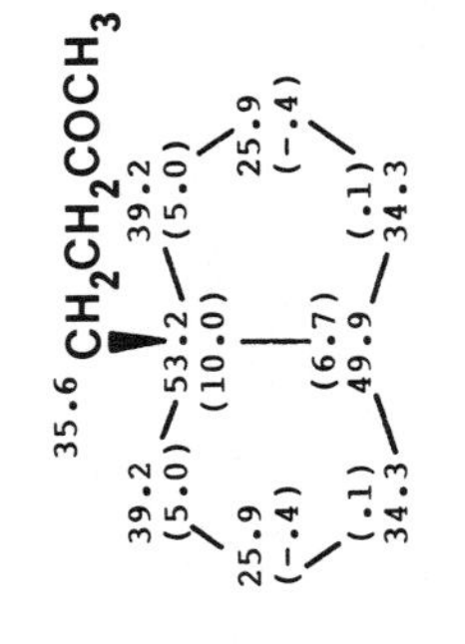
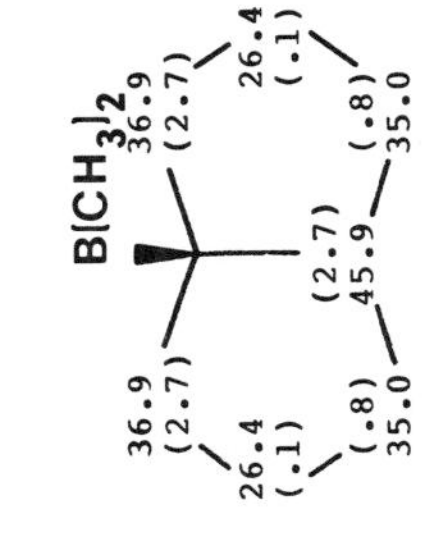
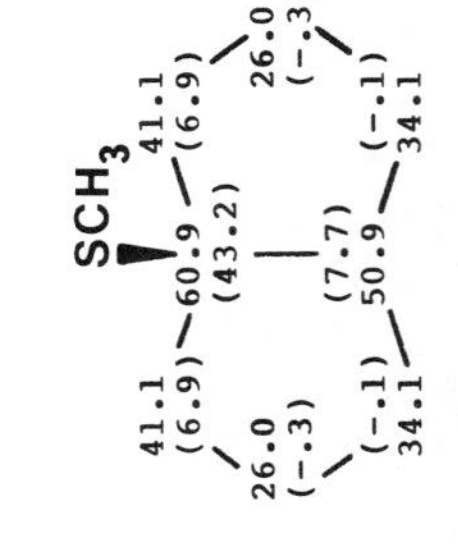
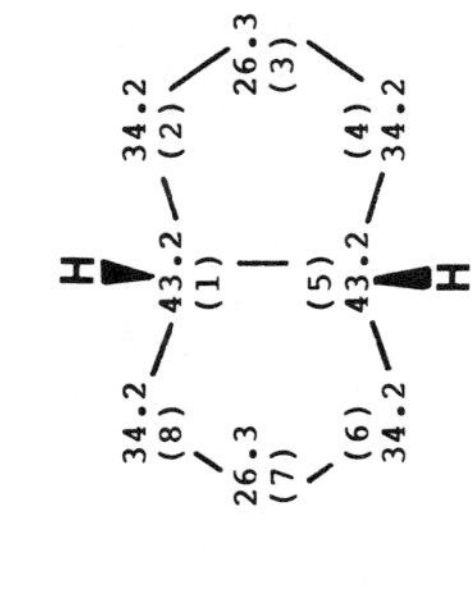
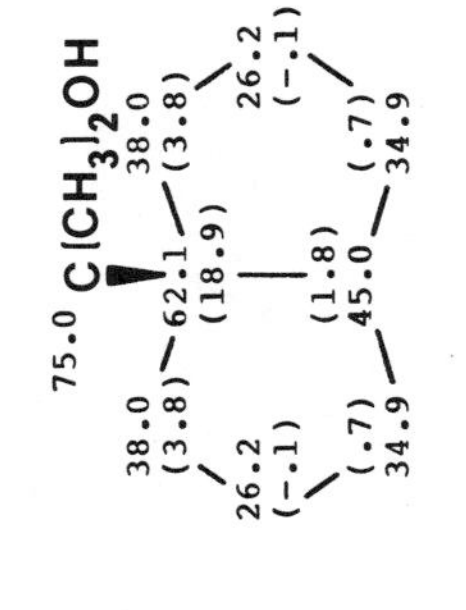
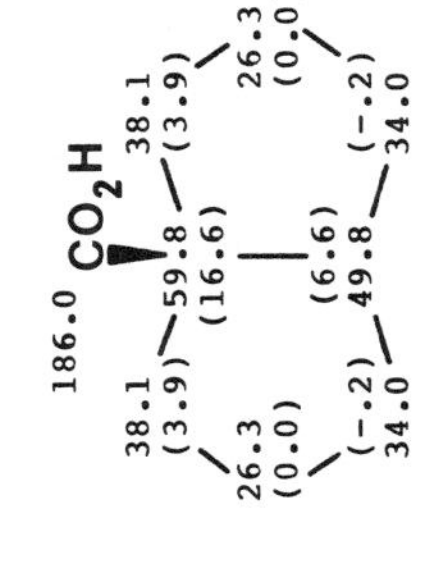
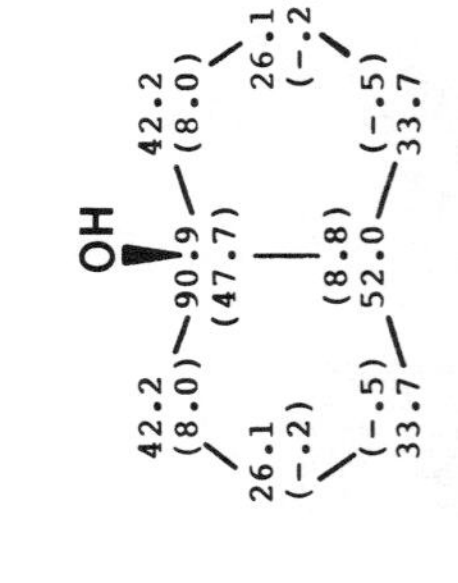

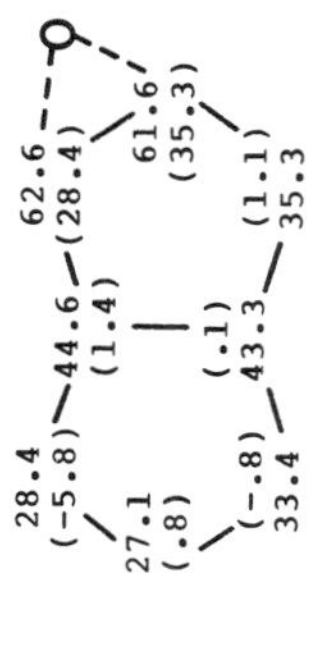

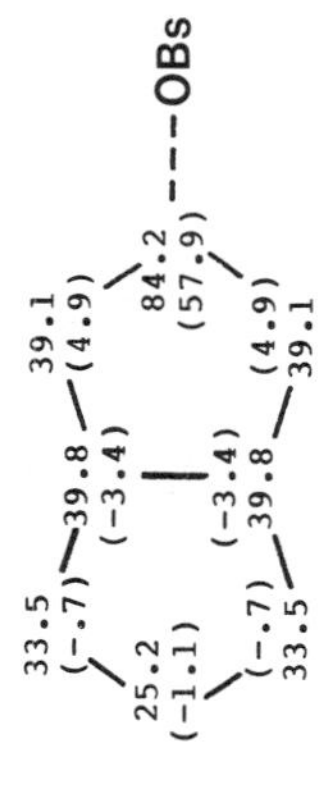

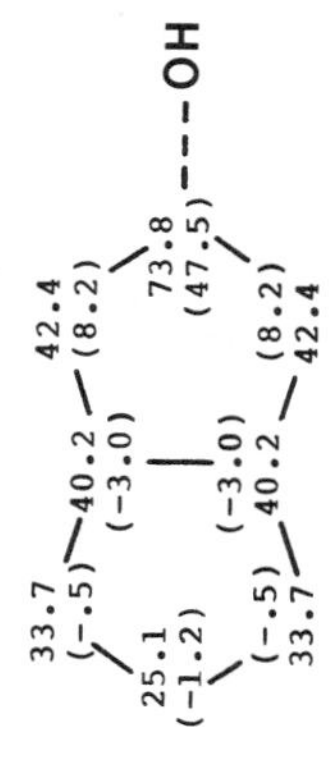

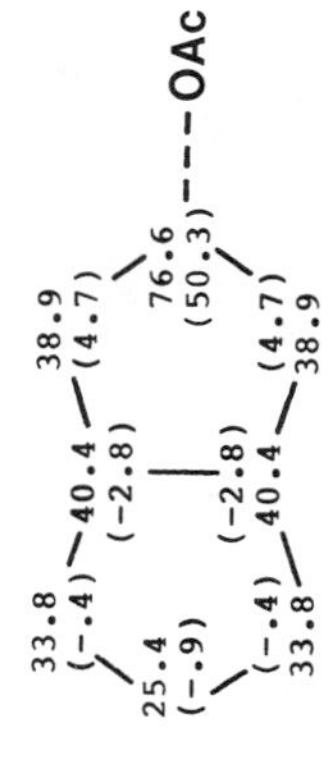

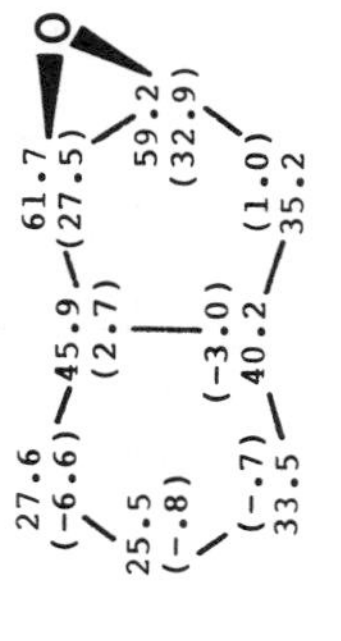

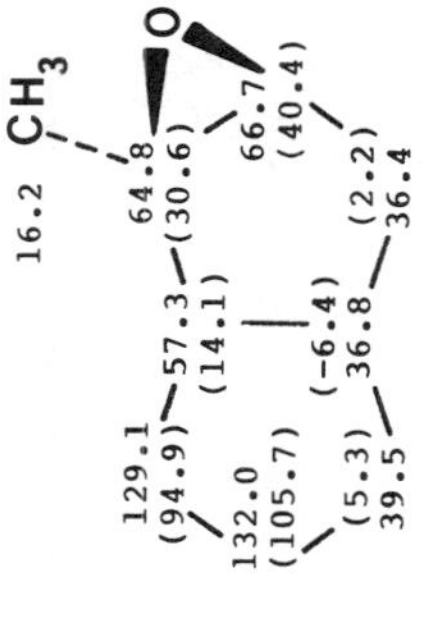

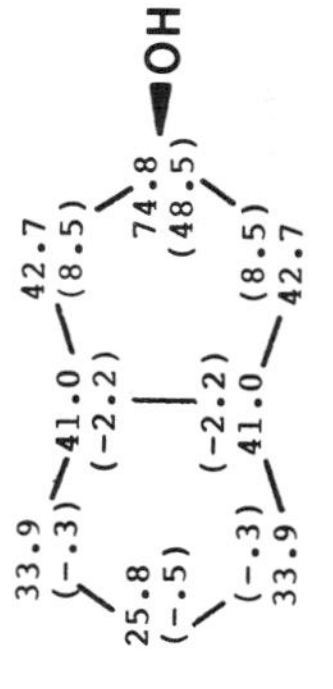

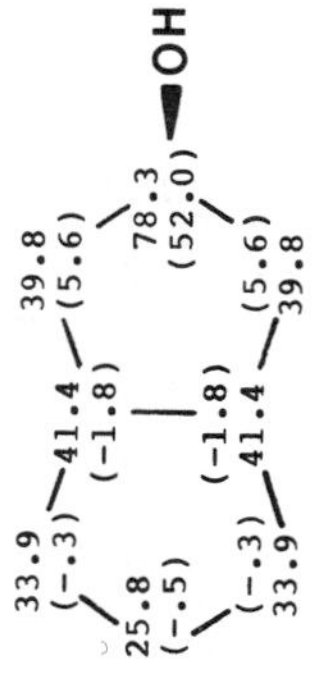

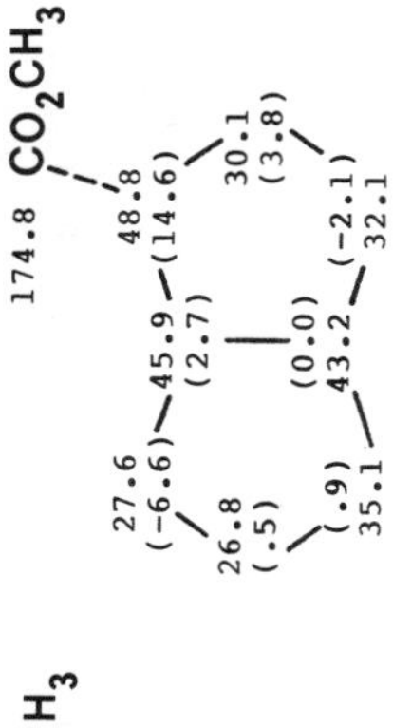

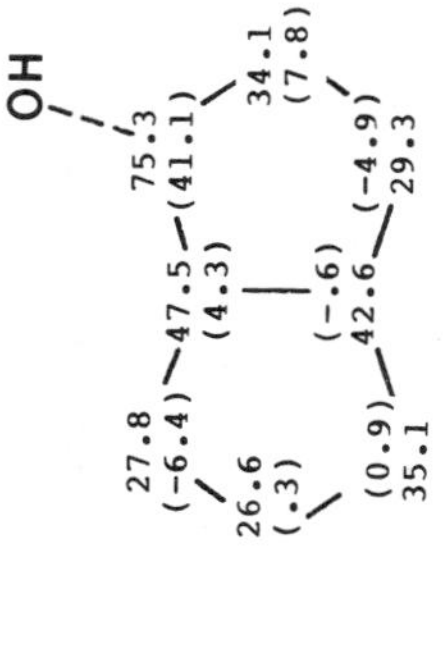

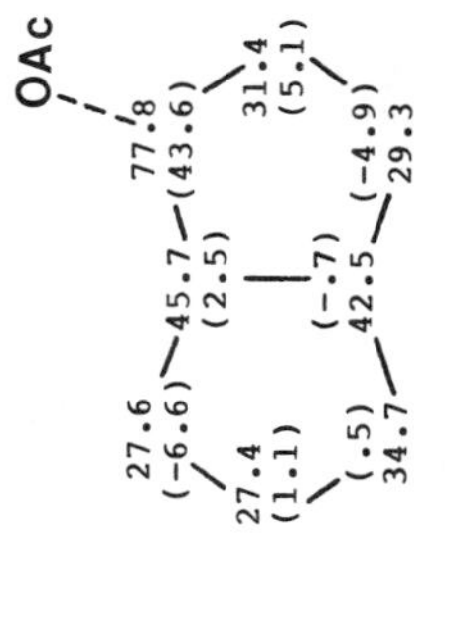

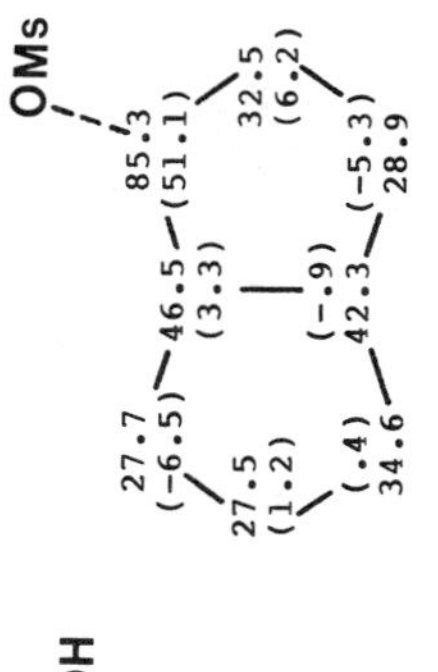

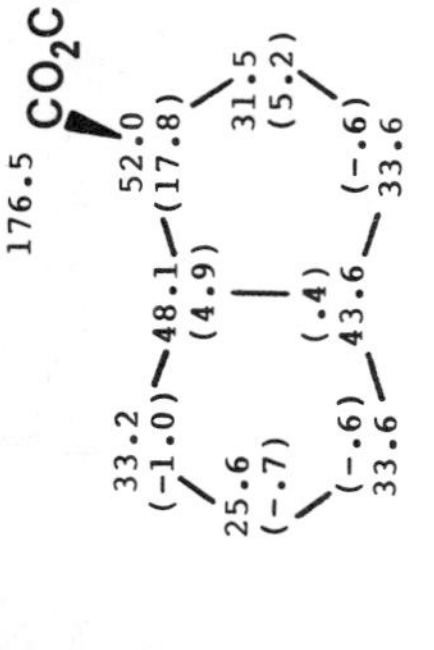

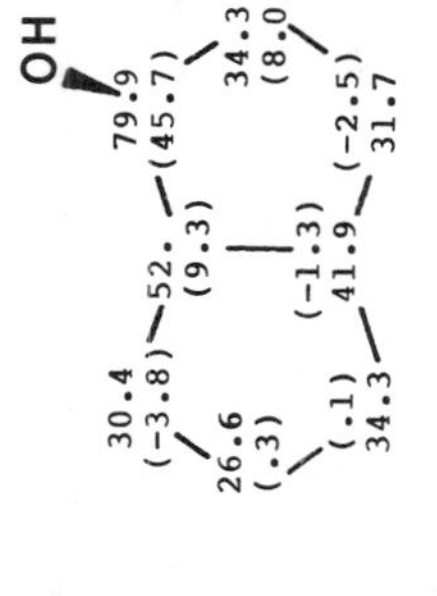

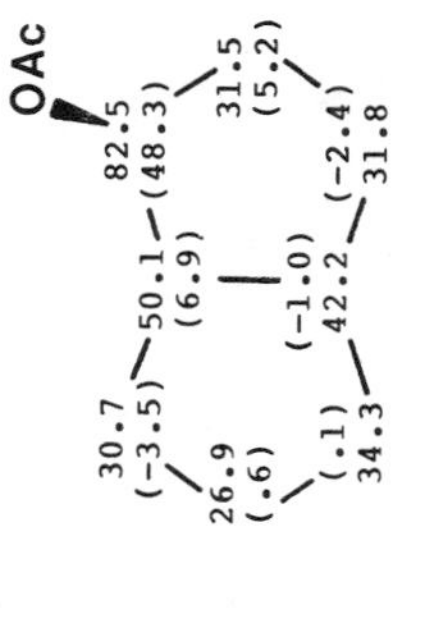

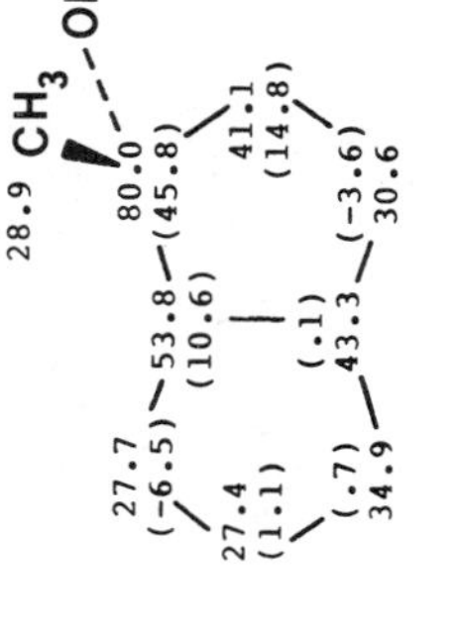

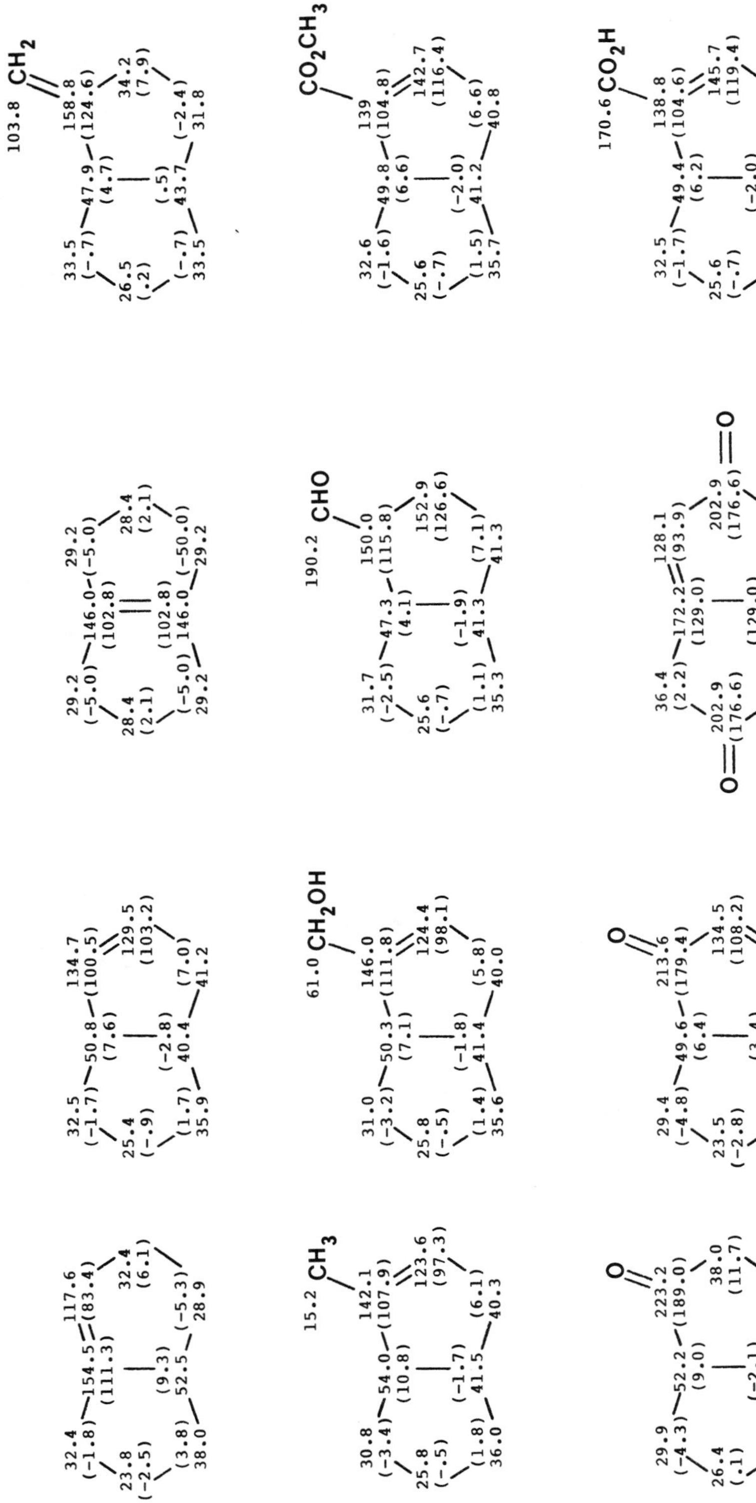

Structure 1 (CH₂):

103.8 CH₂

158.8 (124.6)
47.9 (4.7)
34.2 (7.9)
33.5 (-.7)
(.5)
43.7 (-2.4)
31.8
26.5 (.2)
33.5 (-.7)

Structure 2 (CO₂CH₃):

CO₂CH₃

139 (104.8)
49.8 (6.6)
142.7 (116.4)
32.6 (-1.6)
25.6 (-.7)
(-2.0)
41.2 (1.5)
6.6)
40.8
35.7

Structure 3 (CO₂H):

170.6 CO₂H

138.8 (104.6)
49.4 (6.2)
145.7 (119.4)
32.5 (-1.7)
25.6 (-.7)
(-2.0)
41.2 (1.4)
(6.8)
41.0
35.6

Structure 4:

29.2 (-5.0)
146.0 (102.8)
29.2 (-5.0)
28.4 (2.1)
29.2 (2.1)
(102.8)
146.0 (-50.0)
29.2
28.4 (2.1)

Structure 5 (CHO):

190.2 CHO

150.0 (115.8)
47.3 (4.1)
152.9 (126.6)
31.7 (-2.5)
25.6 (-.7)
(-1.9)
41.3 (1.1)
(7.1)
41.3
35.3

Structure 6 (O / O):

202.9 (176.6) =O
128.1 (93.9)
172.2 (129.0)
36.4
36.4 (2.2)
(129.0)
172.2 (93.9)
128.1
O= 202.9 (176.6)
(2.2)

Structure 7 (=O):

210.9 (184.6) =O
124.8 (90.6)
191.4 (148.2)
46.7 (-7.9)
(8.2)
42.4
31.2 (-3.0)
25.5 (-.8)
(3.5)
46.7
26.3

Structure 8:

134.7 (100.5)
50.8 (7.6)
129.5 (103.2)
32.5 (-1.7)
25.4 (-.9)
(-2.8)
40.4 (1.7)
(7.0)
41.2
35.9

Structure 9 (CH₂OH):

61.0 CH₂OH

146.0 (111.8)
50.3 (7.1)
124.4 (98.1)
31.0 (-3.2)
25.8 (-.5)
(-1.8)
41.4 (1.4)
(5.8)
40.0
35.6

Structure 10 (O):

O
213.6 (179.4)
49.6 (6.4)
134.5 (108.2)
29.4 (-4.8)
23.5 (-2.8)
(3.4)
46.6 (-4.1)
(133.4)
167.6
30.1

Structure 11 (=O):

220.9 (194.6) =O
44.7 (10.5)
39.7 (-3.5)
(10.5)
44.7
33.5 (-.7)
25.6 (-.7)
(-3.5)
39.7
33.5

Structure 12:

117.6 (83.4)
154.5 (111.3)
32.4 (6.1)
32.4 (-1.8)
23.8 (-2.5)
(9.3)
52.5 (-5.3)
28.9
(3.8)
38.0

Structure 13 (CH₃):

15.2 CH₃

142.1 (107.9)
54.0 (10.8)
123.6 (97.3)
30.8 (-3.4)
25.8 (-.5)
(-1.7)
41.5 (1.8)
(6.1)
40.3
36.0

Structure 14 (O):

O
223.2 (189.0)
52.2 (9.0)
38.0 (11.7)
29.9 (-4.3)
26.4 (.1)
(-2.1)
41.1 (-.7)
(-7.7)
26.5
33.5

Structure 15 (O):

O
203.9 (169.7)
148.9 (105.7)
41.2 (14.9)
25.7 (-8.5)
27.9 (1.6)
(144.2)
187.4 (-2.0)
(-9.7)
24.5
32.2

BICYCLO[3.3.0]OCTANES

Str	Substituents	C1	C2	C3	C4	C5	C6	C7	C8	Substituent Shifts	References
1	PARENT	43.2	34.2	26.3	34.2	43.2	34.2	26.3	34.2		2,38,57,66,67
2	1-CH2(CH2)2OH/3-=0/7-=0	47.4	48.2	217.4	43.9	41.9	43.9	217.4	48.2	35.4	51
3	1-CH2C CH/2-=0/X4-Me N4-Me/X8-CH2OMe	59.3	219.9	50.5	34.7	47.6	22.0	28.2	57.0	31.6,31.0,24.9,72.4	45
4	1-CH2CH2COMe	53.2	39.2	25.9	34.3	49.9	34.3	25.9	39.2	35.6	57
5	1-CH2CH2COOH/3-=0/7-=0	47.1	47.8	216.8	43.7	41.8	43.7	216.8	47.8	33.4	50,51
6	1-COOH	59.8	38.1	26.3	34.0	49.8	34.0	26.3	38.1	186.0	73
7	1-COOMe/2-COOMe 2= 3-OH/X4-Me N4-Me/X8-CN	61.8	100.2	181.0	44.4	58.6	29.1	32.6	40.1	173.9,169.7,28.2,19.7,120.0	62
8	1-COOMe/2-COOMe 2= 3-OH/X4-Me N4-Me/X8-Me	62.0	102.1	180.6	44.0	58.0	29.1	36.2	45.8	176.5,171.5,28.0,19.4,16.1	16
9	1-COOMe/2-COOMe 2= 3-OH/X4-Me N4-Me/X8-Br	65.3	99.4	182.4	44.3	56.6	27.4	38.1	55.1	173.7,170.8,27.9,19.9	62
10	1-COOMe/2= 3-COMe/X4-Me N4-Me/X8-Me	67.3	144.0	149.2	46.9	59.9	28.3	35.1	45.3	174.9,197.3,28.1,21.7,15.8	16
11	1-COOMe/3=/X4-Me N4-Me/6=/X8-COOMe	55.2	41.5	218.7	47.9	65.2	128.4	131.4	58.4	175.7,26.6,22.1,171.6	62
12	1-COOMe/3=/X4-Me N4-Me/7= 8-COOMe	58.1	42.9	219.3	48.8	57.0	34.5	145.2	138.8	174.5,26.6,20.6,163.9	62
13	1-COOMe/3=/X4-Me N4-Me/X8-CN	55.5	45.0	217.6	48.6	57.6	29.7	32.1	42.0	174.2,26.9,19.3,119.1	62
14	1-COOMe/3=/X4-Me N4-Me/X8-COOMe	54.1	46.2	219.6	48.5	57.7	29.4	29.0	58.4	176.0,27.3,19.6,172.7	62
15	1-COOMe/3-=0/X4-Me N4-Me/X8-Me	55.6	45.9	220.1	49.0	57.2	28.3	34.6	49.1	176.6,27.4,19.2,15.4	16
16	1-COOMe/3=/X4-Me N4-Me/X8-Ph	57.2	46.4	219.9	49.0	57.8	29.7	31.3	59.9	175.8,27.3,18.9,139.6	62
17	1-COOMe/N3-OH X3-C CH/X4-Me N4-Me/X8-Me	61.3	47.4	81.9	48.0	60.0	29.2	35.3	47.8	177.8,81.9,29.1,17.5,14.6	16
18	1-COOMe/N3-OH X3-COMe/X4-Me N4-Me/X8-Me	61.6	42.3	91.6	48.4	61.9	29.2	35.1	47.6	178.1,211.9,28.7,17.7,14.5	16
19	1-Me	49.8	41.9	25.9	34.5	50.9	34.5	25.9	41.9	29.2	2
20	1-Me/2= 3=	53.9	215.1	133.0	165.7	54.2	29.1	24.2	37.9	21.9	74
21	1-Me/2=0 3=/6-=0	53.0		133.6	158.3	63.5		36.8	29.9	22.0	75
22	1-Me/2=0 3=/N6-OH X6-CH=CH2	53.9		133.5	163.1	62.6	79.2	38.1	32.4	22.4,143.1	75
23	1-Me/2=0 3=/X6-OH N6-CH=CH2	56.6	221.1	134.2	162.8	66.2	82.0	36.7	33.9	22.6,140.9	75
24	1-Me/2=0/5-Me/6-=0	57.1	224.0	35.9	28.8	56.6	221.1	35.9	28.8	16.5,16.5	41
25	1-Me/2=0/5-Me/6= 7-Me	64.7	219.9	36.5	31.4	55.2	133.0	139.1	47.7		47
26	1-Me/2=0/5-Me/7-Me 7=	56.3	212.1	38.1	34.2	48.7	51.6	140.0	128.0	23.0	47
27	1-Me/2=0/6=	54.6	225.3	38.1	24.2	184.5	125.8	207.1	44.6	20.6	42,43
28	1-Me/2=0/6=/N8-Me	56.9	225.9	36.2	23.9	49.3	133.3	130.7	45.0	21.6,16.0	4
29	1-Me/2=0/6=/X8-Me	57.3	223.6	38.7	24.1	55.9	131.7	136.4	51.6	16.4,15.6	4
30	1-Me/2=0/X3-Me N3-Me	56.4	225.1	35.5	22.9	46.3	132.4	138.2	53.1	24.9,25.8,24.4	4
31	1-Me/2-CN 2=/5-Me/6-CN 6=	58.2	115.2	144.2	43.0	44.5	31.8	24.4	37.4	19.9,123.4,19.9,123.4	59
32	1-Me/2-CN 2=/X4-Br/5-Me/6-CN 6=/X8-Br	61.0	113.6	143.9	56.7	58.2	115.2	144.2	43.0	19.4,123.0,19.4,123.0	41
33	1-Me/3-=0/7-=0	43.7	50.4	217.0	43.7	61.0	113.6	143.9	56.7	26.2	41
34	1-OH	90.9	42.2	26.1	33.7	52.0	43.7	217.0	50.4		51
35	1=	154.5	117.6	32.4	28.9	52.5	38.0	23.8	42.2		57,56
36	1= 2-Me 3-=0/X6-CH2CH2CH=CH2/X7-Me N7-Me	181.4	131.4	209.9	42.5	48.6	53.9	42.7	32.4	8.3,28.9,25.0,29.4	66
37	1= 2-Me 3-=0/X6-CH2COOH/X7-Me N7-Me	181.9	132.5	211.7	41.1	48.7	50.7	42.9	42.5	8.3,34.1,25.1,28.9	51
38	1= 3-=0	191.4	124.8	210.9	42.4	46.7	46.7	25.6	42.1		51
39	1= 3-=0/5-Me/X6-OH	191.2	125.2	210.3	49.7	53.2	77.7	31.9	31.2	18.4	48,49
40	1= 3-=0/5= 7-=0	172.2	128.1	202.9	36.4	172.2	128.1	202.9	24.0		42
41	1=/N3-OH/8-=CH2	152.6	122.3	82.8	44.4	49.8	32.1	36.1	142.8	106.9	54
42	1=/N3-OH/N8-OTBS	157.6	123.6	81.9	45.1	45.6	28.8	36.1	69.8		1
43	(5)1=	146.0	29.2	28.4	29.2	146.0	29.2	28.4	29.2		66
44	(5)1= 2-=0	148.9	203.9	41.2	24.5	187.4	32.2	27.9	25.7		2
45	(5)1= 2-=0/3-COOMe	147.2	195.8	57.8	190.5	187.4	31.9	27.8	24.6	169.8	52
46	(5)1= 2-=0/6-Me	147.6	204.2	40.9	23.8	32.1	38.9	37.0	23.9	18.3	52
47	(5)1=/2-OH	151.1	80.0	37.1	28.0	146.7	29.5	28.2	27.6		1
48	(5)1=/3-=0/7-=0	138.5	42.3	214.5	42.3	138.5	42.3	214.5	42.3		54
49	2-=0/X4-Me X8-Me	60.1	221.3	46.9	35.5	49.9	31.1	35.8	38.5	20.7,20.7	1

141

BICYCLO[3.3.0]OCTANES (cont.)

Str	Substituents	C1	C2	C3	C4	C5	C6	C7	C8	Substituent Shifts	References
51	2=CH2	47.9	158.8	34.2	31.8	43.7	33.5	26.5	33.5	103.8	6
52	2=CHOH 3=O/7=	49.4	116.1	208.9	44.7	36.4	39.9	129.8	132.9	161.4	1
53	2=O	52.2	223.2	38.0	26.4	41.1	33.5	26.4	29.9		2,59
54	2=O 3=	49.6	213.6	134.5	167.6	46.6	30.1	23.5	29.4		2
55	2=O 3=/X7-OCH2C(Me)2CH2O-7N		212.0	133.4	166.8	42.7	38.3	107.8	34.0		71
56	2=O 3=X8-Me	57.5	212.9	133.2	167.3	46.6	27.2	32.3	37.4	20.4	1
57	2=O 8=N6-OH	135.5	203.3	42.3	30.0	46.6	46.1	82.1	148.3		40
58	2=O 8=/X7-OH	131.1	203.4	42.5	29.0	46.0	43.8	81.2	153.5		40
59	2=O/5-Me	59.5	223.1	37.9	33.6	47.9	40.3	25.7	29.4	27.2	2
60	2=O/6=O	49.5	220.3	37.6	23.1	49.5	220.3	37.6	23.1		2
61	2=O/6=	49.4	224.0	36.4	25.4	47.8	133.9	131.2	37.4		2,3,39
62	2=O/6=/N8-Me	51.9	220.4	38.5	25.6	48.5	132.1	136.9	43.5	16.4	4
63	2=O/6=/X8-Me	57.6	223.4	36.3	25.2	48.5	132.3	137.3	45.6	21.2	4
64	2=O/7=	59.9	218.2	37.9	28.5	38.2	40.4	127.4	132.6		2
65	2=O/8=O	62.3	209.0	37.7	26.1	39.3	26.1	37.7	209.0		5
66	2=O/N5-OH/N6-OH	48.8	224.0	37.7	19.7	42.2	75.6	74.6	33.5		1
67	2=O/N6-OC(Me)2O-7N	52.6	219.3	37.9	20.0	44.7	83.1	82.5	33.3		1
68	2=O/N7-O-8N	52.8	218.1	38.6	29.7	39.6	34.1	60.3	59.3		40
69	2=O/N7-OAc/N8-OAc	50.6	216.5	38.5	27.9	36.0	35.4	74.2	74.6		2
70	2=O/N8-OTBS	58.4		40.5	28.6	40.9	31.6	38.1	75.9		1
71	2=O/X3-Me N3-Me	50.3	224.6	47.6	43.5	36.3	33.6	25.0	29.4	25.3,23.1	59
72	2=O/X4-Me	52.2	222.1	47.2	35.1	49.4	32.4	25.9	29.6	20.8	2
73	2=O/X6-Me	52.5	223.8	36.5	23.9	49.6	40.4	35.3	27.8	18.6	2
74	2=O/X6-Me N6-Me/7= 8-Me	61.5	218.1	39.6	23.8	51.2	46.4	137.0	133.3	30.0,22.1,14.5	63
75	2=O/X6-Me N6-Me/8-Me 8=	138.8	202.9	44.6	22.0	59.2	43.3	58.6	148.1	28.2,24.2,15.3	63
76	2=O/X6-Me N6-Me/N8-CH2I	54.2	220.3	40.6	24.6	53.9	40.4	47.2	43.2	29.6,23.6,10.4	63
77	2=O/X6-OC(Me)2O-7X	50.5	218.4	38.4	23.8	49.3	86.4	82.2	34.3		1
78	2=O/X6-OH/X7-OH	47.7	222.8	37.3	23.2	45.4	78.8	73.8	33.6		1
79	2=O/X7-O-8X	55.1	218.2	35.4	25.0	36.8	32.8	57.3	57.7		40
80	X2-CH(OH)CCl3/3=	48.2	55.7	126.0	141.0	49.6	31.5	24.9	34.5	86.0	1
81	X2-CH(OH)COOMe/3=	44.0	58.2	126.6	139.1	50.4	31.4	24.9	34.5	73.2	1
82	N2-CH(X-OH)O-8N/6=	49.1	52.0	27.6	32.1	50.0	138.2	131.3	89.0	103.4	1
83	N2-CH(X-OH)O-8N/X7-OH	51.9	51.8	29.1	32.1	43.1	38.1	77.9	89.9	104.3	1
84	2-CH2Br 2=/N6-OC(iPr)2O-7N	51.3	141.2	130.8	31.6	45.5	84.2	83.2	34.4	30.0	1
85	X2-CH2CH2OH/3=/N6-CH=CH2	48.3	52.1	135.5	130.6	54.6	47.4	30.2	34.0	39.7,140.8	1
86	N2-CH2CH2OH/3=/N6-CH=CH2	44.8		135.4	129.8	56.1	47.9	31.2	26.2	34.0,140.4	1
87	N2-CH2CH2OH/3=/N6-OTBS	42.2	44.8	134.7	129.8	56.6	75.3	34.7	22.9	33.7	1
88	X2-CH2CH2OH/3=/N8-CH=CH2	51.2	43.7	134.2	133.9	50.1	28.9	31.3	48.5	40.1,140.3	1
89	X2-CH2CH2OMs/3=/N6-CH=CH2	44.5	43.2	134.1	130.8	56.0	47.6	31.1	26.0	30.4,140.2	1
90	N2-CH2CH2OMs/3=/N6-OTBS	42.1	44.3	133.3	130.8	56.8	75.0	35.0	23.0	30.2	1
91	X2-CH2CH2OTs/3=/N6-CH=CH2	48.2	51.5	134.2	131.4	54.6	47.1	30.1	33.8	35.6,140.6	1
92	N2-CH2CH2OTs/3=/N6-CH=CH2	44.5	43.2	134.1	130.8	56.1	47.6	31.1	26.0	30.2,140.3	1
93	X2-CH2COOTBS/3=/N6-CH=CH2	48.2	52.0	134.7	131.2	54.7	47.1	30.1	33.9	43.1,140.6	1
94	N2-CH2COOTBS/3=/N6-CH=CH2	45.0	43.0	134.6	130.5	56.0	47.8	31.2	26.7	37.6,140.2	1
95	N2-CH2COOTBS/3=/N6-OTBS	44.6	42.0	134.2	130.3	56.7	75.0	35.1	23.3	37.3	1
96	X2-CH2COOTBS/3=/N8-CH=CH2	50.7	44.1	134.5	133.5	50.0	28.8	31.1	48.1	43.3,139.5	1
97	2-CH2Ac 2=/6=	48.0	140.8	126.9	37.8	48.7	134.5	128.1	36.0	62.0	1
98	N2-CH2OAc/3=/N6-CH2OAc/7=	50.5	47.5	131.5	130.8	50.5	47.5	131.5	130.8	65.3,65.3	46
99	2-CH2OCH2OCH2CH2TMS 2=/N7-OC(Me)2O-8N	53.8	140.7	127.8	39.3	43.8	35.7	84.3	82.0	64.9	1
100	2-CH2OCH2Ph 2=/6=	48.6	142.9	125.8	37.7	48.0	134.7	128.2	35.9	67.8	1

BICYCLO[3.3.0]OCTANES (cont.)

Str	Substituents	C1	C2	C3	C4	C5	C6	C7	C8	Substituent Shifts	References
101	2-CH2OCH2Ph 2=/X7-O-8X	58.9	147.7	123.0	43.2	42.5	36.4	86.2	84.7	66.2	1
102	X2-CH2OCH2Ph/3=/6=	43.3	55.0	133.6	131.1	58.2	132.0	129.4	40.0	74.8	1
103	X2-CH2OH	46.5	50.5	30.8	33.3	43.4	33.7	25.8	33.2	66.5	1
104	N2-CH2OH	45.1	46.4	27.9	32.7	42.9	35.2	27.5	27.5	64.1	1
105	2-CH2OH 2=	50.3	146.0	124.4	40.0	41.4	35.6	25.8	31.0	61.0	1
106	2-CH2OH 2=/6=	47.5	145.9	123.8	37.7	48.6	134.9	128.2	35.9	60.9	1
107	2-CH2OH 2=/7=	58.3	145.5	124.5	40.3	39.6	41.0	130.3	131.2	61.5	1
108	2-CH2OH 2=/N6-OC(iPr)2O-7N	51.6	144.8	126.3	31.2	45.7	84.6	83.2	34.9	60.8	1
109	2-CH2OH 2=/N6-OC(Me)2O-7N	51.7	144.3	127.4	31.0	46.0	82.6	82.2	32.4	60.5	1
110	2-CH2OH 2=/N7-C1/X8-OH	56.8	144.2	125.3	39.7	37.0	41.4	63.4	84.2	60.8	1
111	2-CH2OH 2=/N7-OC(Me)2O-8N	55.3	142.6	129.0	39.2	44.4	35.2	84.5	82.2	61.8	1
112	2-CH2OH 2=/X7-O-8X	53.9	142.3	126.8	39.3	36.8	36.2	59.0	58.6	60.6	1
113	X2-CH2OH/3=/6=	42.7	57.4	134.5	130.5	58.4	132.0	129.6	40.2	66.5	1
114	X2-CH2OH/3=/7=	52.3	54.1	129.0	137.6	47.2	37.4	128.2	134.3	66.5	1
115	N2-CH2OH/3=/N5-CH2OH/7=	50.5	51.2	131.7	131.4	50.5	51.2	131.7	131.4	63.7,63.7	46
116	X2-CH2OH/7=	54.4	48.8	29.3	34.0	40.1	40.8	128.8	134.1	64.7	1
117	N2-CH2OH/7=	52.2	46.7	27.9	34.4	40.1	42.0	132.0	129.7	64.4	1
118	N2-CH2OH/N7-O-8N	47.4	41.3	29.2	33.4	44.7	35.3	60.0	59.2	65.0	1
119	N2-CH2OMs/N7-O-8N	47.0	41.5	29.1	33.2	41.7	35.1	59.3	59.1	72.7	1
120	2-CH2OTBS 2=/N7-OC(Me)2O-8N	53.6	143.9	125.2	39.1	44.1	36.1	84.5	82.6	62.6	1
121	2-CH2OTMS 2=/6=	47.5	145.4	123.7	37.7	48.8	134.8	128.1	36.1	60.7	1
122	2-CH2OTMS 2=/7=	58.3	145.1	124.1	40.4	39.8	41.0	129.7	131.3	61.2	1
123	2-CH2OTMS 2=/X7-O-8X	53.9	141.7	126.9	38.2	36.9	36.2	59.0	58.5	60.9	1
124	2-CH2P(Ph)3Br 2=/N6-OH/N7-OH	50.7		133.4	32.6	43.8	75.7	75.3	35.1	25.2	1
125	N2-CH=C(OTBS)2/3=/N6-CH=CH2	45.8	44.2	136.9	128.9	55.4	48.7	31.1	28.3	84.2,140.7	1
126	X2-CHO	44.4	60.1	28.0	33.2	43.7	33.6	26.0	33.1	203.8	1
127	N2-CHO	44.7	56.7	29.4	32.1	43.7	34.3	24.8	27.9	204.2	1
128	2-CHO 2=	47.3	150.0	152.9	41.3	41.3	35.3	25.6	31.7	190.2	1
129	2-CHO 2= N7-O-8N	49.2	144.8	153.5	42.4	40.5	33.8	61.6	59.9	189.5	1
130	2-CHO 2=/6=	44.7	149.9	151.7	39.0	48.5	133.4	129.4	37.0	190.0	3
131	2-CHO 2=/7=	55.4	149.0	151.3	41.8	39.3	40.7	129.7	131.1	189.5	3
132	2-CHO 2=/N6-OC(iPr)2O-7N	48.7	149.2	152.6	45.4	45.4	84.3	83.0	35.7	189.3	3
133	2-CHO 2=/N6-OC(Me)2O-7N	48.8	148.9	152.6	33.0	45.7	82.7	81.9	32.8	189.4	1
134	2-CHO 2=/X7-O-8X	51.0	145.2	153.8	39.6	36.7	35.7	58.1	58.1	189.7	1
135	X2-C1/N3-OAc	50.1	66.4	80.2	35.7	38.3	33.6	25.1	31.2		1
136	N2-C1/X3-SePh	49.8	69.4	53.2	40.8	39.6	33.3	27.4	31.1		1
137	2-CN 2=/6=	49.5	118.8	146.4	36.7	48.3	134.0	128.2	38.9	116.4	58
138	2-CN 2=/7=	60.0	117.0	147.5	41.7	38.9	41.0	129.2	132.1	116.5	3
139	X2-CN N2-OH/6=	52.4	77.9	37.5	28.8	49.3	133.4	128.6	37.3	121.4	3
140	X2-CN N2-OH/7=	63.3	76.2	38.2	32.0	38.7	41.8	128.6	135.3	121.4	3
141	N2-CN X2-OH/6=	50.4	75.4	37.8	28.5	49.5	133.2	129.9	32.4	123.1	3
142	N2-CN X2-OH/7=	61.9	74.3	39.5	31.1	39.8	41.3	125.6	136.0	123.0	3
143	2-COC1 2=/6=	47.2	143.6	152.9	37.5	48.6	133.3	129.0	38.8	163.4	1
144	X2-COC1/3=/6=	43.5	70.3	129.	137.9	46.3	131.0	125.6	39.5	179.0	1
145	N2-COO-8N	50.9	45.5	32.0	32.0	46.3	29.1	34.7	84.8	180.1	1
146	N2-COO-8N/6=	47.3	45.0	31.0	33.0	51.6	140.8	130.1	87.7	179.1	1
147	N2-COO-8N/X6-O-7X	50.6	45.8	32.0	30.0	47.4	55.1	58.0	88.0	180.1	1
148	N2-COO-8N/X7-SePh	50.0	45.2	31.8	32.0	48.0	36.0	45.4	88.5	180.1	1
149	X2-COOBu/6=	45.2	52.7	30.3	31.4	50.8	134.6	128.4	39.2	176.1	1
150	2-COOCH2CH2OH 2= 3-=O/6=	42.3	164.1	135.9	208.7	59.4	131.9	126.8	35.6	165.0	53

Str	Substituents	C1	C2	C3	C4	C5	C6	C7	C8	Substituent Shifts	References
151	X2-COOH	48.1	52.0	31.4	33.6	43.7	33.6	25.6	33.2	182.7	1
152	N2-COOH	45.8	48.9	30.0	32.0	43.1	35.2	26.5	27.6	181.0	1
153	2-COOH 2=	49.4	138.8	145.7	41.0	41.2	35.6	25.6	32.5	170.6	1
154	2-COOH 2=/6=	46.7	138.7	144.9	37.8	48.3	133.6	129.0	38.8	170.7	3
155	2-COOH 2=/7=	57.2	137.6	145.8	41.5	39.2	41.0	129.8	131.5	170.7	3
156	2-COOH 2=/N6-OC(Me)2O-7N	50.5	138.0	145.5	32.2	45.6	82.5	82.0	33.8	169.0	1
157	2-COOH 2=/X6-OC(Me)2O-7X	48.3	139.8	143.3	37.0	50.3	87.4	83.2	37.6	168.5	1
158	X2-COOH/3=/6=	43.3	59.0	129.4	135.5	59.2	132.0	127.9	40.1	179.0	1
159	X2-COOH/3=/7=	52.3	55.9	129.5	139.1	47.7	37.1	125.8	133.0	181.2	1
160	X2-COOH/7=	54.9	50.3	30.2	34.5	40.0	40.8	130.0	132.8	182.2	1
161	N2-COOH/7=	53.2	48.8	26.7	34.2	39.8	42.0	129.5	132.9	180.5	1
162	N2-COOMe	48.1	52.0	31.5	33.6	43.6	33.6	25.6	33.2	176.5	1
163	N2-COOMe	45.9	48.8	30.1	32.1	43.2	35.1	26.8	27.6	174.8	1
164	2-COOMe 2=	49.8	139.0	142.7	40.8	41.2	35.7	25.6	32.6		1
165	2-COOMe 2= 3-OH/X4-COOMe/6-COOMe 6= 7-OH/X8-COOMe	43.7	103.6	170.5	55.2	43.7	103.6	170.5	55.2	168.7,170.2,168.7,170.2	69
166	2-COOMe 2=/6-COOMe 6=	47.2	137.9	142.6	38.6	47.2	137.9	142.6	38.6	165.3,165.3	38
167	2-COOMe 2=/6=	47.0	138.9	141.8	37.8	48.1	133.7	129.0	38.6	165.6	1
168	2-COOMe 2=/7=	57.5	137.8	142.7	41.0	39.1	41.2	129.7	131.7	165.5	1
169	2-COOMe 2=/N6-Br/X7-OAc	43.5	138.3	142.7	37.6	46.2	57.1	79.4	34.1	164.5	1
170	2-COOMe 2=/N6-OC(iPr)2O-7N	50.9	138.3	143.3	32.3	45.2	84.3	83.2	36.6	165.2	1
171	2-COOMe 2=/N6-OC(Me)2O-7N	50.8	138.1	143.2	32.1	45.5	82.5	82.0	34.0	165.3	1
172	2-COOMe 2=/N6-OH/N7-OH	46.2	138.7	144.0	32.6	43.0	75.3	74.4	36.1	165.7	1
173	2-COOMe 2=/N7-OC(Me)2O-8N	53.6	135.8	143.7	40.5	43.6	35.0	81.9	84.2	165.7	1
174	2-COOMe 2=/X6-Br/N7-OAc	45.9	138.9	141.1	38.0	48.9	57.0	80.3	34.6	164.5	1
175	2-COOMe 2=/X6-OC(Me)2O-7X	48.5	139.9	141.4	36.9	50.4	87.3	83.2	37.7	165.0	1
176	2-COOMe 2=/X6-OH/X7-OH	45.8	139.2	142.3	37.3	46.0	80.8	73.9	36.1	165.4	1
177	2-COOMe 2=/X6-OMs/X7-OMs	44.6	138.0	142.0	36.2	44.1	85.1	80.8	34.6	164.5	1
178	2-COOMe 2=/X7-OC(Me)2O-8X	58.3	135.7	144.1	37.8	39.6	38.5	81.6	84.2	165.4	1
179	X2-COOMe/3=/6=	43.3	58.7	129.3	135.2	59.1	131.7	127.9	39.9	174.4	1
180	X2-COOMe/3=/7=	52.3	55.9	129.3	138.6	47.6	37.1	126.1	133.2	174.8	1
181	X2-COOMe/3=/N6-COOMe/7=	52.5	56.8	129.1	135.6	51.4	53.5	128.7	134.2	173.0,173.8	1
182	N2-COOMe/3=/N6-COOMe/7=	47.5	58.3	128.7	135.1	47.5	58.3	128.7	135.1	174.0,174.0	38
183	X2-COOMe/3=/N6-Me/7==O	38.3	58.5	129.5	133.4	51.8	45.6	218.4	43.0	173.5,11.3	1
184	X2-COOMe/3=/N6-OC(Me)2O-7N	47.2	57.5	129.0	132.9	54.1	83.8	81.6	34.8	170.0	1
185	X2-COOMe/3=/X6-COOMe/7=	52.6	55.8	128.1	137.0	52.6	55.8	128.1	137.0	173.9,173.9	38
186	X2-COOMe/3=/X6-Me/7==O	39.3	58	128.7	136.9	54.9	47.7	219.9	42.9	173.8,15.4	1
187	X2-COOMe/X3-O-4X/N6-Me/7-=O	36.4	52.4	60.3	58.4	46.4	43.8	217.8	44.2	172.2,11.4	1
188	X2-I/N6-I	54.5	31.1	39.8	30.6	47.8	32.1	36.9	34.3		61
189	N2-I/N6-I	47.6	33.2	38.1	33.4	47.6	33.2	38.1	33.4		61
190	X2-Me	51.9	42.3	36.0	33.6	43.4	34.0	25.5	32.3	19.8	2
191	N2-Me	47.9	37.9	33.1	32.4	43.2	35.7	27.8	27.8	15.4	2
192	2-Me 2=	54.0	142.1	123.6	40.3	41.5	36.0	25.8	30.8	15.2	2
193	2-Me 2=/5-Me	61.3	141.9	122.9	48.4	49.2	43.2	26.0	30.9	15.6,29.0	2
194	2-Me 2=/7=	62.0	141.5	131.2	40.7	39.7	41.6	123.0	130.1	15.4	2
195	2-Me 2=/X6-Me	53.8	143.2	122.2	37.8	50.2	43.0	35.3	30.0	14.9,19.2	2
196	X2-Me/3-=O	48.6	49.1	222.2	43.6	37.5	33.6	25.5	32.0	14.2	1
197	N2-Me/3-=O	46.3	47.5	221.0	43.6	36.8	33.0	25.3	27.1	10.7	1
198	X2-Me/3-=O/6=	46.1	48.9	220.2	41.3	43.8	134.7	130.3	38.7	13.2	1
199	X2-Me/3-=O/N4-Me/6=	43.6	47.2	218.5	43.8	49.4	129.5	130.3	37.9	13.6,11.2	1
200	X2-Me/3-=O/N7-OC(Me)2O-8N	52.4	42.2		41.6	38.7	36.1	83.2	82.1	16.6	1

Str	Substituents	C1	C2	C3	C4	C5	C6	C7	C8	Substituent Shifts	References
201	X2-Me/3-=O/X4-Me/6=	43.9	46.3	221.3	48.8	53.0	134.3	130.0	38.8	13.3,15.8	1
202	X2-Me/3-=O/X6-Me	48.6	50.5	222.1	41.7	46.2	41.5	35.2	30.9	13.8,18.9	2,70
203	N2-Me/3-=O/X6-Me	44.7	46.8	220.7	43.1	44.9	42.3	34.5	27.3	10.5,21.0	2
204	X2-Me/3-=O/X7-OC(Me)2O-8X	55.7	44.2	219.2	41.9	35.3	38.7	82.3	84.6	13.2	1
205	X2-Me/3-=O/X7-OH/X8-OH	53.4	47.0	221.5	44.1	32.6	38.9	73.8	79.3	15.4	1
206	X2-Me/X3-OCH2C(Me)2CH2O-3N/6=/X8-OCOCH2OMe	53.0	45.0	108.4	33.3	44.8	142.9	127.0	85.2	11.6	1
207	X2-Me/X3-OCH2C(Me)2CH2O-3N/6=/X8-OH	56.1	45.7	108.6	33.3	44.8	140.1	131.1	82.3	11.5	1
208	X2-Me/X3-OCH2C(Me)2CH2O-3N/7=	55.7	48.6	108.1	39.5	36.4	36.6	128.5	133.7	12.1	1
209	N2-Me/X3-OCH2C(Me)2CH2O-3N/7=	52.3	42.5	109.4	40.9	36.4	37.4	130.4	130.9	10.9	1
210	X2-Me/X3-OCH2CH2O-3N/7=	56.1	46.3	118.5	42.9	36.0	39.9	129.0	133.6	12.7	1
211	X2-Me/N3-OH/X6-Me	48.6	49.2	80.6	40.8	46.9	42.5	34.8	30.6	16.6,19.6	2
212	N2-Me/X3-OH/X6-Me	44.3	46.8	77.4	40.1	46.8	44.1	35.8	28.1	12.7,19.6	2
213	N2-O-3N	44.6	62.6	61.6	35.3	43.3	33.4	27.1	28.4		1
214	N2-O-3N/N8-OH	49.8	60.0	60.2	36.3	40.9	31.1	34.6	74.9		1
215	N2-O-3N/N8-OTBS	50.8	57.5	59.2	36.7	37.5	30.6	33.8	75.6		1
216	X2-O-3X	45.9	61.7	59.2	35.2	40.2	33.5	25.5	27.6		2
217	X2-O-3X N2-Me/7=	57.3	64.8	66.7	36.4	36.8	39.5	132.0	129.1	16.2	1
218	X2-O-3X/N8-OH	51.0	60.3	59.0	37.0	39.4	30.1	36.3	73.5		1
219	X2-OAc	50.1	82.5	31.5	31.8	42.2	34.3	26.9	30.7		1
220	N2-OAc	45.7	77.8	31.4	29.3	42.5	34.7	27.4	27.6		2
221	X2-OAc/3=/N6-CH=CH2	48.0	88.3	128.7	138.4	54.2	47.5	31.2	29.4	139.9	1
222	X2-OAc/3=/N8-CH=CH2	51.0	82.2	129.5	138.3	49.2	28.3	29.6	46.5	140.9	1
223	N2-OAc/3=/N8-CH=CH2	47.8	79.5	139.4	129.7	49.9	29.3	30.9	47.3	139.9	1
224	N2-OAc/3=/N8-OTBS	47.0	78.4	129.6	138.5	48.4	26.2	33.6	75.4		1
225	N2-OAc/7=	53.7	77.6	30.9	31.0	31.0	41.7	128.6	132.5		2
226	N2-OAc/N3-OAc	43.8	75.0	74.7	34.7	39.4	33.9	27.0	27.4		1
227	N2-OAc/N6-OAc	44.9	77.2	32.4	23.0	44.9	77.2	32.4	23.0		2
228	X2-OAc/X3-OAc	46.4	79.5	75.0	35.4	39.0	33.8	26.1	31.0		1
229	X2-OAc/X3-OAc/6=	43.6	79.8	73.9	33.7	45.9	134.0	129.2	37.2		1
230	N2-OC(Me)2O-3N	46.9	84.0	82.8	36.5	46.0	32.8	25.6	27.3		1
231	N2-OC(Me)2O-3N/5=	54.4	84.7	78.5	37.2	147.7	121.5	32.7	23.4		1
232	N2-OC(Me)2O-3N/6=	44.7	82.8	82.5	34.8	52.0	134.2	129.5	32.6		1
233	N2-OC(Me)2O-3N/7=	53.8	83.6	82.4	36.8	42.6	39.6	130.3	129.8		1
234	N2-OC(Me)2O-3N/N6-OCS2Me	43.8	83.2	82.3	30.9	46.3	85.8	31.9	21.9		1
235	N2-OC(Me)2O-3N/N6-OH	45.1	84.5	82.0	30.2	50.3	73.9	37.1	22.1		1
236	N2-OC(Me)2O-3N/N7-OCS2Me	45.1	83.7	81.9	34.6	43.2	37.0	86.9	30.4		1
237	N2-OC(Me)2O-3N/X6-I	45.7	83.1	82.1	33.4	59.2	31.2	41.0	24.0		2
238	X2-OC(Me)2O-3X	51.9	87.1	82.9	38.7	42.2	31.3	25.2	29.5		1
239	X2-OC(Me)2O-3X/5=	56.3	85.6	83.3	36.2	148.0	122.1	32.6	30.9		1
240	X2-OC(Me)2O-3X/6=	48.7	88.7	82.7	37.9	49.8	135.0	128.9	37.4		1
241	X2-OC(Me)2O-3X/7=	59.3	84.8	82.0	38.8	39.4	38.4	130.5	131.3		1
242	X2-OC(Me)2O-3X/N6-OCS2Me	49.6	87.0	82.3	32.4	45.6	86.2	30.6	26.0		1
243	X2-OC(Me)2O-3X/N6-OH	49.0	88.5	82.8	31.9	48.0	73.9	33.3	26.3		1
244	X2-OC(Me)2O-3X/N7-OCS2Me	50.1	86.4	82.6	38.7	48.0	37.4	86.9	35.5		1
245	N2-OC(OTBS)=CH2/3=/N8-CH=CH2	49.5	83.1	129.5	138.7	49.8	29.4	32.5	21.9	140.6	1
246	N2-OC(OTBS)=CH2/3=/N8-OTBS	51.8	83.0	130.7	136.7	48.0	28.6	36.2	74.5		1
247	N2-OCH(N-Me)-8N/3=	49.7	87.6	131.7	137.9	52.0	31.0	23.7	49.7	78.3	1
248	N2-OCH(X-Me)-8N/3=	49.7	87.4	130.4	138.9	52.0	30.3	32.7	52.5	78.9	1
249	N2-OCH2-8N/3-=O	50.9	83.5	215.8	37.4	46.7	31.6	34.0	43.0	76.3	1
250	N2-OCH2-8N/3=	50.4	89.0	131.4	138.9	52.5	30.4	33.0	46.2	73.4	1

BICYCLO[3.3.0]OCTANES (cont.)

Str	Substituents	C1	C2	C3	C4	C5	C6	C7	C8	Substituent Shifts	References
251	N2-OCH2-8N/3=/6=	50.6	88.1	129.6	138.8	57.0	131.3	134.4	50.8	75.0	46
252	N2-OCH2-8N/X3-OH	53.7	90.8	77.4	38.1	45.1	31.5	32.7	42.9	74.3	1
253	N2-OCH2-8N/X3-OMs	53.7	88.1	87.3	36.2	45.3	31.4	32.4	42.7	74.5	1
254	X2-OCH2CH2O-2N/7-CH2CH2CO-8 7=	51.4	116.9	35.6	30.6	46.1	39.1	186.4	146.9	25.5,203.3	19
255	X2-OCH2CH2O-2N/X3-CH2CH2CH2-4X	51.8	120.1	51.8	49.2	49.2	33.9	26.4	26.4	26.4,33.9	1
256	N2-OCO2Me/7=	53.6	81.0	30.7	30.9	39.2	41.3	127.6	132.7	132.7	40
257	N2-OCOO-8N	44.7	82.4	35.0	29.6	43.8	29.6	35.0	82.4		40
258	N2-OCOO-8N/X3-Br	42.2	86.3	53.9	40.2	42.6	28.5	35.4	81.3		40
259	N2-OCOPh/6=	42.5	78.0	29.9	28.2	49.5	133.8	130.1	33.1		1
260	N2-OCOPh/7=	54.0	77.9	31.2	31.2	39.4	41.6	128.5	132.3		1
261	N2-OCS2Me/7=	53.6	86.7	30.8	30.8	39.6	41.4	128.3	132.4		1
262	X2-OH	52.5	79.9	34.3	31.7	41.9	34.3	26.6	30.4		2
263	N2-OH	47.5	75.3	34.1	29.3	42.6	35.1	26.6	27.8		2
264	X2-OH N2-Me/3==0/N7-OC(Me)2O-8N	57.0	78.2	214.2	40.2	37.5	36.6	81.3	83.7	18.1	1
265	X2-OH N2-Me/3-=0/X7-OC(Me)2O-8X	58.8	76.4	218.8	39.8	34.2	39.3	82.5	82.6	21.4	1
266	X2-OH N2-Me/X6-Me	56.8	82.2	38.3	30.5	50.7	43.3	36.3	29.4	24.0,19.6	2
267	N2-OH X2-CBr3/6=	53.3	91.6	40.1	30.9	46.9	132.9	129.5	34.0	60.6	1
268	N2-OH X2-CHCl2/6=	51.7	85.3	38.4	29.7	47.5	133.4	129.7	32.8	80.1	1
269	N2-OH X2-Me	53.8	80.0	41.1	30.6	43.3	34.9	27.4	27.7	28.9	2
270	N2-OH X2-Me/7=	62.0	79.5	41.3	32.1	40.1	41.5	133.8	128.8	27.8	2
271	N2-OH X2-Me/X6-Me	54.0	79.7	40.7	28.4	51.4	43.4	36.6	26.1	28.9,19.4	2
272	X2-OH/3=	51.4	85.3	132.4	139.6	49.4	32.4	24.9	30.8		2
273	X2-OH/3=/N8-CH=CH2	55.5	79.1	133.0	138.4	49.2	28.3	29.9	46.6	139.3	1
274	X2-OH/3=/N8-CH=CH2	49.3	79.1	133.3	137.6	49.6	29.3	31.7	47.0	141.7	1
275	X2-OH/3=/N8-OH	48.5	78.6	133.6	137.1	48.7	28.4	35.1	76.4		1
276	N2-OH/3=/N8-OTBS	48.1	78.3	134.1	135.9	47.0	26.8	34.0	79.4		1
277	X2-OH/3=/X8-Me	60.2	83.4	130.8	140.7	49.2	34.6	29.4	39.3	19.7	1
278	X2-OH/6=	49.5	80.9	33.0	28.7	49.2	134.5	128.6	38.0		2
279	N2-OH/6=	44.0	75.2	32.3	28.2	49.4	134.4	130.5	32.4		1
280	X2-OH/7=	60.9	77.6	32.3	33.5	38.7	41.4	131.2	131.7		1
281	N2-OH/7=	55.8	74.8	34.4	31.2	39.5	41.9	128.4	133.8		2
282	N2-OH/N3-OH	45.4	76.1	74.2	37.5	40.1	34.0	26.2	27.7		1
283	N2-OH/N6-OH	49.0	72.6	38.8	20.3	49.0	72.6	38.8	20.3		2
284	N2-OH/N7-OH/8=	158.9	68.6	37.5	29.6	48.1	44.3	82.6	124.3		1
285	N2-OH/N7-OH/X8-I	58.4	73.2	35.1	29.9	40.5	38.4	82.7	31.5		1
286	N2-OH/X3-Cl/N8-OTBS	47.6	82.0	66.0	40.6	39.6	27.3	35.2	77.3		1
287	N2-OH/X3-I/N8-OH	47.2	84.2	32.9	41.2	40.8	30.6	36.3	75.9		1
288	X2-OH/X3-OH	48.7	80.5	74.8	37.8	39.4	34.1	26.2	31.4		1
289	X2-OH/X3-OH/6=	45.7	80.9	74.1	36.0	46.6	134.8	128.6	37.5		1
290	X2-OH/X3-OH/N8-OH	51.7	74.9	76.3	38.0	39.2	29.1	33.4	73.9		1
291	X2-OH/X6-Me	53.3	78.9	33.7	30.9	50.5	42.2	35.9	28.8	19.6	2
292	N2-OH/X6-Me	50.8	75.1	33.4	27.6	47.4	43.5	36.4	25.5	19.4	2
293	X2-OH/X6-OH	51.0	79.9	34.2	27.8	51.0	79.9	34.2	27.8		2
294	2-OMe 2=/7=	56.8	161.8	91.6	36.9	37.6	41.6	130.6	130.7		1
295	X2-OMe N2-OMe/7=	56.5	112.1	31.1	31.1	42.2	38.2	130.1	132.3		1
296	N2-OMe/7=	53.0	84.8	29.8	29.6	38.6	42.4	129.0	132.1		1
297	N2-OMe/X3-SePh	48.1	87.6	53.2	37.0	39.3	33.8	25.1	32.6		58
298	N2-OMs	46.5	85.3	32.5	28.9	42.3	34.6	27.5	27.7		1
299	N2-OMs 7=	53.9	84.2	31.4	30.4	38.7	41.8	127.9	133.2		1
300	N2-OTBS/7-Bu 7=	55.6	76.3	33.5	31.1	38.8	45.1	145.4	122.8	31.0	1

BICYCLO[3.3.0]OCTANES (cont.)

Str	Substituents	C1	C2	C3	C4	C5	C6	C7	C8	Substituent Shifts	References
301	N2-OTBS/7=	55.6	76.1	33.3	31.0	38.3	42.8	130.0	131.3		1
302	N2-OTBS/N7-OTBS/X8-C1	53.9	73.3	34.7	28.7	36.9	40.6	81.5	63.9		1
303	X2-SePh/N3-C1	48.8	56.9	63.2	42.5	40.0	32.2	24.9	33.7		58
304	2=	50.8	134.7	129.5	41.2	40.4	35.9	25.4	32.5		2,67
305	2=/6=	47.8	135.2	128.1	37.9	47.8	135.2	128.1	37.9		1
306	2=7=	58.9	132.4	129.6	41.2	38.8	41.2	129.6	132.4		
307	2=/8-=CH2	56.1	133.4	128.9	40.2	41.2	33.5	33.5	155.3	104.8	2
308	2=/X8-Me	59.5	134.2	128.7	40.6	40.7	34.5	34.3	40.6	20.5	1
309	3-=0/N6-OH	38.3	46.2	221.6	37.9	45.2	74.5	35.3	30.9		2
310	3-=0	39.7	44.7	220.9	44.7	39.7	33.5	25.6	33.5		1
311	3-=0/6=	37.1	44.8	218.8	42.5	46.4	134.3	130.4	40.2		2
312	3-=0/7-=0	36.5	43.7	218.0	43.7	36.5	43.7	218.0	43.7		3,39
313	3-=0/N4-Me/6=	34.5	44.1	220.5	46.4	52.3	129.7	131.7	40.7		2,69
314	3-=0/N6-OC(Me)2O-7N	41.1	42.6	218.8	39.6	44.4	83.4	82.2	35.8	11.3	1
315	3-=0/X4-Me/6=	35.2	44.0	222.0	47.9	55.1	134.0	130.5	40.0		1
316	3-=0/X6-OC(Me)2O-7X	38.0	43.2	217.2	39.0	47.1	85.4	82.1	37.8	15.8	2
317	X3-OAc	41.4	39.8	78.3	39.8	41.4	33.8	25.8	33.9		1
318	N3-OAc	40.4	38.9	76.6	38.9	40.4	33.8	25.4	33.8		1
319	N3-OBS	39.8	39.1	84.2	39.1	39.8	33.5	25.2	33.5		1
320	X3-OCH2CH2O-3N/X8-Me	49.0	41.9	119.6	42.3	40.6	32.8	35.7	40.7	19.6	72
321	N3-OCS2Me/6=	38.0	39.4	86.2	36.6	48.6	133.9	128.2	40.3		1
322	X3-OH	41.0	42.7	74.8	42.7	41.0	33.9	25.8	33.9		2
323	N3-OH	40.2	42.4	73.8	42.4	40.2	33.7	25.1	33.7		1
324	N3-OH/6=	38.1	43.8	74.6	40.6	49.0	135.8	128.7	40.6		1
325	N3-OH/N6-OC(Me)2O-7N	43.3	41.9	75.0	34.9	45.3	84.0	82.4	36.5		1
326	N3-OH/N7-OH	41.2	43.4	76.1	41.2	43.4	41.2	76.1	41.2		2
327	N3-OH/X6-OC(Me)2O-7X	40.4	40.9	73.9	38.8	49.9	86.8	82.8	39.8		1

References to Bicyclo[3.3.0]octanes and related* systems

1. Whitesell, J. K., et al, University of Texas at Austin, unpublished results.
2. Whitesell, J. K. and Matthews, R. S., *J. Org. Chem.* **42**, 3878 (1877).
3. Whitesell, J. K., Minton, M. A. and Flanagan, W. G., Tetrahedron **37**, 4451 (1981).
4. Hudlicky, T., Koszyk, F. J., Kutchan, T. M. and Sheth, J. P., *J. Org. Chem.* **45**, 5020 (1980).
5. Eaton, P. E., Mueller, R. H., Carlson, G. R., Cullison, D. A., Cooper, G. F., Chou, T.-C. and Krebs, E. P., *J. Am. Chem. Soc.* **99**, 2751 (1977).
6. Gassman, P. G., Valcho, J. J., Proehl, G. S. and Cooper, C F., *J. Am. Chem. Soc.* **102**, 6519 (1980).
*7. Weber, R. W. and Cook, J. M., *Can. J. Chem.* **56**, 189 (1978).
*8. Inamoto, Y., Aigami, K., Fujikura, Y., Takaishi, N. and Tsuchihashi, K., *J. Org. Chem.* **44**, 854 (1979).
*9. Cargill, R. L., Dalton, J. R., O'Connor, S. and Michels, D. G., *Tetrahedron Lett.* 4465 (1978).
*10. Takaishi, N., Inamoto, Y., Tsuchihashi, K., Yashima, K. and Aigami, K., *J. Org. Chem.* **40**, 2929 (1975).
*11. Salomon, R. G., Ghosh, S., Zagorski, M. G. and Reitz, M., *J. Org. Chem.* **47**, 829 (1982).
*12. Fujikura, Y., Takaishi, N. and Inamoto, Y., *Tetrahedron* **37**, 4465 (1981).
*13. Codding, P. W., Kerr, K. A., Oudeman, A. and Sorensen, T. S., *J. Organomet. Chem.* **232**, 193 (1982).
*14. Bosse, D. and de Meijere, A., *Chem. Ber.* **111**, 2243 (1978).
*15. Paquette, L. A., Kearney, F. R., Drake, A. F. and Mason, S. F., *J. Am. Chem. Soc.* **103**, 5064 (1981).
16. Yates, P. and Stevens, K. E., *Tetrahedron* **37**, 4401 (1981).
*17. Eaton, P. E., Giordano, C., Schloemer, G. and Vogel, U., *J. Org. Chem.* **41**, 2238 (1976).
*18. Eaton, P. E., Sidhu, R. S., Langford, G. E., Cullison, D. A. and Pietruszewski, C., *Tetrahedron* **37**, 4479 (1981).
19. Mehta, G., Srikrishna, A., Reddy, A. V. and Nair, M. S., *Tetrahedron* **37**, 4543 (1981).
*20. Takaishi, N., Inamoto, Y., Tsuchihashi, K., Yashima, K. and Aigami, K., *J. Org. Chem.* **43**, 5026 (1978).
*21. Seto, H., Sasaki, T., Uzawa, J., Takeuchi, S. and Yonehara, H., *Tetrahedron Lett.* 4411 (1978).
*22. Seto, H. and Yonehara, H., *J. Antibiot.* **33**, 92 (1980).
*23. Jaggi, F. J. and Ganter, C., *Helv. Chim. Acta* **63**, 2087 (1980).
*24. Majerski, Z., Djigas, S. and Vinkovic, V., *J. Org. Chem.* **44**, 4064 (1979).
*25. Jaggi, F. J. and Ganter, C., *Helv. Chim. Acta* **63**, 866 (1980).
*26. Tobe, Y., Terashima, K., Sakai, Y. and Odaira, Y., *J. Am. Chem. Soc.* **103**, 2307 (1981).
*27. Bishop, R. and Landers, A. E., *Aust. J. Chem.* **32**, 2675 (1979).
*28. Klester, A. M. and Ganter, C., *Helv. Chim. Acta* **66**, 1200 (1983).
*29. Bohlmann, F., Van, N. L. and Pickardt, J., *Chem. Ber.* **110**, 3777 (1977).
*30. Boglmann, F. and Jakupovic, J., *Phytochem.* **19**, 259 (1980).
*31. Bohlmann, F., Van, N. L., Pham, T. V. C., Jacupovic, J., Schuster, A., Zabel, V. and Watson, W. H., *Phytochem.* **18**, 1831 (1979).

*32. Feline, T. C., Mellows, G., Jones, R. B. and Phillips, L., *J. Chem. Soc Chem. Commun.* 63 (1974).

*33. Zalkow, L. H., Harris, III, R. N. and Burke, N. I., *J. Nat. Prod.* **42**, 96 (1979).

*34. Schostarez, H. and Paquette, L. A., *Tetrahedron* **37**, 4431 (1981).

*35. Pirrung, M. C., *J. Am. Chem. Soc.* **103**, 82 (1981).

*36. Inamoto, Y., Aigami, K., Takaishi, N., Fujikura, Y., Tsuchihashi, K. and Ikeda, H., *J. Org. Chem.* **42**, 3833 (1977).

*37. Inamoto, Y., Takaishi, N., Fujikura, Y., Aigami, K. and Tsuchihashi, K., *Chem. Lett.* 631 (1976).

38. Goldstein, M. J., Wenzel, T. T., Whittaker, G. and Yates, S. F., *J. Am. Chem. Soc.* **104**, 2669 (1982).

39. Berger, S., *Org. Magn. Reson.* **14**, 65 (1980).

40. Haslanger, M. F. and Ahmed, S., *J. Org. Chem.* **46**, 4808 (1981).

41. Quast, H., Christ, J., Gorlach, Y. and von der Saal, W., *Tetrahedron Lett.* **23**, 3653 (1982).

42. Trost, B. M. and Curran, D. P., *Tetrahedron Lett.* **22**, 4929 (1981).

43. Trost, B. M. and Curran, D. P., *J. Am. Chem. Soc.* **102**, 5699 (1980).

*44. Venegas, M. and Little, R. D., *Tetrahedron Lett.* 309 (1979).

45. Takeda, K., Shimono, Y. and Yoshii, E., *J. Am. Chem. Soc.* **105**, 563 (1983).

46. Farnum, D. G. and Monego, T., *Tetrahedron Lett.* **24**, 1361 (1983).

47. Smith, I. I. I. A. B., Toder, B. H., Branca, S. J. and Dieter, R. K., *J. Am. Chem. Soc.* **103**, 1996 (1981).

48. Klipa, D. K. and Hart, H., *J. Org. Chem.* **46**, 2815 (1981).

49. Aristoff, P. A., *Synth. Commun.* **13**, 145 (1983).

50. Oehldrich, J., Cook, J. M. and Weiss, U., *Tetrahedron Lett.* 4549 (1976).

51. Cooper, K. and Pattenden, G., *J. Chem. Soc. Perkin Trans. I* 799 (1984).

52. Horton, M. and Pattenden, G., *J. Chem. Soc. Perkin Trans. I* 811 (1984).

53. Sakkers, P. J. D., Vankan, J. M. J., Klunder, A. J. H. and Zwanenburg, B., *Tetrahedron Lett.* 897 (1979).

54. Docken, A. M., *J. Org. Chem.* **46**, 4096 (1981).

*55. Mehta, G. and Rao, K. S., *Tetrahedron Lett.* **24**, 809 (1983).

56. Crandall, J. K., Magaha, H. S., Henderson, M. A., Widener, R. K. and Tharp, G. A., *J. Org. Chem.* **47**, 5372 (1982).

57. Kramer, G. W. and Brown, H. C., *J. Org. Chem.* **42**, 2832 (1977).

58. Garratt, D. G. and Kabo, A., *Can. J. Chem.* **58**, 1030 (1980).

59. Gover, S. H., Marr, D. H., Stothers, J. B. and Tan, C. T., *Can. J. Chem.* **53**, 1351 (1975).

*60. Geneste, P., Durand, R., Kamenka, J.-M., Beierbeck, H., Martino, R. and Saunders, J. K., *Can. J. Chem.* **56**, 1940 (1978).

61. Uemura, S., Fukuzawa, S., Toshimitsu, A., Okano, M., Tezuka, H. and Sawada, S., *J. Org. Chem.* **48**, 270 (1983).

62. Yates, P. and Stevens, K. E., *Can. J. Chem.* **60**, 825 (1982).

63. Tsunoda, T., Kodama, M. and Ito, S., *Tetrahedron Lett.* **24**, 83 (1983).

*64. Mikhail, G. and Demuth, M., *Helv. Chim. Acta* **66**, 2362 (1983).

*65. Little, R. D., Muller, G. W., Venegas, M. G., Carroll, G. L., Bukhari, A., Patton, L. and Stone, K., *Tetrahedron* **37**, 4371 (1981).

66. Becker, K. B., *Helv. Chim. Acta* **60**, 68 (1977).

67. Kelly, D. P., Underwood, G. R. and Barron, F. F., *J. Am. Chem. Soc.* **98**, 3106 (1976).

*68. Wender, P. A. and Dreyer, G. B., *J. Am. Chem. Soc.* **104**, 5805 (1982).

69. Bertz, S. H., Rihs, G. and Woodward, R. B., *Tetrahedron* **38**, 63 (1982).

70. Callant, P., Ongena, R. and Vandewalle, M., *Tetrahedron* **37**, 2085 (1981).

71. Carceller, E., Moyano, A. and Serratosa, F., *Tetrahedron Lett.* **25**, 2031 (1984).

72. Callant, P., De Wilde, H. and Vandewalle, M., *Tetrahedron* **37**, 2079 (1981).

73. Warner, P. and Lu, S.-L., *J. Am. Chem. Soc.* **98**, 6752 (1976).

74. Koller, M., Karpf, M. and Dreiding, A. S., *Helv. Chim. Acta* **66**, 2760 (1983).

75. Tice, C. M. and Heathcock, C. H., *J. Org. Chem.* **46**, 9 (1981).

Chapter 13

Bicyclo[3.3.1]nonanes and Tricyclo[3.3.1.1³,⁷]decanes

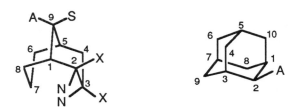

Bicyclo[3.3.1]nonanes can be specified with exo/endo and syn/anti, where syn always refers to the orientation of a group on C-9 toward the lower numbered, three-carbon bridge. Adamantanes, or tricyclo[3.3.1.1³,⁷]decanes, can serve as quite useful models for chair cyclohexane rings that have no conformational freedom. For C-2 monosubstituted compounds, the lower numbered ring has the substituent axial while C-7, C-8, and C-10 are anti to the substituent. For 2,2-disubstituted compounds, the lower numbered ring has the substituent of highest priority (in the Cahn, Ingold, Prelog system) axial. With substituents on C-2 and C-4, stereochemistry for each is specified as either axial (A) or equitorial (E) to the ring containing both of these carbons. Finally, with substituents on C-2 and C-6, C-1 is syn to the C-6 group, C-9 is syn to both, C-3 is anti to the C-6 substituent and C-10 is anti to both.

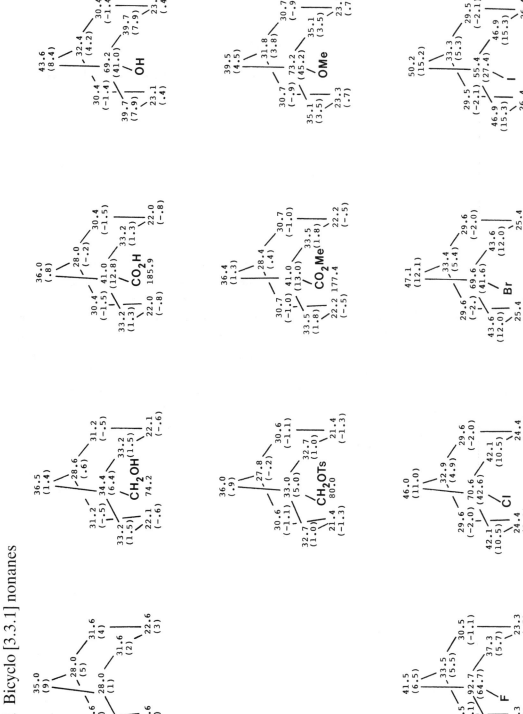

13.1 Bicyclo [3.3.1] nonanes

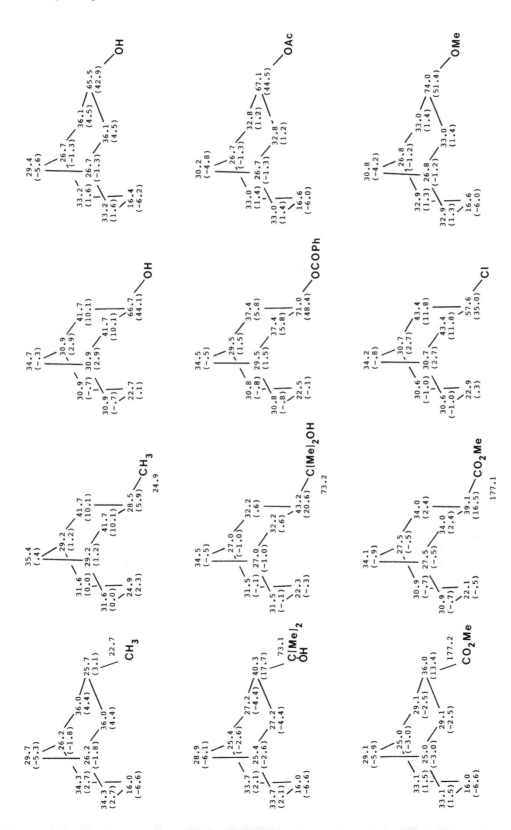

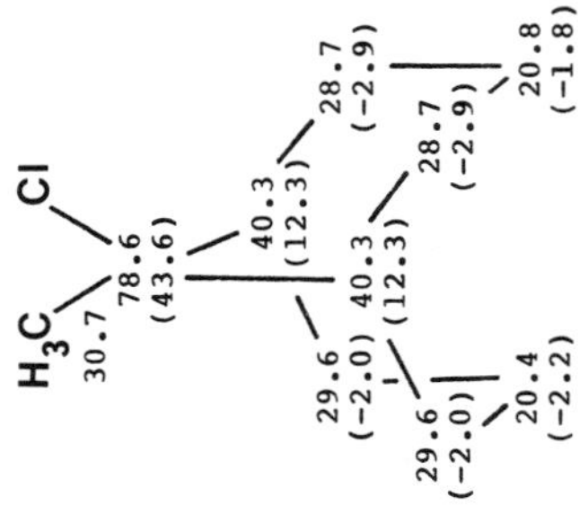
H3C
30.7
Cl
78.6
(43.6)
40.3
(12.3)
28.7
(-2.9)
40.3
(12.3)
28.7
(-2.9)
20.8
(-1.8)
29.6
(-2.0)
29.6
(-2.0)
20.4
(-2.2)

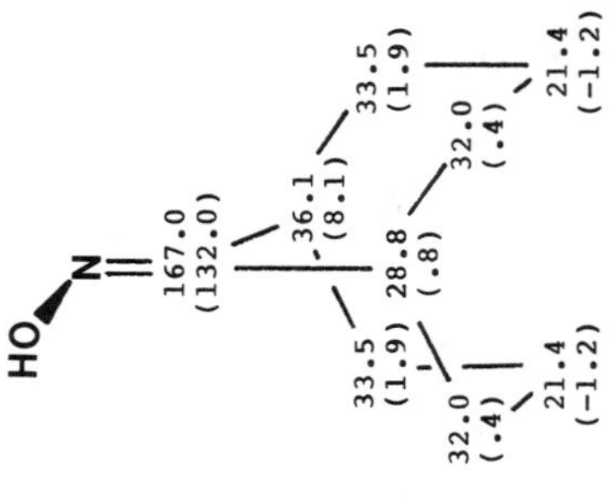
HO
N
167.0
(132.0)
36.1
(8.1)
33.5
(1.9)
28.8
(.8)
32.0
(.4)
21.4
(-1.2)
33.5
(1.9)
32.0
(.4)
21.4
(-1.2)

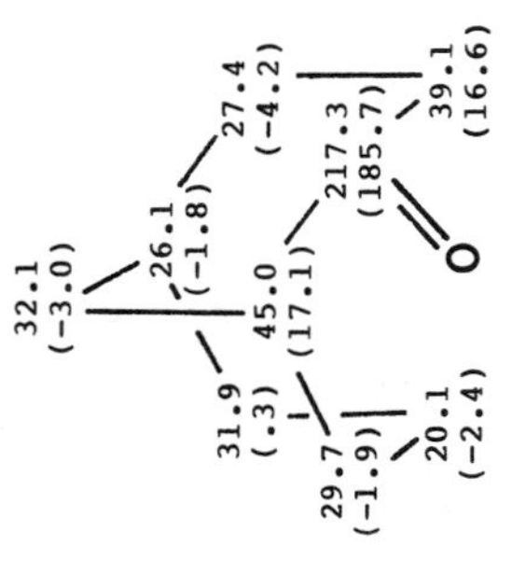
32.1
(-3.0)
26.1
(-1.8)
27.4
(-4.2)
45.0
(17.1)
217.3
(185.7)
O
39.1
(16.6)
31.9
(.3)
29.7
(-1.9)
20.1
(-2.4)

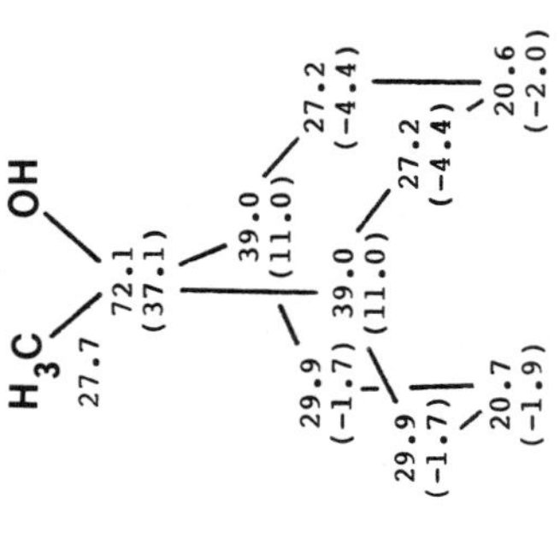
H3C
27.7
OH
72.1
(37.1)
39.0
(11.0)
27.2
(-4.4)
39.0
(11.0)
27.2
(-4.4)
20.6
(-2.0)
29.9
(-1.7)
29.9
(-1.7)
20.7
(-1.9)

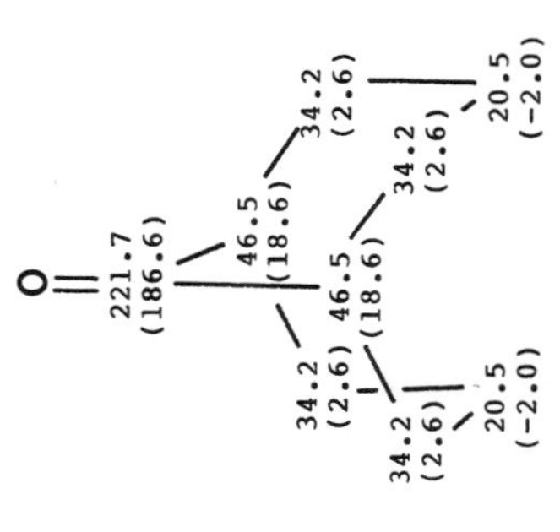
O
221.7
(186.6)
46.5
(18.6)
34.2
(2.6)
46.5
(18.6)
34.2
(2.6)
20.5
(-2.0)
34.2
(2.6)
34.2
(2.6)
20.5
(-2.0)

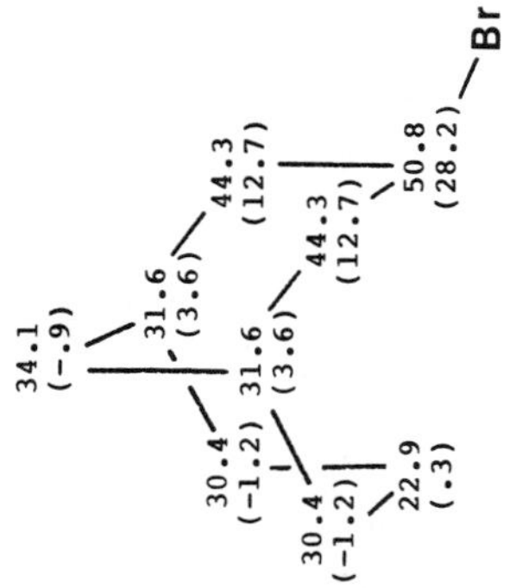
34.1
(-.9)
31.6
(3.6)
44.3
(12.7)
31.6
(3.6)
44.3
(12.7)
50.8
(28.2)
Br
30.4
(-1.2)
30.4
(-1.2)
22.9
(.3)

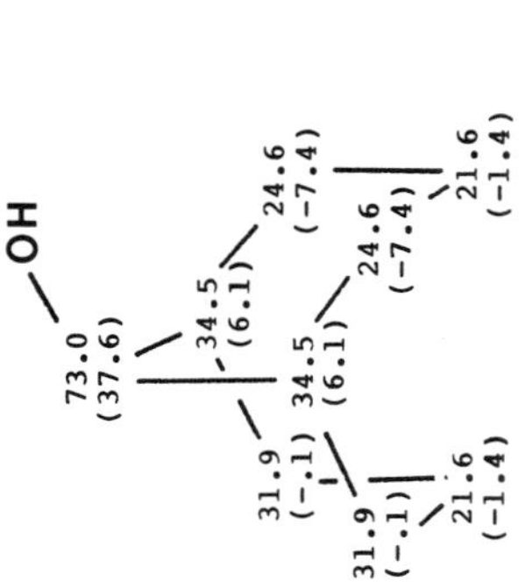
OH
73.0
(37.6)
34.5
(6.1)
24.6
(-7.4)
34.5
(6.1)
24.6
(-7.4)
21.6
(-1.4)
31.9
(-.1)
31.9
(-.1)
21.6
(-1.4)

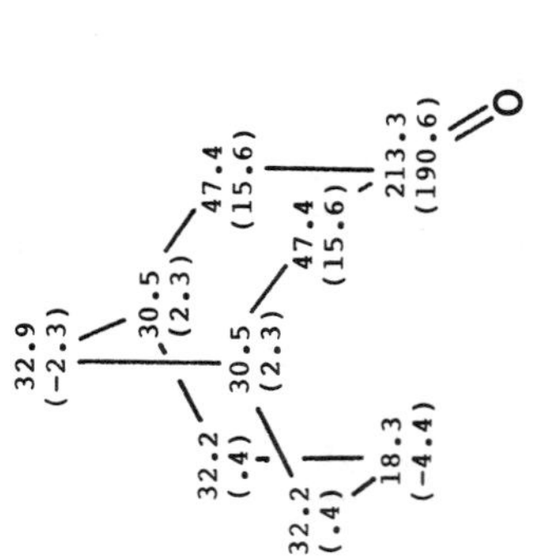
32.9
(-2.3)
30.5
(2.3)
47.4
(15.6)
30.5
(2.3)
47.4
(15.6)
213.3
(190.6)
O
32.2
(.4)
32.2
(.4)
18.3
(-4.4)

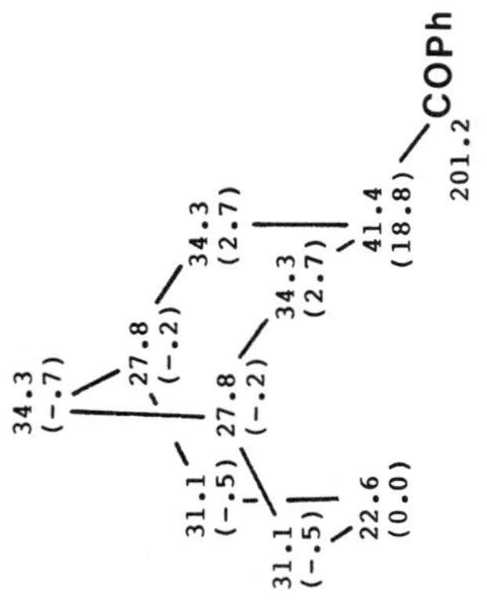
34.3
(-.7)
27.8
(-.2)
34.3
(2.7)
27.8
(-.2)
34.3
(2.7)
41.4
(18.8)
COPh
201.2
31.1
(-.5)
31.1
(-.5)
22.6
(0.0)

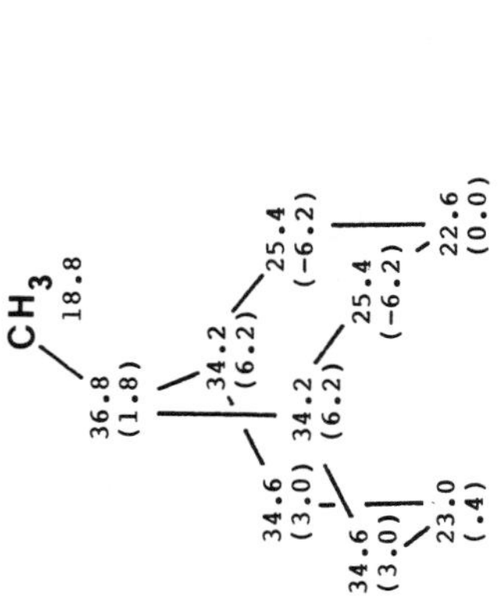
CH3
18.8
36.8
(1.8)
34.2
(6.2)
25.4
(-6.2)
34.2
(6.2)
25.4
(-6.2)
22.6
(0.0)
34.6
(3.0)
34.6
(3.0)
23.0
(.4)

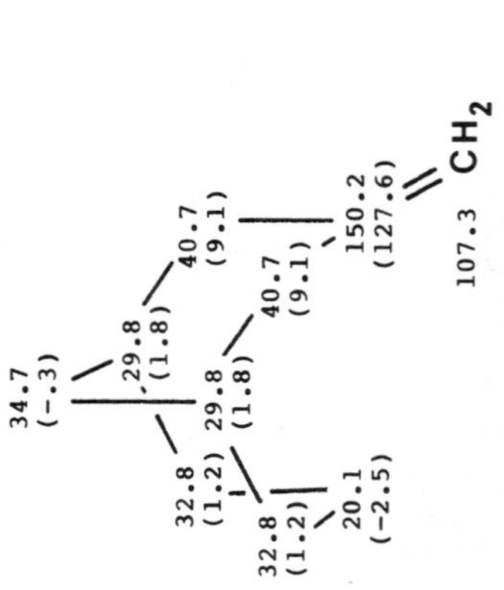
34.7
(-.3)
29.8
(1.8)
40.7
(9.1)
29.8
(1.8)
40.7
(9.1)
150.2
(127.6)
CH2
107.3
32.8
(1.2)
32.8
(1.2)
20.1
(-2.5)

BICYCLO[3.3.1]NONANES

Str	Substituents	C1	C2	C3	C4	C5	C6	C7	C8	C9	Substituent Shifts	References
1	PARENT	28.0	31.6	22.6	31.6	28.0	31.6	22.6	31.6	35.0		6,1,2,3,9,35,83
2	1-Br	69.6	43.6	25.4	29.6	33.4	29.6	25.4	43.6	47.1		4,83
3	1-Br/3-=O	62.8	58.2	207.5	45.2	32.0	30.1	21.6	44.4	44.6		83,108
4	1-Br/3-=O/N4-Ph	62.5	58.0	204.4	59.1	40.4	24.9	21.8	44.5	46.6	136.1	112
5	1-Br/N2-Me/3-=O	70.3	57.6	207.3	47.0	32.2	30.2	21.3	39.1	45.8	10.9	111
6	1-Br/N2-Ph/3-=O	69.0	69.7	204.6	47.4	32.2	30.2	22.1	39.2	46.0	133.7	112
7	1-CH(COOH)2/3-=O	38.1	47.7	211.9	45.6	29.4	35.3	18.3	30.5	35.3	60.9	112
8	1-CH(COOMe)(COOH)/3-=O	39.2	48.0	212.4	46.1	30.0	35.8	18.7	31.0	36.1	61.1	112
9	1-CH(COOMe)2/3-=O	38.9	48.2	210.4	46.0	29.9	31.0	18.6	31.0	36.1	61.1	110
10	1-CH2OH	34.4	33.2	22.1	31.2	28.6	31.2	22.1	33.2	36.5	74.2	3
11	1-CH2OTs	33.0	32.7	21.4	30.6	27.8	30.6	21.4	32.7	36.0	80.0	3
12	1-Cl	70.6	42.1	24.4	29.6	32.9	29.6	24.4	42.1	46.0		4,9
13	1-Cl/3-=O	67.5	56.9	208.2	45.3	31.5	30.2	21.0	42.9	43.5		83,108
14	1-CN/3-=O	33.3	48.2	205.9	45.6	29.1	30.0	17.5	35.4	35.6	123.0	110
15	1-COOH	41.0	33.2	22.0	30.4	28.0	30.4	22.0	33.2	36.0	185.9	9,3
16	1-COOH/5-Br	45.8	31.4	24.4	42.2	65.6	42.2	24.4	31.4	47.3	182.7	3
17	1-COOMe	41.0	33.5	22.2	30.7	28.4	30.7	22.2	33.5	36.4	177.4	3
18	1-COOMe/2-=O/3-COOMe/5-COOMe/6-=O/7-COOMe	51.8	199.4	47.5	29.6	51.8	199.4	47.5	29.6	35.2	170.7,168.3,170.7,168.3	3
19	1-COOMe/5-COOH	41.6	32.3	21.4	32.1	41.6	32.1	21.4	32.3	36.9	177.4,184.1	3
20	1-COOMe/5-COOMe	41.6	32.4	21.4	32.4	41.6	32.4	21.4	32.4	37.2	176.8,176.8	3
21	1-F	92.7	37.3	23.3	30.5	33.5	30.5	23.3	37.3	41.5		4
22	1-F/3-=O	93.6	53.2	208.8	45.6	30.5	30.5	20.0	37.8	39.0		83
23	1-I	55.4	46.9	26.4	29.5	33.3	29.5	26.4	46.9	50.2		4
24	1-I/3-=O	44.2	61.1	207.4	45.5	32.5	29.5	22.2	47.2	47.8		83
25	1-NHAc/5-Br	54.7	34.3	23.6	42.0	65.9	42.0	23.6	34.3	50.7		3
26	1-NHNHCOOEt/3-=O	58.4	50.3	211.0	45.7	29.9	31.1	19.1	35.7	36.3		108
27	1-NMe2/3-=O	57.5	46.7	211.4	46.0	30.2	31.1	19.3	32.5	36.5		108
28	1-O)2/3-=O	80.5	51.0	209.3	45.8	30.2	31.1	19.5	35.4	36.6		110
29	1-OH	69.2	39.7	23.1	30.4	32.4	30.4	23.1	39.7	43.6		83,4,3,9
30	1-OH/3-=O	70.8	55.4	212.1	45.7	30.7	30.8	20.1	40.3	41.2		83,108
31	1-OH/N2-Me/3-=O	72.9	55.5	212.0	45.7	30.4	30.0	19.8	34.6	43.0	7.5	111
32	1-OH/N2-Ph/3-=O	73.0	68.2	207.5	45.9	30.2	31.2	20.1	35.7	41.8	132.5	112
33	1-OH/X2-OH N2-Ph/3-=O	74.1	82.5	207.2	42.4	29.3	30.7	19.8	34.9	36.6	135.4	113
34	1-OH/X2-SePh/3-=O	70.6	66.1	206.5	41.3	29.5	30.5	19.5	38.0	39.1		108
35	1-OMe	73.2	35.1	23.3	30.7	31.8	30.7	23.3	35.1	39.5		4
36	1-OMe/3-=O	74.9	51.3	210.9	46.1	30.4	31.1	19.7	35.9	37.6		108
37	1-OMe/N2-Ph/3-=O	77.5	64.4	208.2	46.1	30.0	31.2	20.1	32.3	39.1	133.8	113
38	1-OOH/3-=O	81.9	50.0	211.7	45.7	30.9	30.9	19.5	34.8	35.7		110
39	1-SePh/3-=O	45.6	53.9	210.0	46.2	31.4	30.8	20.6	34.8	40.1		108
40	1-SMe/3-=O	44.5	52.1	209.4	45.7	30.8	30.8	19.7	37.7	39.1		108
41	1-SPh/3-=O	49.2	53.0	209.9	45.6	30.9	30.9	20.1	38.4	39.3		108
42	1= 2-Ph 3-=O	163.5	137.9	199.6	50.3	33.3	29.1	27.7	36.2	36.8	135.7	113
43	2-=CH2/6-=CH2	37.6	152.8	31.0	32.7	37.6	152.8	31.0	29.7	36.1	107.5,107.5	41
44	2-=O	45.0	217.3	39.1	27.4	26.1	31.9	20.1	29.7	32.1		35
45	2-=O 3= 4-CH2OCH2CH=CH2/9-=O	62.3	197.8	127.9	158.8	50.3	32.9	17.1	30.5	207.4	70.1	119
46	2-=O 3= 4-CH2OCH2CH2CH2/9-OCH2CH2O-9	52.3	201.1	127.1	160.0	42.0	25.9	16.0	24.2	110.2	70.7	119
47	2-=O 3= 4-CH2OMe/9-=O	62.3	197.8	158.7	158.7	50.2	32.8	17.0	30.3	207.4	72.7	119
48	2-=O 3= 4-Me/9-=O	61.6	197.8	130.2	159.8	54.6	32.8	17.1	29.7	208.0	22.6	119
49	2-=O 3= 4-Me/9-OCH2CH2O-9	46.5	200.9	128.7	161.1	51.5	25.8	16.0	23.6	110.3	23.1	119
50	2-=O 3/9-=O	63.3	198.9	132.7	147.9	33.4	33.4	16.4	29.9	207.8		119

BICYCLO[3.3.1]NONANES (cont.)

Str	Substituents	C1	C2	C3	C4	C5	C6	C7	C8	C9	Substituent Shifts	References
51	2-=O 3=/9-OCH_2CH_2O-9	52.9	201.4	131.8	149.2	41.0	26.3	15.4	23.7	110.0	74.2	119
52	2-=O/4-CH_2OCH_2CH=CH_2/9-=O	63.4	209.0	43.6		49.4	36.0	18.9	35.5	211.5	76.7	119
53	2-=O/4-CH_2OMe/9-=O	63.3	208.7	43.3	35.4	49.2	35.9	18.8	35.7	211.2	23.2	119
54	2-=O/4-Me/9-=O	62.7	209.5	48.3	30.0	53.3	35.3	18.6	34.9	211.9		119
55	2-=O/6-=O	43.5	212.6	37.0	26.6	43.5	212.6	37.0	26.6	31.7		35,3
56	2-=O/7-OOCPh/9-=O	59.9	211.1	39.5	22.6	42.6	39.8	68.0	37.9	209.2		35
57	2-=O/9-=O	64.2	211.3	39.0	22.2	44.3	35.6	18.6	34.9	209.7		35
58	2-=O/N6-OH	43.6	217.4	38.2	19.6	33.1	72.0	28.7	25.3	30.4		115
59	2-=O/X4-CH=CH_2 N4-CH_2OCH_2CH=CH_2/9-OCH_2CH_2O-9		211.3			42.6	27.0	19.1	25.3	210.4	147.1,75.9	119
60	2-=O/X4-CH=CH_2 N4-CH_2OCH_2CH=CH_2/9-=O	65.9	207.9	45.7	42.8	52.5	35.2	19.9	30.5	210.4	142.1,74.7	119
61	2-=O/X4-CH=CH_2 N4-Me/9-=O	65.4	208.0	50.1	38.6	55.4	35.0	19.7	30.1	210.8	145.4,24.9	119
62	2-=O/X4-CH=CH_2 N4-Me/9-OCH_2CH_2O-9	54.4	212.0	51.3	38.3	45.7	27.2	18.6	25.9	110.4	150.5,26.9	119
63	X2-Cl/9-=O	54.6	65.5	28.4	29.9	45.5	34.0	19.6	32.7	216.1		1
64	N2-Cl/9-=O	53.9	61.7	31.5	29.8	44.9	34.0	20.4	28.8	215.8		1
65	2-CN 2=/6-CN 6=	28.7	115.9	142.4	30.7	28.7	115.9	142.4	30.7	26.3	118.3,118.3	44
66	2-CN 2=/X4-Br/6-CN 6=	28.8	113.3	144.0	41.9	37.0	116.4	140.2	30.6	22.0	117.2,117.5	44
67	2-CN 2=/X4-Br/6-CN 6=/X8-OH	37.1	141.0	141.5	40.9	37.1	113.7	113.7	40.9	17.8	116.6,116.6	44
68	2-COOEt 2= 3-OH/4-Ph/5-OH	30.0	100.7	172.1	58.4	70.6	35.5	18.6	27.6	41.0	171.6,135.9	112
69	2-COOEt 2= 3-OH/5-OH	30.5	99.3	172.1	43.8	68.4	40.8	19.3	28.7	41.2	170.8	80
70	2-COOMe 2= 3-OH/N4-Me/5-OH	30.1	98.3	175.5	45.0	69.4	36.2	18.9	28.4	42.0	171.3,51.2	111
71	2-COOtBu 2= 3-OH/N4-Ph/5-OH	30.7	101.6	175.8	58.5	80.8	35.8	19.0	28.2	41.3	175.8,135.8	112
72	2-COOtBu 2= 3-OH/X4-Ph/5-OH	30.5	102.4	175.8	57.0	81.0	37.0	18.8	28.2	40.8	175.8,137.4	112
73	2-Me 2= 3-OTMS/6-Me 6=	32.8	114.5	142.3	34.8	34.0	135.8	119.3	30.6	30.1	15.1,22.4	47
74	2-Me 2=/6-=CH_2	34.2	136.2	121.8	28.5	36.8	155.1	29.9	32.6	33.3	22.2,105.3	41
75	2-Me 2=/6-Me 6=	32.4	136.6	118.8	30.4	32.4	136.6	118.8	30.4	29.7	22.2,22.2	47
76	X2-Me/3-=O/6-Me 6=	35.7	51.9	210.3	40.0	35.7	135.6	118.8	33.4	25.8	18.0,21.2	47
77	N2-Me/3-=O/6-Me 6=	36.2	48.4	210.3	45.0	36.5	135.6	118.8	32.6	25.5	11.6,21.2	47
78	2-OAc 2=/6-OAc 6=	30.2	149.7	111.4	28.3	30.2	149.7	111.4	28.3	28.7		3
79	X2-OAc/9-=O	51.1	78.2	26.2	28.8	45.8	34.7	19.5	30.8	217.8		1
80	N2-OAc/9-=O	51.2	74.7	28.8	27.7	45.5	34.5	20.9	26.7	216.6		1
81	X2-OH N2-Et/X6-OH N6-Et	36.3	71.4	25.4	23.2	36.3	71.4	25.4	23.2	33.5	33.3,33.3	42
82	X2-OH/9-=O	54.7	76.8	28.5	29.1	46.2	34.5	19.6	30.6	220.1		1
83	N2-OH/9-=O	54.4	73.4	29.5	27.7	45.2	34.0	20.8	27.2	219.3		1
84	X2-OH/X6-OH	34.3	69.3	23.3	20.8	34.3	69.3	23.3	20.8	28.4		42
85	X2-OTs/9-=O	51.7	86.4	27.1	27.8	45.5	34.8	19.3	30.9	215.5		1
86	N2-OTs/9-=O	51.9	81.7	28.2	27.5	44.8	34.0	18.5	26.7	215.0		1
87	N2-Ph/3-=O	38.9	60.9	209.6	47.2	31.3	32.3	18.5	26.7	35.1	137.4	113
88	2=/6=	28.2	131.4	124.7	30.6	28.2	131.4	124.7	30.6	28.5		18
89	3-=CH_2	29.8	40.7	150.2	40.7	29.8	32.8	20.1	32.8	34.7	107.3	20
90	3-=CH_2/7-=CH_2	29.4	40.8	145.2	40.8	29.4	40.8	145.2	40.8	33.7	110.4,110.4	20
91	3-=O	30.5	47.4	213.3	47.4	30.5	32.2	18.3	32.2	32.9		83,20
92	3-=O/6= 7-Me	30.4	46.3	208.9	49.1	31.4	125.2	132.5	37.7	30.4	23.2	3
93	3-=O/6=/X8-Br	30.4	45.6	208.6	46.8	38.9	127.3	133.2	50.6	25.3		46
94	3-=O/7-=CH_2	31.0	47.3	209.3	47.3	31.0	41.6	141.8	41.6	32.2	114.2	20,3
95	3-=O/7-=O	32.7	47.8	208.8	47.8	32.7	47.8	208.8	47.8	31.3		35
96	3-=O/N7-Me	28.9	50.4	211.8	50.4	28.9	36.5	23.7	36.5	28.5	22.1	20
97	3-=O/X7-Me	30.9	47.6	52.8	47.6	30.9	41.1	24.1	41.1	32.5	23.0	20
98	X3-Br	31.6	44.3	50.8	44.3	31.6	30.4	30.4	30.4	34.1		4
99	X3-C(Me)$_2$OH	27.0	32.2	43.2	32.2	27.0	31.5	22.3	31.5	34.5	73.2	5
100	N3-C(Me)$_2$OH	25.4	27.2	40.3	27.2	25.4	33.7	16.0	33.7	28.9	73.1	5

BICYCLO[3.3.1]NONANES (cont.)

Str	Substituents	C1	C2	C3	C4	C5	C6	C7	C8	C9	Substituent Shifts	References
101	N3-C(Me)2OH/N7-C(Me)2OH	24.5	32.0	41.4	32.0	24.5	32.0	41.4	32.0	23.7	72.7,72.7	5
102	N3-C(Me)2OH/N7-C(Me)3/9-=O	42.4	34.4	42.7	34.4	42.4	34.8	42.2	34.8	224.7	0,32.1	5
103	X3-C(Me)3/9-=O	46.1	35.9	40.7	35.9	46.1	34.2	21.3	34.2	221.3	32.8	5
104	N3-C(Me)3/9-=O	44.6	30.7	41.3	30.7	44.6	36.1	15.2	36.1	222.2	32.4	5
105	X3-C(Me)3/9-=)2anti	32.2	35.2	42.1	35.2	32.2	22.6	33.9	33.9	131.7	32.8	106
106	X3-C(Me)3/9-=)2syn	32.2	53.4	42.2	53.4	32.2	33.7	22.6	33.7	131.7	32.9	106
107	X3-Cl	30.7	43.4	57.6	43.4	30.7	30.6	22.9	30.6	34.2		4
108	N3-COOEt	25.0	29.2	36.1	29.2	25.0	33.0	16.0	33.0	29.1	176.5	4
109	N3-COOH X3-Me/N7-COOH	25.0	40.9	38.8	40.9	25.0	27.3	35.6	27.3	28.8	180.9,33.2,178.0	7,17
110	X3-COOH/9-=O	46.0	37.0	38.0	37.0	46.0	34.1	21.0	34.1	218.9	177.1	10
111	N3-COOH/9-=O	43.7	31.8	36.9	31.8	43.7	35.9	15.1	35.9	220.9	179.5	10
112	N3-COOH/N7-COOH	24.1	29.8	35.1	29.8	24.1	29.8	35.1	29.8	29.0	0	7
113	X3-COOMe	27.5	34.0	39.1	34.0	27.5	30.9	22.1	30.9	34.1	177.1	5
114	N3-COOMe	25.0	29.1	36.0	29.1	25.0	33.1	16.0	33.1	29.1	177.2	5
115	N3-COOMe X3-Me/N7-COOMe	24.9	41.2	39.4	41.2	24.9	27.2	35.5	27.2	28.7	0,0,32.9	17
116	X3-COOMe/9-=O	45.4	36.4	37.4	36.4	45.4	34.0	21.0	34.0		0	5
117	N3-COOMe/9-=O	43.8	32.0	37.0	32.0	43.8	35.8	15.1	35.8	219.8	174.6	5
118	X3-COOMe/N7-C(Me)3	25.2	36.4	33.9	36.4	25.2	27.8	38.8	27.8	27.2	177.0,32.5	5
119	X3-COOMe/N7-C(Me)3/9-=O	43.1	37.9	33.2	37.9	43.1	32.4	40.9	32.4	219.9	184.9,32.4	5
120	X3-COOMe/X7-C(Me)3	28.1	34.0	39.2	34.0	28.1	33.3	42.3	33.3	33.8	176.9,32.0	5
121	N3-COOMe/X7-C(Me)3	25.4	29.6	35.8	29.6	25.4	34.0	36.5	34.0	29.0	177.2,32.1	5
122	X3-COOMe/X7-C(Me)3/9-=O	45.2	36.4	38.1	36.4	45.2	35.7	41.5	35.7		0,31.0	5
123	N3-COOMe/X7-C(Me)3/9-=O	43.0	32.5	36.8	32.5	43.0	36.8	36.0	36.8	220.6	174.6,31.9	5
124	N3-COOMe/X7-CH(Me)2/9-=O	43.0	32.6	36.8	32.6	43.0	39.1	31.7	39.1	220.6	174.7,32.3	5
125	N3-COOMe/X7-Me	25.5	29.6	35.3	29.6	25.5	42.3	22.8	42.3	28.9	21.7,0	17
126	X3-COPh	27.8	34.3	41.4	34.3	27.8	31.1	22.6	31.1	34.3	201.2	81
127	X3-COPh/9-=O	45.6	36.8	39.1	36.8	45.6	34.1	21.3	34.1	218.5	201.0	81
128	N3-COPh/9-=O	43.8	32.2	38.8	32.2	43.8	35.8	15.1	35.8	220.0	201.5	81
129	X3-Me	29.2	41.7	28.5	41.7	29.2	31.6	24.9	31.6	35.4	24.9	20
130	N3-Me	26.2	36.0	25.7	36.0	26.2	34.3	16.0	34.3	29.7	22.7	4
131	N3-Me/N7-Me	25.7	40.0	25.4	40.0	25.7	40.0	25.4	40.0	23.4	25.0,25.0	100
132	X3-Me/X7-Me	29.7	41.1	28.4	41.1	29.7	41.1	28.4	41.1	35.3	24.8,24.8	100
133	N3-Me/X7-Me	26.4	35.9	24.7	35.9	26.4	42.9	21.9	42.9	28.8	22.9,22.2	100
134	N3-OAc	26.7	32.8	67.1	32.8	26.7	33.0	16.6	33.0	30.2		4
135	X3-OCOPh	29.5	37.4	71.0	37.4	29.5	30.8	22.5	30.8	34.5		82
136	X3-OH	30.9	41.7	66.7	41.7	30.9	30.9	22.7	30.9	34.7		4
137	N3-OH	26.7	36.1	65.5	36.1	26.7	33.2	16.4	33.2	29.4		4
138	X3-OH/6= 7-Me	28.4	38.6	64.8	43.2	30.1	125.4	134.7	37.2	31.3	23.1	3
139	X3-OH/7-=CH2	30.9	42.0	63.8	42.0	30.9	40.2	148.8	40.2	33.9	108.8	20
140	N3-OH/7-=CH2	28.6	37.9	63.3	37.9	28.6	40.6	147.9	40.6	32.2	112.2	20
141	X3-OH/9-=O	45.2	41.6	63.8	41.6	45.2	35.2	19.6	35.2	218.6		11
142	X3-OH/N7-Me	26.6	43.2	64.1	43.2	26.6	35.7	24.2	35.7	28.2	22.0	20
143	N3-OH/N7-Me	25.3	36.2	68.0	36.2	25.3	40.9	28.1	40.9	25.0	22.6	20
144	X3-OH/X7-Me	30.2	41.4	66.7	41.4	30.2	40.5	28.8	40.5	34.7	24.3	20
145	N3-OH/X7-Me	26.9	36.2	65.2	36.2	26.9	42.3	22.8	42.3	28.9	22.1	20
146	N3-OMe	26.8	33.0	74.0	33.0	26.8	32.9	16.6	32.9	30.8		4
147	X3-OMe/7-=CH2	30.9	38.7	72.9	38.7	30.9	40.6	148.8	40.6	34.2	109.2	20
148	N3-OMe/7-=CH2	28.3	32.5	74.0	32.5	28.3	43.0	144.0	43.0	29.4	111.9	20
149	9-=NOH	28.8	32.0	21.4	33.5	36.1	33.5	21.4	32.0	167.0		14
150	9-=O	46.5	34.2	20.5	34.2	46.5	34.2	20.5	34.2	221.7		1,2,19

157

BICYCLO[3.3.1]NONANES (cont.)

Str	Substituents	C1	C2	C3	C4	C5	C6	C7	C8	C9	Substituent Shifts	References
151	9-Cl 9-Me	40.3	28.7	20.8	28.7	40.3	29.6	20.4	29.6	78.6	30.7	12
152	9-Me	34.2	25.4	22.6	25.4	34.2	34.6	23.0	34.6	36.8	18.8	8
153	9-OH	34.5	24.6	21.6	24.6	34.5	31.9	21.6	31.9	73.0		2,13
154	9-OH 9-Me	39.0	27.2	20.6	27.2	39.0	29.9	20.7	29.9	72.1	27.7	13,12

13.2 Tricyclo [3.3.1.1³,⁷] decanes

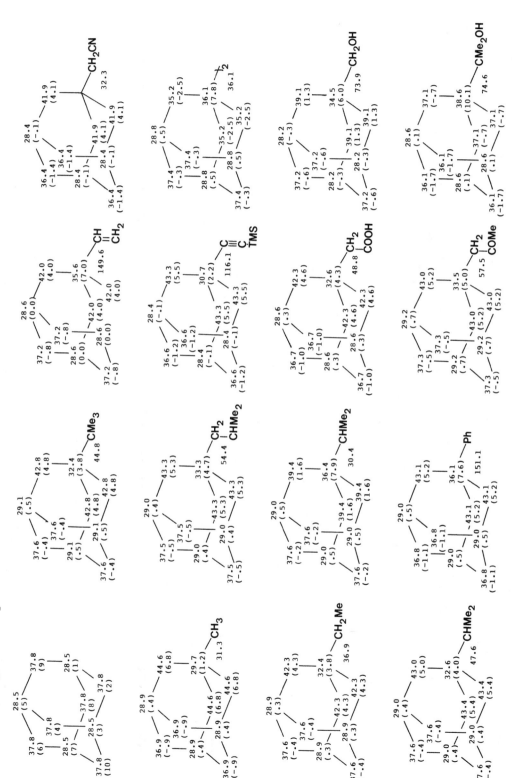

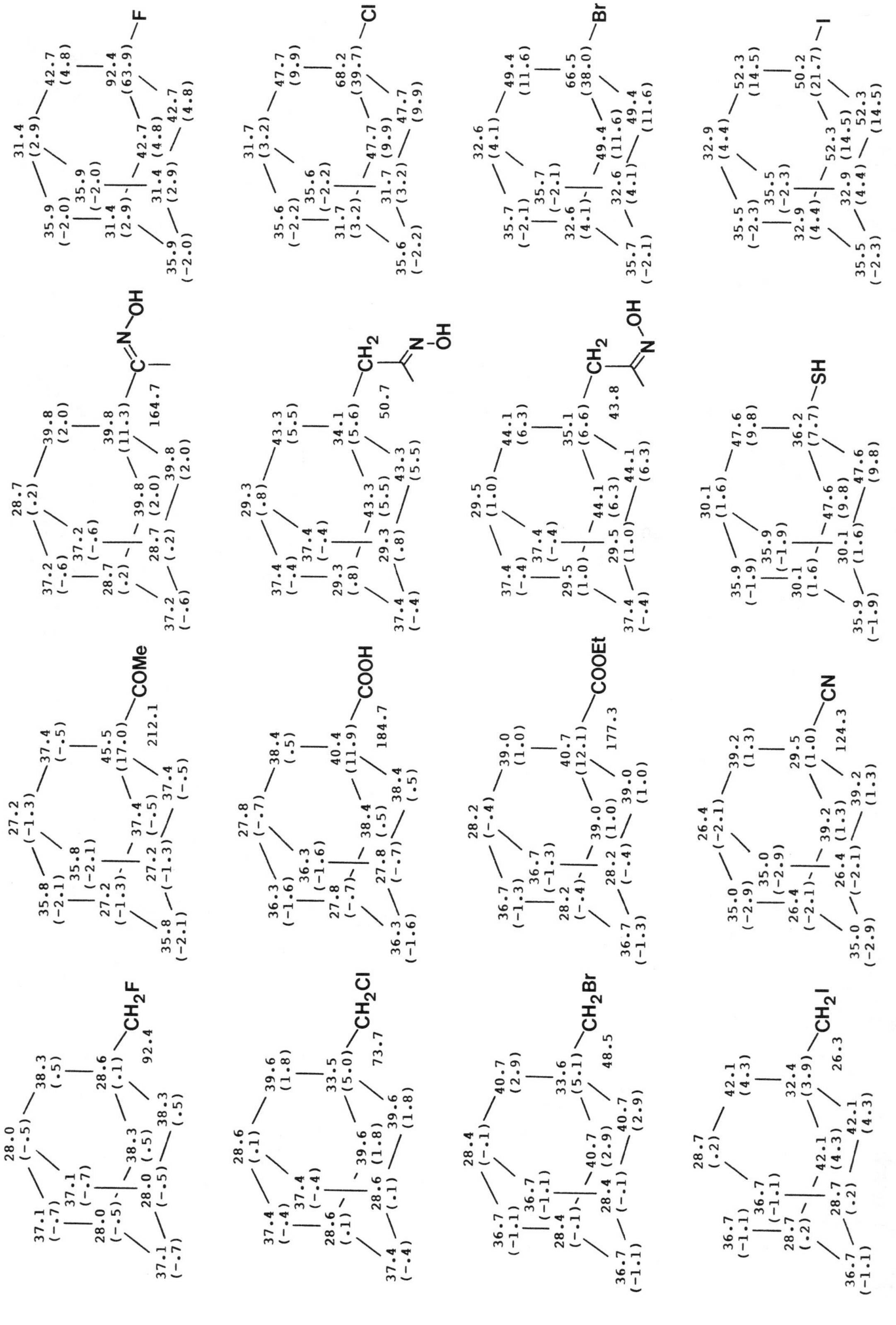

F

42.7 (4.8) 92.4 (63.9) 42.7 (2.9) 31.4 (2.9) 35.9 (-2.0) 35.9 (-2.0) 31.4 (2.9) 42.7 (4.8) 31.4 (4.8) 35.9 (-2.0)

Cl

47.7 (9.9) 68.2 (39.7) 47.7 (9.9) 31.7 (3.2) 35.6 (-2.2) 35.6 (-2.2) 31.7 (3.2) 47.7 (9.9) 31.7 (3.2) 35.6 (-2.2)

Br

49.4 (11.6) 66.5 (38.0) 49.4 (11.6) 32.6 (4.1) 35.7 (-2.1) 35.7 (-2.1) 32.6 (4.1) 49.4 (11.6) 32.6 (4.1) 35.7 (-2.1)

I

52.3 (14.5) 50.2 (21.7) 52.3 (14.5) 32.9 (4.4) 35.5 (-2.3) 35.5 (-2.3) 32.9 (4.4) 52.3 (14.5) 32.9 (4.4) 35.5 (-2.3)

C N—OH

39.8 (2.0) 164.7 39.8 (2.0) 28.7 (.2) 37.2 (-.6) 37.2 (-.6) 28.7 (.2) 39.8 (11.3) 28.7 (.2) 37.2 (-.6)

CH₂ N—OH

43.3 (5.5) 50.7 43.3 (5.5) 29.3 (.8) 37.4 (-.4) 37.4 (-.4) 29.3 (.8) 34.1 (5.6) 29.3 (.8) 37.4 (-.4)

CH₂ N—OH

44.1 (6.3) 43.8 44.1 (6.3) 29.5 (1.0) 37.4 (-.4) 37.4 (-.4) 29.5 (1.0) 35.1 (6.6) 29.5 (1.0) 37.4 (-.4)

SH

47.6 (9.8) 36.2 (7.7) 47.6 (9.8) 30.1 (1.6) 35.9 (-1.9) 35.9 (-1.9) 30.1 (1.6) 47.6 (9.8) 30.1 (1.6) 35.9 (-1.9)

COMe

37.4 (-.5) 212.1 37.4 (-.5) 27.2 (-1.3) 35.8 (-2.1) 35.8 (-2.1) 27.2 (-1.3) 45.5 (17.0) 27.2 (-1.3) 35.8 (-2.1)

COOH

38.4 (.5) 184.7 38.4 (.5) 27.8 (-.7) 36.3 (-1.6) 36.3 (-1.6) 27.8 (-.7) 40.4 (11.9) 27.8 (-.7) 36.3 (-1.6)

COOEt

39.0 (1.0) 177.3 39.0 (1.0) 28.2 (-.4) 36.7 (-1.3) 36.7 (-1.3) 28.2 (-.4) 40.7 (12.1) 28.2 (-.4) 36.7 (-1.3)

CN

39.2 (1.3) 124.3 39.2 (1.3) 26.4 (-2.1) 35.0 (-2.9) 35.0 (-2.9) 26.4 (-2.1) 29.5 (1.0) 26.4 (-2.1) 35.0 (-2.9)

CH₂F

38.3 (.5) 92.4 38.3 (.5) 28.0 (-.5) 37.1 (-.7) 37.1 (-.7) 28.0 (-.5) 28.6 (.1) 28.0 (-.5) 37.1 (-.7)

CH₂Cl

39.6 (1.8) 73.7 39.6 (1.8) 28.6 (.1) 37.4 (-.4) 37.4 (-.4) 28.6 (.1) 33.5 (5.0) 28.6 (.1) 37.4 (-.4)

CH₂Br

40.7 (2.9) 48.5 40.7 (2.9) 28.4 (-.1) 36.7 (-1.1) 36.7 (-1.1) 28.4 (-.1) 33.6 (5.1) 28.4 (-.1) 36.7 (-1.1)

CH₂I

42.1 (4.3) 26.3 42.1 (4.3) 28.7 (.2) 36.7 (-1.1) 36.7 (-1.1) 28.7 (.2) 32.4 (3.9) 28.7 (.2) 36.7 (-1.1)

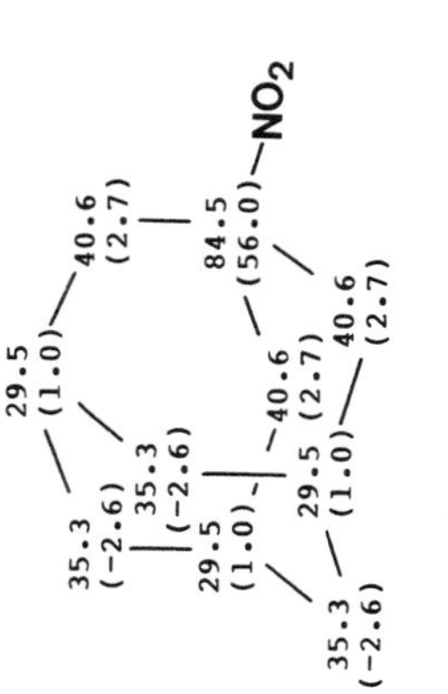

CHMe₂

28.5 (−.1) · 32.0 (−6.0) · 32.0 (3.4) · 39.8 · 44.4 (6.4) —CHMe₂
35.0 (−3) · 32.0 (−6.0) · 39.6 (1.6) · 32.0 (3.4)
28.3 (−.3) · 39.6 (1.6)

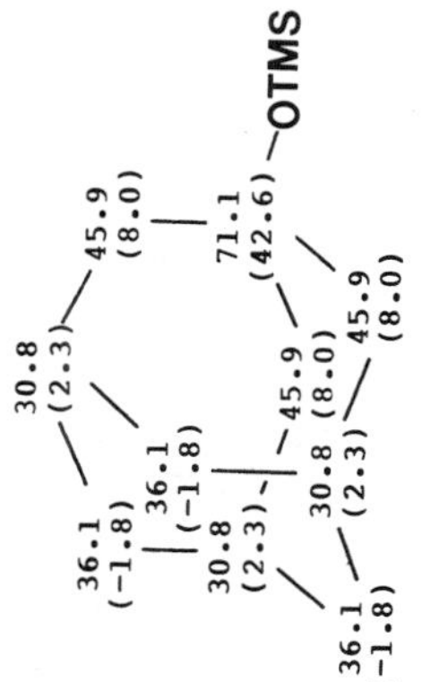

CH₂ / CH₂OH

28.8 (.3) · 42.9 (5.1) · 32.0 (3.5) · 47.2 · 42.9 (5.1) —CH₂OH
37.2 (−.6) · 37.2 (−.6) · 42.9 (5.1) · 28.8 (.3)
28.8 (.3) · 37.2 (−.6)

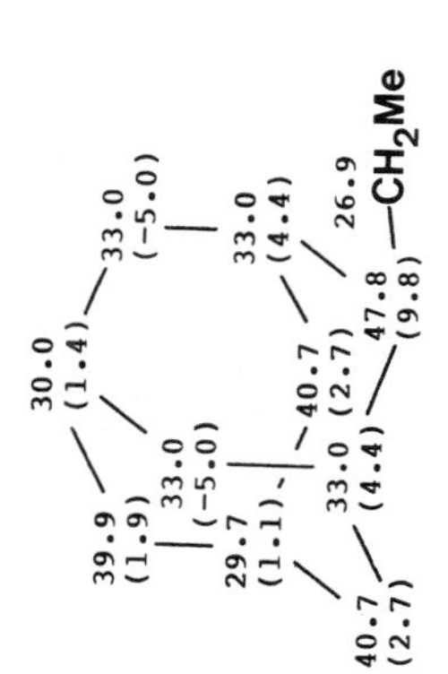

Ph

28.0 (−.5) · 31.9 (−5.9) · 31.1 (2.6) · 143.9 · 46.8 (9.0) —Ph
37.9 (.1) · 31.9 (−5.9) · 39.2 (1.4) · 31.1 (2.6)
27.8 (−.7) · 31.1 (2.6) · 39.2 (1.4)

F₂

28.3 (−.2) · 31.7 (−6.2) · 28.1 (−.4) · 42.5 · 42.5 (4.6) F₂
38.5 (.6) · 31.7 (−6.2) · 39.7 (−.4) · 28.1 (−.4)
28.0 (−.5) · 28.1 (−.4) · 39.7 (1.8)

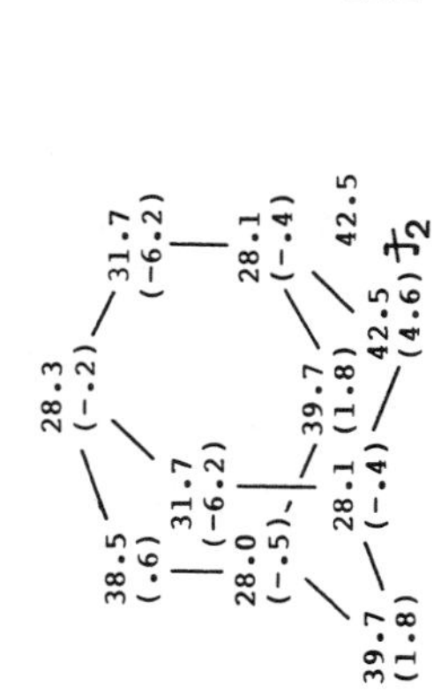

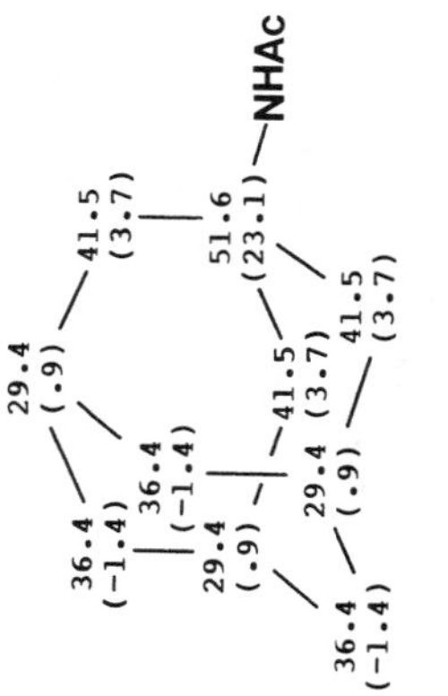

SI₂

30.1 (1.6) · 43.2 (5.4) · 47.3 (18.8) —SI₂
36.2 (−1.6) · 36.2 (−1.6) · 43.2 (5.4) · 43.2 (5.4)
30.1 (1.6) · 30.1 (1.6) · 30.1 (1.6)
36.2 (−1.6) · 36.2 (−1.6)

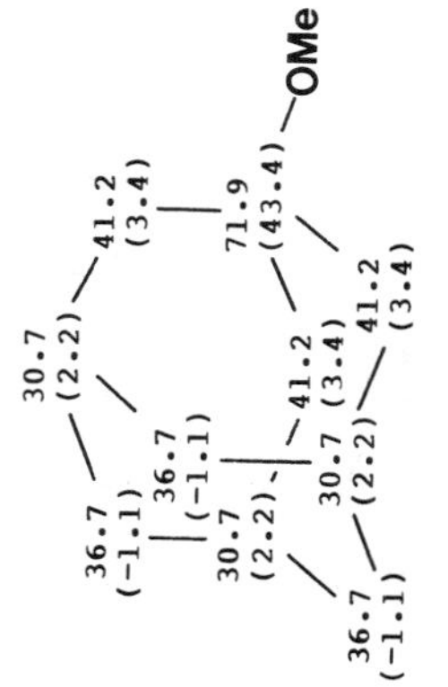

SnMe₃

28.9 (.4) · 41.7 (3.8) · 27.8 (−.7) —SnMe₃
37.5 (−.4) · 37.5 (−.4) · 41.7 (3.8) · 41.7 (3.8)
28.9 (.4) · 28.9 (.4)
37.5 (−.4)

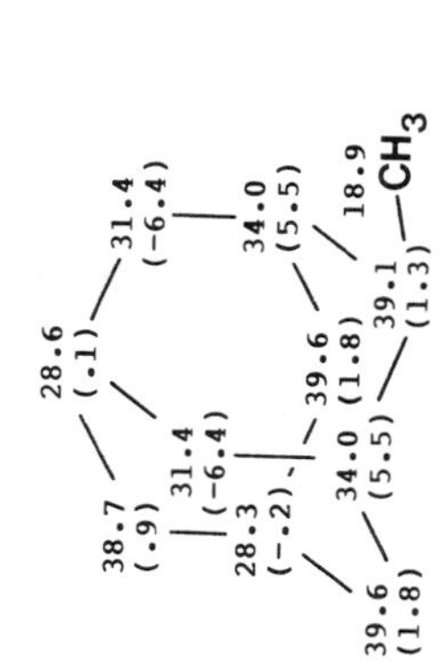

CH₃

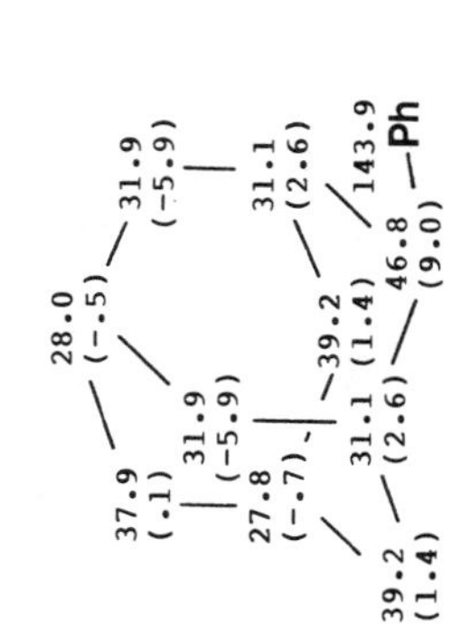

28.6 (.1) · 31.4 (−6.4) · 34.0 (5.5) · 18.9 —CH₃
38.7 (.9) · 31.4 (−6.4) · 39.6 (1.8) · 39.1 (1.3)
28.3 (−.2) · 34.0 (5.5) · 34.0 (5.5)
39.6 (1.8)

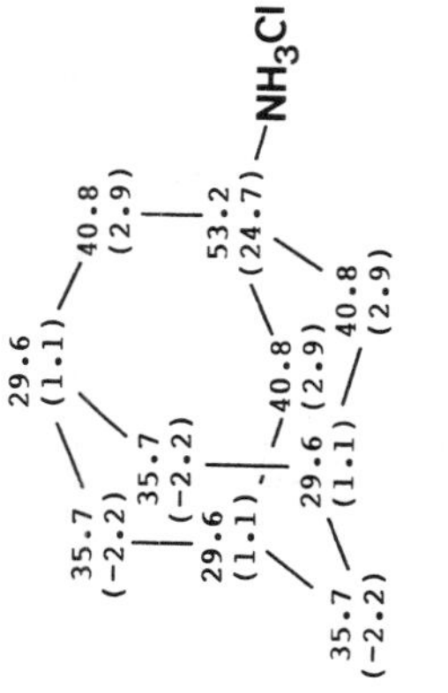

CH₂Me

30.0 (1.4) · 33.0 (−5.0) · 33.0 (4.4) · 26.9 —CH₂Me
39.9 (1.9) · 33.0 (−5.0) · 40.7 · 47.8 (9.8)
29.7 (1.1) · 33.0 (4.4)
40.7 (2.7) · 33.0 (4.4)
40.7 (2.7)

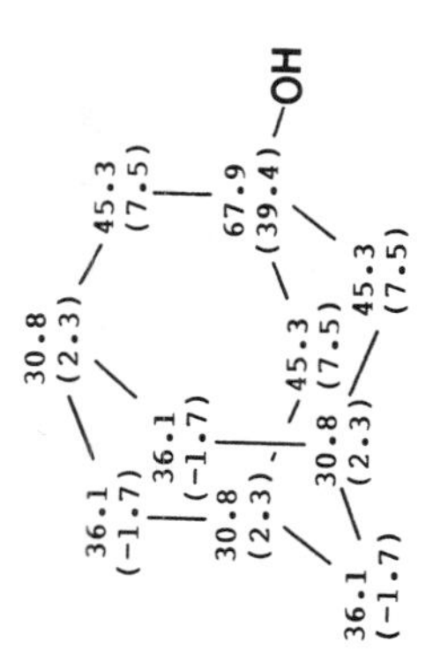

NCS

29.3 (.8) · 43.8 (6.0) · 58.4 (29.9) —NCS
35.6 (−2.2) · 35.6 (−2.2) · 43.8 (6.0) · 43.8 (6.0)
29.3 (.8) · 29.3 (.8)
35.6 (−2.2)

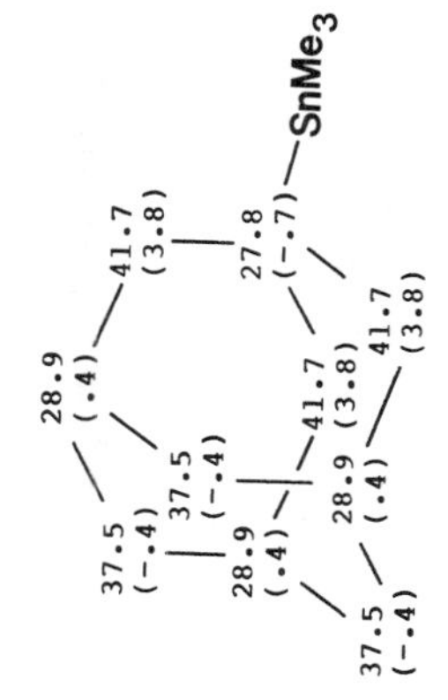

OH

30.8 (2.3) · 45.3 (7.5) · 67.9 (39.4) —OH
36.1 (−1.7) · 36.1 (−1.7) · 45.3 (7.5) · 45.3 (7.5)
30.8 (2.3) · 30.8 (2.3)
36.1 (−1.7)

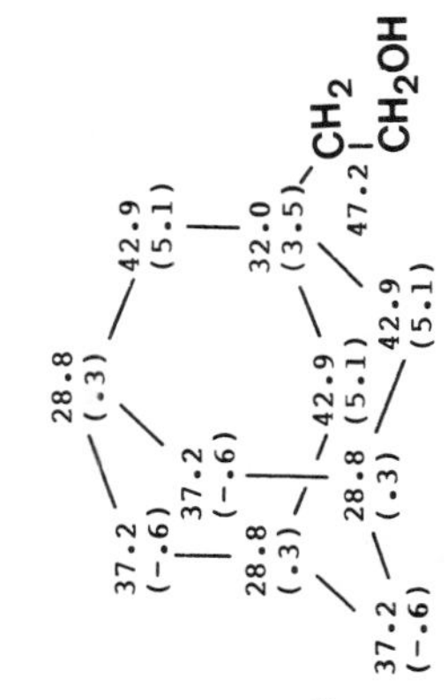

OMe

30.7 (2.2) · 41.2 (3.4) · 71.9 (43.4) —OMe
36.7 (−1.1) · 36.7 (−1.1) · 41.2 (3.4) · 41.2 (3.4)
30.7 (2.2) · 30.7 (2.2)
36.7 (−1.1)

OTMS

30.8 (2.3) · 45.9 (8.0) · 71.1 (42.6) —OTMS
36.1 (−1.8) · 36.1 (−1.8) · 45.9 (8.0) · 45.9 (8.0)
30.8 (2.3) · 30.8 (2.3)
36.1 (−1.8)

NH₂

29.8 (1.3) · 46.1 (8.3) · 47.1 (18.6) —NH₂
36.2 (−1.6) · 36.2 (−1.6) · 46.1 (8.3) · 46.1 (8.3)
29.8 (1.3) · 29.8 (1.3)
36.2 (−1.6)

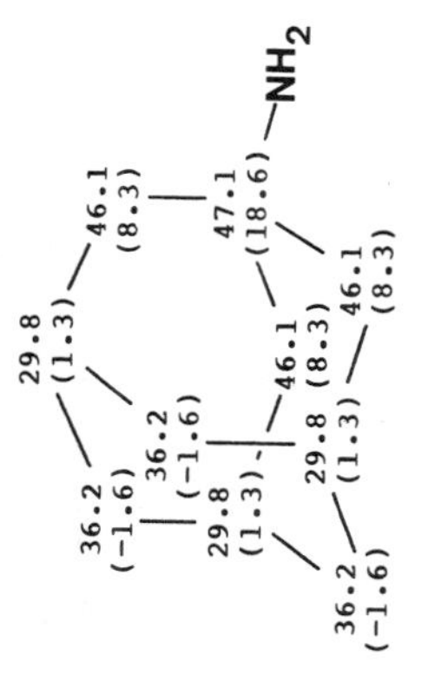

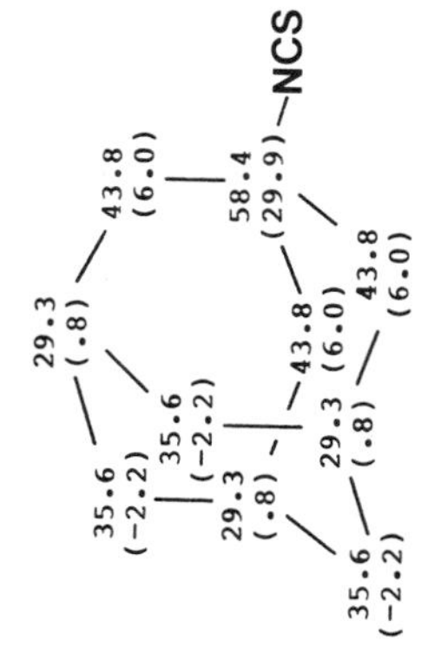

NH₃Cl

29.6 (1.1) · 40.8 (2.9) · 53.2 (24.7) —NH₃Cl
35.7 (−2.2) · 35.7 (−2.2) · 40.8 (2.9) · 40.8 (2.9)
29.6 (1.1) · 29.6 (1.1)
35.7 (−2.2)

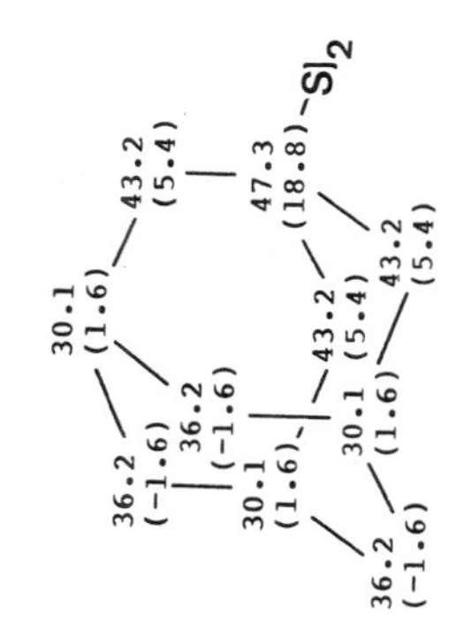

NHAc

29.4 (.9) · 41.5 (3.7) · 51.6 (23.1) —NHAc
36.4 (−1.4) · 36.4 (−1.4) · 41.5 (3.7) · 41.5 (3.7)
29.4 (.9) · 29.4 (.9)
36.4 (−1.4)

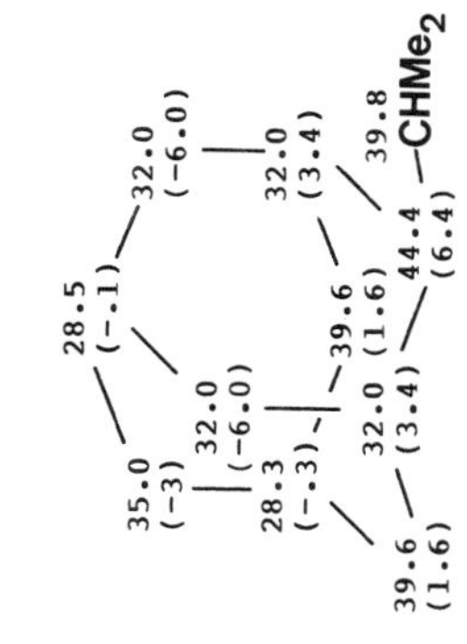

NO₂

29.5 (1.0) · 40.6 (2.7) · 84.5 (56.0) —NO₂
35.3 (−2.6) · 35.3 (−2.6) · 40.6 (2.7) · 40.6 (2.7)
29.5 (1.0) · 29.5 (1.0)
35.3 (−2.6)

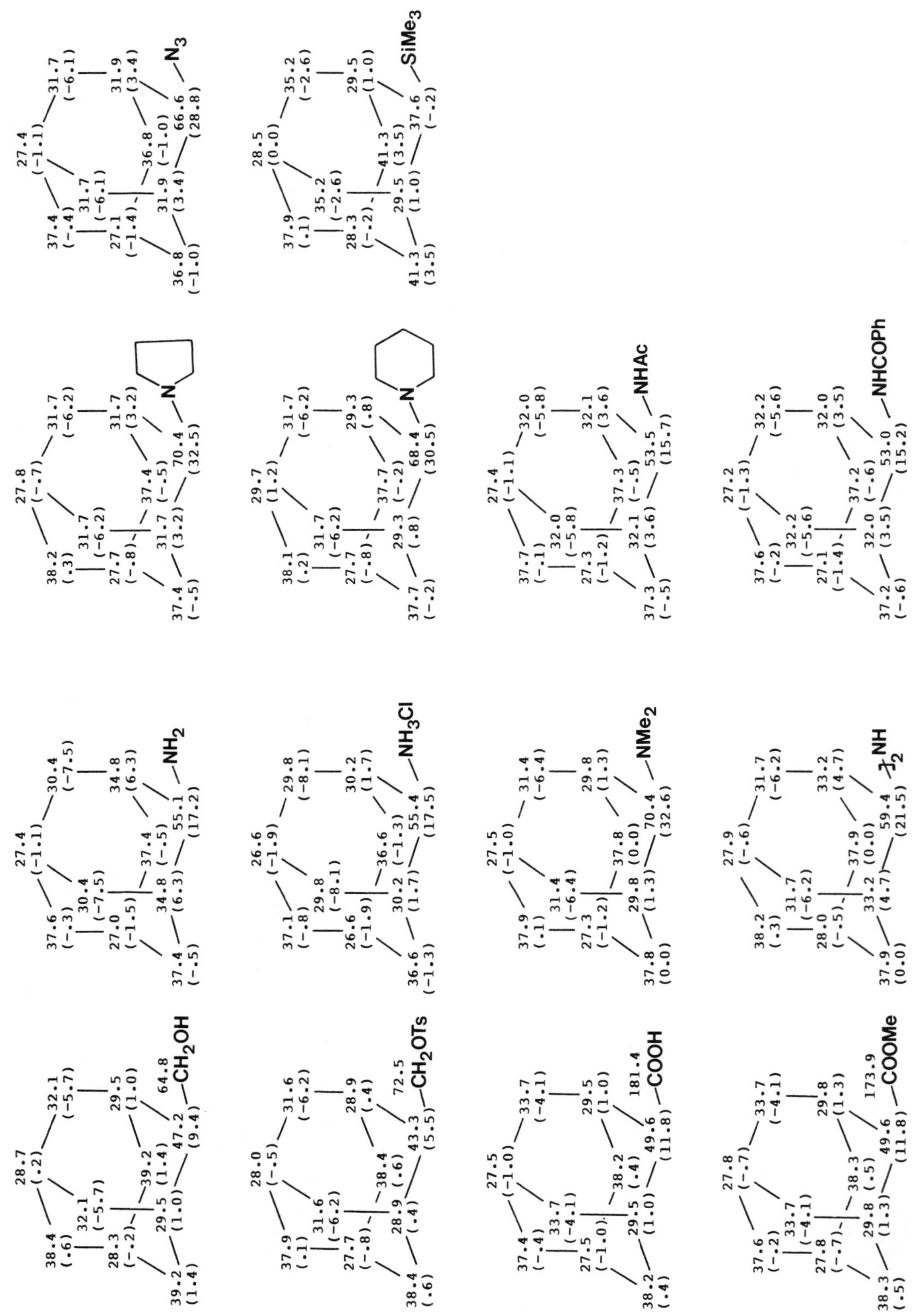

Row 1:

N₃ substituted adamantane:
31.7 (−6.1) 31.9 (3.4) 66.6 (28.8) 36.8 (−1.0) 31.9 (3.4) 36.8 (−1.0) 27.1 (−1.4) 31.7 (−6.1) 37.4 (−.4) 27.4 (−1.1)

N₃

SiMe₃ substituted adamantane:
35.2 (−2.6) 29.5 (1.0) 37.6 (−.2) 41.3 (3.5) 29.5 (1.0) 41.3 (3.5) 28.3 (−.2) 35.2 (−2.6) 37.9 (.1) 28.5 (0.0)

SiMe₃

Row 2:

Pyrrolidino adamantane (N in 5-ring):
31.7 (−6.2) 31.7 (3.2) 70.4 (32.5) 37.4 (−.5) 31.7 (3.2) 37.4 (−.5) 27.7 (−.8) 31.7 (−6.2) 38.2 (.3) 27.8 (−.7)

Piperidino adamantane (N in 6-ring):
31.7 (−6.2) 29.3 (.8) 68.4 (30.5) 37.7 (−.2) 29.3 (.8) 37.7 (−.2) 27.7 (−.8) 31.7 (−6.2) 38.1 (.2) 29.7 (1.2)

NHAc
32.0 (−5.8) 32.1 (3.6) 53.5 (15.7) 37.3 (−.5) 32.1 (3.6) 37.3 (−.5) 27.3 (−1.2) 32.0 (−5.8) 37.7 (−.1) 27.4 (−1.1)

NHCOPh
32.2 (−5.6) 32.0 (3.5) 53.0 (15.2) 37.2 (−.6) 32.0 (3.5) 37.2 (−.6) 27.1 (−1.4) 32.2 (−5.6) 37.6 (−.2) 27.2 (−1.3)

Row 3:

NH₂
30.4 (−7.5) 34.8 (6.3) 55.1 (17.2) 34.8 (6.3) 37.4 (−.5) 37.4 (−.5) 27.0 (−1.5) 30.4 (−7.5) 37.6 (−.3) 27.4 (−1.1)

NH₃Cl
29.8 (−8.1) 30.2 (1.7) 55.4 (17.5) 36.6 (−1.3) 30.2 (1.7) 36.6 (−1.3) 26.6 (−1.9) 29.8 (−8.1) 37.1 (−.8) 26.6 (−1.9)

NMe₂
31.4 (−6.4) 29.8 (1.3) 70.4 (32.6) 37.8 (0.0) 29.8 (1.3) 37.8 (0.0) 27.3 (−1.2) 31.4 (−6.4) 37.9 (.1) 27.5 (−1.0)

NH₂ (with extra line)
31.7 (−6.2) 33.2 (4.7) 59.4 (21.5) 37.9 (0.0) 33.2 (4.7) 37.9 (0.0) 28.0 (−.5) 31.7 (−6.2) 38.2 (.3) 27.9 (−.6)

Row 4:

CH₂OH
32.1 (−5.7) 29.5 (1.0) 64.8 (9.4) 47.2 (1.4) 39.2 (1.0) 29.5 (1.0) 28.3 (−.2) 32.1 (−5.7) 38.4 (.6) 28.7 (.2) 39.2 (1.4)

CH₂OTs
31.6 (−6.2) 28.9 (.4) 72.5 (5.5) 43.3 (.4) 38.4 (.6) 28.9 (.4) 27.7 (−.8) 31.6 (−6.2) 37.9 (.1) 28.0 (−.5) 38.4 (.6)

COOH
33.7 (−4.1) 29.5 (1.0) 181.4 (11.8) 49.6 (.4) 38.2 (.4) 29.5 (1.0) 27.5 (−1.0) 33.7 (−4.1) 37.4 (−.4) 27.5 (−1.0) 38.2 (.4)

COOMe
33.7 (−4.1) 29.8 (1.3) 173.9 (11.8) 49.6 (.5) 38.3 (1.3) 29.8 (1.3) 27.8 (−.7) 33.7 (−4.1) 37.6 (−.2) 27.8 (−.7) 38.3 (.5)

F
31.5 (-6.3) · 32.9 (4.4) · 27.7 (-.8) · 35.8 (-2.0) · 95.6 (57.8) · 37.3 (-.5) · 31.5 (-6.3) · 32.9 (4.4) · 27.3 (-1.2) · 35.8 (-2.0)

Cl
31.0 (-6.8) · 36.0 (7.5) · 27.6 (-.9) · 38.3 (.5) · 68.1 (30.3) · 37.9 (.1) · 31.0 (-6.8) · 36.0 (7.5) · 27.0 (-1.5) · 38.3 (.5)

Br
31.7 (-6.1) · 36.5 (8.0) · 27.7 (-.8) · 38.8 (1.0) · 63.7 (25.9) · 38.0 (.2) · 31.7 (-6.1) · 36.5 (8.0) · 27.0 (-1.5) · 38.8 (1.0)

I
33.0 (-4.8) · 37.6 (9.1) · 27.8 (-.7) · 38.8 (1.0) · 46.4 (8.6) · 38.2 (.4) · 33.0 (-4.8) · 37.6 (9.1) · 27.1 (-1.4) · 38.8 (1.0)

SH
31.0 (-6.8) · 35.7 (7.2) · 27.9 (-.6) · 38.9 (1.1) · 46.5 (8.7) · 37.9 (.1) · 31.0 (-6.8) · 35.7 (7.2) · 27.1 (-1.4) · 38.9 (1.1)

SMe
32.0 (-5.8) · 32.7 (4.2) · 28.0 (-.5) · 38.9 (1.1) · 55.0 (17.2) · 38.0 (.2) · 32.0 (-5.8) · 32.7 (4.2) · 27.9 (-.6) · 38.9 (1.1)

SEt
32.1 (-5.7) · 33.5 (5.0) · 27.9 (-.6) · 38.9 (1.1) · 52.8 (15.0) · 38.0 (.2) · 32.1 (-5.7) · 33.5 (5.0) · 27.7 (-.8) · 38.9 (1.1)

SnPh₃
36.9 (-.9) · 32.4 (3.9) · 28.6 (.1) · 40.7 (2.9) · 43.3 (5.5) · 38.0 (.2) · 36.9 (-.9) · 32.4 (3.9) · 28.2 (-.3) · 40.7 (2.9)

OMs
31.1 (-6.7) · 33.0 (4.5) · 26.9 (-1.6) · 36.3 (-1.5) · 85.6 (47.8) · 37.1 (-.7) · 31.1 (-6.7) · 33.0 (4.5) · 26.6 (-1.9) · 36.3 (-1.5)

OTMS
31.2 (-6.7) · 35.3 (6.8) · 27.7 (-.8) · 36.6 (-1.3) · 74.9 (37.0) · 37.8 (-.1) · 31.2 (-6.7) · 35.3 (6.8) · 27.2 (-1.3) · 36.6 (-1.3)

OAc
31.9 (-5.9) · 32.0 (3.5) · 27.4 (-1.1) · 36.5 (-1.3) · 77.0 (39.2) · 37.5 (-.3) · 31.9 (-5.9) · 32.0 (3.5) · 27.2 (-1.3) · 36.5 (-1.3)

SnMe₃
36.5 (-1.3) · 32.1 (3.6) · 28.7 (.2) · 40.9 (3.1) · 40.0 (2.2) · 38.2 (.4) · 36.5 (-1.3) · 32.1 (3.6) · 28.3 (-.2) · 40.9 (3.1)

CN
32.9 (-5.3) · 30.3 (1.5) · 26.8 (-2.0) · 36.4 (-1.8) · 122.0 (-1.4) · 36.8 (-1.4) · 36.6 (-1.6) · 32.9 (-5.3) · 30.3 (1.5) · 26.6 (-2.2) · 36.4 (-1.8)

OH
31.2 (-6.7) · 34.7 (6.2) · 27.8 (-.7) · 36.7 (-1.2) · 74.7 (36.8) · 37.8 (-.1) · 31.2 (-6.7) · 34.7 (6.2) · 27.3 (-1.2) · 36.7 (-1.2)

OMe
31.5 (-6.3) · 31.5 (3.0) · 27.6 (-.9) · 36.6 (-1.2) · 83.3 (45.5) · 37.7 (-.1) · 31.5 (-6.3) · 31.5 (3.0) · 27.6 (-.9) · 36.6 (-1.2)

Structure 1 (OTMS, CH₃):
27.3 (−1.2); 33.3 (−4.6); 39.8 (11.3); OTMS; 38.6 (.7); 33.3 (−4.6); 34.9 (−3.0); 77.3 (39.4); 27.6 (−.9); 39.8 (11.3); 26.8 CH₃; 34.9 (−3.0)

Structure 2 (OH, Ph):
27.1 (−1.4); 33.1 (−4.7); 35.8 (7.3); OH; 37.9 (.1); 33.1 (−4.7); 35.0 (−2.8); 75.6 (37.8); 27.6 (−.9); 35.8 (7.3); 145.5 Ph; 35.0 (−2.8)

Structure 3 (OTMS, Ph):
27.0 (−1.5); 33.1 (−4.8); 36.0 (7.5); OTMS; 37.8 (−.1); 33.1 (−4.8); 34.4 (−3.5); 77.9 (40.0); 27.4 (−1.1); 36.0 (7.5); 143.6 Ph; 34.4 (−3.5)

Structure 4 (OH, CH₃):
27.2 (−1.3); 33.1 (−4.7); 39.3 (10.8); OH; 38.5 (.7); 33.1 (−4.7); 35.2 (−2.6); 73.8 (36.0); 27.7 (−.8); 39.3 (10.8); 27.5 CH₃; 35.2 (−2.6)

Structure 5 (Cl, CH₃):
26.9 (−1.6); 34.1 (−3.8); 40.7 (12.2); Cl; 38.7 (.8); 34.1 (−3.8); 34.7 (−3.2); 79.9 (42.0); 27.1 (−1.4); 40.7 (12.2); 30.3 CH₃; 34.7 (−3.2)

Structure 6 (Cl, Ph):
26.6 (−1.9); 33.6 (−4.3); 36.9 (8.4); Cl; 37.9 (0.0); 33.6 (−4.3); 34.7 (−3.2); 79.1 (41.2); 26.9 (−1.6); 36.9 (8.4); 143.8 Ph; 34.7 (−3.2)

Structure 7 (CH₃, CH₃, CH₃):
27.7 (−.8); 33.3 (−4.5); 37.4 (8.9); 27.7 CH₃; 39.3 (1.5); 33.3 (−4.5); 33.3 (−4.5); 27.7 (−.8); 37.4 (8.9); CH₃; 27.7; 33.3 (−4.5)

Structure 8 (CN, CH₂CN):
26.5 (−2.0); 33.5 (−4.3); 30.7 (2.2); CN 121.8; 37.5 (−.3); 33.5 (−4.3); 35.1 (−2.7); 43.3 (5.5); 26.5 (−2.0); 30.7 (2.2); 25.1 CH₂CN; 35.1 (−2.7)

Structure 9 (CH₂OH, NH₂):
29.1 (.6); 34.8 (−3.0); 35.6 (7.1); CH₂OH 67.3; 39.9 (2.1); 34.8 (−3.0); 33.7 (−4.1); 57.0 (19.2); 28.7 (.2); 35.6 (7.1); NH₂; 33.7 (−4.1)

26.9 (-1.6) — 32.8 (-5.0) — 32.8 (4.3) — OMe
37.7 (-.1) | 32.8 (-5.0) | 34.5 (-3.3) — 79.8 (42.0)
27.5 (-1.0) — 32.8 (4.3) — 141.1 Ph
34.5 (-3.3)

27.1 (-.3) — 35.0 (-2.9) — 34.6 (7.2) — OH
39.6 (1.7) | 35.0 (-2.9) | 35.6 (-2.3) — 77.5 (39.6)
27.2 (-.2) — 34.6 (7.2) — 39.8 CMe₃
35.6 (-2.3)

27.3 (-1.2) — 34.6 (-3.2) — 42.2 (13.7) — Br
39.4 (1.6) | 34.6 (-3.2) | 36.0 (-1.8) — 82.3 (44.5)
27.5 (-1.0) — 42.2 (13.7) — 32.8 CH₃
36.0 (-1.8)

27.0 (-1.5) — 34.9 (-2.9) — 38.2 (9.7) — Br
38.7 (.9) | 34.9 (-2.9) | 35.4 (-2.4) — 80.2 (42.4)
27.4 (-1.1) — 38.2 (9.7) — 145.3 Ph
35.4 (-2.4)

26.7 (-1.8) — 33.8 (-4.0) — 34.0 (5.5) — PO(OMe)₂
38.2 (.4) | 33.8 (-4.0) | 32.2 (-5.6) — 77.3 (39.5)
27.2 (-1.3) — 34.0 (5.5) — OH
32.2 (-5.6)

27.3 (-1.2) — 33.9 (-3.9) — 33.1 (4.6) — OMe
37.5 (-.3) | 33.9 (-3.9) | 33.9 (-3.9) — 102.2 (64.4)
27.3 (-1.2) — 33.1 (4.6) — OMe
33.9 (-3.9)

26.6 (-1.9) — 34.1 (-3.7) — 36.0 (7.5) — F
36.7 (-1.1) | 34.1 (-3.7) | 34.1 (-3.7) — 127.5 (89.7)
26.6 (-1.9) — 36.0 (7.5) — F
34.1 (-3.7)

26.4 (-2.1) — 34.8 (-3.0) — 44.8 (16.3) — Cl
38.2 (.4) | 34.8 (-3.0) | 34.8 (-3.0) — 100.6 (62.8)
26.4 (-2.1) — 44.8 (16.3) — Cl
34.8 (-3.0)

26.8 (-1.7) — 35.5 (-2.3) — 47.0 (18.5) — Br
39.1 (1.3) | 35.5 (-2.3) | 35.5 (-2.3) — 85.6 (47.8)
26.8 (-1.7) — 47.0 (18.5) — Br
35.5 (-2.3)

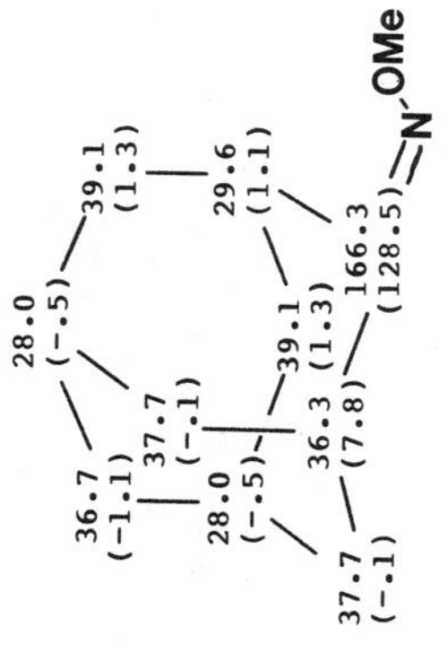

=N–OMe

39.1 (1.3) · 28.0 (−.5) · 29.6 (1.1) · 36.7 (−1.1) · 37.7 (−.1) · 39.1 (1.3) · 166.3 (128.5) · 28.0 (−.5) · 36.3 (7.8) · 37.7 (−.1)

=O

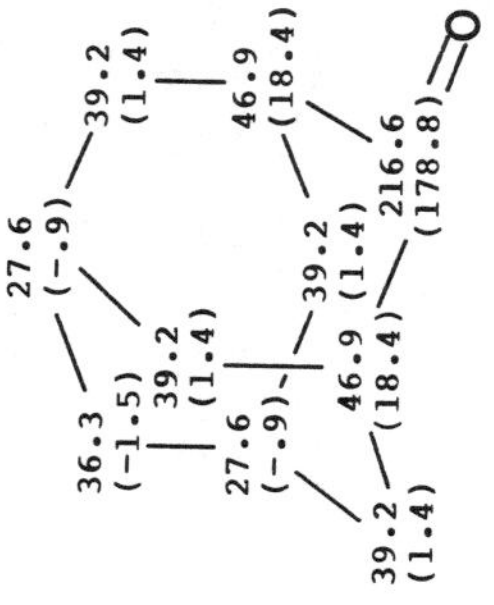

39.2 (1.4) · 27.6 (−.9) · 46.9 (18.4) · 36.3 (−1.5) · 39.2 (1.4) · 39.2 (1.4) · 216.6 (178.8) · 27.6 (−.9) · 46.9 (18.4) · 39.2 (1.4)

=S

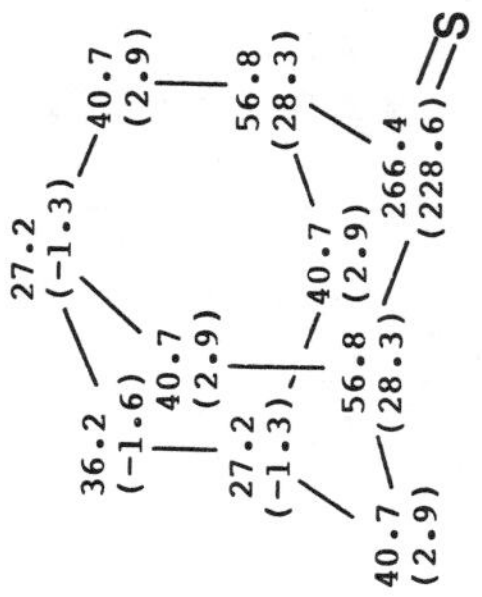

40.7 (2.9) · 27.2 (−1.3) · 56.8 (28.3) · 36.2 (−1.6) · 40.7 (2.9) · 40.7 (2.9) · 266.4 (228.6) · 27.2 (−1.3) · 56.8 (28.3) · 40.7 (2.9)

=C(CN)₂ 77.3

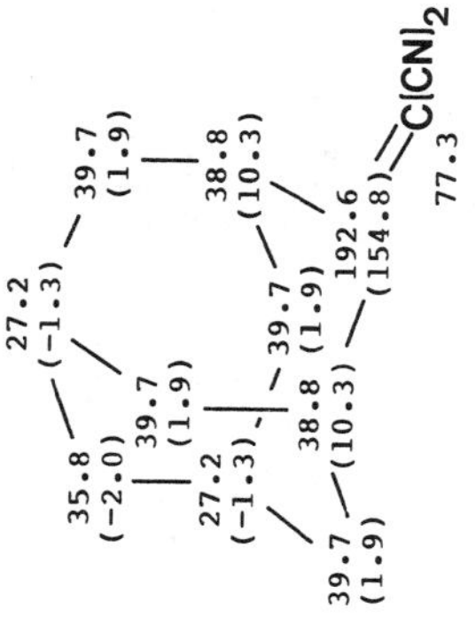

39.7 (1.9) · 27.2 (−1.3) · 38.8 (10.3) · 35.8 (−2.0) · 39.7 (1.9) · 39.7 (1.9) · 192.6 (154.8) · 27.2 (−1.3) · 38.8 (10.3) · 39.7 (1.9)

=C(COOEt)CN 98.5

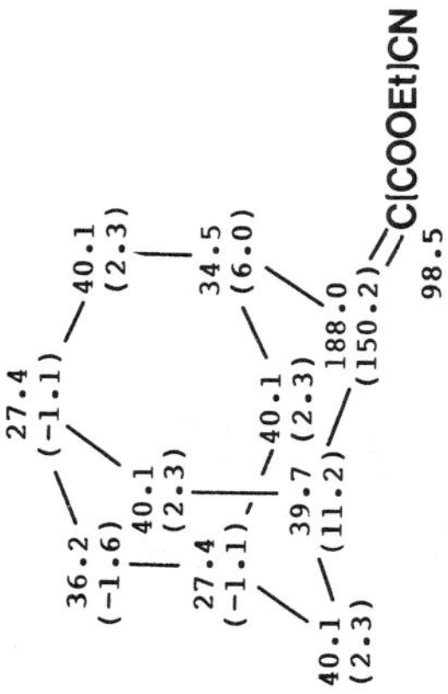

40.1 (2.3) · 27.4 (−1.1) · 34.5 (6.0) · 36.2 (−1.6) · 40.1 (2.3) · 40.1 (2.3) · 188.0 (150.2) · 27.4 (−1.1) · 39.7 (11.2) · 40.1 (2.3)

=N–OH

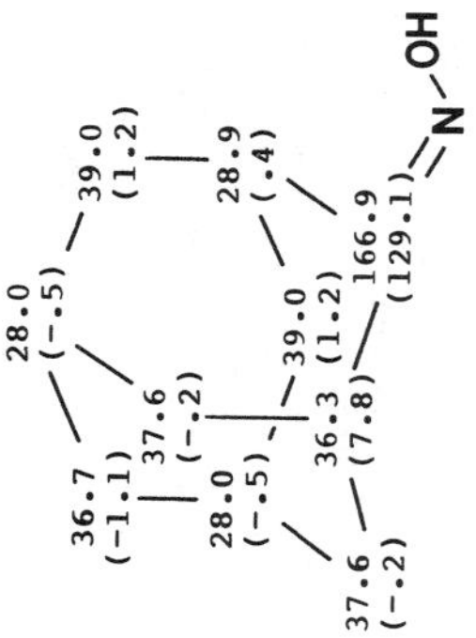

39.0 (1.2) · 28.0 (−.5) · 28.9 (.4) · 36.7 (−1.1) · 37.6 (−.2) · 39.0 (1.2) · 166.9 (129.1) · 28.0 (−.5) · 36.3 (7.8) · 37.6 (−.2)

=CH₂ 100.8

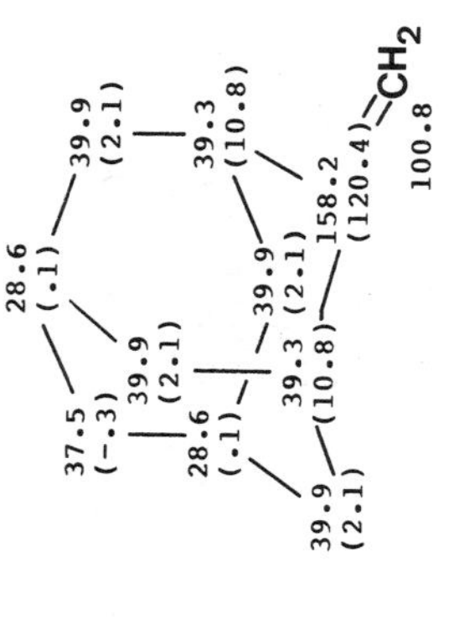

39.9 (2.1) · 28.6 (.1) · 39.3 (10.8) · 37.5 (−.3) · 39.9 (2.1) · 39.9 (2.1) · 158.2 (120.4) · 28.6 (.1) · 39.3 (2.1) · 39.9 (2.1)

=C–Me, H 110.2

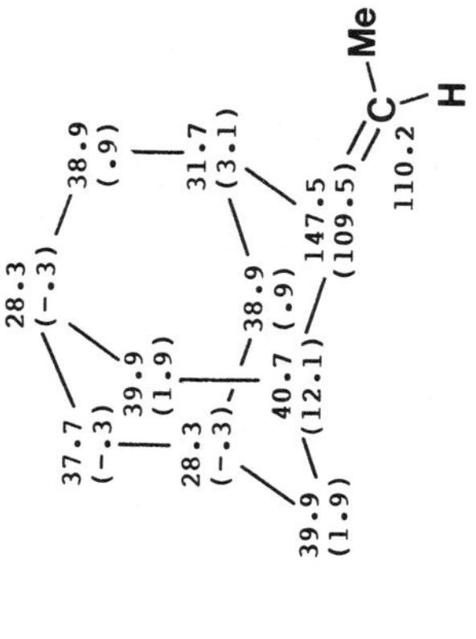

38.9 (.9) · 28.3 (−.3) · 31.7 (3.1) · 37.7 (−.3) · 39.9 (1.9) · 38.9 (.9) · 147.5 (109.5) · 28.3 (−.3) · 40.7 (12.1) · 39.9 (1.9)

=C=₂ 131.6

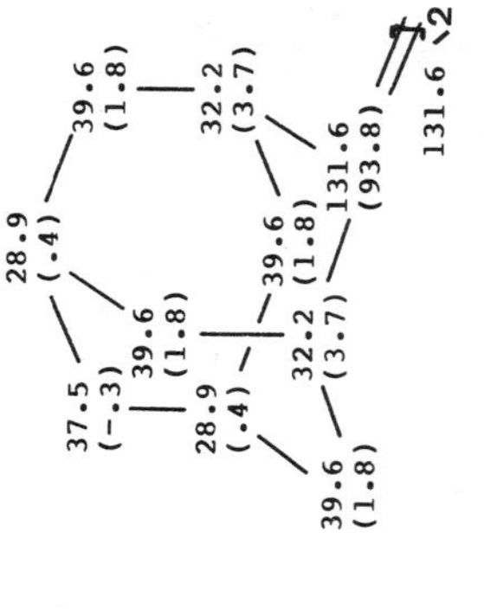

39.6 (1.8) · 28.9 (.4) · 32.2 (3.7) · 37.5 (−.3) · 39.6 (1.8) · 39.6 (1.8) · 131.6 (93.8) · 28.9 (.4) · 39.6 (1.8) · 32.2 (3.7)

TRICYCLO[3.3.1.1^{3,7}]DECANES

Str	Substituents	C1	C2	C3	C4	C5	C6	C7	C8	C9	C10	Substituent Shifts	References
1	PARENT	28.5	37.8	28.5	37.8	28.5	37.8	28.5	37.8	37.8	37.8		6,21,37-8,40,49-53,55-6,58,60-62,67,69,70,88,95,97,120-1,126-7
2	1-(1-AD)	36.1	35.2	28.8	37.4	28.8	37.4	28.8	35.2	35.2	37.4	36.1	21
3	1-Br	66.5	49.4	32.6	35.7	32.6	35.7	32.6	49.4	49.4	35.7		56,21,40,52,60,88,127
4	1-Br 3-Me	62.0	59.1	62.0	46.9	34.9	33.6	34.9	46.9	46.9	46.9		52
5	1-Br/3-Br	65.7	55.9	35.0	42.5	32.5	34.7	32.5	48.4	48.4	42.5	30.1	60
6	1-Br/3-Br/5-Br	58.1	56.8	58.1	56.8	58.1	44.8	36.3	44.8	56.8	44.8		52
7	1-Br/3-Br/5-Br/7-Br	54.2	54.9	54.2	54.9	54.2	54.9	54.2	54.9	54.9	54.9		52
8	1-Br/3-Cl	62.2	57.8	66.8	45.6	34.2	33.6	34.2	46.9	46.9	45.6		52
9	1-Br/3-F	62.0	53.8	92.2	40.9	33.3	34.2	33.3	47.3	47.3	40.9		52
10	1-Br/3-Me/5-Me	65.2	55.0	35.2	49.6	35.2	41.7	32.5	47.5	55.0	41.7	29.7,29.7	127
11	1-Bu	32.4	42.8	29.1	37.6	29.1	37.6	29.1	42.8	42.8	37.6	44.8	62
12	1-C CTMS	30.7	43.3	28.4	36.6	28.4	36.6	28.4	43.3	43.3	36.6	116.1	102
13	1-$C(Me)_2OH$	38.6	37.1	28.6	36.1	28.6	36.1	28.6	37.1	37.1	36.1	74.6	68
14	1-$C(Me)_2OH$/4-OCH_2CH_2O-4	35.8	34.3	36.5	111.2	36.5	33.9	27.3	34.3	37.1	33.9		77
15	1-$C(Me)_2OH=CH_2$/4-OCH_2CH_2O-4	36.7	38.4	36.7	111.2	36.7	34.3	27.4	40.6	38.4	34.3	153.4	77
16	1-$C(Me)=$[E]NOH	39.8	39.8	28.7	37.2	28.7	37.2	28.7	39.8	39.8	37.2	164.7	123
17	1-CH_2Br	33.6	40.7	28.4	36.7	28.4	36.7	28.4	40.7	40.7	36.7	48.5	85,21
18	1-CH_2Br/3-Ph	34.4	46.3	36.7	42.3	29.0	35.7	29.0	40.0	40.7	42.3	46.7,149.7	21
19	1-$CH_2C(Me)=$[E]NOH	34.1	43.3	29.3	37.4	29.3	37.4	29.3	43.3	43.3	37.4	50.7	123
20	1-$CH_2C(Me)=$[Z]NOH	35.1	44.1	29.5	37.4	29.5	37.4	29.5	44.1	44.1	37.4	43.8	123
21	1-CH_2CH_2OH	32.0	42.9	28.8	37.2	28.8	37.2	28.8	42.9	42.9	37.2	47.2	127
22	1-CH_2Cl	33.5	39.6	28.6	37.4	28.6	37.4	28.6	39.6	39.6	37.4	73.7	91
23	1-CH_2CN	33.5	41.9	28.4	36.4	28.4	36.4	28.4	41.9	41.9	36.4	32.3	89,91
24	1-CH_2COMe	33.5	43.0	29.2	37.2	29.0	35.6	28.6	41.9	43.0	37.3	57.5	123
25	1-CH_2COOH	32.6	42.3	28.6	36.7	28.6	36.7	28.6	42.3	42.3	36.7	48.8	128,21,127
26	1-CH_2COOH/3-Br	37.0	53.6	60.6	48.7	32.6	34.8	32.6	40.3	40.3	48.7	47.4	21
27	1-CH_2COOH/3-CH_2Br	32.9	45.3	34.0	39.8	28.4	35.6	28.4	41.1	41.1	39.8	47.6,46.1	21
28	1-CH_2COOH/3-CH_2COOH	32.7	46.6	36.7	41.3	28.6	35.8	28.6	41.1	41.3	41.3	48.2,48.2	127
29	1-CH_2COOH/3-OH	35.5	47.7	68.2	44.0	30.5	35.5	28.0	41.1	41.1	44.0	47.9	21
30	1-CH_2F	28.6	38.3	28.0	37.1	28.0	37.1	28.7	38.3	38.3	37.1	92.4	89
31	1-CH_2I	32.4	42.1	28.7	37.2	28.7	36.7	28.7	42.1	42.1	36.7	26.3	89,91
32	1-CH_2OH	34.5	39.1	28.2	37.2	28.2	37.4	28.2	39.1	39.1	37.2	73.9	85,21,90,127
33	1-CH_2OH/3-Me/5-Me	36.9	46.1	31.3	52.0	31.3	44.0	30.1	38.8	46.1	44.0	72.8	89
34	1-$CH=CH_2$	35.6	47.2	28.6	37.2	28.7	37.2	28.6	42.0	46.1	37.2	149.6	121
35	1-Cl	68.2	47.7	31.7	35.6	31.7	35.6	31.7	47.7	47.7	35.6		6,21,40,52,56,88,97
36	1-Cl/3-Cl	66.9	56.6	66.9	45.7	33.8	33.5	33.8	45.7	45.7	45.7		52
37	1-Cl/3-Cl/5-Cl	64.5	54.4	64.5	54.4	64.5	43.8	33.5	43.8	54.4	43.8		52
38	1-Cl/3-Cl/5-Cl/7-Cl	62.3	52.8	62.3	62.3	62.3	52.8	62.3	52.8	52.8	52.8		52
39	1-Cl/3-F	66.8	52.2	92.4	40.7	32.5	33.9	32.5	45.8	45.5	40.7		52
40	1-Cl/3-I	66.3	60.6	43.5	49.8	34.7	33.6	34.7	45.5	45.5	49.8		52
41	1-CN	29.5	39.2	26.4	35.0	26.4	35.0	26.4	39.2	39.2	35.0	124.3	88,40,70,105,127
42	1-COMe	45.5	37.4	27.2	35.8	35.8	35.8	27.2	37.4	37.4	35.8	212.1	88,123,127
43	1-COOEt	40.7	39.0	28.2	36.7	28.2	36.7	28.2	39.0	39.0	36.7	177.3	127,121
44	1-COOH	40.4	38.4	27.8	36.3	27.8	36.3	27.8	38.4	38.4	36.3	184.7	88,21,40,89,92,127
45	1-COOH/3-Br	44.4	49.1	60.9	48.0	31.6	34.5	31.6	37.0	37.0	48.0		21
46	1-COOH/3-Me/5-Me	42.6	45.5	31.2	51.2	36.5	43.3	34.7	38.5	36.2	35.8	180.2	89
47	1-COOMe/4-OCH_2CH_2O-4	39.9	36.2	36.5	36.5	28.9	33.9	29.8	42.3	42.3	33.9	177.6	77
48	1-Et	32.4	42.3	28.9	37.6	29.5	37.6	26.7	41.6	41.6	37.6	36.9	62
49	1-Et/3-Me	33.3	49.5	30.8	44.6	29.5	36.6	28.9	41.6	41.6	44.6	36.8,31.6	62
50	1-Et/3-Me/5-Me	34.2	48.8	31.5	51.7	31.5	43.8	30.0	40.9	48.8	43.8	36.0,30.9,30.9	62

TRICYCLO[3.3.1.1³,⁷]DECANES (cont.)

Str	Substituents	C1	C2	C3	C4	C5	C6	C7	C8	C9	C10	Substituent Shifts	References
51	1-F	92.4	42.7	31.4	35.9	31.4	35.9	31.4	42.7	42.7	35.9		52,21,40,84,88
52	1-F/3-F	93.4	48.0	93.4	41.2	31.5	34.3	31.5	41.2	41.2	41.2		52
53	1-F/3-F/5-F	92.4	46.9	92.4	46.9	92.4	39.9	28.1	39.9	46.9	39.9		52
54	1-F/3-F/5-F/7-F	90.4	46.2	90.4	46.2	90.4	46.2	90.4	46.2	46.2	46.2		52
55	1-I	50.2	52.3	32.9	35.5	32.9	35.5	32.9	52.3	52.3	35.5		56,40,52,88
56	1-I/3-I	44.8	64.6	44.8	49.9	36.6	33.7	36.6	49.9	49.9	49.9		52
57	1-I/3-I/5-I	39.3	61.7	39.3	61.7	39.3	47.4	39.5	47.4	61.7	47.4		52
58	1-I/3-I/5-I/7-I	31.1	59.7	31.1	59.7	31.1	59.7	31.1	59.7	59.7	59.7		52
59	1-iBu	33.3	43.3	29.0	37.5	29.0	37.5	29.0	39.4	39.4	37.6	54.4	62
60	1-iPr	36.4	39.4	29.0	37.6	29.0	37.6	29.0	39.4	39.4	37.6	30.4	76,121
61	1-Me	29.7	44.6	28.9	36.9	28.9	36.9	28.9	44.6	44.6	36.9	31.3	60,21,40,88,89,126
62	1-Me/2=O/3-Me/5-Me/7-Me	45.7	45.7	45.5	52.7	32.0	49.5	32.0	52.7	52.7	52.7	22.5,22.5,28.7,28.7	120
63	1-Me/2-Me	32.4	43.9	35.5	30.8	29.6	38.1	39.0	47.4	38.2	39.5	28.7,15.4	121
64	1-Me/2-OH 2-Me/3-Me/5-Me/7-Me	38.3	76.4	38.3	47.0	31.2	51.2	31.7	48.8	47.0	48.8	23.2,27.9,23.2,29.8,30.0	120
65	1-Me/2-OH 2-tBu/3-Me/5-Me/7-Me	42.8	81.6	42.8	48.6	30.4	52.0	31.1	52.8	48.6	52.8	29.4,44.1,29.4,29.6,29.9	120
66	1-Me /2-OH/3-Me/5-Me/7-Me	36.0	81.8	36.0	43.9	31.6	50.6	31.5	50.8	43.9	50.8	26.2,26.2,29.7,30.0	120
67	1-Me/3-Me	30.6	51.8	30.6	43.9	29.5	36.2	29.5	43.9	43.9	43.9	31.0,31.0	70,62,127
68	1-Me/3-Me/5-Me	31.6	51.4	31.6	51.4	31.6	43.4	30.2	43.4	51.4	43.4	31.3,31.0,31.0	62
69	1-Me/3-Me/5-Me/7-Me	32.2	50.5	32.2	50.5	32.2	50.5	32.2	50.5	50.5	50.5	4*30.3	120,62
70	1-Me/3-Me/6-Me	30.9	52.7	30.6	45.7	34.9	37.6	34.9	37.8	37.8	45.7	30.9,31.3,18.6	121
71	1-Me/A4-Me	29.9	38.5	34.3	38.3	34.3	38.8	29.0	45.7	38.5	38.8	31.4,18.7	121
72	1-Me/E2-Me/5-Me	32.9	43.2	36.0	46.3	31.3	44.9	29.4	37.5	54.4	30.0	28.5,15.0,31.1	121
73	1-Me/E4-Me	30.2	46.5	34.5	38.5	34.5	30.7	28.7	45.6	46.5	30.7	31.7,19.0	121
74	1-N=C=S	58.4	43.8	29.3	35.6	29.3	35.6	29.3	43.8	43.8	35.6		127
75	1-NCF3	71.4	39.8	29.1	36.3	29.1	36.3	29.1	39.8	39.8	36.3		96
76	1-N=NCF3/3-Cl	73.6	48.9	66.4	46.4	31.2	34.3	31.2	45.7	38.1	46.4		96
77	1-NH2	47.1	46.1	29.8	36.2	29.8	36.2	29.8	46.1	46.1	36.2		6,21,40,88,124,127
78	1-NH3Cl	53.2	40.8	29.6	35.7	29.6	35.7	29.6	40.8	40.8	35.7		127,21,88
79	1-NHAc	51.6	41.5	29.4	36.4	29.4	36.4	29.4	41.5	41.5	36.4		60
80	1-NO2	84.5	40.6	29.5	35.3	29.5	35.3	29.5	40.6	40.6	35.3		88,21
81	1-OAc/A4-NMe2	79.8	35.7	32.7	68.8	32.7	35.9	30.1	41.0	35.7	35.9		97
82	1-OAc/A4-OAc	78.8	35.9	34.8	74.9	34.8	35.0	29.7	40.8	35.9	35.0		97
83	1-OAc/A4-OH	79.3	35.1	37.4	72.6	37.4	35.1	30.1	41.0	35.1	35.1		97
84	1-OAc/E4-NMe2	79.2	40.6	32.2	69.7	32.2	29.9	30.1	41.4	40.6	29.9		97
85	1-OAc/E4-OAc	78.5	39.7	33.8	75.3	33.8	30.6	29.7	41.1	39.7	30.6		97
86	1-OAc/E4-OH	79.6	39.7	36.4	73.1	36.4	29.8	29.6	41.5	39.7	29.8		97
87	1-OH	67.9	45.3	30.8	45.3	36.1	36.1	30.8	45.3	45.3	36.1		6,21,49,56,60,88,97,124,127
88	1-OH/3-Me	68.9	52.2	33.4	43.3	31.0	35.4	31.0	44.6	44.6	43.3	30.3	60,121
89	1-OH/2-OH	69.8	77.3	36.3	29.5	30.2	35.8	29.7	43.1	38.0	35.8		49
90	1-OH/3-Me/5-Me/7-Me	70.4	50.7	34.0	49.9	34.0	49.9	34.0	50.7	50.7	49.9	29.6,29.6,29.6	127
91	1-OH/3-OH	69.5	51.9	69.5	43.4	31.1	34.7	31.1	43.4	43.4	43.4		49
92	1-OH/3-OH/5-Me/7-Me	69.8	51.4	69.8	50.5	34.3	49.5	34.3	50.5	50.5	50.5	29.2,29.2	127
93	1-OH/A4-NH2	67.5	38.8	37.7	54.0	37.7	36.2	29.7	45.6	38.8	36.2		97
94	1-OH/A4-NMe2	67.6	39.4	37.4	69.1	32.5	35.9	29.8	39.4	39.4	29.9		97
95	1-OH/A4-OH	66.9	38.3	36.7	72.0	36.7	35.0	29.3	44.5	38.3	35.0		49
96	1-OH/E4-NH2	67.4	44.9	36.9	54.5	36.9	29.5	30.1	45.8	44.9	29.5		97
97	1-OH/E4-NMe2	67.6	44.5	36.0	69.6	32.2	30.0	30.2	45.6	44.5	30.0		97
98	1-OH/2-OH	66.9	43.0	35.8	72.7	35.8	29.8	29.8	44.9	43.0	29.8		49
99	1-OMe	71.9	41.2	30.7	36.7	30.7	36.7	30.7	41.2	41.2	36.7		60,61
100	1-OTMS	71.1	45.9	30.8	36.1	30.8	36.1	30.8	45.9	45.9	36.1		88,50

TRICYCLO[3.3.1.1^{3,7}]DECANES (cont.)

Str	Substituents	C1	C2	C3	C4	C5	C6	C7	C8	C9	C10	Substituent Shifts	References
101	1-OTMS/2-OTMS	73.2	78.6	37.5	29.9	29.9	36.5	30.3	45.1	38.9	35.7		50
102	1-OTMS/3-OTMS	72.8	54.2	72.8	44.7	31.1	34.9	31.1	44.7	44.7	44.7		50
103	1-OTMS/A4-OTMS	70.2	39.4	37.9	72.9	37.0	35.1	29.5	46.1	39.4	35.1		50
104	1-OTMS/E4-OTMS	70.2	44.4	37.0	73.7	37.0	29.8	29.0	46.2	44.4	29.8		50
105	1-Ph	36.1	43.1	29.0	36.8	29.0	36.8	29.0	43.1	43.1	36.8	151.1	88
106	1-Pr	32.6	43.4	29.0	37.6	29.0	37.6	29.0	43.4	43.0	37.6	47.6	62
107	1-Pr/3-Pr	33.3	47.9	33.3	42.5	29.0	37.3	29.4	43.4	42.5	42.5	47.3,47.3	62
108	1-SH	36.2	47.6	30.1	35.9	30.1	35.9	30.1	47.6	47.6	35.9		127
109	1-SnMe3	27.8	41.7	28.9	37.5	28.9	37.5	28.9	41.7	41.7	37.5		95,75,94
110	1-SS(1-AD)	47.3	43.2	30.1	36.2	36.2	36.2	30.1	43.2	43.2	36.2		122
111	1-tBu/2-=O	54.0	215.0	48.9	38.8	28.2	36.1	28.2	43.2	39.3	38.8	34.4	78
112	2-(2-AD)	28.1	42.5	28.1	31.7	28.3	38.5	28.0	39.7	31.7	39.7	42.5	88
113	2-=AD	32.2	131.6	32.2	39.6	28.9	37.5	28.9	39.6	39.6	39.6	131.6	104
114	2-=C(CN)2	38.8	192.6	38.8	39.7	27.2	35.8	27.2	39.6	39.6	41.0	77.3	83,103
115	2-=C(CN)2/A4-Br	38.0	186.1	46.5	59.1	35.1	36.8	25.9	39.7	34.3	41.0	81.5	83
116	2-=C(CN)2/A4-Cl	38.1	187.1	46.4	66.8	34.8	36.4	25.9	39.6	33.7	40.2	81.8	83
117	2-=C(CN)2/A4-F	38.2	188.5	43.7	96.4	32.9	34.1	25.8	39.1	33.7	38.0	81.1	64
118	2-=C(CN)2/A4-I	38.0	186.2	47.6	38.3	35.7	35.7	26.0	40.1	36.1	41.7	81.4	83
119	2-=C(CN)2/E4-Br	37.5	187.5	45.4	56.0	34.6	30.6	26.5	40.2	38.0	34.6	79.6	83
120	2-=C(CN)2/E4-Cl	37.5	187.4	45.1	63.2	34.1	29.3	26.4	39.8	37.0	33.7	79.7	83
121	2-=C(CN)2/E4-F	37.7	187.1	43.6	92.2	31.5	30.0	26.3	39.5	33.7	34.5	80.6	83
122	2-=C(CN)2/E4-I	37.5	187.5	46.2	37.5	35.7	31.4	26.7	40.8	39.0	35.9	78.9	64
123	2-=C(CN)2/E4-Me	38.4	193.6	44.1	41.1	32.2	29.6	27.1	40.0	40.3	33.0	77.4,17.5	83
124	2-=C(CN)2/E4-Ph	38.4	192.3	42.0	48.5	28.7	30.4	26.9	39.5	40.0	33.8	78.2,139.9	83
125	2-=C(CN)2/E4-SiMe3	39.0	194.3	39.9	41.2	28.7	33.2	26.8	39.4	45.2	36.1	76.0	73
126	2-=C(CN)2/E4-SMe	38.1	190.7	42.7	55.2	32.0	30.5	26.7	39.4	39.3	34.0	78.6,15.3	83
127	2-=C(CN)2/E4-SnMe3	38.9	195.0	42.2	41.8	31.0	34.5	27.0	39.6	45.6	37.6	75.8	83
128	2-=C(COOEt)CN	34.5	188.0	39.7	40.1	27.4	36.2	27.4	40.1	40.1	40.1	98.5	74
129	2-=CH2	39.3	158.2	39.3	39.9	28.6	37.5	28.6	39.9	39.9	39.9	100.8	127
130	2-=CH2/6-=CH2	38.2	156.9	38.8	41.3	38.8	156.9	38.8	41.3	41.3	41.3	101.4,101.4	83,69,87
131	2-=CH2/A4-F	38.2	152.8	43.8	96.0	32.8	35.4	26.9	39.0	33.4	37.7	104.9	69
132	2-=CH2/E4-Br	37.4	154.7	46.8	62.1	36.0	31.2	27.5	39.3	39.3	33.4	103.9	64
133	2-=CH2/E4-Cl	37.5	155.1	46.4	67.3	35.6	30.6	27.5	39.4	38.7	32.8	104.3	83
134	2-=CH2/E4-F	37.5	153.7	43.9	94.9	32.5	31.1	27.2	38.9	35.5	33.3	104.7	83,72
135	2-=CH2/E4-I	37.8	155.2	48.0	44.4	37.7	32.7	27.8	39.9	34.9	34.9	103.8	64
136	2-=CH2/E4-Me	38.7	159.3	44.5	40.7	33.6	31.4	28.4	40.2	40.7	32.8	100.5,18.1	83
137	2-=CH2/E4-SnMe3	39.2	160.0	42.5	41.6	32.0	35.9	28.2	39.3	43.4	38.1	99.4	83
138	2-=CHMe	31.7	147.5	40.7	39.9	28.3	37.7	28.0	38.9	38.9	39.9	110.2	74
139	2-=NOMe	29.6	166.3	36.3	37.7	28.0	36.7	26.6	39.1	39.1	37.7		121
140	E2-=NOMe/A4-Br	28.7	161.0	44.6	60.9	36.0	37.5	26.6	37.7	32.0	40.3		83
141	E2-=NOMe/A4-Cl	28.9	161.2	44.2	67.0	35.6	37.0	27.0	37.5	31.3	39.6		64
142	E2-=NOMe/A4-I	28.7	160.9	54.7	42.2	37.0	36.8	26.8	37.4	33.4	40.8		83
143	E2-=NOMe/A4-OH	29.3	164.0	43.5	75.8	34.1	35.4	26.7	37.4	31.7	33.2		83
144	E2-=NOMe/E4-Br	28.4	163.5	59.3	59.3	35.7	30.7	27.3	37.6	36.8	33.2		83
145	Z2-=NOMe/E4-Br	34.9	162.6	37.3	58.1	35.8	30.7	27.3	39.2	38.1	32.0		83
146	E2-=NOMe/E4-Cl	28.3	163.4	43.4	65.3	35.2	29.9	27.2	38.0	36.0	32.4		83
147	Z2-=NOMe/E4-Cl	34.8	162.7	36.7	64.0	35.2	29.9	27.7	38.9	37.4	31.1		83
148	E2-=NOMe/E4-I	28.6	163.1	44.8	40.3	36.7	32.1	27.4	38.1	37.5	34.5		83
149	Z2-=NOMe/E4-I	35.1	162.2	38.2	39.2	36.8	32.1	27.4	39.5	38.7	33.4		83
150	E2-=NOMe/E4-OH	28.4	165.0	42.9	73.7	33.9	30.0	27.2	37.2	34.1	32.2		83

TRICYCLO[3.3.1.1³,⁷]DECANES (cont.)

Str	Substituents	C1	C2	C3	C4	C5	C6	C7	C8	C9	C10	Substituent Shifts	References
151	S2-=NOMe/E4-OH	35.1	164.4	36.5	72.2	34.0	30.0	27.2	38.7	35.4	31.0		83
152	A2-=NOMe/E4-OMs	28.2	162.0	41.1	83.2	32.7	30.1	26.7	36.8	33.6	32.8		83
153	S2-=NOMe/E4-OMs	34.8	161.2	34.5	81.9	32.9	30.2	26.8	35.5	34.8	31.6		83
154	2==O	46.9	216.6	46.9	39.2	27.6	36.3	27.6	39.2	39.2	39.2		51,15,19,21,53-4,69 98,103,120,125,127
155	2==O/4-==CH2	45.6	209.7	57.9	152.5	38.0	37.7	27.5	37.2	41.7	38.8	104.9	71
156	2==O/4-==O	45.0	207.6	68.3	207.6	45.0	38.2	26.0	38.2	30.0	44.0		76
157	2==O/5-Me/7-Me	46.0	214.5	46.0	44.9	31.1	51.1	31.1	44.9	44.9	44.9	29.4	121
158	2==O/5-t-Bu	46.6	219.4	46.6	38.9	35.4	34.4	28.2	38.6	38.9	38.6	36.6	77
159	2==O/6-==O	45.3	211.7	45.3	39.6	45.3	211.7	45.3	39.6	39.6	39.6		66,69
160	2==O/A4-Br	45.8	210.8	54.8	60.1	35.1	36.9	26.1	39.6	33.6	40.6		51,53,101
161	2==O/A4-Br/6-==O	44.6	207.7	54.3	53.0	53.2	207.7	43.9	40.1	35.3	37.3		66
162	2==O/A4-Cl	46.1	212.4	54.6	67.6	34.9	36.6	26.1	39.2	32.9	39.7		53
163	2==O/A4-F	46.4	214.0	52.2	98.0	32.4	34.6	26.3	38.9	32.6	37.7		83,64
164	2==O/A4-I	45.8	211.7	56.2	40.2	36.2	36.8	26.1	39.5	35.4	41.5		53
165	2==O/A4-I/6-==O	44.4	208.6	54.1	32.4	54.1	207.6	44.3	40.0	37.4	39.3		66
166	2==O/A4-Me	46.9	217.7	53.4	43.4	33.1	37.9	27.1	39.6	33.9	39.8	18.8	83,125
167	2==O/A4-Me/6-==O	45.4	213.1	51.7	43.7	51.0	213.1	44.4	39.8	38.9	38.9	17.2	66
168	2==O/A4-N3	46.3	214.1	50.9	69.2	31.6	35.4	26.4	38.9	33.3	38.2		83
169	2==O/A4-NMe2	46.4	216.7	49.3	73.7	29.6	35.9	26.5	39.2	33.8	38.1		83
170	2==O/A4-NMe2/6-==O	44.8	212.1	47.3	72.4	48.3	212.4	43.8	39.7	33.9	35.3		66
171	2==O/A4-OAc	46.1	213.4	50.8	79.3	31.3	35.1	26.3	38.9	37.9	37.9		51,53
172	2==O/A4-OH	46.2	215.9	53.8	77.2	33.3	34.9	26.2	38.7	32.9	37.4		51,53
173	2==O/A4-OMe	46.3	214.7	50.9	86.1	30.9	35.0	26.6	38.8	32.4	37.6		53,43
174	2==O/A4-OMs	45.7	211.5	51.3	86.5	34.2	34.8	25.7	38.5	32.4	38.0		51
175	2==O/A4-Ph	46.6	210.7	49.8	52.8	34.2	38.0	27.4	40.2	32.3	41.2	141.8	83,101
176	2==O/A4-SiMe3	47.0	218.3	48.0	39.8	28.9	40.3	28.0	39.6	37.1	42.6		73
177	2==O/E4-Br	45.0	210.4	54.5	56.2	34.7	30.2	26.7	39.2	35.5	33.9		51,53,101
178	2==O/E4-Cl	45.1	212.2	54.2	63.2	34.5	30.0	26.6	39.1	34.9	33.2		53
179	2==O/E4-F	45.2	212.7	52.6	92.2	31.8	30.1	26.6	38.9	30.9	33.9		64,83
180	2==O/E4-I	45.7	210.7	55.9	37.3	35.7	31.8	27.0	39.8	36.8	35.7		53
181	2==O/E4-Me	46.1	217.5	52.3	40.2	32.7	30.0	27.5	39.3	39.4	32.3	17.0	83,125
182	2==O/E4-N3	45.5	213.3	50.8	64.7	31.2	30.2	26.6	38.7	34.1	33.5		83
183	2==O/E4-NMe2	45.4	216.8	49.4	70.0	29.2	30.0	26.8	39.1	35.5	32.9		83
184	2==O/E4-OAc	45.4	212.3	51.1	73.9	30.9	30.4	26.7	39.0	33.8	32.7		51,53
185	2==O/E4-OH	45.0	210.8	53.7	72.2	33.2	29.4	26.6	38.6	32.8	32.7		51,53
186	2==O/E4-OMe	45.5	214.3	50.5	81.5	30.9	30.0	26.6	38.6	33.1	32.8		53,43
187	2==O/E4-OMs	44.7	209.6	51.7	80.3	31.9	29.6	26.2	38.4	33.3	31.6		51
188	2==O/E4-Ph	45.6	214.6	49.2	47.2	29.7	30.3	26.9	38.6	38.3	32.8	139.2	83
189	2==O/E4-SEt	45.9	214.7	51.8	52.2	33.1	31.1	27.2	39.1	38.1	33.7		83
190	2==O/E4-SiMe3	46.9	218.1	47.4	39.0	28.7	33.4	27.1	39.2	44.2	35.5		73
191	2==O/E4-SMe	45.9	214.5	50.7	54.1	32.2	30.8	27.1	39.0	37.9	33.5		83
192	2==O/E4-SnMe3	47.2	218.3	50.1	40.4	35.0	27.4	27.4	39.2	44.5	37.3		74
193	2==S	56.8	266.4	56.8	40.7	27.2	36.2	27.2	40.7	40.7	40.7		83,98,125
194	2==S/A4-F	56.8	263.9	61.4	98.0	32.0	34.9	26.4	40.9	34.4	40.0		64
195	2==S/A4-Me	57.7	269.5	63.5	44.6	32.7	38.2	27.2	41.9	35.6	42.2	18.7	125
196	2==S/E4-Br	55.4	259.5	63.4	57.4	34.5	30.4	26.6	41.1	37.9	35.3		83
197	2==S/E4-Cl	55.3	259.7	63.9	63.0	34.1	29.8	26.6	40.8	36.9	34.6		83
198	2==S/E4-F	56.0	262.0	62.2	93.2	31.8	30.6	26.7	40.8	33.5	35.9		64
199	2==S/E4-I	55.5	259.2	64.4	38.4	35.5	31.8	26.9	41.6	39.1	36.7		83
200	2==S/E4-Me	56.9	270.6	62.9	42.4	32.6	30.3	27.5	42.0	41.4	35.5	17.5	83,125

170

TRICYCLO[3.3.1.1^{3,7}]DECANES (cont.)

Str	Substituents	C1	C2	C3	C4	C5	C6	C7	C8	C9	C10	Substituent Shifts	References
201	2-=S/E4-OMS	55.2	257.7	60.7	81.1	31.6	29.9	26.1	40.2	35.1	33.9		83
202	2-Br	36.5	63.7	36.5	31.7	27.7	38.0	27.0	38.8	31.7	38.8		6,40,51,53,55,60,88,127
203	2-Br 2-Br	47.0	85.6	47.0	35.5	26.8	39.1	26.8	35.5	35.5	35.5		6
204	2-Br 2-Me	42.2	82.3	42.2	34.6	27.3	39.4	27.5	36.0	34.6	36.0	32.8	60,16,88
205	2-Br 2-Ph	38.2	80.2	38.2	34.9	27.0	34.9	27.4	35.4	34.9	35.4	145.3	60,88
206	E2-Br/4-Me 4-Me	36.1	61.7	45.7	38.9	36.4	33.6	27.1	33.2	34.6	28.1	27.4,27.4	76
207	A2-Br/E4-Br	35.5	62.3	43.7	57.2	35.8	31.8	26.2	39.1	33.2	33.4		57
208	E2-Br/E4-Br	34.8	58.0	43.5	58.0	34.8	32.1	26.9	32.1	38.0	26.1		57
209	A2-Br/E4-F	35.4	60.9	40.4	91.6	32.4	31.5	25.8	38.4	38.0	26.1		64
210	E2-Br/E4-F	35.5	56.6	40.6	93.2	31.6	32.0	26.5	31.4	34.1	26.1		64
211	E2-Br/E4-SiMe₃	37.4	67.4	37.4	39.5	28.1	35.1	28.1	32.0	42.6	28.7		73
212	2-CH₂CN 2-CN	30.7	43.3	30.7	33.5	26.5	37.5	26.5	35.1	35.1	35.1	25.1,121.8	127
213	2-CH₂OH	29.5	47.2	29.5	32.1	28.7	38.4	28.3	39.2	32.1	39.2	64.8	83,89,93
214	2-CH₂OH 2-NH₂	35.6	57.0	35.6	34.8	29.1	39.9	28.7	33.7	34.8	33.7	67.3	114
215	E2-CH₂OH/4-=O	28.3	47.9	47.6	216.6	46.5	39.0	27.0	30.5	39.2	32.8	62.2	83
216	2-CH₂OTs	28.9	43.3	28.9	31.6	28.0	37.9	27.7	38.4	31.6	38.4	72.5	83
217	E2-CH₂OTs/4-=O	28.0	43.7	47.0	214.3	46.1	38.8	26.8	30.2	38.4	32.7	69.2	83
218	2-Cl	36.0	68.1	36.0	31.0	27.6	37.9	27.0	38.3	31.0	38.3		6,21,40,53,55,88
219	2-Cl 2-Cl	44.8	100.6	44.8	34.8	26.4	38.2	26.4	34.8	34.8	34.8		6
220	2-Cl 2-Me	40.7	79.9	40.7	34.1	26.9	38.7	27.1	34.7	34.1	34.7	30.3	88,12
221	2-Cl 2-Ph	36.9	79.1	36.9	33.6	26.6	37.9	26.9	34.7	33.6	34.7	143.8	88
222	A2-Cl/E4-Cl	34.8	67.9	42.8	62.0	35.2	30.9	26.1	38.2	31.9	31.9		57
223	E2-Cl/E4-Cl	34.3	64.6	42.8	64.6	34.3	31.1	26.7	31.1	36.6	24.8		57
224	A2-Cl/E4-F	34.9	67.2	40.1	91.1	32.9	31.4	25.8	37.7	29.9	32.5		64
225	E2-Cl/E4-F	34.4	62.8	40.4	93.4	31.5	31.6	26.5	30.6	33.5	25.3		64
226	A2-Cl/E4-SiMe₃	35.8	69.5	37.1	31.0	28.5	35.1	27.5	38.2	34.1	37.7		73
227	E2-Cl/E4-SiMe₃	36.2	71.1	36.9	38.6	28.1	35.5	27.2	31.0	42.0	28.1		73
228	2-CN	30.3	36.8	30.3	32.9	26.8	36.6	26.6	36.4	32.9	36.4	122.0	55,40,83
229	A2-CN/4-=O	30.2	39.6	48.7	212.4	44.8	38.9	26.3	35.4	35.0	38.7	119.9	83
230	E2-CN/4-=C(CN)₂	29.8	37.7	39.5	186.5	37.7	39.3	26.2	31.7	35.0	36.0	119.0,80.4	83
231	E2-CN/4-=O	30.1	37.4	47.6	212.3	45.8	38.7	26.4	32.0	36.7	35.3	119.3	83
232	2-COOH	29.5	49.6	29.5	33.7	27.5	37.4	27.5	38.2	33.7	38.2	181.4	83,40,54,89,92
233	E2-COOH/4-=O	29.0	49.8	47.6	215.7	46.0	39.0	26.8	32.4	37.7	35.3	177.6	83
234	2-COOMe	29.8	49.6	29.8	33.7	27.8	37.6	27.8	38.3	33.7	35.3	173.9	83,89
235	E2-COOMe/4-=O	29.1	51.8	47.8	215.6	46.1	39.0	26.8	32.4	37.9	35.3	172.6	83
236	A2-COOMe/4-=O	29.9	53.3	47.9	214.9	46.5	40.0	27.2	37.2	34.5	40.2	173.6	83
237	A2-COOMe/4-=O/8-=O	46.7	52.1	46.5	209.5	44.8	40.4	44.7	209.4	35.0	35.0	170.3	66
238	2-Et	33.0	47.8	33.0	33.0	30.0	39.9	29.7	40.7	33.0	40.7	26.9	62
239	2-F	32.9	95.6	32.9	31.5	27.7	37.3	27.3	35.8	31.5	35.8		64,40,83,88
240	2-F 2-F	36.0	127.5	36.0	34.1	26.6	36.7	26.6	34.1	34.1	34.1		65
241	E2-F/E4-F	31.3	92.2	37.7	92.2	31.3	31.0	26.2	31.0	29.6	25.7		64
242	A2-F/E4-F	32.0	96.5	37.7	91.4	31.8	31.0	25.8	35.0	33.0	29.8		64
243	2-I	37.6	46.4	37.6	33.0	27.8	38.2	27.1	38.8	33.0	38.8		53,40,55
244	E2-I/4-Me 4-Me	37.4	44.6	47.0	39.0	36.7	34.1	28.5	34.9	34.7	28.8	27.5,27.5	76
245	E2-I/E4-I	36.2	38.1	45.3	38.1	36.2	33.9	27.4	33.9	39.0	28.6		57
246	2-Me	34.0	39.1	34.0	31.4	28.6	38.7	28.3	39.6	31.4	39.6	18.9	60,8,40,55,69,83,88,89,93,121
247	2-Me 2-Me	37.4	39.1	37.4	33.4	27.4	39.3	27.7	33.3	33.3	33.3	27.7,27.7	59
248	2-Me/6-Me	39.3	73.8	39.3	33.1	27.2	38.5	27.7	35.2	35.2	35.8	27.5	60,12,16,88,120-1
249	A2-Me/E4-F	33.8	39.6	33.3	92.3	33.8	39.6	33.3	32.9	24.8	40.9	18.5,18.5	69
250	A2-Me/E4-F	33.6	40.5	38.4	92.3	32.6	27.0	38.7	38.8	30.5	32.1	18.5	64

TRICYCLO[3.3.1.1³,⁷]DECANES (cont.)

Str	Substituents	C1	C2	C3	C4	C5	C6	C7	C8	C9	C10	Substituent Shifts	References
251	2-N(CH₂)₄	32.9	36.8	38.3	96.4	32.7	32.3	27.3	31.0	36.6	25.4	17.6	64
252	2-N(CH₂)₅	31.7	70.4	31.7	31.7	29.7	38.2	27.7	37.7	31.7	37.7		97
253	2-N₃	29.3	68.4	29.3	31.7	29.7	38.1	27.7	37.7	31.7	37.7		97
254	2-NH(2-Ad)	31.9	66.6	31.9	31.7	27.4	37.4	27.1	36.8	31.7	36.8		83
255	2-NH₂	33.2	59.4	33.2	31.7	27.9	38.2	28.0	37.9	31.7	37.9		97
256	2-NH₃Cl	34.8	55.1	34.8	30.4	27.4	37.6	27.0	37.4	30.4	37.4		88,40,54
257	2-NHAc	30.2	55.4	30.2	29.8	26.6	37.1	26.6	36.6	29.8	36.6		127,54,88
258	E2-NHAc/E4-Br	32.1	53.5	32.1	32.0	27.4	37.7	27.3	37.3	32.0	37.3		57,60
259	E2-NHAc/E4-Cl	30.8	53.3	40.0	59.4	35.2	31.8	26.7	32.4	36.7	26.4		57
260	2-NHCOPh	30.6	52.9	39.3	64.8	34.6	30.9	26.5	32.0	36.0	25.7		57
261	2-NMe₂	32.0	53.0	32.0	32.2	27.2	37.6	27.1	37.2	32.2	37.2		54
262	A2-NMe₂/5-Cl	29.8	70.4	29.8	31.4	68.1	48.0	30.9	35.3	41.6	35.3		83,8,97
263	E2-NMe₂/5-Cl	33.4	68.1	33.4	41.6	67.5	48.0	31.2	29.5	47.0	29.5		97
264	2-OAc	33.4	68.7	33.4	47.0	27.4	37.5	26.5	36.5	31.9	36.5		97
265	2-OAc/6-OAc	32.0	77.0	32.0	31.9	27.4	76.0	30.9	30.6	26.3	34.7		51,53,69,97
266	2-OCH₂CH₂O-2/A4-OH	31.3	76.0	30.6	30.6	34.5	36.3	25.7		29.1	34.2		69
267	2-OCH₂CH₂O-2/A4-OMe	36.3	111.9	41.4	76.3	31.0	36.7	26.3	34.6	29.4	34.5		43
268	2-OCH₂CH₂O-2/E4-OH	35.6	110.8	39.2	85.6	32.0	30.5	26.3	34.5	33.4	28.2		43
269	2-OCH₂CH₂O-2/E4-OMe	35.6	111.6	43.1	71.3	30.1	30.8	26.3	34.6	31.8	28.7		43
270	2-OH	35.9	111.4	40.4	80.5	27.8	37.8	27.3	36.7	31.2	36.7		51,21,40,48-9,53-55
271	2-OH 2-Me	34.7	74.7	34.7	31.2	27.1	37.9	27.3	35.0	31.2	36.7		60,69,88,127
272	2-OH 2-Ph	35.8	75.6	35.8	33.1	27.1	39.6	27.6	35.0	33.1	35.0	145.5	60,79,88
273	2-OH 2-t-Bu	34.6	77.5	34.6	35.0	37.7	39.3	27.2	35.6	35.0	35.6	39.8	120
274	E2-OH/4-=C(CN)₂	33.7	73.0	45.6	189.2	38.6	38.6	26.5	29.4	35.1	33.5	78.9	83
275	A2-OH/4-=C(CN)₂	33.7	77.2	46.3	190.8	38.6	38.6	25.9	34.9	34.0	38.2	80.0	83
276	E2-OH/4-=CH₂	34.4	74.8	45.7	155.6	37.8	39.3	27.6	38.6	36.8	32.8	103.3	83
277	2-OH/6-OH	33.7	72.4	33.1	29.7	33.7	72.4	33.1	30.7	24.4	34.6		69,49
278	A2-OH/A4-Br	34.4	75.8	41.0	60.1	36.6	39.2	26.0	29.7	25.6	39.2		57
279	E2-OH/A4-Br	34.5	69.5	42.4	62.3	35.8	38.9	26.3	36.8	31.2	33.0		57
280	A2-OH/A4-Cl	34.3	75.7	40.5	67.1	35.8	38.4	25.8	31.4	24.8	38.0		57
281	E2-OH/A4-Cl	34.3	68.9	41.9	68.0	35.1	38.1	26.2	36.6	30.6	32.1		57
282	E2-OH/A4-F	33.9	69.3	39.5	96.6	32.0	32.0	26.2	31.0	30.6	29.3		65
283	E2-OH/A4-Me	35.0	69.8	40.2	40.1	32.7	39.1	27.3	30.6	31.1	32.6	18.6	65
284	A2-OH/A4-OH	34.6	76.0	38.7	76.0	34.6	36.3	25.8	31.6	25.3	35.2		57,49
285	A2-OH/A4-SiMe₃	35.0	72.4	35.9	37.9	28.3	40.9	28.0	36.3	34.6	35.0		73
286	E2-OH/A4-SiMe₃	34.8	74.5	34.6	34.2	36.3	41.0	28.0	31.3	29.1	38.7		73
287	A2-OH/A4-SnMe₃	34.9	73.8	38.4	34.2	31.4	41.0	28.0	36.1	31.1	34.3		74
288	E2-OH/A4-SnMe₃	33.7	75.2	38.3	37.1	31.3	36.5	27.2	31.3	35.6	31.0		74
289	A2-OH/E4-Br	33.3	76.2	43.0	58.6	36.0	31.7	26.7	36.7	32.5	31.0		57
290	E2-OH/E4-Br	33.7	73.8	42.9	58.9	35.1	31.8	26.9	31.4	35.5	25.6		57
291	A2-OH/E4-Cl	33.2	76.0	42.5	63.3	35.5	31.0	26.5	36.5	32.0	30.2		57
292	E2-OH/E4-Cl	33.6	73.6	42.4	64.7	34.6	31.0	26.9	31.2	34.9	24.9		57
293	A2-OH/E4-F	33.3	76.0	39.5	92.2	32.6	31.3	26.1	35.9	29.7	30.5		64
294	E2-OH/E4-F	33.8	71.9	39.7	93.6	31.7	30.7	26.6	31.4	31.8	25.3		64
295	A2-OH/E4-I	34.4	75.7	44.2	40.6	36.3	33.1	26.7	36.1	32.5	32.5		57
296	E2-OH/E4-I	34.3	73.6	40.3	40.2	33.1	33.3	27.2	37.1	35.9	30.3		57
297	A2-OH/E4-Me	33.5	75.5	40.3	31.3	33.1	31.9	27.3	32.0	33.1	30.3	18.1	65
298	E2-OH/E4-Me	33.6	75.7	37.8	37.8	31.5	31.9	27.8	37.0	37.6	25.0	17.8	65
299	A2-OH/E4-OH	33.5	74.0	40.9	67.3	34.0	30.9	26.3	32.6	29.9	30.2		57,49
300	E2-OH/E4-OH	33.3	71.4	41.2	71.4	33.3	31.1	27.0	31.1	33.1	24.8		57,49

172

TRICYCLO[3.3.1.1^{3,7}]DECANES (cont.)

Str	Substituents	C1	C2	C3	C4	C5	C6	C7	C8	C9	C10	Substituent Shifts	References
301	A2-OH/E4-SiMe3	33.9	74.8	35.9	33.9	28.8	35.1	27.0	36.9	34.9	34.6		73
302	E2-OH/E4-SiMe3	34.6	76.8	35.5		28.3	35.1	27.4	31.5	40.3	28.6		73
303	A2-OH/E4-SnMe3	34.9	75.2	39.1	31.3	31.2	36.4	27.1	38.1	34.2	35.5		74
304	E2-OH/E4-SnMe3	34.9	76.5	38.0	37.3	30.7	36.5	27.5	31.4	39.8	29.8		74
305	2-OMe	31.5	83.3	31.5	31.5	27.6	37.7	27.6	36.6	31.5	36.6		6,53,60
306	2-OMe 2-OMe	33.1	102.2	33.1	33.9	27.3	37.5	27.3	33.9	33.9	33.9		6
307	2-OMe 2-Ph	32.8	79.8	32.8	32.8	26.9	37.7	27.5	34.5	32.8	34.5	141.1	79
308	2-OMS	33.0	85.6	33.0	31.1	26.9	37.1	26.6	36.3	31.1	36.3		51
309	E2-OMS/4-=C(CN)$_2$	31.9	80.3	43.1	185.5	37.4	39.4	26.1	29.6	34.4	34.2		83
310	2-OTMS	35.3	74.9	35.3	31.2	27.7	37.8	27.2	36.6	31.2	36.6	80.6	88,50
311	2-OTMS 2-Me	39.8	77.3	39.8	33.3	27.3	38.6	27.6	34.9	33.3	34.9	26.8	88
312	2-OTMS 2-Ph	36.0	77.9	36.0	33.1	27.0	37.8	27.4	34.4	33.1	34.4	143.6	88
313	2-OTMS/6-OTMS	34.5	74.6	34.0	30.1	34.5	74.6	34.0	30.1	24.5	35.1		50
314	A2-OTMS/A4-OTMS	35.4	75.3	41.0	75.3	35.4	36.9	26.3	36.9	25.0	35.8		50
315	E2-OTMS/E4-OTMS	34.0	73.2	42.6	73.2	34.0	31.2	26.8	31.2	33.1	24.8		50
316	A2-OTMS/E4-OTMS	34.3	76.1	42.4	69.6	34.6	31.1	26.3	36.4	30.4	30.2		50
317	2-Ph	31.1	46.8	31.1	31.9	28.0	37.9	27.2	39.2	31.9	39.2	143.9	60,83,88
318	2-PO(OMe)$_2$ 2-OH	34.0	77.3	34.0	33.8	26.7	38.2	27.2	32.2	33.8	32.2		107
319	2-Pr	32.0	44.4	32.0	32.0	28.5	35.0	28.3	39.6	32.0	39.6		62
320	2-SEt	33.5	52.8	33.5	32.1	27.9	38.0	27.7	38.9	32.1	38.9	39.8	83
321	2-SH	35.7	46.5	35.7	31.0	27.9	37.9	27.1	38.9	31.0	38.9		48,55
322	2-SiMe3	29.5	37.6	29.5	35.2	28.5	37.9	28.3	41.3	35.2	41.3		73
323	A2-SiMe3/A4-SiMe3	29.6		26.9		29.6	41.3	28.3	41.3	33.5	52.6		73
324	E2-SiMe3/E4-SiMe3	29.2	41.3		41.3	29.2	35.2	27.9	35.2		33.5		73
325	2-SMe	32.7	55.0	32.7	32.0	28.0	38.0	27.9	38.9	32.0	38.9		83
326	2-SnMe3	32.1	40.0	32.1	36.5	28.7	38.2	28.3	40.9	36.5	40.9		74,75,94
327	A2-SnMe3/E4-SnMe3	32.2	43.5	36.0	39.0	32.2	36.8	28.8	41.8	39.6	39.5		74
328	E2-SnMe3/E4-SnMe3	32.7	44.0	36.0	44.0	32.7	37.1	28.5	37.1	44.3	35.4		74
329	2-SnPh3	32.4	43.3	32.4	36.9	28.6	38.0	28.2	40.7	36.9	40.7		75
330	2=NOH	28.9	166.9	36.3	37.6	28.0	36.7	28.0	39.0	39.0	37.6		14,15

References to Bicyclo[3.3.1]nonanes and Tricyclo[3.3.1.1^{3,7}]decanes and related* systems

1. Heumann, A. and Kolshorn, H., *Tetrahedron* **31**, 1571 (1975).
2. Schneider, H.-J., Lonsdonfer, M. and Weigand, E. F., *Org. Magn. Reson.* **8**, 363 (1976).
3. Pehk, T., Lippmaa, E., Baklan, V. F., Utochka, T. and Yurchenko, A. G., *Eesti NSV Tead. Akad. Toim., Fuus., Mat.* **23**, 425 (1974).
4. Schneider, H.-J. and Ansorge, W., *Tetrahedron* **33**, 265 (1977).
5. Peters, J. A., van der Toorn, J. M. and van Bekkum, H., *Tetrahedron* **33**, 349 (1977).
6. Duddeck, H. and Wolff, P., *Org. Magn. Reson.* **8**, 593 (1976).
7. Peters, J. A., van der Toorn, J. M. and van Bekkum, H., *Tetrahedron* **31**, 2273 (1975).
8. Nelson, S. F., Weisman, G. R., Clennan, E. L. and Peacock, V. E., *J. Am. Chem. Soc.* **98**, 6893 (1976).
9. Wiseman, J. R. and Krabbenhoft, H. O., *J. Org. Chem.* **42**, 2240 (1977).
10. van Oosterhout, H., Kruk, C. and Speckamp, W. N., *Tetrahedron Lett.* 653 (1978).
11. Bok, T. R., Kruk, C. and Speckamp, W. N., *Tetrahedron Lett.* 657 (1978).
12. Kirchen, R. P. and Sorensen, T. S., *J. Am. Chem. Soc.* **100**, 1487 (1978).
13. Ayer, W. A., Browne, L. M., Fung, S. and Stothers, J. B., *Org. Magn. Reson.* **11**, 73 (1978).
14. Geneste, P., Durand, R., Kamenka, J.-M., Beierbeck, H., Martino, R. and Saunders, J. K., *Can. J. Chem.* **56**, 1940 (1978).
15. Hawkes, G., Herwig, K. and Roberts, J. D., *J. Org. Chem.* **39**, 1017 (1974).
16. Beierbeck, H. and Saunders, J. K., *Can. J. Chem.* **55**, 3161 (1977).
17. Peters, J. A., Baas, J. M. A., van de Graaf, B., van der Toorn, J. M. and van Bekkum, H., *Tetrahedron* **34**, 3313 (1978).
18. Bishop, R., *Aust. J. Chem.* **31**, 1485 (1978).
19. Wasylishen, R. E. and Friesen, K. J., *Org. Magn. Reson.* **13**, 343 (1980).
20. Senda, Y., Ishiyama, J. and Imaizumi, S., *J. Chem. Soc. Perkin Trans. II* 90 (1981).
21. Pehk, T., Lippmaa, E., Sevostjanova, V. V., Krayusfkin, M. M. and Tarasova, A. I., *Org. Magn. Reson.* **3**, 783 (1971).
*22. Pekhk, T. I., Lippmaa, E. T., Sokolova, I. M., Vorob'eva, N. S., Gervits, E. S., Bobyleva, A. A., Kalinichenko, A. N. and Belikova, N. A., *J. Org. Chem. USSR* **12**, 1207 (1976).
*23. Takaishi, N., Inamoto, Y. and Aigami, K., *Chem. Lett.* 1185 (1973).
*24. Takaishi, N., Inamoto, Y. and Aigami, K., *J. Org. Chem.* **40**, 276 (75).
*25. Takaishi, N., Inamoto, Y. and Aigami, K., *Synth. Commun.* **4**, 225 (1974).
*26. Majerski, K. M. and Majerski, Z., *Tetrahedron Lett.* 4915 (1973).
*27. Fujikura, Y., Aigami, K., Takaishi, N., Ikeda, H. and Inamoto, Y., *Chem. Lett.* 507 (1976).
*28. Takaishi, N., Fujikura, Y., Inamoto, Y., Ikeda, H. and Aigami, K., *J. Chem. Soc. Chem. Commun.* 371 (1975).
*29. Aigami, K., Inamoto, Y., Takaishi, N., Fujikura, Y., Takatsuki, A. and Tamura, G., *J. Med. Chem.* **19**, 536 (1976).
*30. Aigami, K., Inamoto, Y., Takaishi, N. and Fujikura, Y., *Phytochem.* **16**, 41 (1977).
*31. Takaishi, N., Fujikura, Y., Inamoto, Y. and Aigami, K., *J. Org. Chem.* **42**, 1737 (1977).
*32. Klester, A. M. and Ganter, C., *Helv. Chim. Acta* **66**, 1200 (1983).
*33. Fujikura, Y., Ohsugi, M., Inamoto, Y., Takaishi, N. and Aigami, K., *J. Org. Chem.* **43**, 2608 (1978).

*34. Ohsugi, M., Inamoto, Y., Takaishi, N., Fujikura, Y. and Aigami, K., *Synthesis* 632 (1977).

35. Kessenikh, A. V., Aredova, E. N., Sevost'yanova, V. V. and Krayushkin, M. M., *Bull. Acad. Sci. USSR, Div. Chem. Sci.* **28**, 2209 (1979).

*36. Takaishi, N., Fujikura, Y. and Inamoto, Y., *Chem. Lett.* 825 (1978).

37. Lippmaa, E., Pehk, T. and Past, J., *Eesti NSV Tead. Akad. Toim., Fuus., Mat.* **16**, 345 (1967).

38. Lippmaa, E. and Pehk, T., *Kemian Teollisuus* **24**, 1001 (1967).

*39. Berman, S. S., Denisov, Y. V., Murskhovskaya, A. S., Stepanyants, G. U. and Petrov, A. A., *Neftekhimiya* **14**, 3 (1974).

40. Maciel, G. E., Dorn, H. C., Greene, R. L., Kleschick, W. A., Peterson, J. M. R. and Wahl, J. G. H., *Org. Magn. Reson.* **6**, 178 (1974).

41. Bishop, R. and Landers, A. E., *Aust. J. Chem.* **32**, 2675 (1979).

42. Bishop, R., Choudhury, S. and Dance, I., *J. Chem. Soc. Perkin Trans. II* 1159 (1982).

43. Henkel, J. G. and Spector, J. H., *J. Org. Chem.* **48**, 3657 (1983).

44. Quast, H., Gorlach, Y. and Stawitz, J., *Angew. Chem., Int. Ed. Engl.* **20**, 93 (1981).

*45. Engdahl, C. and Ahlberg, P., *J. Am. Chem. Soc.* **101**, 3940 (1979).

46. Henkel, J. G. and Hane, J. T., *J. Org. Chem.* **48**, 3858 (1983).

47. Bessiere, Y., Grison, C. and Boussac, G., *Tetrahedron* **34**, 1957 (1978).

48. Duddeck, H. and Dietrich, W., *Tetrahedron Lett.* 2925 (1975).

49. Schraml, J., Jancke, H., Engelhardt, G., Vodecka, L. and Hlavaty, J., *Coll. Czech. Chem. Commun.* **44**, 2230 (1979).

50. Schraml, J., Vcelak, J, Chvalovsky, V., Engelhardt, G., Jancke, H., Vodicka, L. and Hlavaty, J., *Coll. Czech. Chem. Commun.* **43**, 3179 (1978).

51. Duddeck, H., *Org. Magn. Reson.* **7**, 151 (1975).

52. Perkins, R. R. and Pincock, R. E., *Org. Magn. Reson.* **8**, 165 (1976).

53. Duddeck, H. and Wolff, P., *Org. Magn. Reson.* **9**, 528 (1977).

54. Berger, S. and Zeller, K.-P., *J. Chem. Soc. Chem. Commun.* 649 (1976).

55. Gerhards, R., Dietrick, W., Bergmann, G. and Duddeck, H., *J. Magn. Reson.* **36**, 189 (1979).

56. Duddeck, H. and Klein, H., *Tetrahedron* **33**, 1971 (1977).

57. Duddeck, H., *Tetrahedron* **34**, 247 (1978).

58. Dheu, M.-L., Gagnaire, D., Duddeck, H., Hollowood, F. and McKervey, M. A., *J. Chem. Soc. Perkin Trans. II* 357 (1979).

59. Graham, W. D., Schleyer, P. R., Hagaman, E. W. and Wenkert, E., *J. Am. Chem. Soc.* **95**, 5785 (1973).

60. Duddeck, H., Hollowood, F., Karim, A. and McKervey, M. A., *J. Chem. Soc. Perkin Trans. II* 360 (1979).

61. Poindexter, G. S. and Kropp, P. J., *J. Org. Chem.* **41**, 1215 (1976).

62. Murakhovsksya, A. S., Stepanyants, A. U., Bagrii, Y. I., Frid, T. Y., Zimina, K. I. and Sanin, P. I., *Petroleum Chemistry* **14**, 39 (1974).

*63. Vorob'eva, N. S., Aref'ev, O. A., Pekhk, T. I., Denisov, Y. V. and Petrov, A. A., *Neftekhimiya* **15**, 659 (1975).

64. Duddeck, H. and Islam, M. R., *Tetrahedron* **37**, 1193 (1981).

65. Duddeck, H. and Islam, M. R., *Org. Magn. Reson.* **16**, 32 (1981).

66. Duddeck, H., *Tetrahedron* **39**, 1365 (1983).

67. Boaz, H., *Tetrahedron Lett.* 55 (1973).

68. Fuchs, W., Kalbacher, H. and Voelter, W., *Org. Magn. Reson.* **17**, 157 (1981).

69. Majerski, Z., Vinkovic, V. and Meic, Z., *Org. Magn. Reson.* **18**, 169 (1981).
70. Pincock, R. E. and Fung, F.-N., *Tetrahedron Lett.* **21**, 19 (1980).
71. Mlinaric-Majerski, K. and Majerski, Z., *J. Am. Chem. Soc.* **102**, 1418 (1980).
72. Sasaki, T., Eguchi, S. and Suzuki, T., *J. Org. Chem.* **45**, 3824 (1980).
73. Duddeck, H. and Islam, M. R., *Chem. Ber.* **117**, 554 (1984).
74. Duddeck, H. and Islam, M. R., *Chem. Ber.* **117**, 565 (1984).
75. Kuivila, H. G., Considine, J. L., Sarma, R. H. and Mynott, R. J., *J. Orgmet. Chem.* **111**, 179 (1976).
76. Duddeck, H., Wiskamp, V. and Rosenbaum, D., *J. Org. Chem.* **46**, 5332 (1981).
77. le Noble, W. J., Srivastava, S. and Cheung, C. K., *J. Org. Chem.* **48**, 1099 (1983).
78. Raber, D. J. and Janks, C. M., *J. Org. Chem.* **48**, 1101 (1983).
79. Peoples, III, P. R., *The Chemistry and Geometry of Carbanions*, PhD Thesis, Purdue University, 1979).
80. Momose, T. and Muraoka, O., *Chem. Pharm. Bull.* **26**, 288 (1978).
81. Momose, T. and Muraoka, O., *Chem. Pharm. Bull.* **26**, 2217 (1978).
82. Momose, T. and Muraoka, O., *Chem. Pharm. Bull.* **26**, 2589 (1978).
83. Duddeck, H. and Feuerhelm, H.-T., *Tetrahedron* **36**, 3009 (1980).
84. Olah, G. A., Shih, J. G., Singh, B. P. and Gupta, B. G. B., *Synthesis* 713 (1983).
85. Israel, R. J. and Murray, J. R. K., *J. Org. Chem.* **48**, 4701 (1983).
*86. Osawa, E., Tahara, Y., Togashi, A., Iizuka, T., Tanaka, N., Kan, T., Farcasiu, O., Kent, G. J., Engler, E. M. and Schleyer, P. v. R., *J. Org. Chem.* **47**, 1923 (1982).
87. Saba, J. A. and Fry, J. L., *J. Am. Chem. Soc.* **105**, 533 (1983).
88. Krishnamurthy, V. V., Iyer, P. S. and Olah, G. A., *J. Org. Chem.* **48**, 3373 (1983).
89. Walter, S. R., Marshall, J. L., McDaniel, J. C. R., Canada, J. E. D. and Barfield, M., *J. Am. Chem. Soc.* **105**, 4185 (1983).
90. Marshall, J. L., Conn, S. A. and Barfield, M., *Org. Magn. Reson.* **7**, 404 (1977).
91. Barfield, M., Conn, S. A., Marshall, J. L. and Muller, D. E., *J. Am. Chem. Soc.* **98**, 6253 (1976).
92. Marshall, J. L. and Miiller, D. E., *J. Am. Chem. Soc.* **95**, 8305 (1973).
93. Barfield, M., Marshall, J. L. and Canada, J. E. D., *J. Am. Chem. Soc.* **102**, 7 (1980).
94. Doddrell, D., Burfitt, I., Kitching, W., Bullpitt, M., Lee, C.-H., Mynott, R. J., Considine, J. L., Kuivila, H. G. and Sarma, R. H., *J. Am. Chem. Soc.* **96**, 1640 (1974).
95. Della, E. W. and Patney, H. K., *Aust. J. Chem.* **32**, 2243 (1979).
96. Golitz, P. and de Meijere, A., *Angew. Chem., Int. Ed. Engl.* **20**, 298 (1981).
97. Pekhk, T. I., Lippmaa, E. T., Lavrova, L. N., Vinogradova, N. N., Klimova, N. V., Shmar'yan, M. I. and Skoldinov, A. P., *J. Org. Chem. USSR* **14**, 1526 (1978).
98. Andrieu, C. G., Debruyne, D. and Paquer, D., *Org. Magn. Reson.* **11**, 528 (1978).
99. Duddeck, H. and Islam, M. R., *Org. Magn. Reson.* **21**, 140 (1983).
100. Peters, J. A., van Ballegoyen-Eekhout, G. W. M., van de Graaf, B., Bovee, W. M. M. J., Baas, J. M. A. and van Bekkum, H., *Tetrahedron* **39**, 1649 (1983).
101. Hoffmann, G. G. and Klein, H., *Tetrahedron* **40**, 199 (1984).
102. Sekiguchi, A. and Ando, W., *J. Am. Chem. Soc.* **106**, 1486 (1984).
103. Dekkers, A. W. J. D., Verhoeven, J. W. and Speckamp, W. N., *Tetrahedron* **29**, 1691 (1973).
104. Olah, G. A., Schilling, P., Westerman, P. W. and Lin, H. C., *J. Am. Chem. Soc.* **96**, 358 (1974).
105. Ajisaka, K. and Kainosho, M., *J. Am. Chem. Soc.* **97**, 330 (1975).
106. Ando, W., Kabe, Y. and Takata, T., *J. Am. Chem. Soc.* **104**, 7314 (1982).

107. Buchanan, G. W. and Benezra, C., *Can. J. Chem.* **54**, 231 (1976).
108. House, H. O., Kleshnick, W. A. and Zaiko, E. J., *J. Org. Chem.* **43**, 3653 (1978).
109. House, H. O., DeTar, M. B. and Van Derveer, D., *J. Org. Chem.* **44**, 3973 (1979).
110. House, H. O., Sieloff, R. F., Lee, T. V. and DeTar, M. B., *J. Org. Chem.* **45**, 1800 (1980).
111. House, H. O., DeTar, M. B., Sieloff, R. F. and Van Derveer, D., *J. Org. Chem.* **45**, 3545 (1980).
112. House, H. O., Outcalt, R. J. and Cliffton, M. D., *J. Org. Chem.* **47**, 2413 (1982).
113. House, H. O., Outcalt, R. J., Haack, J. L. and VanDerveer, D., *J. Org. Chem.* **48**, 1654 (1983).
114. Morat, C. and Rassat, A., *Tetrahedron Lett.* 4561 (1979).
115. Hoffmann, G. and Wiartalla, R., *Tetrahedron Lett.* **23**, 3887 (1982).
*116. Sasaki, T., Eguchi, S. and Okano, T., *Tetrahedron Lett.* **23**, 4969 (1982).
*117. Sasaki, T., Eguchi, S., Okano, T. and Wakata, Y., *J. Org. Chem.* **48**, 4067 (1983).
*118. Duddeck, H. and Wiskamp, V., *Org. Magn. Reson.* **15**, 361 (1981).
119. Harding, K. E., Clement, B. A., Moreno, L. and Peter-Katalinic, J., *J. Org. Chem.* **46**, 940 (1981).
120. Loomes, D. J. and Robinson, M. J. T., *Tetrahedron* **33**, 1149 (1977).
121. Murakhovskaya, A. S., Stepanyants, A. V., Bagru, E. I., Dolgopolova, T. N., Frid, T. Y., Simina, K. I. and Sanin, P. I., *Bull. Acad. Sci. USSR, Div. Chem. Sci.* 1203 (1975).
122. Jorgensen, F. S. and Snyder, J. P., *J. Org. Chem.* **45**, 1015 (1980).
123. Grava, I. Y., Polis, Y. Y., Lidak, M. Y., Liepin'sh, E. E., Shatts, V. D., Dipan, I. V., Gavars, M. P. and Sekatsis, I. P., *J. Org. Chem. USSR* **17**, 679 (1981).
124. Chadwick, D. J. and Williams, D. H., *J. Chem. Soc. Perkin Trans. II* 1202 (1974).
125. Lightner, D. A., Bouman, T. D., Wijekoon, W. M. D. and Hansen, A. E., *J. Am. Chem. Soc.* **106**, 934 (1984).
126. Della, E. W. and Pigou, P. E., *J. Am. Chem. Soc.* **106**, 1085 (1984).
127. Sadtler Standard Carbon-13 NMR Spectra, Sadtler Research Laboratories, Inc., Philadelphia, Pa.
128. Johnson, L. F. and Jankowski, W. C., *Carbon-13 NMR Spectra*, Wiley-Interscience, New York (1972).

Chapter 14

Bicyclo[4.1.0]heptanes

Simple endo/exo terminology can be applied to the bicyclo[4.1.0]heptanes. No significant amount of data on trans-fused compounds was found.

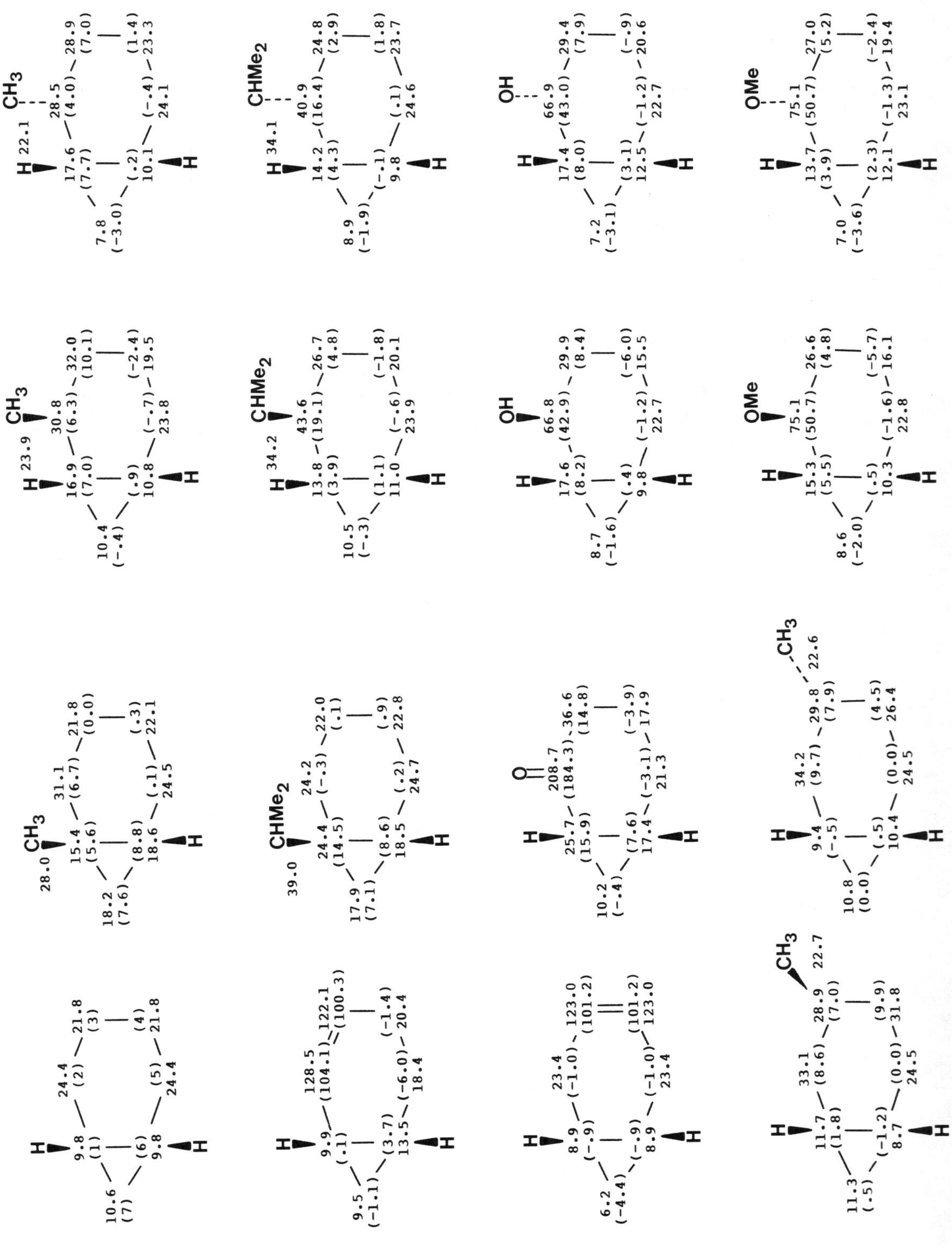

Structure 1 (COOH):
```
            COOH 183.1
      27.0    40.1    (1.1)
H  7.8 (2.6)(18.3)(−1.7) 22.9
  (−2.0)      9.9   22.7
   10.6    (.1)
   (0.0)         H
```

Structure 2 (COOH):
```
            COOH 183.1
      26.0    37.7    (3.2)
H  8.4 (1.6)(15.9)(−1.3) 25.0
  (−1.4)     10.0  23.1
   10.9    (.2)
   (.3)          H
```

Structure 3 (CHMe₂):
```
            CHMe₂ 33.1
      27.7    38.1    (5.2)
H 11.9 (3.2)(16.2)(−.8) 27.1
  (2.0)      9.3   25.3
   11.4    (−.6)
   (.6)          H
```

Structure 4 (CHMe₂):
```
            CHMe₂ 32.9
      28.4    41.2    (1.6)
H  9.4 (3.9)(19.3)(−.4) 23.5
  (−.5)     10.9   24.9
   10.4    (1.0)
   (−.4)         H
```

Structure 5 (Ph):
```
      20.4    21.9    (.1)
H 12.9(−4.0)( )(−4.0) 21.9
  (3.1)     12.9  20.4
   22.4    (3.1)
  (11.8)        H
Ph 138.6
```

Structure 6 (Ph):
```
      23.8    21.6    (−.2)
H 22.5 (−.6)( )(−.6) 21.6
  (12.7)    22.5  23.8
   28.3   (12.7)
  (18.3)        H
Ph 144.5
```

Structure 7 (H₃C):
```
      24.0    22.0    (.2)
H 18.4 (−.4)( )(−.4) 22.0
  (8.6)     18.4  24.0
   18.0    (8.6)
   (7.4)         H
H₃C 19.0
```

Structure 8 (H₃C):
```
      18.9    22.9    (1.1)
H 10.4(−5.5)( )(−5.5) 22.9
  (.6)      10.4  18.9
   12.2    (.6)
  (1.6)         H
H₃C 8.3
```

Structure 9 (Cl/Ph):
```
      20.7    20.6    (−1.2)
H 24.7 (−3.7)( )(−3.7) 20.6
  (14.9)    24.7  20.7
   48.5   (14.9)
  (37.9)        H
Cl  Ph 138.0
```

Structure 10 (Ph/Cl):
```
      19.3    21.2    (−.6)
H 21.4 (−5.1)( )(−5.1) 21.2
  (11.6)    21.4  19.3
   55.2   (11.6)
  (44.6)        H
Ph  Cl 145.1
```

Structure 11 (MeOOC):
```
      23.1    21.3    (−.5)
H 22.1 (−1.3)( )(−1.3) 21.3
  (12.3)    22.1  23.1
   25.9   (12.3)
  (15.3)        H
MeOOC 174.7
```

Structure 12 (MeOOC):
```
      18.9    21.5    (−.3)
H 16.6(−5.5)( )(−5.5) 21.5
  (6.8)     16.6  18.9
   22.1    (6.8)
  (11.5)        H
MeOOC 172.1
```

Structure 13 (MeO):
```
      18.3    22.6    (.8)
H 11.6 (−6.1)( )(−6.1) 22.6
  (1.8)     11.6  18.3
   60.4    (1.8)
  (49.8)        H
MeO
```

Structure 14 (MeO):
```
      22.7    22.0    (.2)
H 17.7 (−1.7)( )(−1.7) 22.0
  (7.9)     17.7  22.7
   66.4    (7.9)
  (55.8)        H
MeO
```

Structure 15 (MeOOC/Cl):
```
      18.8    21.0    (−.8)
H 24.6 (−5.6)( )(−5.6) 21.0
  (14.8)    24.6  18.8
   56.9   (14.8)
  (46.3)        H
MeOOC  Cl 179.8
```

Structure 16 (Br/MeOOC):
```
      19.6    20.7    (−.3)
H 25.9(−4.8)( )(−4.8) 20.7
  (16.1)    25.9  19.6
   32.2   (16.1)
  (21.6)        H
Br  MeOOC 172.7
```

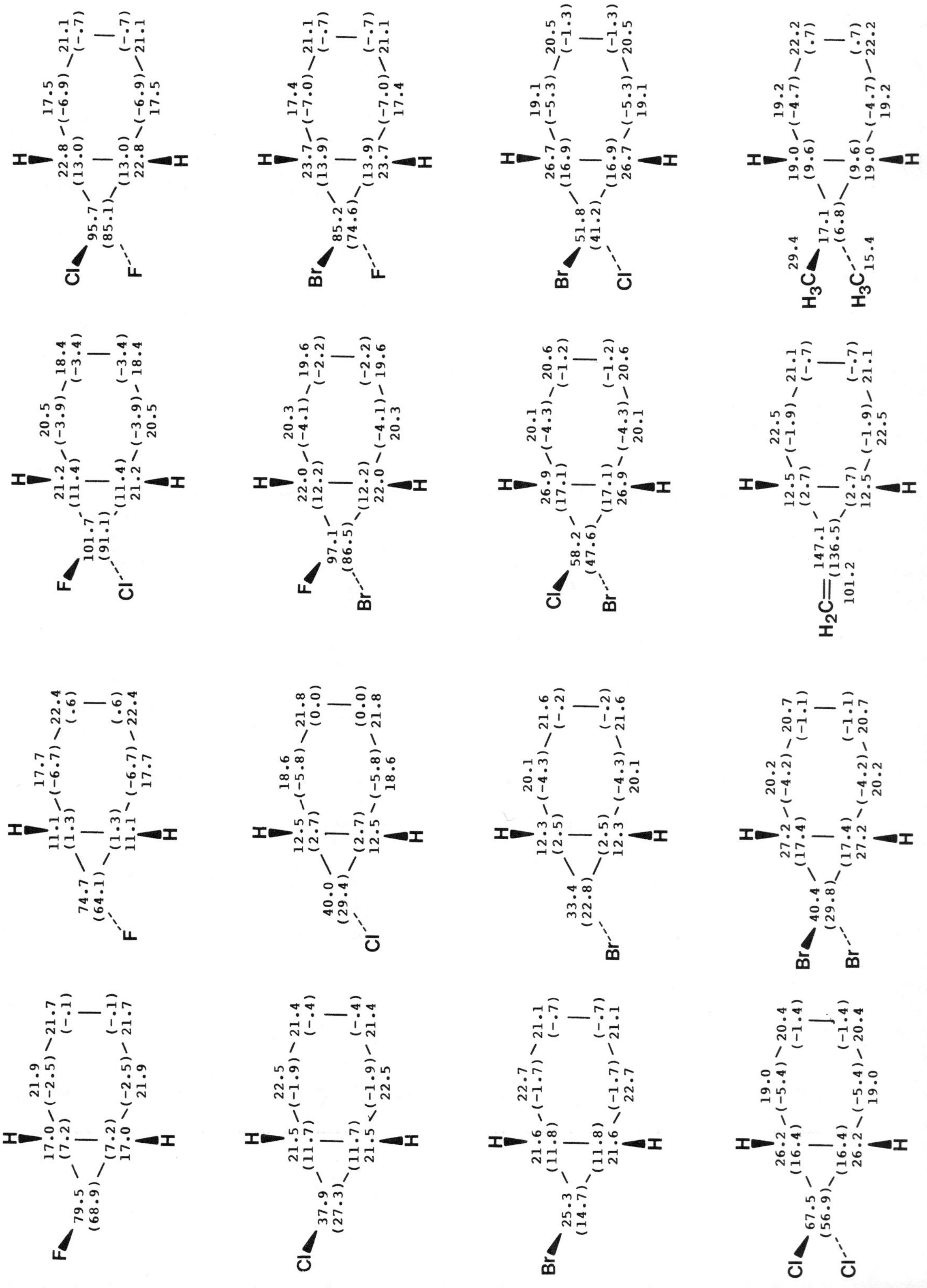

Row 1:

Structure 1 (Cl, F):
17.5 (−6.9) 21.1 (−.7) — (−.7) 21.1 (−6.9) 17.5
22.8 (13.0) — (13.0) 22.8
95.7 (85.1)
Cl — F

Structure 2 (Br, F):
17.4 (−7.0) 21.1 (−.7) — (−.7) 21.1 (−7.0) 17.4
23.7 (13.9) — (13.9) 23.7
85.2 (74.6)
Br — F

Structure 3 (Br, Cl):
19.1 (−5.3) 20.5 (−1.3) — (−1.3) 20.5 (−5.3) 19.1
26.7 (16.9) — (16.9) 26.7
51.8 (41.2)
Br — Cl

Structure 4 (H₃C, H₃C):
19.2 (−4.7) 22.2 (.7) — (.7) 22.2 (−4.7) 19.2
19.0 (9.6) — (9.6) 19.0
29.4 17.1 (6.8) 15.4
H₃C — H₃C

Row 2:

Structure 5 (F, Cl):
20.5 (−3.9) 18.4 (−3.4) — (−3.4) 18.4 (−3.9) 20.5
21.2 (11.4) — (11.4) 21.2
101.7 (91.1)
F — Cl

Structure 6 (F, Br):
20.3 (−4.1) 19.6 (−2.2) — (−2.2) 19.6 (−4.1) 20.3
22.0 (12.2) — (12.2) 22.0
97.1 (86.5)
F — Br

Structure 7 (Cl, Br):
20.1 (−4.3) 20.6 (−1.2) — (−1.2) 20.6 (−4.3) 20.1
26.9 (17.1) — (17.1) 26.9
58.2 (47.6)
Cl — Br

Structure 8 (H₂C=):
22.5 (−1.9) 21.1 (−.7) — (−.7) 21.1 (−1.9) 22.5
12.5 (2.7) — (2.7) 12.5
H₂C= 147.1 (136.5)
101.2

Row 3:

Structure 9 (F):
17.7 (−6.7) 22.4 (.6) — (.6) 22.4 (−6.7) 17.7
11.1 (1.3) — (1.3) 11.1
74.7 (64.1)
F

Structure 10 (Cl):
18.6 (−5.8) 21.8 (0.0) — (0.0) 21.8 (−5.8) 18.6
12.5 (2.7) — (2.7) 12.5
40.0 (29.4)
Cl

Structure 11 (Br):
20.1 (−4.3) 21.6 (−.2) — (−.2) 21.6 (−4.3) 20.1
12.3 (2.5) — (2.5) 12.3
33.4 (22.8)
Br

Structure 12 (Br, Br):
20.2 (−4.2) 20.7 (−1.1) — (−1.1) 20.7 (−4.2) 20.2
27.2 (17.4) — (17.4) 27.2
40.4 (29.8)
Br — Br

Row 4:

Structure 13 (F):
21.9 (−2.5) 21.7 (−.1) — (−.1) 21.7 (−2.5) 21.9
17.0 (7.2) — (7.2) 17.0
79.5 (68.9)
F

Structure 14 (Cl):
22.5 (−1.9) 21.4 (−.4) — (−.4) 21.4 (−1.9) 22.5
21.5 (11.7) — (11.7) 21.5
37.9 (27.3)
Cl

Structure 15 (Br):
22.7 (−1.7) 21.1 (−.7) — (−.7) 21.1 (−1.7) 22.7
21.6 (11.8) — (11.8) 21.6
25.3 (14.7)
Br

Structure 16 (Cl, Cl):
19.0 (−5.4) 20.4 (−1.4) — (−1.4) 20.4 (−5.4) 19.0
26.2 (16.4) — (16.4) 26.2
67.5 (56.9)
Cl — Cl

BICYCLO[4.1.0]HEPTANES

Str	Substituents	C1	C2	C3	C4	C5	C6	C7	Substituent Shifts	References
1	PARENT	9.8	24.4	21.8	21.8	24.4	9.8	10.6		27,2,5,43
2	1-COOH/3=/X7-Me N7-Me	27.6	22.2	124.0	124.4	20.1	24.8	28.4	181.5,22.1,15.3	33
3	1-COOMe/3=/X7-Me N7-Me	27.9	22.4	124.2	124.3	20.5	23.7	26.7	175.0,22.0,15.0	33
4	1-Et/X7-C1 N7-C1	31.3	24.3	20.3	21.3	19.1	32.3	73.3	31.8	42
5	1-iPr	24.4	24.2	22.0	22.8	24.7	18.5	17.9	39.0	43
6	1-iPr/N4-Me	24.5	23.1	29.6	30.2	34.4	18.4	18.1	38.2,22.3	43
7	1-iPr/N4-Me/X7-C1 N7-C1	33.2	19.0	27.8	27.9	29.5	33.7	73.5	35.5,21.6	43
8	1-iPr/X4-Me	23.7	24.8	32.5	26.6	33.4	19.9	18.1	38.9,22.3	43
9	1-iPr/X4-Me/X7-C1 N7-C1	34.8	20.1	28.3	25.9	31.5	33.4	73.4	36.4,22.8	43
10	1-Me	15.4	31.1	21.8	22.1	24.5	18.6	18.2	28.0	1,43
11	1-Me/2-=0	29.8	210.2	21.3	19.3	21.8	26.3	18.1	19.8	6
12	1-Me/2-=0/6-Me	35.4	209.9	36.0	18.3	29.4	28.5	24.1	14.3,21.2	6
13	1-Me/2-=0/X4-C(Me)=CH2/X7-SPh	35.1	207.4	42.2	37.3	26.7	28.9	30.1	14.5,146.4	30
14	1-Me/2-=0/X7-CH=CH2	36.1	209.3	36.8	20.4	21.9	31.7	32.4	15.1,135.2	28
15	1-Me/2-=0/X7-CH=[E]CHMe	35.2	208.0	36.1	19.9	21.4	31.7	30.3	14.4,128.0	28
16	1-Me/6-Me/X7-C1 N7-C1	28.5	28.5	21.3	21.3	28.5	28.5	78.1	20.2,20.2	42
17	1-Me/N4-C(Me)=CH2	15.8	31.1	26.4	42.8	31.6	18.6	18.5	27.5,150.1	43
18	1-Me N4-C(Me)=CH2/X7-C1 N7-C1	25.7	28.4	25.0	40.3	26.6	33.4	72.6	25.5,148.9	43
19	1-Me/N4-1Pr	16.1	31.8	24.5	41.3	28.7	18.5	18.2	27.4,32.8	43
20	1-Me/N4-1Pr/X7-C1 N7-C1	25.8	28.5	22.9	38.9	25.4	32.2	72.7	24.4,33.6	43
21	1-Me/X4-C(Me)=CH2	14.5	30.9	27.6	38.6	29.5	20.3	18.5	27.9,149.4	43
22	1-Me/X4-C(Me)=CH2/X7-C1 N7-C1	27.3	27.6	24.3	38.2	26.0	32.9	72.8	24.9,148.8	43
23	1-Me/X4-1Pr	14.7	31.6	26.5	37.6	27.7	20.5	18.5	27.9,32.2	43
24	1-Me/X4-1Pr/X7-C1 N7-C1	27.6	27.9	22.8	37.5	24.9	32.9	72.9	24.9,32.5	43
25	1-Me/X7-Br N7-Br	26.4	28.8	20.8	20.6	20.0	33.4	50.3	28.3	1
26	1-Me/X7-C1 N7-C1	26.9	27.4	20.2	20.9	18.9	32.1	73.0	25.5	1
27	2-=0	25.7	208.7	36.6	17.9	21.3	17.4	10.2		6,2
28	2-=0 3-Me 3=/X7-Me N7-Me	33.6	195.3	134.2	141.8	22.3	27.8	20.9	13.6,25.6,15.3	29
29	2-=0/6-Me	34.3	208.7	35.8	18.1	28.1	23.9	17.5	24.8	6
30	2-=0/N3-Me X3-0-4X/X7-Me N7-Me	29.8	200.6	57.0	60.4	18.9	20.5	22.7	14.3,27.6,15.1	29
31	2-=0/N3-Me/X7-Me N7-Me	33.9	210.0	42.5	28.3	19.1	25.5	23.4	17.6,29.7,14.0	29
32	2-=0/X3-Me/X7-Me N7-Me	34.8	210.0	42.9	35.1	18.4	31.4	18.8	15.2,28.9,16.1	2
33	2-=0/X7-CH=CH2	34.4	206.5	36.9	18.7	21.0	25.1	27.5	137.9	28
34	2-=0/X7-CH=[E]CHMe	34.0	206.4	36.4	18.3	20.7	24.6	24.6	130.1	28
35	X2-1Pr	13.8	43.6	26.7	20.1	23.9	11.0	10.5	34.2	43
36	N2-1Pr	14.2	40.9	24.8	23.7		9.8	8.9	34.1	43
37	X2-1Pr/N5-Me	14.8	43.3	28.8	27.5	29.3	18.4	7.8	34.1,21.9	43
38	N2-1Pr/X5-Me	15.0	41.1	24.7	34.1	31.7	17.6	8.6	33.8,24.0	43
39	X2-Me	16.9	30.8	32.0	19.5	23.8	10.8	10.4	23.9	43
40	N2-Me	17.6	28.5	28.9	23.3	24.1	10.1	7.8	22.1	43
41	2-Me 2= 3-Me 4=/X7-COOMe N7-Me	49.0	128.4	127.9	131.4	122.0	45.8	16.7	20.0,18.7,178.3,9.1	16
42	X2-0-3X N3-Me/X7-Me N7-Me	22.4	56.3	56.6	27.6	16.8	21.3	20.4	23.7,29.0,16.5	43
43	N2-OAc/X3-Me/X7-Me N7-Me	23.8	78.6	33.5	31.9	18.9	25.7	19.7	20.6,29.5,16.5	2
44	X2-OH	17.6	66.8	29.9	15.5	22.7	9.8	8.7		2
45	N2-OH	17.4	66.9	29.4	20.6	22.7	12.5	7.2		2
46	N2-OH/X3-Me/X7-Me N7-Me	26.8	75.5	36.0	32.2	18.8	25.2	19.8	19.8,29.7,16.4	2
47	X2-OMe	15.3	75.1	26.6	16.1	23.8	10.3	8.6		7
48	N2-OMe	13.7	75.1	27.0	19.4	23.1	12.1	7.0		7
49	X2-OMe/X7-COOMe	26.5	74.9	27.5	16.7	22.7	22.5	23.9	174.2	7
50	N2-OMe/X7-COOMe	25.0	72.6	27.8	18.6	22.6	23.3	21.6	173.8	7

BICYCLO[4.1.0]HEPTANES (cont.)

Str	Substituents	C1	C2	C3	C4	C5	C6	C7	Substituent Shifts	References
51	2=	9.9	128.5	122.1	20.4	18.4	13.5	9.5		23
52	2= 3-Me/X7-Me N7-Me	23.3	119.7	133.3	27.7	18.2	21.3	24.1	23.8,28.6,15.5	43
53	2=/7-=CH$_2$	17.3	125.7	123.6	20.9	16.7	14.7	135.2	102.8	23,24
54	2=/N4-Me/X7-Me N7-Me	22.3	125.3	136.2	27.3	28.6	20.6	23.5	21.6,27.8,15.2	2,43
55	3-=O 4-Me 4=/7-Me N7-Me	23.2	33.1	196.3	132.9	144.4	23.9	24.7	15.8,27.4,13.2	29
56	3-=O/N4-Me/X7-Me N7-Me	22.6	36.4	215.3	41.5	29.4	20.0	19.1	13.7,27.6,14.4	2,3
57	X3-COOH	8.4	26.0	37.7	25.0	23.1	10.0	10.9	183.1	36
58	N3-COOH	7.8	27.0	40.1	22.9	22.7	9.9	10.6	183.1	36
59	X3-iPr	11.9	27.7	38.1	27.1	25.3	9.3	11.4	33.1	43
60	N3-iPr	9.4	28.4	41.2	23.5	24.9	10.9	10.4	32.9	43
61	X3-Me	11.7	33.1	28.9	31.8	24.5	8.7	11.3	22.7	43
62	N3-Me	9.4	34.2	29.8	26.4	24.5	10.4	10.8	22.6	43
63	3-Me 3=/X7-Me N7-Me	16.8	24.9	131.2	119.6	20.9	18.7	16.7	23.6,28.4,13.2	22,2,4,40,43
64	X3-Me/X7-Me N7-Me	19.8	28.7	28.4	31.3	19.3	20.0	17.5	23.0,29.6,15.6	43
65	N3-Me/X7-Me N7-Me	21.3	28.8	29.0	31.5	20.1	18.4	17.5	22.3,29.5,15.5	43
66	N3-O-4N X3-Me/X7-Me N7-Me	17.3	23.9	54.5	56.9	19.8	18.3	17.3	24.6,29.3,14.8	43
67	X3-O-4X N3-Me/X7-Me N7-Me	14.3	23.5	54.2	56.7	19.4	16.4	14.6	23.1,27.9,14.6	43
68	N3-OAc/4-Me 4=/X7-Me N7-Me	18.5	26.7	70.8	135.3	122.4	22.4	24.9	19.8,27.3,14.9	29
69	X3-OH N3-Me/X7-Me N7-Me	17.8	34.5	66.6	32.6	15.9	18.1	16.6	30.4,29.3,14.8	43
70	N3-OH X3-Me/X7-Me N7-Me	17.6	35.8	69.6	33.6	18.6	20.6	17.1	26.2,29.1,15.0	43
71	X3-OH/4-Me 4=/X7-Me N7-Me	17.2	28.2	66.6	137.0	122.7	22.8	24.7	20.8,27.6,14.8	29
72	N3-OH/4-Me 4=/X7-Me N7-Me	18.7	30.1	68.4	139.9	120.7	22.6	24.7	20.5,27.4,15.2	29
73	X3-OH/N4-Me-Me/X7-Me N7-Me	20.2	30.5	74.5	36.4	28.5	21.9	17.5	17.2,28.8,15.9	2,3,43
74	N3-OH/N4-Me/X7-Me N7-Me	21.3	28.0	68.0	33.1	28.7	22.1	17.7	17.2,28.7,15.0	2,3
75	3=	8.9	23.4	123.0	123.0	23.4	8.9	6.2		23,43
76	3=/7-=CH$_2$	12.1	22.4	123.0	123.0	22.4	12.1	138.2	102.3	23
77	3=/7-Br 7-Me	20.9	19.3	123.9	123.9	19.3	20.9	34.6	18.6	23
78	3=/N7-CH$_2$OH	9.9	19.8	125.1	125.1	19.8	9.9	20.5	59.3	26
79	7-=CH$_2$	12.5	22.5	21.1	21.1	22.5	12.5	147.1	101.2	23
80	X7-Br	21.6	22.7	21.1	21.1	22.7	21.6	25.3		27
81	N7-Br	12.3	20.1	21.6	21.6	20.1	12.3	33.4		27
82	X7-Br N7-Br	27.2	20.2	20.7	20.7	20.2	27.2	40.4		27,1,43
83	X7-Br N7-Cl	26.7	19.1	20.5	20.5	19.1	26.7	51.8		27
84	N7-Br X7-Cl	26.9	20.1	20.6	20.6	20.1	26.9	58.2		27
85	X7-Br N7-F	23.7	17.4	21.1	21.1	17.4	23.7	85.2		27
86	N7-Br X7-F	22.0	20.3	19.6	19.6	20.3	22.0	97.1		27
87	X7-C(OMe)(OCH$_2$CH$_2$O)	14.0	23.2	21.7	21.7	23.2	14.0	28.1	123.0	31,41
88	X7-Cl	21.5	22.5	21.4	21.4	22.5	21.5	37.9		27
89	N7-Cl	12.5	18.6	21.8	21.8	18.6	12.5	40.0		27
90	X7-Cl N7-Cl	26.2	19.0	20.4	20.4	19.0	26.2	67.5		27,1,40,43
91	X7-Cl N7-F	22.8	17.5	21.1	21.1	17.5	22.8	95.7		27
92	N7-Cl X7-F	21.2	20.5	18.4	18.4	20.5	21.2	101.7		27
93	X7-Cl N7-Ph	24.7	20.7	20.6	20.6	20.7	24.7	48.5	138.0	27
94	N7-Cl X7-Ph	21.4	19.3	21.2	21.2	19.3	21.4	55.2	145.1	27
95	X7-COOCH$_2$CH$_2$OH	22.6	22.8	21.0	21.0	22.8	22.6	25.6	175.4	31,41
96	X7-COOEt	22.1	22.8	20.9	20.9	22.8	22.1	25.8	175.0	31,41
97	X7-COOH	22.8	23.4	20.9	20.9	23.4	22.8	25.7	181.7	31,41
98	X7-COOMe	22.1	23.1	21.3	21.3	23.1	22.1	25.9	174.7	27,7,31,41
99	N7-COOMe	16.6	18.9	21.5	21.5	18.9	16.6	22.1	172.1	27
100	X7-COOMe N7-Cl	24.6	18.8	21.0	21.0	18.8	24.6	56.9	179.8	27

184

BICYCLO[4.1.0]HEPTANES (cont.)

Str	Substituents	C1	C2	C3	C4	C5	C6	C7	Substituent Shifts	References
101	N7-COOMe X7-Br	25.9	19.6	20.7	20.7	19.6	25.9	32.2	172.7	27
102	X7-F	17.0	21.9	21.7	21.7	21.9	17.0	79.5		27
103	N7-F	11.1	17.7	22.4	22.4	17.7	11.1	74.7		27
104	X7-Me	18.4	24.0	22.0	22.0	24.0	18.4	18.0	19.0	27, 2
105	N7-Me	10.4	18.9	22.9	22.9	18.9	10.4	12.2	8.3	27
106	X7-Me N7-Me	19.0	19.2	22.2	22.2	19.2	19.0	17.1	29.4,15.4	2,43
107	X7-OMe	17.7	22.7	22.0	22.0	22.7	17.7	66.4		27
108	N7-OMe	11.6	18.3	22.6	22.6	18.3	11.6	60.4		27
109	X7-Ph	22.5	23.8	21.6	21.6	23.8	22.5	28.9	144.5	27
110	N7-Ph	12.9	20.4	21.9	21.9	20.4	12.9	22.4	138.6	27

185

References to bicyclo[4.1.0]heptanes and related* systems

1. Brun, P., Casanova, J., Hatem, J., Vincent, E.-J., Waegell, B. and Zahra, J.-P., *Compt. Rend., Ser. C* **288**, 201 (1979).
2. Fringuelli, F., Gottlieb, H. E., Hagaman, E. W., Taticchi, A., Wenkert, E. and Wovkulich, P. M., *Gazz. Chim. Ital.* **105**, 1215 (1975).
3. Fringuelli, F., Hagaman, E. W., Moreno, L. N., Taticchi, A. and Wenkert, E., *J. Org. Chem.* **42**, 3168 (1977).
4. Johnson, L. F. and Jankowski, W. C., *Carbon-13 NMR Spectra*, Wiley-Interscience, New York (1972), spectrum no. 394.
5. Christl, M., *Chem. Ber.* **108**, 2781 (1975).
6. Grover, S. H., Marr, D. H., Stothers, J. B. and Tan, C. T., *Can. J. Chem.* **53**, 1351 (1975).
7. Razin, V. V. and Eremenko, M. V., *J. Org. Chem. USSR* **14**, 1378 (1978).
*8. Gunther, H. and Jikeli, G., *Chem. Ber.* **106**, 1863 (1973).
*9. Gunther, H. and Keller, T., *Chem. Ber.* **103**, 3231 (1970).
*10. Warner, P. M., Ah-King, M. and Palmer, R. F., *J. Am. Chem. Soc.* **104**, 7166 (1982).
*11. Gunther, H., Schmickler, H., Bremser, W., Straube, F. A. and Vogel, E., *Angew. Chem., Int. Ed. Engl.* **12**, 570 (1973).
*12. Warner, P. M., Palmer, R. F. and Lu, S.-L., *J. Am. Chem. Soc.* **99**, 3773 (1977).
*13. Olah, G. A., Liang, C., Ledlie, D. B. and Costopoulos, M. G., *J. Am. Chem. Soc.* **99**, 4196 (1977).
*14. Neidlein, R. and Wesch, K. F., *Helv. Chim. Acta* **66**, 891 (1983).
*15. Seebach, D., Hassig, R. and Gabriel, J., *Helv. Chim. Acta* **66**, 308 (1983).
16. Mukherjee-Muller, G., Winkler, T., Zsindely, J. and Schmid, H., *Helv. Chim. Acta* **59**, 1763 (1976).
*17. Kaupp, G. and Rosch, K., *Angew. Chem., Int. Ed. Engl.* **15**, 163 (1976).
*18. Gooding, K. R., Jackson, W. R., Pincombe, C. F. and Rash, D., *Tetrahedron Lett.* **263** (1979).
*19. Minter, D. E., Fonken, G. J. and Cook, F. T., *Tetrahedron Lett.* **711** (1979).
*20. Dauben, W. G., Michno, D. M. and Olsen, E. G., *J. Org. Chem.* **46**, 687 (1981).
*21. Yates, P. and Stevens, K. E., *Tetrahedron* **37**, 4401 (1981).
22. Bohlmann, F., Zeisberg, R. and Klein, E., *Org. Magn. Reson.* **7**, 426 (1975).
23. Mil'vitskaya, E. M., Babushkina, O. Y., Ranneva, Y. I., Shapiro, I. O., Luzikov, Y. N., Plate, A. F. and Shatenshtein, A. I., *J. Org. Chem. USSR* **18**, 1243 (1982).
24. Tarakanova, A. V., Grishin, Y. K., Vashakidze, A. G., Mil'vitskaya, E. M. and Plate, A. F., *J. Org. Chem. USSR* **8,** 1655 (1972).
*25. Smith, R. A. J. and Hannah, D. J., *Tetrahedron* **35**, 1183 (1979).
26. Neidlein, R. and Wesch, K. F., *Chem. Ber.* **116**, 2466 (1983).
27. Ishihara, T., Ando, T., Muranaka, T. and Saito, K., *J. Org. Chem.* **42**, 666 (1977).
28. Hudlicky, T., Koszyk, F. J., Dochwat, D. M. and Cantrell, G. L., *J. Org. Chem.* **46**, 2911 (1981).
29. Maas, D. D., Blagg, M. and Wiemer, D. F., *J. Org. Chem.* **49**, 853 (1984).
30. Cohen, T. and Yu, L.-C., *J. Am. Chem. Soc.* **105**, 2811 (1983).
31. Krabbenhoft, H. O., *J. Org. Chem.* **44**, 4285 (1979).
*32. Banwell, M. G. and Halton, B., *Aust. J. Chem.* **33**, 2673 (1980).
33. Galloway, N., Dent, B. R. and Halton, B., *Aust. J. Chem.* **36**, 593 (1983).
34. Halton, B. and Officer, D. L., *Aust. J. Chem.* **36**, 1291 (1983).

*35. Durr, H. and Albert, K.-H., *Org. Magn. Reson.* **11**, 69 (1978).

36. Collum, D. B., Mohamadi, F. and Hallock, J. S., *J. Am. Chem. Soc.* **105**, 6882 (1983).

*37. Izac, R. R., Fenical, W. and Wright, J. M., *Tetrahedron Lett.* **25**, 1325 (1984).

*38. Hamon, D. P. G. and Richards, K. R., *Aust. J. Chem.* **36**, 2243 (1983).

*39. Brun, P., Casanova, J., Hatem, J., Zahra, J. P. and Waegell, B., *Org. Magn. Reson.* **12**, 537 (1979).

40. *Sadtler Standard Carbon-13 NMR Spectra*, Sadtler Research Laboratories, Inc., Philadelphia, Pa.

41. Krabbenhoft, H. O., *J. Org. Chem.* **43**, 4556 (1978).

42. Anke, L., Reinhard, D. and Weyerstahl, P., *Liebigs Ann. Chem.* 591 (1981).

43. Pehk, T. I., Kooskora, H. E., Lippmaa, E. T., Lysenkov, V. I., Deshchits, G. V. and Bardyshev, I. I., *Vestsi Akad. Navuk BSSR, Ser. Khim. Navuk* 96 (1977).

*44. Dailey Jr., O. D. and Fuchs, P. L., *J. Org. Chem.* **45**, 216 (1980).

Chapter 15

Bicyclo[4.2.0]octanes

As with the bicyclo[3.2.2]nonanes, the problem of nomenclature is avoided here. Because of the scarcity of information, all data can be displayed as three-dimensional representations. Too little information is available to be able to make definitive deductions concerning conformational biases in these systems.

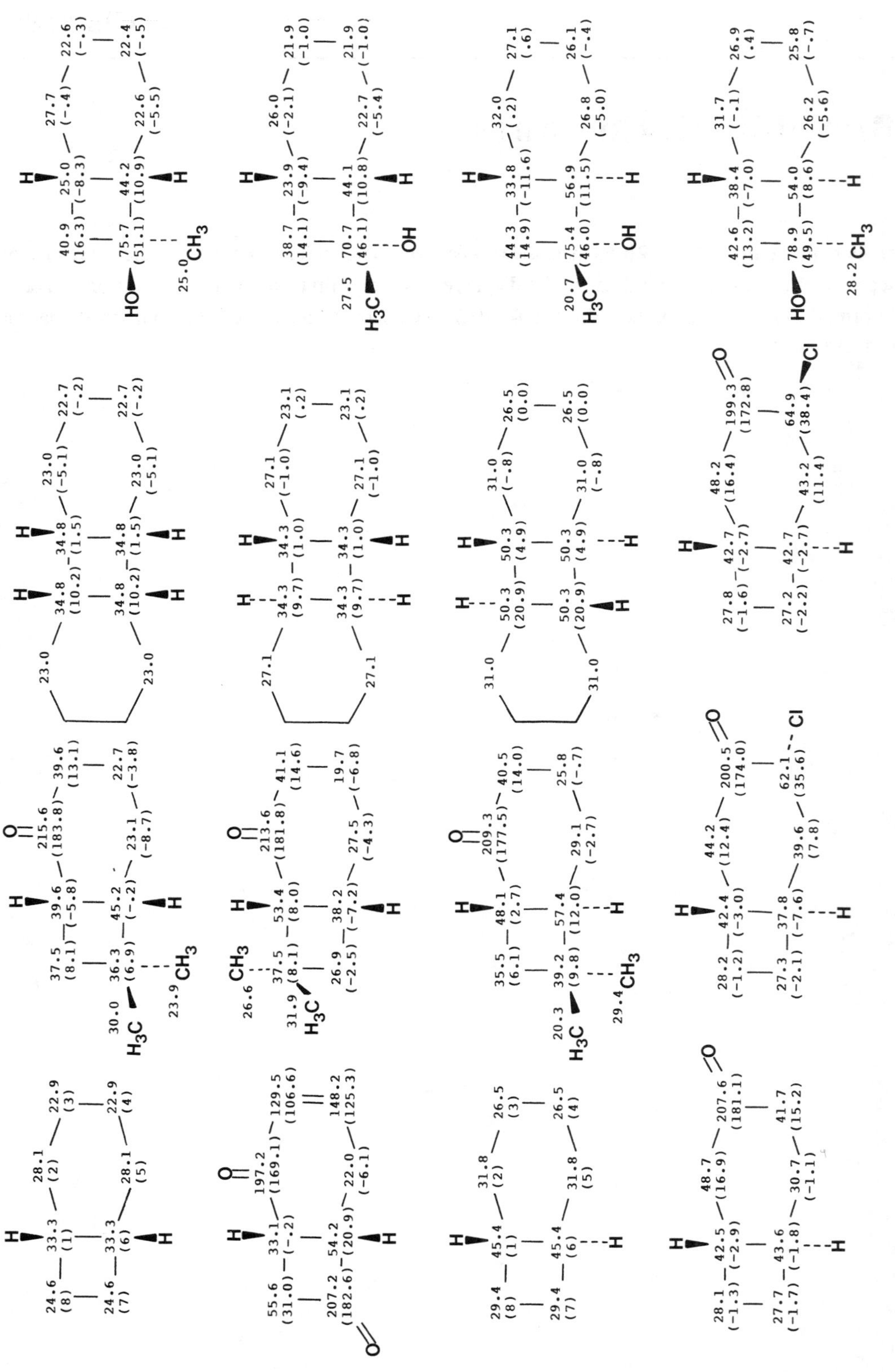

References to bicyclo[4.2.0]octanes and related* systems

1. Bouquant, J., Chuche, J., Convert, O. and Furth, B., *Org. Magn. Reson.* **12**, 5 (1979).
2. Salomon, R. G., Folting, K., Streib, W. E. and Kochi, J. K., *J. Am. Chem. Soc.* **96**, 1145 (1974).
3. Schippers, P. H. and Dekkers, H. P. J. M., *J. Chem. Soc. Perkin Trans. II* 1429 (1982).
4. Metzger, P., Cabestaing, C., Casadevall, E. and Casadevall, A., *Org. Magn. Reson.* **19**, 144 (1982).
5. Fitzer, L., Kliebisch, U., Wehle, D. and Modaressi, S., *Tetrahedron Lett.* **23**, 1661 (1982).
6. Wexler, A. J., Hyatt, J. A., Raynolds, P. W., Cottrell, C. and Swenton, J. S., *J. Am. Chem. Soc.* **100**, 512 (1978).
*7. Yates, P. and Stevens, K. E., *Tetrahedron* **37**, 4401 (1981).
*8. Mehta, G., Srikrishna, A. and , *Tetrahedron Lett.* 3187 (1979).
*9. Graham, C. R., Scholes, G. and Brookhart, M., *J. Am. Chem. Soc.* **99**, 1180 (1977).
*10. Scholes, G., Graham, C. R. and Brookhart, M., *J. Am. Chem. Soc.* **96**, 5665 (1974).
11. Metzger, P., Casadevall, E., Casadevall, A. and Pouet, M.-J., *Can. J. Chem.*, **58**, 1503 (1980).

Chapter 16

Bicyclo[4.3.0]nonanes

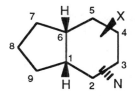 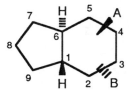

Substantial data are available for both the cis- and the trans-fused systems. Exo and endo are used for cis-bicyclo[4.3.0]nonanes, as well as for those having a double bond emanating from a bridgehead (though these are listed with the trans-fused compounds). For the trans-fused comounds, the orientation of the hydrogen on C-1 is defined as alpha (A), while beta (B) refers to the orientation of the C-6 hydrogen. Groups are then defined based on the side of the molecule that they are on.

16.1 cis – Bicyclo [4.3.0] nonanes

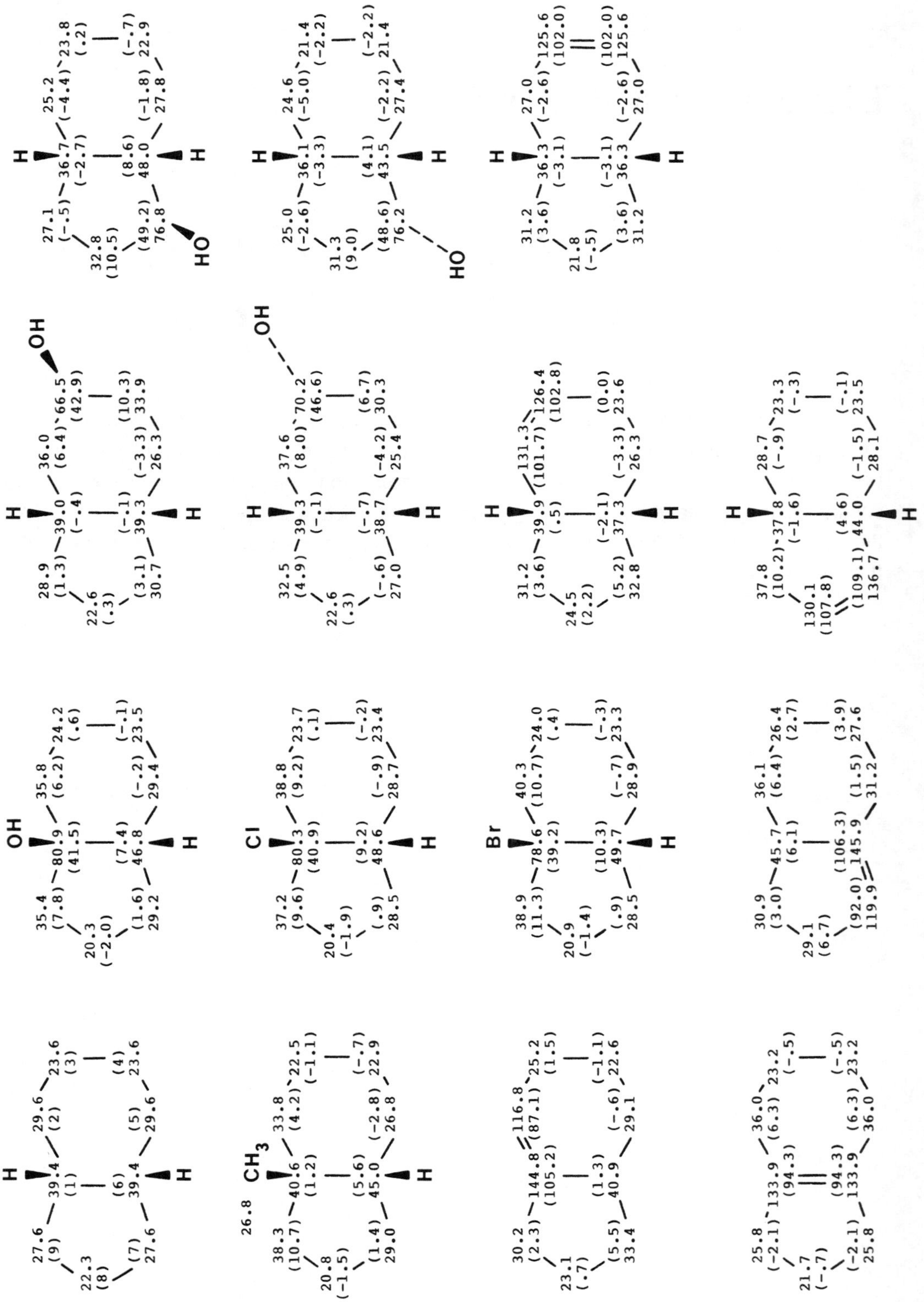

Row 1 (top)

Structure 1:
```
        25.2   23.8      (-.7)
   27.1 (-4.4)        22.9
  (-.5) 36.7  (8.6) (-1.8)
        (-2.7) 48.0  27.8
   32.8        (49.2) 76.8
  (10.5)              HO
   H                    H
```

Structure 2:
```
        24.6   21.4      (-2.2)
   25.0 (-5.0)        21.4
  (-2.6) 36.1 (4.1) (-2.2)
        (-3.3) 43.5  27.4
   31.3        (48.6) 76.2
  (9.0)              HO
   H                    H
```

Structure 3:
```
        27.0   125.6      ══  (102.0)
   31.2 (-2.6)       125.6
  (-3.6) 36.3 (-3.1) (-2.6)
        (-3.1) 36.3  27.0
   21.8        (3.6) 31.2
  (-.5)
   H                    H
```

Row 2

Structure 4 (OH top):
```
   OH
        36.0   66.5      (10.3)
   28.9 (6.4)        33.9
  (1.3) 39.0 (-.1) (-3.3)
        (-.4) 39.3  26.3
   22.6        (3.1) 30.7
  (.3)
   H                    H
```

Structure 5 (OH top):
```
   OH
        37.6   70.2      (6.7)
   32.5 (8.0)        30.3
  (4.9) 39.3 (-.7) (-4.2)
        (-.1) 38.7  25.4
   22.6        (-.6) 27.0
  (.3)
   H                    H
```

Structure 6:
```
        131.3  126.4      (0.0)
   31.2 (101.7)      23.6
  (3.6) 39.9 (-2.1) (-3.3)
        (.5) 37.3  26.3
   24.5        (5.2) 32.8
  (2.2)
   H                    H
```

Structure 7:
```
        28.7   23.3      (-.1)
   37.8 (-.9)        23.5
  (10.2) 37.8 (4.6) (-1.5)
        (-1.6) 44.0  28.1
   130.1       (109.1) 136.7
  (107.8)
   H                    H
```

Row 3

Structure 8 (OH):
```
   OH
        35.8   24.2      (-.1)
   35.4 (6.2)        23.5
  (7.8) 80.9 (7.4) (-.2)
        (41.5) 46.8  29.4
   20.3        (1.6) 29.2
  (-2.0)
                        H
```

Structure 9 (Cl):
```
   Cl
        38.8   23.7      (-.2)
   37.2 (9.2)        23.4
  (9.6) 80.3 (9.2) (-.9)
        (40.9) 48.6  28.7
   20.4        (.9) 28.5
  (-1.9)
                        H
```

Structure 10 (Br):
```
   Br
        40.3   24.0      (-.3)
   38.9 (10.7)       23.3
  (11.3) 78.6 (10.3) (-.7)
        (39.2) 49.7  28.9
   20.9        (.9) 28.5
  (-1.4)
                        H
```

Structure 11:
```
        36.1   26.4      (3.9)
   30.9 (6.4)        27.6
  (3.0) 45.7 (106.3) (1.5)
        (6.1) 145.9  31.2
   29.1        (92.0) 119.9
  (6.7)
                        H
```

Row 4 (bottom)

Structure 12:
```
        29.6   23.6      (4)
   27.6 (2)          23.6
  (9) 39.4 (6) (5)
        (1) 39.4  29.6
   22.3        (7) 27.6
  (8)
   H                    H
```

Structure 13 (CH3):
```
   26.8  CH3
        33.8   22.5      (-.7)
   38.3 (4.2)        22.9
  (10.7) 40.6 (5.6) (-2.8)
        (1.2) 45.0  26.8
   20.8        (1.4) 29.0
  (-1.5)
                        H
```

Structure 14:
```
        116.8  25.2      (-1.1)
   30.2 (87.1)       22.6
  (2.3) 144.8 (1.3) (-.6)
        (105.2) 40.9  29.1
   23.1        (5.5) 33.4
  (.7)
                        H
```

Structure 15:
```
        36.0   23.2      (-.5)
   25.8 (6.3)        23.2
  (-2.1) 133.9 (94.3) (6.3)
        (94.3) 133.9  36.0
   21.7        (-2.1) 25.8
  (-.7)
```

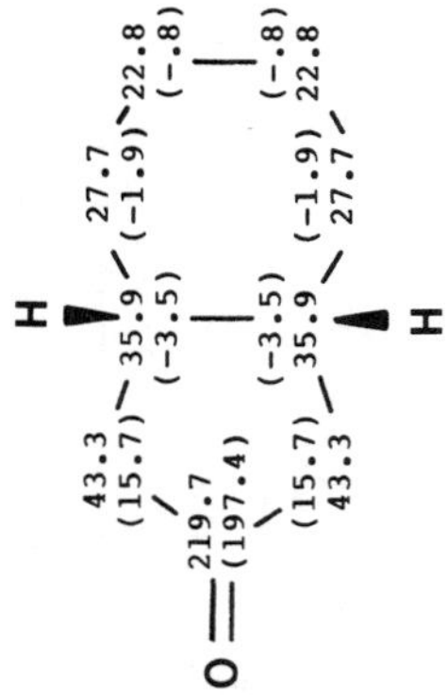

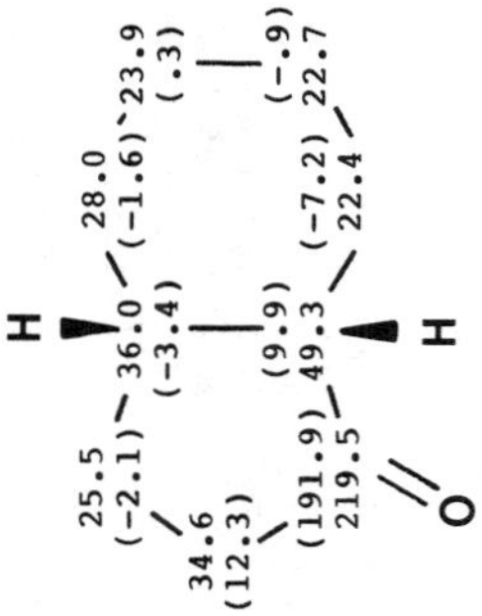

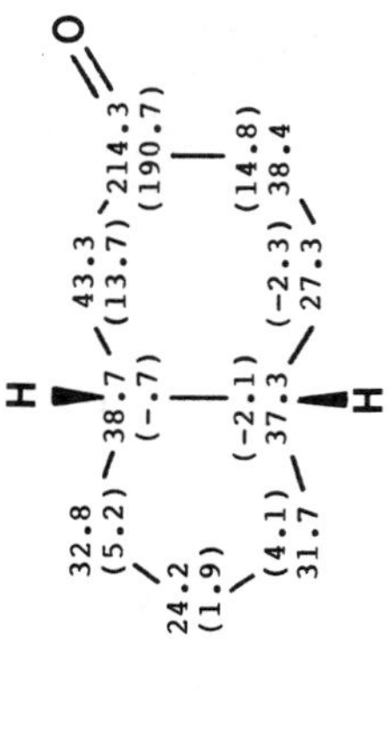

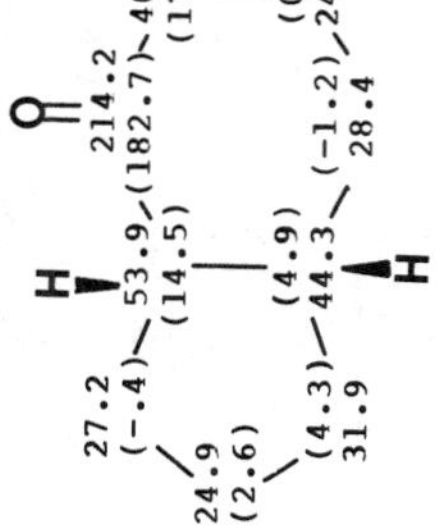

cis-BICYCLO[4.3.0]NONANES

Str	Substituents	C1	C2	C3	C4	C5	C6	C7	C8	C9	Substituent Shifts	References
1	PARENT	39.4	29.6	23.6	23.6	29.6	39.4	27.6	22.3	27.6		1,8,25,42,57
2	1-Br	78.6	40.3	24.0	23.3	28.9	49.7	28.5	20.9	38.9		1
3	1-Cl	80.3	38.8	23.7	23.4	28.7	48.6	28.5	20.4	37.2		1
4	1-COOH/X5-SEt/6-OH	55.1	29.6	20.2	30.0	51.0	82.9	36.1	18.9	32.8	181.3	44
5	1-Me	40.6	33.8	22.5	22.9	26.8	45.0	29.0	20.8	38.3	26.8	4
6	1-Me/2-=O/7=	54.5	216.5	38.4	21.3	27.4	54.1	134.4	129.6	43.5	24.2	6
7	1-Me/2-=O/7=/N9-Me	54.8	216.1	41.7	22.1	25.0	51.9	132.2	133.9	54.2	26.9,18.2	6
8	1-Me/2-=O/7=/X9-Me	56.2		38.5	21.6	26.6	51.9	132.9	136.2	52.7	18.6,13.3	6
9	1-Me/2=/X5-Me N5-Me/N8-OH/X9-Me N9-Me	49.4	135.3	125.1	36.8	31.9	44.1	33.9	81.1	46.0	22.3,29.0,29.9,23.9,17.9	3
10	1-Me/2=/X5-Me N5-Me/X8-OH/X9-Me N9-Me	45.8	132.7	125.0	33.8	32.3	48.8	30.5	51.1	45.8	18.8,29.5,30.2,65.3,21.1,22.2	3
11	1-Me/2=/X5-Me N5-Me/X8-OH/X9-Me N9-Me	46.9	132.9	124.4	33.7	32.2	49.7	34.0	78.5	45.9	20.8,29.2,30.5,16.3,22.0	3
12	1-OH	80.9	35.8	24.2	23.5	29.4	46.8	29.2	20.3	35.4		1,56
13	2-=O	53.9	212.3	40.9	24.3	28.4	44.3	31.9	24.9	27.2		54,5
14	2-=O 3= 3-Me/7=	47.7	201.1	135.7	142.5	36.7	42.4	134.8	130.7	27.3	16.4	24
15	2-=O/6-Me	60.4	214.0	38.9			47.2			27.1	27.4	5
16	2-=O/6-Me N9-Me	62.9	214.8	41.0			44.9			37.7	28.8,18.1	5
17	2-=O/6-Me/X9-Me	69.3	214.0	37.8			47.3			39.0	27.0,20.0	5
18	2-=O/7=	50.1		39.8	21.0	28.0	46.8	134.5	130.4	34.7		6
19	2-=O/7=/N9-Me	52.4	215.2	42.3	22.8	30.6	46.0	132.9	134.8	42.7	18.1	6
20	2-=O/7=/X9-Me	58.8	213.9	39.5	22.1	28.6	46.7	133.5	135.6	43.0	19.4	6
21	2-=O/N9-Me	55.6	214.7	42.5	38.5	25.5	41.3	133.3	129.8	37.5	18.0	5
22	2-=O/X9-Me	61.2	213.8	39.3			42.8			36.1	20.0	5
23	2-COOH 2=/8=	44.0	130.9	142.7		24.7	35.2	38.1	128.7	132.9	172.2	36
24	X2-Me N2-Me/3-=O/7=	49.3	46.3	215.8	36.4	27.0	40.6	135.2	129.8	36.3	27.7,23.4	50
25	X2-Me N2-Me/3-=O/X4-Me N4-Me/7=	44.1	46.4	217.6	42.7	39.2	40.8	135.2	129.1	35.3	26.9,21.9,27.8,25.8	50
26	X2-OH/3-=O/7=	44.9	76.6	213.0	38.5	25.5	43.0	133.3	129.8	36.3		50
27	X2-OH/X3-OMe N3-OMe/7=	41.8	71.3	101.0	27.1	25.0	45.1	135.5	130.1	35.1		50
28	N2-OH/X3-OMe N3-OMe/7=	37.4	69.8	100.1	32.3	31.2	40.5	135.5	130.1	36.3		50
29	2=	39.9	131.3	126.4	23.6	26.3	37.3	32.8	24.5	31.2		7
30	2= 3-OMe/7=	37.6	96.4	156.5	26.5	26.5	43.8	135.1	130.4	40.8		50
31	2=/8=	43.3	130.8	126.9	23.4	25.8	37.5	38.8	130.0	135.5		17
32	3-=O	38.7	43.3	214.3	38.4	27.3	37.3	31.7	24.2	32.8		50,7,51
33	3-=O/7=	34.0	43.6	212.6	37.1	25.6	43.3	133.6	130.3	40.5		50,7
34	3-=O/8=	43.4	42.5	212.9	37.6	27.2	33.8	39.7	130.2	133.5		50,7
35	3-Me 3=/6-Me/7-=O	40.5	29.6	130.7	116.4	27.9	45.3	220.6	35.1	24.7	23.4,19.4	45
36	3-Me 3=/7-=O	32.6	30.6	131.8	118.4	21.5	46.2	219.1	33.9	26.2	23.6	45
37	N3-OAc/N4-OAc/7=	36.4	29.5	70.8	70.4	28.5	41.8	135.6	129.0	38.7		50
38	N3-OCMe2O-4 N/7=	32.6	32.1	74.3	74.0	26.3	39.3	132.9	132.9	40.0		50
39	X3-OH	39.0	36.0	66.5	33.9	26.3	39.3	30.7	22.6	28.9		7
40	N3-OH	39.3	37.6	70.2	30.3	25.4	38.7	27.0	22.6	32.5		7
41	X3-OH/X4-OH/7=	33.9	32.2	69.1	69.0	30.9	41.9	135.6	130.0	38.1		48,40
42	N3-OH/N4-OH/7=	36.6	31.9	70.3	70.2	31.1	42.0	136.0	129.3	39.1		48
43	X3-OH/X4-OH/X7-OH/X8-OH	36.0	30.5	71.5	70.1	28.7	43.2	76.7	73.7	39.1		40
44	3-OMe 3=/7=	36.7	32.6	157.5	91.4	27.2	44.3	136.0	129.7	40.4		50
45	X3-OMe N3-OMe/7=	34.6	35.9	100.3	29.3	24.3	43.6	134.9	130.3	39.2		50
46	3=	36.3	27.0	125.6	125.6	27.0	36.3	31.2	21.8	31.2		7
47	3= 4-Me/6-Me/7-=O	39.5	25.4	118.0	129.4	32.2	46.6	221.2	35.2	24.5	23.6,19.7	45
48	3=/4-Me/7=	34.9	29.1	121.4	135.5	32.8	43.5	135.9	129.5	40.1	23.8	18
49	3=/6-Me/7-=O	39.7	24.6	123.9	122.4	27.2	45.4	220.1	35.0	24.6	19.4	45
50	3=/7-=O	31.6	25.3	124.8	124.3	20.9	46.5	218.5	33.6	26.1		45,7

cis-BICYCLO[4.3.0]NONANES (cont.)

Str	Substituents	C1	C2	C3	C4	C5	C6	C7	C8	C9	Substituent Shifts	References
51	3=/7=	35.2	28.6	128.2	128.2	27.6	42.8	136.2	129.7	40.2		48,17,18,57
52	3=/8==O	32.4	26.4	124.7	124.7	26.4	32.4	44.6	218.7	44.6		7
53	3=/N5-Me/6-Me/7-==O	40.9	25.1	122.3	129.9	34.9	48.9	222.2	37.7	25.7		45
54	3=/N7-OAc/N8-OH	31.2	27.3	125.5	125.3	21.2	36.6	75.7	72.3	35.6	23.0,18.7	50
55	3=/N7-OCMe$_2$O-8N	34.6	27.4	125.4	125.3	22.6	38.6	83.6	80.6	37.1		50
56	3=/N7-OH/N8-OH	31.2	27.6	126.0	125.8	21.2	38.4	75.5	71.8	38.6		48
57	3=/X5-Me/7=	31.1	28.8	127.3	134.8	35.4	50.2	131.5	131.2	41.7	18.7	39
58	3=/X7-OH/X8-OH	31.2	28.3	126.6	126.3	24.6	41.0	77.3	71.4	39.5		48,40
59	N5-Me/6-Me/7-==O	44.5	29.7	23.7	31.6	38.1	51.5	221.7	35.2	24.0	21.9,16.0	45
60	6-Me/7-==O	42.2	26.7	22.3	22.1	29.4	48.1	222.0	34.8	22.9	21.2	45
61	7-==CH$_2$	39.6	28.6	24.2	22.3	27.5	44.1	146.7	30.1	26.4	104.0	53
62	7-==O	36.0	28.0	23.9	22.7	22.4	49.3	219.5	34.6	25.5		45,7,46,55
63	7-Me 7=	38.5	28.3	23.4	23.6	27.0	46.4	144.4	123.5	36.5	14.9	53
64	X7-OH	36.7	25.2	23.8	22.9	27.8	48.0	76.8	32.8	27.1		1
65	N7-OH	36.1	24.6	21.4	21.4	27.4	43.5	76.2	31.3	25.0		1
66	7=	37.8	28.7	23.3	23.5	28.1	44.0	136.7	130.1	37.8		7
67	8-==O	35.9	27.7	22.8	22.8	27.7	35.9	43.3	219.7	43.3		7

16.2 trans – Bicyclo [4.3.0] nonanes

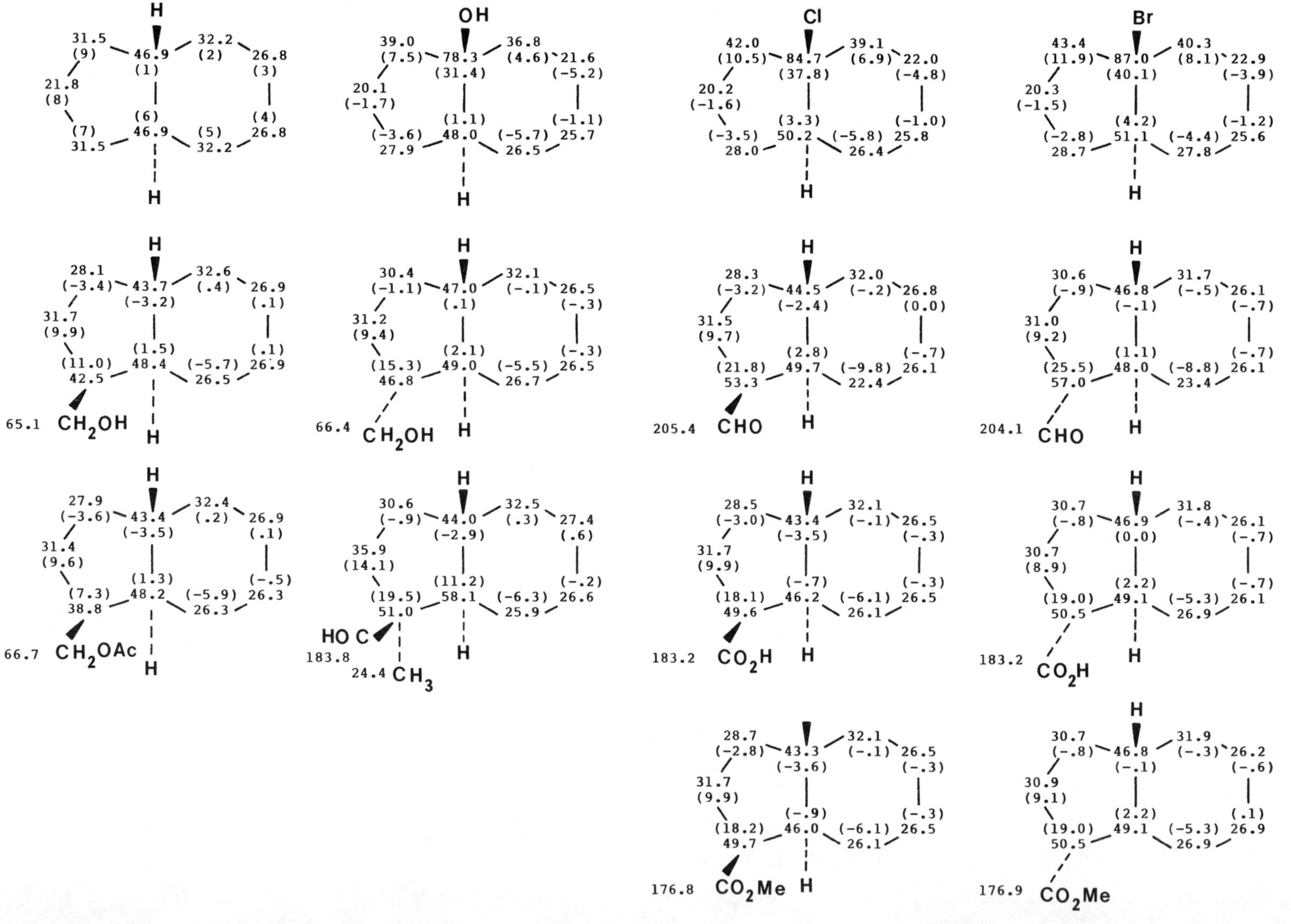

Structure 1 (Br):
40.3 (8.1) 22.9 (-3.9)
(-1.2) 25.6
Br 87.0 (40.1) (4.2) 51.1 (-4.4) 27.8 ---- H
43.4 (11.9)
20.3 (-1.5)
28.7

Structure 2 (CHO):
31.7 (-.5) 26.1 (-.7)
(-.7) 26.1
46.8 (-.1) (1.1) 48.0 (-8.8) 23.4 ---- H
30.6 (-.9)
31.0 (9.2)
(25.5) 57.0 CHO
204.1

Structure 3 (CO$_2$H):
31.8 (-.4) 26.1 (-.7)
(-.7) 26.1
46.9 (0.0) (2.2) 49.1 (-5.3) 26.9 ---- H
30.7 (-.8)
30.7 (8.9)
(19.0) 50.5 CO$_2$H
183.2

Structure 4 (CO$_2$Me):
31.9 (-.3) 26.2 (-.6)
(.1) 26.9
46.8 (-.1) (2.2) 49.1 (-5.3) 26.9 ---- H
30.7 (-.8)
30.9 (9.1)
(19.0) 50.5 CO$_2$Me
176.9

Structure 5 (Cl):
39.1 (6.9) 22.0 (-4.8)
(-1.0) 25.8
Cl 84.7 (37.8) (3.3) 50.2 (-5.8) 26.4 ---- H
42.0 (10.5)
20.2 (-1.6)
28.0

Structure 6 (CHO):
32.0 (-.2) 26.8 (0.0)
(-.7) 26.1
44.5 (-2.4) (2.8) 49.7 (-9.8) 22.4 ---- H
28.3 (-3.2)
31.5 (9.7)
(21.8) 53.3 CHO
205.4

Structure 7 (CO$_2$H):
32.1 (-.1) 26.5 (-.3)
(-.3) 26.5
43.4 (-3.5) (-.7) 46.2 (-6.1) 26.1 ---- H
28.5 (-3.0)
31.7 (9.9)
(18.1) 49.6 CO$_2$H
183.2

Structure 8 (CO$_2$Me):
32.1 (-.1) 26.5 (-.3)
(-.3) 26.5
43.3 (-3.6) (-.9) 46.0 (-6.1) 26.1 ---- H
28.7 (-2.8)
31.7 (9.9)
(18.2) 49.7 CO$_2$Me
176.8

Structure 9 (OH):
36.8 (4.6) 21.6 (-5.2)
(-1.1) 25.7
OH 78.3 (31.4) (1.1) 48.0 (-5.7) 26.5 ---- H
39.0 (7.5)
20.1 (-1.7)
27.9

Structure 10 (CH$_2$OH):
32.1 (-.1) 26.5 (-.3)
(-.3) 26.5
47.0 (.1) (2.1) 49.0 (-5.5) 26.7 ---- H
30.4 (-1.1)
31.2 (9.4)
(15.3) 46.8 CH$_2$OH
66.4

Structure 11 (CH$_3$):
32.5 (.3) 27.4 (.6)
(-.2) 26.6
44.0 (-2.9) (11.2) 58.1 (-6.3) 25.9 ---- H
30.6 (-.9)
35.9 (14.1)
(19.5) 51.0 CH$_3$ 24.4
HO C
183.8

Structure 12:
31.5 (9) 32.2 (2) 26.8 (3)
(4) 26.8
H 46.9 (1) (6) 46.9 (5) 32.2 ---- H
21.8 (8)
(7) 31.5

Structure 13 (CH$_2$OH):
28.1 (-3.4) 32.6 (.4) 26.9 (.1)
(.1) 26.9
H 43.7 (-3.2) (1.5) 48.4 (-5.7) 26.5 ---- H
31.7 (9.9)
(11.0) 42.5 CH$_2$OH
65.1

Structure 14 (CH$_2$OAc):
27.9 (-3.6) 32.4 (.2) 26.9 (.1)
(-.5) 26.3
H 43.4 (-3.5) (1.3) 48.2 (-5.9) 26.3 ---- H
31.4 (9.6)
(7.3) 38.8 CH$_2$OAc
66.7

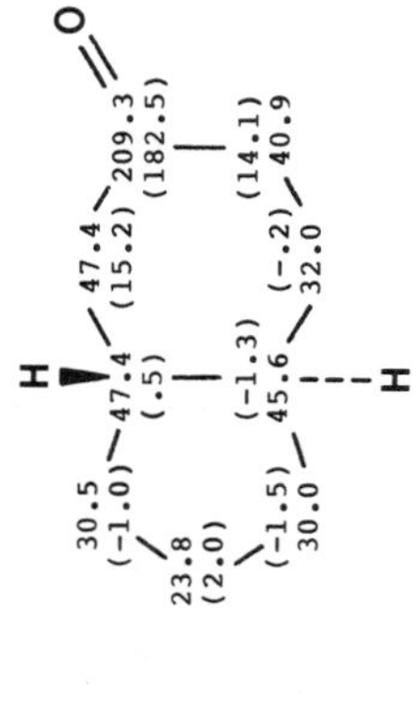

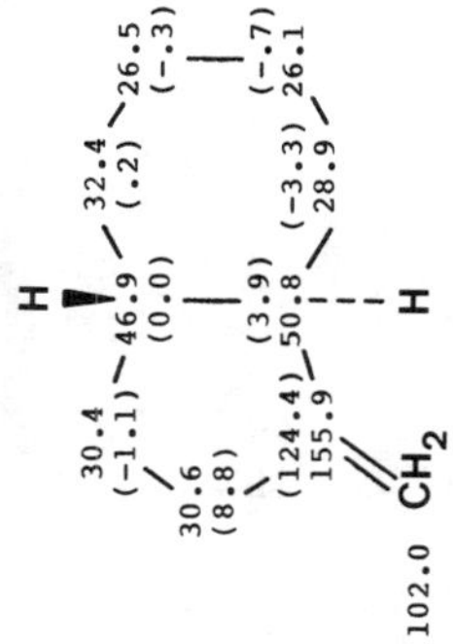

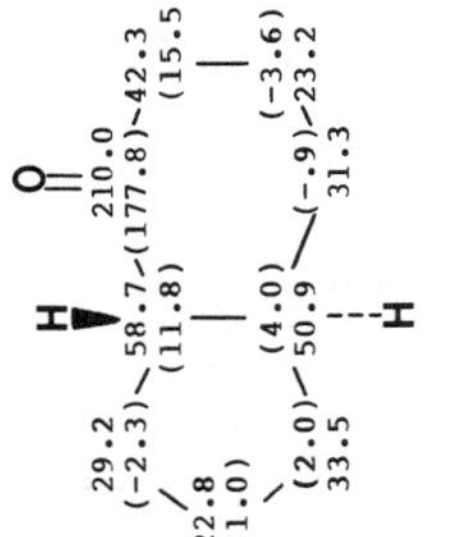

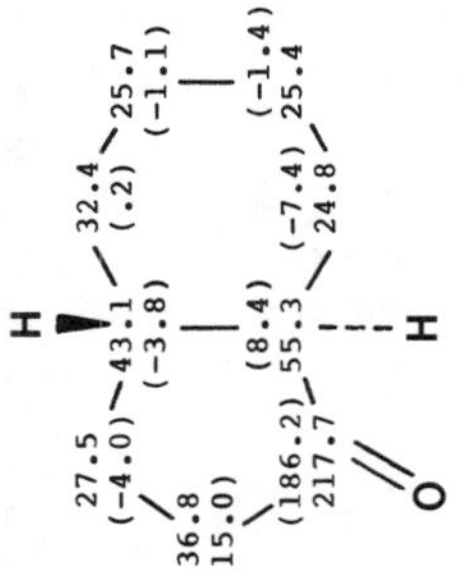

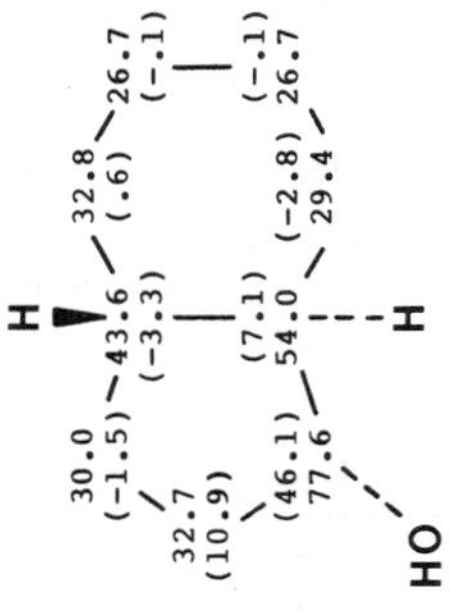

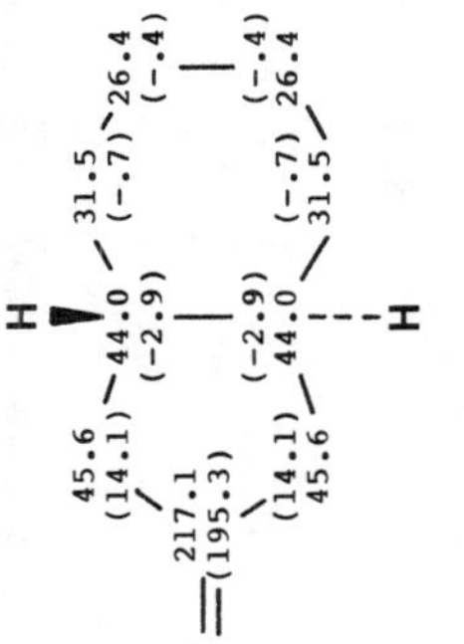

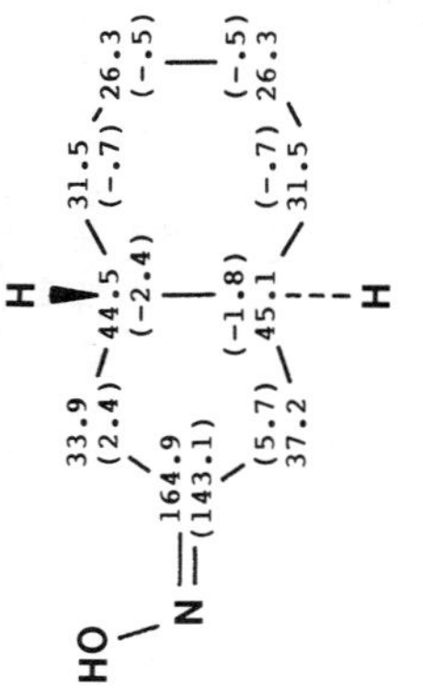

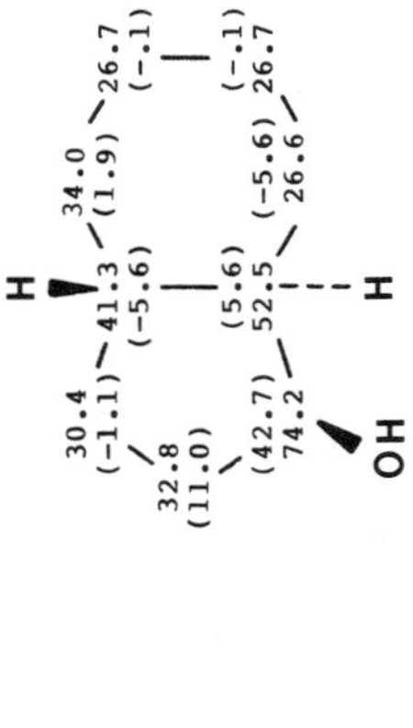

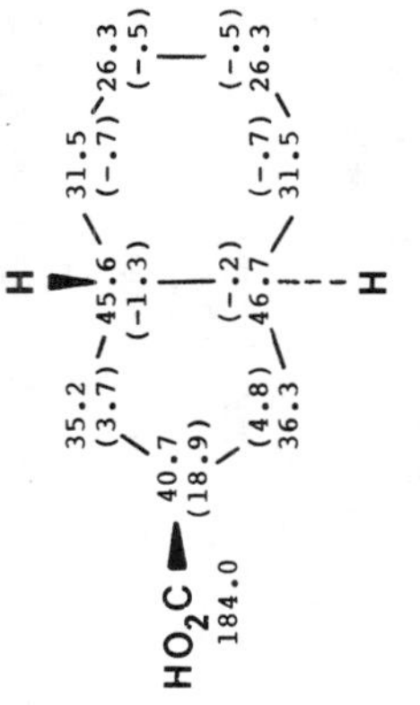

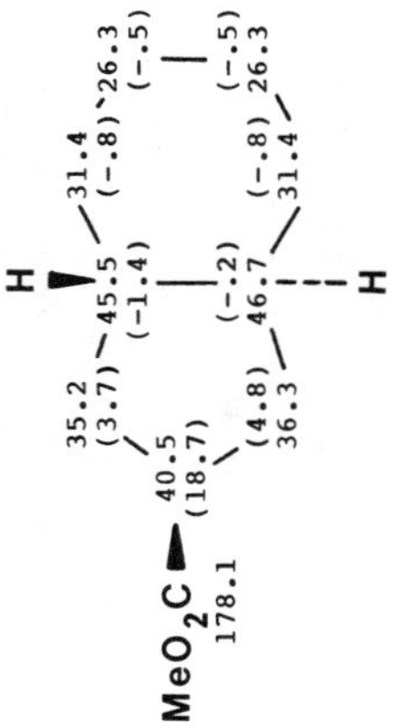

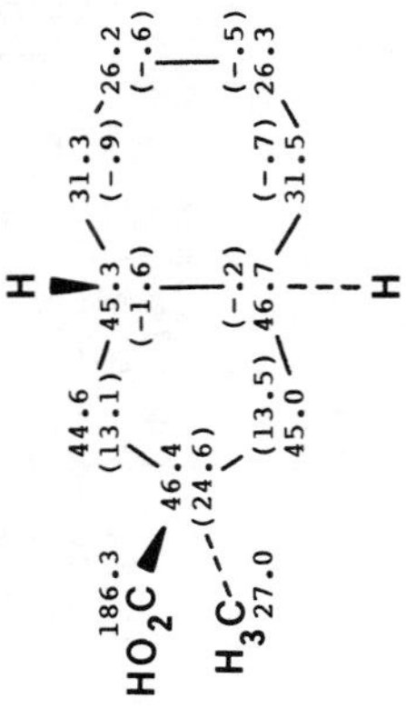

O

47.4 209.3
30.5 (15.2) (182.5) (14.1)
(−1.0) 47.4 (−1.3) (−.2) 40.9
H 47.4 45.6 32.0
23.8 (.5)
(2.0) (−1.5) H
30.0

26.5 (−.7)
32.4 (.2) (−.3) 26.1
30.4 (−3.3) 28.9
(−1.1) 46.9 (3.9)
H (0.0) 50.8
30.6 (124.4) H
(8.8) 155.9
102.0 CH₂

26.7 (−.1)
30.2 (−2.0) 26.4
49.2 37.6 (−.4)
(17.7) (−9.3) (−2.8) 29.4
153.5 (3.3)
(131.7) 50.2
H (119.0) H
150.5
189.8 CHO

O
210.0 42.3
29.2 (177.8) (15.5) (−3.6)
(−2.3) 58.7 (4.0) (−.9) 23.2
H (11.8) 50.9 31.3
22.8 (2.0)
(1.0) 33.5
H

32.4 25.7 (−1.4)
27.5 (.2) (−1.1) 25.4
(−4.0) 43.1 (8.4) (−7.4) 24.8
H (−3.8) 55.3
36.8 (186.2)
(15.0) 217.7 H
O

32.8 26.7 (−.1)
30.0 (.6) (−.1) 26.7
(−1.5) 43.6 (7.1) (−2.8) 29.4
H (−3.3) 54.0
32.7 (46.1)
(10.9) 77.6 H
HO

31.5 26.4 (−.4)
45.6 (−.7) (−.4) 26.4
(14.1) 44.0 (−2.9) (−.7) 31.5
H (−2.9) 44.0
217.1 (14.1)
O (195.3) 45.6 H

31.5 26.3 (−.5)
33.9 (−.7) (−.5) 26.3
(2.4) 44.5 (−1.8) (−.7) 31.5
H (−2.4) 45.1
164.9 (5.7)
HO N (143.1) 37.2 H

34.0 26.7 (−.1)
30.4 (1.9) (−.1) 26.7
(−1.1) 41.3 (5.6) (−5.6) 26.6
H (−5.6) 52.5
32.8 (42.7)
(11.0) 74.2 H
OH

31.5 26.3 (−.5)
35.2 (−.7) (−.5) 26.3
(3.7) 45.6 (−.2) (−.7) 31.5
H (−1.3) 46.7
40.7 (4.8)
HO₂C (18.9) 36.3 H
184.0

31.4 26.3 (−.5)
35.2 (−.8) (−.5) 26.3
(3.7) 45.5 (−.2) (−.8) 31.4
H (−1.4) 46.7
40.5 (4.8)
MeO₂C (18.7) 36.3 H
178.1

31.3 26.2 (−.5)
44.6 (−.9) (−.6) 26.3
(13.1) 45.3 (−.2) (−.7) 31.5
H (−1.6) 46.7
186.3 (13.5)
HO₂C 46.4 45.0 H
H₃C (24.6)
27.0

trans-BICYCLO[4.3.0]NONANES

Str	Substituents	C1	C2	C3	C4	C5	C6	C7	C8	C9	Substituent Shifts	References
1	PARENT (A1,B6)	46.9	32.2	26.8	26.8	32.2	46.9	31.5	21.8	31.5		1,8,25,42
2	1-Br	87.0	40.3	22.9	25.6	27.8	51.1	28.7	20.3	43.4		1
3	1-Cl	84.7	39.1	22.0	25.8	26.4	50.2	28.0	20.2	42.0		1
4	1-iPr/6= 7-Me	28.5	39.0	22.4	23.8	28.0	139.4	127.1	30.5	37.3	17.3,13.3	31
5	1-Me/2=/X5-Me N5-Me/6= 8-=O	45.4	132.9	125.1	42.3	35.4	192.6	123.9	206.5	53.4	27.1,28.9,29.1	3
6	1-Me/2=/X5-Me N5-Me/6= 8-=O/X9-Me N9-Me	51.9	129.7	125.0	40.9	35.3	190.2	122.1	213.1	55.4	17.5,27.8,29.1,27.2,27.6	3
7	1-Me/2=/X5-Me N5-Me/6=/X8-CH2OH/X9-Me N9-Me	52.9	131.8	124.3	41.1	33.5	157.0	119.3	54.1	48.5	19.9,27.8,28.8,62.6,20.1,20.3	3
8	1-Me/2=/X5-Me N5-Me/6=/X8-OH	47.5	135.2	122.5	41.8	33.3	159.3	124.6	75.8	53.8	26.8,27.0,28.8	3
9	Al-Me/A2-Br/A5-OH B5-Me/7-CHO 7=	52.7	61.4	31.0	41.3	69.6	60.9	150.4	156.4	47.4	15.5,31.9,189.6	43
10	Al-Me/A2-Br/A5-OH B5-Me/B7-CHO	48.2	63.0	31.0	41.1	71.0	56.8	49.9	22.2	42.2	16.2,31.3,203.6	43
11	Al-Me/A2-Br/A5-OH B5-Me/B7-CHO/A8-OAc	46.8	62.5	30.9	41.8	70.0	58.6	57.6	75.3	47.8	17.4,31.9,201.5	43
12	Al-Me/A2-Br/A5-OH B5-Me/B7-CHO/B8-OAc	46.8	62.2	30.8	41.9	69.9	55.0	52.2	73.2	48.7	17.5,31.2,199.9	43
13	Al-Me/A8-COOH B8-Me	41.7	39.4	21.5	25.0	26.4	46.8	40.4	46.1	53.0	17.1,186.0,28.8	21
14	Al-Me/B8-COOH	41.5	38.8	21.6	25.4	26.5	47.5	32.2	39.4	45.3	16.6,184.3	21
15	1-Me/X5-Me N5-Me/6= 8-=O/X9-Me N9-Me	45.1	130.8	124.7	34.5	31.6	47.3	38.5	221.4	53.4	17.2,29.7,30.5,21.6,21.8	3
16	1-Me/X5-Me N5-Me/6=/X8-CH2OH/X9-Me N9-Me	51.1	31.0	19.2	40.6	33.6	157.3	119.4	75.3	47.2	20.1,29.8,31.5,63.1,20.9,22.4	3
17	1-Me/X5-Me N5-Me/6=/X8-OH	46.3	41.8	19.4	41.2	33.9	160.3	124.6	75.3	55.1	27.0,27.6,31.2	3
18	1-OH/3=	78.3	36.8	21.6	25.7	26.5	48.0	27.9	20.1	55.1		1
19	1-OH/3=	96.2	37.9	125.0	127.3	27.4	43.2	28.3	20.8	39.0		58
20	Al-OH/B2-CH=CMe2/B3-Me/B7-Me	83.9	47.8	29.6	34.9	24.0	30.3	30.3	34.9	38.7	121.2,19.4,19.0	15
21	Al-OH/B2-COOMe/B3-Me/B7-Me	80.2	55.4	34.6	29.5	23.2	48.7	30.6	35.9	25.8	173.2,19.3,18.9	15
22	1=	144.8	116.8	25.2	22.6	29.1	40.9	33.4	23.1	30.0		8
23	1= 2-COOMe 3-=O/6-NO2	161.5	130.9	191.7	38.8	32.7	95.0	31.4	21.5	30.2	164.3	35
24	1=/N3-COOH/4=	145.4	111.9	43.2	122.4	130.3	40.5	31.9	22.9	30.4	180.0	58
25	(6)1=	133.9	36.0	23.2	23.2	36.0	133.9	25.8	21.7	29.5		8,23,29
26	(6)1= 2-=O/X4-Me N4-Me	136.4	197.4	51.8	35.1	41.0	163.4	37.8	21.8	28.8	28.6	26
27	(6)1=/X3-SePh/N4-Cl	131.2	35.7	44.3	58.9	35.5	129.6	29.1	21.6	32.7	23.8	9
28	2-=O	58.7	210.0	42.3	23.2	31.3	50.9	33.5	22.8	29.2		54,5
29	2-=O/6-Me		211.6	41.0			48.3			31.9		5
30	2-=O/A9-Me	64.9	211.6	41.9			50.1			29.2	20.4	5
31	2-=O/B6-Me/A9-Me	67.4	211.3	41.3			49.2			29.4	18.8,20.5	5
32	X2-Me N2-Me/6-Me/N7-OH X7-Me/9=	156.8	33.8	40.3	18.8	29.3	51.7	82.5	44.7	117.4	29.2,31.1,20.4,22.4	22
33	X2-Me N2-Me/6-Me/X7-OH N7-Me/9=	156.0	34.0	40.7	19.2	31.5	49.6	82.7	44.9	117.3	29.1,30.9,19.7,25.2	22
34	B2-OH A2-Me/B5-Br/B6-Me/A9-CH=CMe2	60.7	71.6	40.4	31.2	63.3	47.7	42.0	28.5	36.4	30.6,16.3,132.0	49,43
35	A2-OH B2-CH=CMe2/B3-Me/B7-Me	59.0	79.0	46.0	33.3	30.6	50.3	40.1	32.8	24.9	121.9,15.4,19.2	14
36	N2-OH X2-Me/N5-Me/6=	55.9	71.8	39.3	31.4	33.3	147.5	121.6	31.0	27.4	22.4,18.4	30
37	N2-OH/6=	50.9	69.6	32.8	20.4	31.2	141.3	123.6	28.3	24.2		30
38	N2-OH/N5-Me/6=	51.4	69.8	32.8	31.5	33.6	146.4	121.7	29.4	24.2	18.5	30
39	2=/9=	142.8	130.1	123.8	26.4	32.0	43.4	31.5	30.1	123.1		27
40	3-=O	47.4	47.4	209.3	40.9	32.0	45.6	30.0	23.8	30.5		2
41	3-=O/A4-Cl	47.6	45.9	200.7	64.4	41.6	45.8	29.9	24.0	31.3		2
42	3-=O/B4-Cl	47.4	42.3	203.2	60.5	38.3	39.2	29.8	23.5	31.5		2
43	3-Me 3=/7-=O	39.1	37.1	133.8	119.9	24.6	50.8	218.2	37.4	27.5	23.3	45
44	3=/6-Me/7-=O	41.0	27.7	125.9	125.2	33.4	45.8	220.8	35.4	24.1	13.1	37
45	3=/7-=O	38.3	32.0	126.3	125.6	24.8	50.5	217.7	36.8	27.4		45,7
46	3=/B5-Me/7-=O	38.3	31.9	124.8	133.4	31.9	56.7	217.4	36.9	27.0	19.4	45
47	B5-Me/7-=O	42.9	32.2	25.7	35.9	32.8	59.9	217.4	36.9	26.9	19.0	45
48	6-Me/7-=O	45.8	26.2	25.5	20.9	32.0	47.5	220.8	35.3	24.2	12.6	37
49	6=	45.7	36.1	26.4	27.6	31.2	145.9	119.9	29.1	30.9		8,30
50	7-=CH2	46.5	32.2	26.3	25.9	28.8	50.5	145.7	30.5	30.3	101.8	52,53

200

trans-BICYCLO[4.3.0]NONANES (cont.)

Str	Substituents	C1	C2	C3	C4	C5	C6	C7	C8	C9	Substituent Shifts	References
51	7-=O	43.1	32.4	25.7	25.4	24.8	55.3	217.7	36.8	27.5		45,46,55
52	A7-CH$_2$OAc	43.4	32.4	26.9	26.3	26.3	48.2	38.8	31.4	27.9	66.7	20
53	A7-CH$_2$OH	43.7	32.6	26.9	26.9	26.5	48.4	42.5	31.7	28.1	65.1	20
54	B7-CH$_2$OH	47.0	32.1	26.5	26.5	26.7	49.0	46.8	31.2	30.4	66.4	19
55	A7-CHO	44.5	32.0	26.8	26.1	22.4	49.7	53.3	31.5	28.3	205.4	20
56	B7-CHO	46.8	31.7	26.1	26.1	23.4	48.0	57.0	31.0	30.6	204.1	20
57	7-CHO 7=	37.6	30.2	26.7	26.4	29.4	50.2	150.5	153.5	49.2	189.8	19
58	A7-COOH	43.4	32.1	26.5	26.5	26.1	46.2	49.6	31.7	28.5	183.2	19
59	B7-COOH	46.9	31.8	26.1	26.1	26.9	49.1	50.5	30.7	30.7	183.2	19
60	A7-COOH B7-Me	44.0	32.5	27.4	26.6	25.9	58.1	51.0	35.9	30.6	183.8,24.4	21
61	A7-COOMe	43.3	32.1	26.5	26.5	26.1	46.0	49.7	31.7	28.7	176.8	19
62	B7-COOMe	46.8	31.9	26.2	26.9	26.9	49.1	50.5	30.9	30.7	176.9	19
63	7-Me 7=	50.3	30.6	26.7	26.6	29.1	53.1	144.4	123.8	36.5	14.3	53
64	A7-OH	41.3	32.8	26.7	26.7	26.6	52.5	74.2	34.0	30.4		1
65	B7-OH	43.6	32.8	26.7	26.7	29.4	54.0	77.6	32.7	30.0		1
66	8-=NOH	44.5	31.5	26.3	26.3	31.5	45.1	37.2	164.9	33.9		34
67	8-=O	44.0	31.5	26.4	26.4	31.5	44.0	45.6	217.1	45.6		34
68	A8-COOH	45.6	31.5	26.3	26.3	31.5	46.7	36.3	40.7	35.2	184.0	21
69	A8-COOH B8-Me	45.3	31.3	26.2	26.3	31.5	46.7	45.0	46.4	44.6	186.3,27.0	21
70	A8-COOMe	45.5	31.4	26.3	26.3	31.4	46.7	36.3	40.5	35.2	178.1	21

References to Bicyclo[4.3.0]nonanes and related* systems

1. Schneider, H.-J. and Nguyen-Ba, N., *Org. Magn. Reson.* **18**, 38 (1982).
2. Metzger, P., Casadevall, E., Casadevall, A. and Pouet, M.-J., *Can. J. Chem.* **58**, 1503 (1980).
3. Thomas, A. F., Ozainne, M. and Guntz-Dubini, R., *Can. J. Chem.* **58**, 1810 (1980).
4. Gooding, K. R., Jackson, W. R., Pincombe, C. F. and Rash, D., *Tetrahedron Lett.* 263 (1979).
5. Lo Cicero, B., Weisbuch, F. and Dana, G., *J. Org. Chem.* **46**, 914 (1981).
6. Hudlicky, T., Koszyk, F. J., Dochwat, D. M. and Cantrell, G. L., *J. Org. Chem.* **46**, 2911 (1981).
7. Arbuzov, V. A., Pekhk, T. I., Belikova, N. A., Bobyleva, A. A. and Plate, A. F., *J. Org. Chem. USSR* **19**, 273 (1983).
8. Becker, K. B., *Helv. Chim. Acta* **60**, 68 (1977).
9. Garratt, D. G. and Kabo, A., *Can. J. Chem.* **58**, 1030 (1980).
*10. Inamoto, Y., Takaishi, N., Fujikura, Y., Aigami, K. and Tsuchihashi, K., *Chem. Lett.* 631 (1976).
*11. Inamoto, Y., Aigami, K., Takaishi, N., Fujikura, Y., Tsuchihashi, K. and Ikeda, H., *J. Org. Chem.* **42**, 3833 (1977).
*12. Dauben, W. G., Michno, D. M. and Olsen, E. G., *J. Org. Chem.* **46**, 687 (1981).
*13. Minter, D. E., Fonken, G. J. and Cook, F. T., *Tetrahedron Lett.* 711 (1979).
14. Connolly, J. D., Harrison, L. J. and Rycroft, D. S., *Tetrahedron Lett.* **25**, 1401 (1984).
15. Izac, R. R., Poet, S. E., Fenical, W., Van Engen, D. and Clardy, J., *Tetrahedron Lett.* **23**, 3743 (1982).
*16. Davies, S. G. and Whitham, G. H., *J. Chem. Soc. Perkin Trans. II* 861 (1975).
17. Bobyleva, A. A., Belikova, N. A., Baranova, S. V., Arbuzov, V. A., Pekhk, T. I. and Plate, A. F., *J. Org. Chem. USSR* **17**, 1183 (1981).
18. Iwase, S., Nakata, M., Hamanaka, S. and Ogawa, M., *Bull. Chem. Soc. Japan* **49**, 2017 (1976).
19. Galteri, M., Lewis, P. H., Middleton, S. and Stock, L. E., *Aust. J. Chem.* **33**, 101 (1980).
20. Lewis, P. H., Middleton, S., Rosser, M. J. and Stock, L. E., *Aust. J. Chem.* **33**, 1049 (1980).
21. Middleton, S. and Stock, L. E., *Aust. J. Chem.* **33**, 2467 (1980).
22. Zink, M. P., Wolf, H. R., Muller, E. P., Schweizer, W. B. and Jeger, O., *Helv. Chim. Acta* **59**, 32 (1976).
23. Kagayama, T., Okabayashi, S., Amaike, Y., Matsukawa, Y., Ishii, Y. and Ogawa, M., *Bull. Chem. Soc. Japan* **55**, 2297 (1982).
24. Huston, R., Rey, M. and Dreiding, A. S., *Helv. Chim. Acta* **65**, 1563 (1982).
25. Metzger, P., Cabestaing, C., Casadevall, E. and Casadevall, A., *Org. Magn. Reson.* **19**, 144 (1982).
26. Helferty, P. H. and Yates, P., *Org. Magn. Reson.* **21**, 352 (1983); Yates, P. and Helferty, P. H. *Can. J. Chem.* **61**, 936 (1983).
27. Turecek, F., Antropiusova, H., Mach, K., Hanus, V. and Sedmera, P., *Tetrahedron Lett.* **21**, 637 (1980).
*28. Turecek, F., Hanus, V., Sedmera, P., Antropiusova, H. and Mach, K., *Tetrahedron* **35**, 1463 (1979).
29. Benkeser, R. A., Belmonte, F. G., Yang, M.-H., *Synth. Commun.* **13**, 1103 (1983); Benkeser, R. A., Belmonte, F. G., Kang, J., *J. Org. Chem.* **48**, 2796 (1983).

30. Snider, B. B. and Deutsch, E. A., *J. Org. Chem.* **48**, 1822 (1983).
31. Epstein, W. W., Grua, J. R. and Gregonis, D., *J. Org. Chem.* **47**, 1128 (1982).
*32. Gunther, H. and Jikeli, G., *Chem. Ber.* **106**, 1863 (1973).
*33. Gunther, H. and Keller, T., *Chem. Ber.* **103**, 3231 (1970).
34. Geneste, P., Durand, R., Kamenka, J.-M., Beierbeck, H., Martino, R. and Saunders, J. K., *Can. J. Chem.* **46**, 1940 (1976).
35. Nakashita, Y. and Hesse, M., *Helv. Chim. Acta* **66**, 845 (1983).
36. Miyano, M. and Stealey, M. A., *J. Org. Chem.* **47**, 3184 (1982).
37. Snider, B. B. and Kirk, T. C., *J. Am. Chem. Soc.* **105**, 2364 (1983).
*38. Ghatak, U. R., Sanyal, B., Ghosh, S., Sarkar, M., Raju, M. S. and Wenkert, E., *J. Org. Chem.* **45**, 1081 (1980).
39. Ishii, Y., Nakagawa, K., Yuki, H., Iwase, S., Hamanaka, S. and Ogawa, M., *J. Japan Petrol. Inst. (Sekiyu Gakkaishi)* **25**, 58 (1982).
40. Matoba, Y., Ohnishi, M., Kagohashi, M., Ishii, Y. and Ogawa, M., *J. Japan Petrol. Inst. (Sekiyu Gakkaishi)* **26**, 15 (1983).
*41. Gadek, T. R. and Vollhardt, K. P. C., *Angew. Chem., Int. Ed. Engl.* **20**, 802 (1981).
42. Weisman, G. R. and Nelsen, S. F., *J. Am. Chem. Soc.* **98**, 7007 (1976).
43. Howard, B. M. and Fenical, W., *J. Org. Chem.* **43**, 4401 (1978).
44. Garst, M. E., McBride, B. J. and Johnson, A. T., *J. Org. Chem.* **48**, 8 (1983).
45. Fringuelli, F., Pizzo, F., Taticchi, A., Halls, T. D. J. and Wenkert, E., *J. Org. Chem.* **47**, 5056 (1982).
46. Larock, R. C., Oertle, K. and Potter, G. F., *J. Am. Chem. Soc.* **102**, 190 (1980).
*47. Schippers, P. H. and Dekkers, H. P. J. M., *J. Am. Chem. Soc.* **105**, 145 (1983).
48. Whitesell, J. K., Minton, M. A. and Flanagan, W. G., *Tetrahedron* **37**, 4451 (1981).
49. Wratten, S. J. and Faulkner, D. J., *J. Org. Chem.* **42**, 3343 (1977).
50. Whitesell, J. K. and Minton, M. A., The University of Texas at Austin, unpublished results.
51. Coustard, J.-M., Douteau, M.-H., Jacquesy, R., Longevialle, P. and Zimmermann, D., *J. Chem. Res. Miniprint* 337 (1978).
52. Lehmkuhl, H. and Tsien, Y.-L., *Chem. Ber.* **116**, 2437 (1983).
53. Mach, K., Sedmera, P., Petrusova, L., Antropiusova, H., Hanus, V. and Turecek, T., *Tetrahedron Lett.* **23**, 1105 (1982).
54. Calmes, D., Gorrichon-Guigon, L., Maroni, P., Accary, A., Barret, R. and Huet, J., *Tetrahedron* **37**, 879 (1981).
55. Snider, B. B. and Cartaya-Marin, C. P., *J. Org. Chem.* **49**, 1688 (1984).
56. Crandall, J. K. and Magaha, H. S., *J. Org. Chem.* **47**, 5368 (1982).
57. *Sadtler Standard Carbon-13 NMR Spectra*, Sadtler Research Laboratories, Inc., Philadelphia, Pa.
58. Davis, S. G. and Whitham, G. H., *J. Chem. Soc. Perkin Trans. I* 1479 (1978).

Chapter 17

Bicyclo[4.4.0]decanes

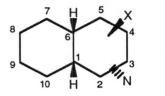

Both the cis- and trans-fused compounds represent valuable models for other, cyclohexane-based systems. Stereochemistry in the cis compounds is specified simply as exo and endo, while the orientations of groups in the trans series and those with a bridgehead double bond are specified based on their orientation as axial (A) or equatorial (E) substituents. Note that this system is distinct from that used with the trans-fused bicyclo[4.3.0]nonanes but it was felt that this inconsistency would be more than repaid here by the ease with which the spatial orientation of the groups could be visualized. Corresponding use of axial and equatorial for the [4.3.0]nonanes would have required new definitions of orientations of substituents on the five-membered ring.

17.1 cis – Bicyclo [4.4.0] decanes

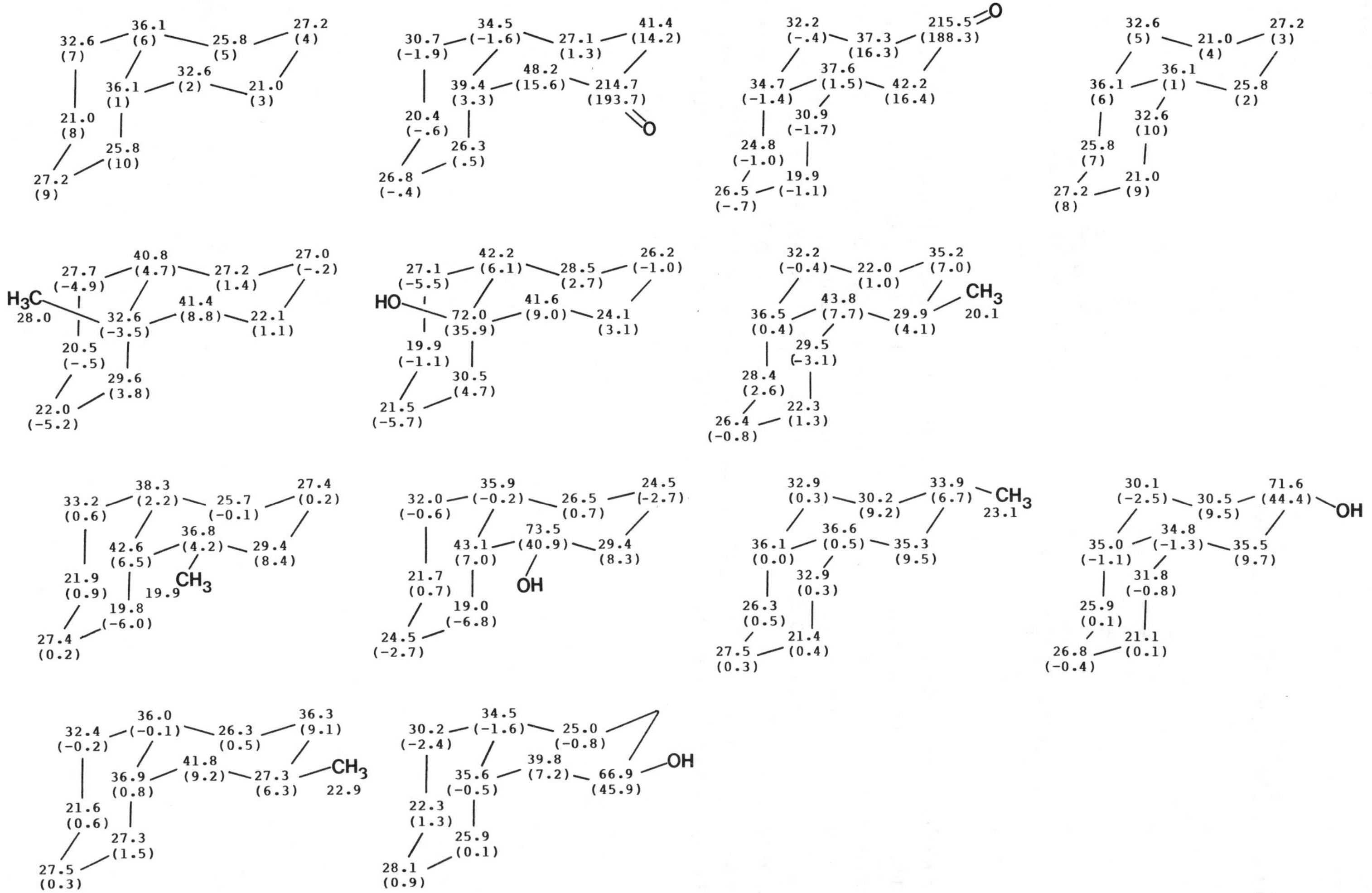

Structure 1

27.2 (3)
21.0 (4)
25.8 (2)
36.1 (1)
32.6 (5)
32.6 (10)
36.1 (6)
25.8 (7)
21.0 (9)
27.2 (8)

Structure 2 (—OH)

71.6 (44.4)
30.5 (9.5)
35.5 (9.7)
30.1 (−2.5)
34.8 (−1.3)
31.8 (−0.8)
35.0 (−1.1)
25.9 (0.1)
21.1 (0.1)
26.8 (−0.4)

Structure 3 (=O)

215.5 (188.3)
37.3 (16.3)
42.2 (16.4)
32.2 (−.4)
37.6 (1.5)
30.9 (−1.7)
34.7 (−1.4)
24.8 (−1.0)
19.9 (−1.1)
26.5 (−.7)

Structure 4 (—CH₃)

35.2 (7.0)
22.0 (1.0)
29.9 (4.1) CH₃ 20.1
32.2 (−0.4)
43.8 (7.7)
29.5 (−3.1)
36.5 (0.4)
28.4 (2.6)
22.3 (1.3)
26.4 (−0.8)

Structure 5 (—CH₃)

33.9 (6.7) CH₃ 23.1
30.2 (9.2)
35.3 (9.5)
32.9 (0.3)
36.6 (0.5)
32.9 (0.3)
36.1 (0.0)
26.3 (0.5)
21.4 (0.4)
27.5 (0.3)

Structure 6 (=O)

41.4 (14.2)
27.1 (1.3)
214.7 (193.7) =O
34.5 (−1.6)
48.2 (15.6)
26.3 (.5)
30.7 (−1.9)
39.4 (3.3)
20.4 (−.6)
26.8 (−.4)

Structure 7 (HO)

26.2 (−1.0)
28.5 (2.7)
24.1 (3.1)
42.2 (6.1)
41.6 (9.0)
30.5 (4.7)
27.1 (−5.5)
72.0 (35.9)
19.9 (−1.1)
HO
21.5 (−5.7)

Structure 8 (—OH)

24.5 (−2.7)
26.5 (0.7)
29.4 (8.3)
35.9 (−0.2)
73.5 (40.9) OH
19.0 (−6.8)
32.0 (−0.6)
43.1 (7.0)
21.7 (0.7)
24.5 (−2.7)

Structure 9 (—OH)

—OH
25.0 (−0.8)
66.9 (45.9)
34.5 (−1.6)
39.8 (7.2)
25.9 (0.1)
30.2 (−2.4)
35.6 (−0.5)
22.3 (1.3)
28.1 (0.9)

Structure 10

27.2 (4)
25.8 (5)
21.0 (3)
36.1 (6)
32.6 (2)
36.1 (1)
25.8 (10)
32.6 (7)
21.0 (8)
27.2 (9)

Structure 11 (H₃C)

27.0 (−.2)
27.2 (1.4)
22.1 (1.1)
40.8 (4.7)
41.4 (8.8)
27.7 (−4.9)
32.6 (−3.5)
29.6 (3.8)
20.5 (−.5)
H₃C 28.0
22.0 (−5.2)

Structure 12 (CH₃)

27.4 (0.2)
25.7 (−0.1)
29.4 (8.4)
38.3 (2.2)
36.8 (4.2)
CH₃ 19.9 (−6.0)
33.2 (0.6)
42.6 (6.5)
21.9 (0.9)
27.4 (0.2)

Structure 13 (—CH₃)

36.3 (9.1)
26.3 (0.5)
27.3 (6.3) CH₃ 22.9
36.0 (−0.1)
41.8 (9.2)
32.4 (−0.2)
36.9 (0.8)
21.6 (0.6)
27.3 (1.5)
27.5 (0.3)

cis–BICYCLO[4.4.0]DECANES

Str	Substituents	A/E	C1	C2	C3	C4	C5	C6	C7	C8	C9	C10	Substituent Shifts	References
1	PARENT		37.0	29.8	24.6	24.6	29.8	37.0	29.8	24.6	24.6	29.8		12,2,4,13,28,43,
1	PARENT	A	36.1	32.6	21.0	27.2	25.8	36.1	32.6	21.0	27.2	25.8		9
1	PARENT	E	36.1	25.8	27.2	21.0	32.6	36.1	25.8	27.2	21.0	32.6		52,72,82,103
2	1-C1/6-Me	A	81.8	39.5	24.2	21.2	33.8	38.9	36.3	22.3	21.0	34.8	25.1	49
3	1-CN/3-=O			46.6	205.6	37.9	26.2	36.7	25.9	21.3	20.6	31.2	122.5	121
4	1-CN/3-=O/X7-Me N7-Me		39.8	51.1	206.1	39.9	24.1	45.9	34.2	33.6	19.9	30.9	125.0,29.0,30.9	120
5	1-COCH=CHMe/3-=O		54.3	43.9	209.6	36.9	27.8	34.4	26.4	23.8	21.5	30.9	201.7	121
6	1-COMe/3-=O		56.0	43.8	209.3	36.9	27.8	34.3	26.3	23.7	21.4	31.5	211.4	121
7	1-COOH		47.5	31.7	22.9	23.3	28.3	36.0	28.3	23.3	22.9	31.7	185.0	70
8	1-COOMe/2-=O 3-=/X5-Me N5-Me/8-=/8-Me 8-=		56.6	197.7	123.9	157.9	36.2	41.3	29.0	132.2	118.5	28.7	173.0,30.4,23.5,23.4	88
9	1-COOMe/2-=O 3-=/X5-Me N5-Me/8-=/8-Me 9-Me		57.5	197.4	124.0	158.0	36.0	40.4	24.2	119.3	131.1	33.1	173.0,30.4,23.6,23.4	88
10	1-COPh/3-=O		56.2	45.1	209.3	37.2	27.6	34.9	26.7	23.8	21.7	31.5	207.5	121
11	1-Et/3-=/6-Et/8-=		31.4	36.3	125.1	125.1	36.3	31.4	36.3	125.1	125.1	36.3	23.8,23.8	77
12	1-Me		32.9	36.4	22.8	24.5	28.1	41.8	28.1	24.5	22.8	36.4	28.2	2
12	1-Me	A	32.6	41.4	22.1	22.1	27.7	40.8	27.7	22.1	22.0	29.6	28.0	9
13	1-Me/2-=O	A	49.0	215.0	37.6	22.6	28.6	44.4	26.2	24.9	22.6	33.8	25.5	78
14	1-Me/2-=O/6-Me/8-=O		51.5	209.3	38.6	21.6	34.3	44.6	50.5	214.4	36.9	31.3	20.9,23.1	18
15	1-Me/2-=O/6-O-7X/8-=O		50.1	210.1	37.0	20.4	28.9	67.6	61.4	204.7	34.3	25.5	22.7	110
16	1-Me/2-=O/6-O-7X/E8-=CHCOOMe		50.2	210.8	37.2	20.3	29.6	65.8	63.6	153.2	23.0	26.3	22.8,121.6	110
17	1-Me/2-=O/6-O-7X/Z8-=CHCOOMe		50.3	211.5	37.3	20.1	30.2	65.8	58.3	153.4	27.8	28.8	22.6,121.7	110
18	1-Me/2-=O/6-OH/N7-OH/E8-=CHCOOMe		54.0	213.8	36.6	20.1	25.5	80.3	75.2	161.8	23.6	33.0	15.1,111.4	110
19	1-Me/2-=O/8-=O		48.5	211.0	38.4	24.9	26.7	46.0	43.7	214.0	37.5	33.6	23.9	111
20	1-Me/2-=O/8-Me 8-=		47.0	214.7	36.8	24.9	27.6	41.3	32.7	130.3	116.4	31.3	20.1,23.4	78
21	1-Me/2-=O/8-Me 8-= 9-Me		48.2	214.7	36.9	24.9	27.5	41.4	34.5	122.1	120.9	37.4	20.4,18.7,18.7	78
22	1-Me/2-=O/8-=		47.2	213.5	36.6	24.7	27.5	40.4	27.5	123.4	122.3	30.6	19.8	78
23	1-Me/2-=O/8-=/N10-Me		50.2	213.6	38.9	23.0	25.4	42.0	27.3	122.0	130.9	38.5	17.7,23.4	78
24	1-Me/2-=O/8-=/X10-Me		51.1	214.3	37.2	25.7	27.5	42.9	28.2	123.3	128.9	31.4	14.3,14.6	78
25	1-Me/2-=O/N10-Me		51.9	215.0	39.4	30.8	22.9	47.5	26.7	26.5	29.0	42.9	16.7,23.9	78
26	1-Me/2-=O/X5-Me N5-Me		48.3	215.7	34.7	34.1	33.7	51.5	24.5	23.8	21.1	34.2	26.9,31.6,27.4	78
27	1-Me/2-=O/X5-Me N5-Me/8-=		46.2	214.6	34.4	40.3	40.3	47.6	22.5	125.1	122.8	31.3	22.8,31.7,20.5	78
28	1-Me/3-=O/6-Me	E	40.5	49.0	215.3	38.0	37.2	34.8	32.5	21.7	21.7	35.4	24.1,22.5	9
29	1-Me/3-=O/6-Me	A	41.0	52.6	214.3	38.3	33.5	34.7	34.5	21.5	20.9	34.4	22.7,23.4	9
30	1-Me/6-Me	A	35.1	37.0	22.7	22.1	33.1	35.1	37.0	22.7	22.1	33.1	24.1,24.1	9
31	1-Me/N4-t-Bu		32.2	42.1	22.8	48.7	28.4	41.5	28.1	20.6	22.2	33.1	27.5,32.2	10
32	1-Me/X2-COOMe/X4-OCH2CH2O-4N/6-OH		40.0	43.5	34.1	48.7	45.0	73.4	33.3	21.6	20.9	33.8	17.4	112
33	1-Me/X2-COOMe/X4-OCH2CH2O-4N/X7-OH		37.0	51.4	33.6	33.6	36.0	42.3	68.2	29.9	19.9	37.9	23.8	112
34	1-Me/X3-COOMe/4-=/7-=O/X10-Me		38.1	35.6	39.5	126.1	126.4	59.7	212.7	39.3	30.1	30.1	20.8,173.5,14.8	114
35	1-Me/Z2-=CHMe 3-=O/6-Me		45.3	146.9	206.7	34.1	36.0	36.0	34.7	21.8	21.8	30.7	21.1,127.1,23.9	17
36	1-O-2X/3-=O/6-Me/N10-Me	E	69.2	61.9	207.5	34.1	30.1	34.0	33.0	20.8	25.9	29.8	22.3	9
37	1-O-2X/3-=O/6-Me/N10-Me		71.6	57.0	206.9	33.0	31.4	35.2	35.1	21.3	35.2	30.8	22.7,13.6	110
38	1-O-2X/3-=O/6-Me/X7-OAc		67.4	62.0	205.0	32.8	25.7	39.3	71.9	26.5	20.8	29.0	17.6	110
39	1-O-2X/3-=O/6-Me/X7-OH		68.3	62.2	206.8	32.7	25.2	40.2	70.5	29.1	29.8	29.8	16.6	110
40	1-O-2X/3-=O/6-Me/X7-OTs		67.2	61.9	204.9	32.3	25.4	40.0	80.4	27.4	20.6	28.6	17.4	110
41	1-O-2X/E3-=CHCOOMe/6-Me		66.5	63.4	154.9	22.3	31.7	33.9	34.6	21.3	26.1	30.8	120.9,22.5	110
42	1-O-2X/E3-=CHCOOMe/6-Me/N10-Me		68.8	57.6	154.9	22.3	34.7	33.5	36.3	21.5	34.8	30.9	121.0,22.9,13.9	110
43	1-O-2X/E3-=CHCOOMe/6-Me/X7-OTs		65.0	63.3	152.7	21.2	25.8	39.7	80.7	27.5	29.1	29.1	121.7,17.5	110
44	1-O-2X/X3-CH2COOMe/6-Me		65.2	64.0	30.0	22.1	33.3	33.5	36.3	21.6	25.8	31.9	36.9,22.8	110
45	1-O-2X/X3-OH/6-Me/X7-OAc		66.5	63.5	64.0	25.5	26.0	38.3	73.4	26.5	20.6	30.3	17.7	110
46	1-O-2X/Z3-=CHCOOMe/6-Me		66.1	58.1	154.5	27.5	33.7	38.3	35.8	21.5	25.8	31.3	121.2,22.7	110
47	1-O-2X/Z3-=CHCOOMe/6-Me/N10-Me		68.4	51.8	154.8	27.2	33.9	34.5	35.1	21.5	34.9	31.5	121.2,23.0,14.2	110

cis-BICYCLO[4.4.0]DECANES (cont.)

Str	Substituents	A/E	C1	C2	C3	C4	C5	C6	C7	C8	C9	C10	Substituent Shifts	References
48	1-O-2X/Z3-=CHCOOMe/6-Me/X7-OTs		64.7	57.7	152.5	26.1	27.1	39.7	81.5	27.5	20.4	29.5	122.3,17.4	110
49	1-O-6		62.1	31.1	20.6	20.6	31.1	62.1	31.1	20.6	20.6	31.1		41
50	1-O-6/N2-O-3N/N4-O-5N/N7-O-8N/N9-O-10N		61.5	50.2	48.0	48.0	50.2	61.5	50.2	48.0	48.0	50.2		96
51	1-OAc/6-Me	A	86.8	30.6	23.3	21.2	33.4	37.1	36.3	21.0	21.7	26.3	22.6	48
52	1-OAc/N2-OAc/6-Me		88.6	73.3	29.5	19.4	32.7	39.8	36.4	20.8	21.5	22.7	23.2	48
52	1-OAc/N2-OAc/6-Me	A	84.5	73.1	27.5	16.6	32.8	37.2	37.1	20.7	21.1	27.0	23.7	48
53	1-OAc/X2-OAc/6-Me	E	88.5	70.6	26.8	19.1	35.2	39.6	33.6	20.5	21.1	27.1	23.3	48
54	1-OH	A	72.0	41.6	24.1	26.2	28.5	42.2	27.1	19.9	21.5	30.5		9
55	1-OH/2-=O/6-Me		80.3	215.4	36.4	22.0	33.3	41.6	35.2	20.9	20.9	32.4	22.4	48
56	1-OH/2-Me 2=/6-Me	E	73.4	135.9	126.4	22.1	31.2	36.7	32.6	21.0	23.8	30.4	19.2,22.1	9
56	1-OH/2-Me 2=/6-Me	A	75.1	138.8	121.6	21.5	30.8	35.8	36.2	21.0	22.5	34.0	17.8,22.1	9
57	1-OH/2=/6-Me	E	71.2	133.3	130.0	22.7	31.2	36.3	32.9	21.4	24.1	35.4	22.1	9
58	1-OH/2=/6-Me/N10-Me	E	72.9	131.0	127.0	22.3	31.3	36.7	32.8	21.1	32.4	35.6	21.7,14.9	47
59	1-OH/3-=O/6-Me		76.8	51.1	211.4	37.7	35.8	37.2	34.1	20.9	23.9	34.3	22.9	9
59	1-OH/3-=O/6-Me	E	78.7	49.0	215.8	37.8	36.5	36.8	33.2	20.7	23.9	36.1	20.9	9
59	1-OH/3-=O/6-Me	A	75.0	53.0	211.7	37.6	34.4	36.8	34.1	20.4	20.9	31.2	22.0	9
60	1-OH/6-Me		73.5	36.1	23.6	22.3	33.6	28.2	36.0	22.3	23.6	36.1	23.0	20
60	1-OH/6-Me	A	73.4	36.2	23.8	21.8	33.6	37.1	36.5	21.1	21.6	32.4	22.5	9,20,48
61	1-OH/N2-C1/6-Me		74.8	69.8	34.1	21.8	32.7	39.5	36.8	21.1	21.1	26.9	21.8	44
62	1-OH/N2-Me/6-Me		75.2	37.5	32.5	21.1	34.2	38.0	37.0	21.6	21.5	25.9	14.9,22.3	9
63	1-OH/N2-OAc/6-Me		74.8	77.0	28.8	19.8	33.1	38.7	36.2	20.9	21.0	26.7	22.3	44,48
64	1-OH/N2-OH/6-Me		76.3	73.6	30.9	19.9	33.1	38.1	36.4	21.0	21.0	25.3	22.3	9,44,48
65	1-OH/N2-OH/E3-=CHCOOMe/6-Me		76.5	75.2	163.0	23.5	32.8	37.7	35.5	20.9	20.7	26.0	110.7,22.6	110
66	1-OH/N2-OH/E3-=CHCOOMe/6-Me/N10-Me		76.5	79.0	160.9	21.2	37.5	37.8	34.3	20.0	31.8	37.5	115.5,23.4,15.5	110
67	1-OH/X2-Me/3-=O/6-Me	E	80.7	46.5	213.1	37.9	36.9	38.2	33.6	21.0	23.4	32.4	6.7,21.6	9
68	1-OH/X2-Me/6-Me		74.8	31.7	30.7	21.3	36.8	38.1	33.5	21.6	23.1	31.3	15.1,22.7	9
69	1-OH/X2-OAc/6-Me	E	74.4	71.3	26.7	19.2	35.1	38.4	33.1	20.8	22.8	30.5	22.2	48
70	1-OH/X2-OH/6-Me	E	75.4	67.8	30.2	19.3	35.6	38.1	33.4	21.2	23.2	29.9	22.5	48
71	1-OMe/2=/6-Me	E	74.2	132.5	130.3	21.3	31.0	36.1	33.4	21.4	23.6	31.0	22.5	9
72	1-OMe/6-Me	A	76.8	29.9	23.3	21.3	33.9	37.2	36.1	21.2	21.2	25.1	22.5	9
73	2-=O		50.7	212.7	39.5	23.2	29.3	40.9	29.6	23.6	25.0	25.4		25,24
74	2-=O 3=/5-=O/N7-Ph/8=		43.2	200.9	138.8	139.3	197.6	51.4	42.3	127.2	126.3	21.1	138.5	89
75	2-=O 3=/N5-OH/8=		44.6	199.4	128.4	150.1	70.6	40.3	21.0	125.7	125.2	21.1		108
76	2-=O/5-=O		48.0	210.4	36.4	36.4	210.4	48.0	25.8	23.4	23.4	25.8		73
77	2-=O/5-=O/8-Me 8=		44.8	208.8	36.0	36.0	209.2	45.5	28.2	131.7	118.4	24.1	23.4	73
78	2-=O/5-=O/8=		44.8	209.2	35.8	35.8	209.2	44.8	23.5	124.4	124.4	23.5		73
79	2-=O/5-=O/N7-Me		50.5	210.8	36.4	37.2	209.8	51.2	33.2	29.4	23.5	26.0	18.5	73
80	2-=O/5-=O/N7-Me/8=		45.2	210.0	35.3	37.1	209.6	48.7	31.0	131.0	123.0	22.6	17.6	89
81	2-=O/5-=O/N7-Ph		52.2	210.2	36.7	37.0	207.6	52.4	43.1	26.8	24.8	25.8	143.5	89
82	2-=O/5-=O/N7-Ph/8=		42.9	210.2	33.3	37.1	208.4	49.6	40.9	126.9	126.9	21.2	138.1	89
83	2-=O/5-=O/N8-Me		46.0	208.9	36.9	36.1	210.6	50.0	36.2	32.2	30.4	23.7	22.2	73
84	2-=O/6-Me	A	57.6	38.5	21.5	32.8	37.0	37.1	21.6	24.1	24.6		28.0	20
84	2-=O/6-Me		59.1	36.8	36.8	21.1	28.1	36.2	38.9	21.7	25.2	26.7	28.0	20
84	2-=O/6-Me	E	54.0	41.8	21.1	39.6	38.7	32.5	21.3	21.3	20.3		28.0	20
85	2-=O/7=		48.1	212.1	40.6	22.3	23.1	37.4	129.7	128.3	29.7	23.4		24
86	2-=O/8=/N10-Me		53.2	210.2	42.8	24.4	29.5	39.1	25.8	123.0	130.6	32.2	17.4	78
87	2-=O/N5-OH		47.5	211.0	38.4	29.3	71.1	47.5	22.0	25.0	22.5	25.5		108
88	2-=O/X4-Me		48.7	212.1	49.8	29.4	39.1	39.1	29.2	24.9	21.9	26.1	22.3	78
89	2-=O/X4-Me/8=		47.0	210.5	49.3	30.7	38.4	34.9	26.7	124.5	124.4	23.3	22.3	78
90	2-=O/X5-Me N5-Me/8=/N10-Me		49.2	210.6	39.3	36.6	33.2	49.5	24.6	123.0	130.7	32.6	27.2,26.6,17.7	78

208

cis-BICYCLO[4.4.0]DECANES (cont.)

Str	Substituents	A/E	C1	C2	C3	C4	C5	C6	C7	C8	C9	C10	Substituent Shifts	References
91	2-=O/X5-Me N5-Me/N10-Me		50.9	212.6	39.8	35.7	33.4	53.4	24.1	26.7	28.7	35.5	28.0,26.9,19.9	78
92	N2-COO-7N/10-=O		53.0	40.5	31.3	19.2	29.0	34.3	79.6	30.6	33.0	208.7	172.3	73
93	N2-COO-7N/3=/N5-Me/N10-OAc		39.6	37.2	125.4	135.9	34.8	37.8	75.1	30.3	22.1	72.4	172.0,16.5	73
94	N2-COO-7N/N5-Me/10-=O		54.1	40.2	30.9	27.8	33.8	40.2	74.1	31.4	33.4	208.6	172.4,19.2	105,73
95	N2-COO-7N/N5-Me/6-Me/X9-Me/10-=O		59.5	41.6	31.1	29.6	41.6	38.6	78.8	37.2	36.6	211.3	172.9,15.4,24.1,13.6	118
96	N2-COO-7N/N5-Me/6-Me/X9-Me/X10-OH		48.4	42.2	31.9	30.3	43.9	35.4	80.8	32.3	27.1	75.0	175.4,16.3,23.4,15.1	105
97	N2-COO-7N/N5-Me/N10-OAc		42.2	37.2	31.5	28.8	35.0	39.8	74.2	30.6	22.4	72.7	174.4,19.1	73
98	N2-COO-7N/N5-Me/N10-OH		44.8	36.4	31.6	28.9	34.8	39.8	74.7	30.6	25.7	70.4	175.8,19.2	73,105
99	N2-COO-7N/N5-Me/N8-Me/10-=O		53.6	40.3	31.0	28.0	33.8	40.4	78.4	37.9	41.7	208.3	172.4,19.2,18.0	105
100	N2-COO-7N/N5-Me/N8-Me/N10-OH		44.4	36.6	31.6	29.1	34.8	40.2	79.1	36.2	34.2	70.2	175.6,19.2,17.8	118
101	N2-COOMe/3=/7-=O/10-=O		46.6	40.0	125.0	123.2	23.8	45.0	208.8	35.2	36.3	207.0	172.0	73
102	N2-COOMe/3=/N5-Me/7-=NOH/10-=O		45.1	38.3	121.4	130.7	28.1	37.6	155.4	26.1	35.9	209.5	172.0,17.4	73
103	N2-COOMe/3=/N5-Me/7-=O/10-=O		46.0	39.4	121.4	131.3	30.6	48.3	207.8	37.6	34.7	210.3	172.1,17.4	73,74
104	N2-COOMe/7-=O/10-=O		49.2	41.2	24.0	23.5	26.6	48.4	209.6	35.8	36.6	207.8	173.6	73,74
105	N2-COOMe/N5-Me/6-Me/7-=O/N9-Me/10-=O		52.6	38.4	18.4	27.4	34.6	51.9	213.1	44.4	49.3	211.2	174.2,16.8,26.0,16.8	118
106	N2-COOMe/N5-Me/6-Me/N7-OH/N9-Me/10-=O		52.9	35.2	19.2	30.7	30.7	46.5	77.1	41.3	41.3	211.6	175.3,18.5,28.8,14.4	118
107	N2-COOMe/N5-Me/7-=NOH/10-=O		46.3	40.5	17.5	30.9	26.8	40.9	155.5	26.8	36.2	211.3	173.4,15.2	73
108	N2-COOMe/N5-Me/7-=O/10-=O		47.7	41.4	20.1	30.5	31.4	51.1	208.9	37.4	35.1	208.8	173.5,15.8	73,74,105
109	N2-COOMe/N5-Me/7-=O/X8-Me/10-=O		47.4	41.1	17.9	31.4	30.8	51.0	211.9	40.8	42.2	208.8	173.7,15.0,13.3	118
110	N2-COOMe/N5-Me/N7-OAc/10-=O		47.2	42.5	19.0	34.0	28.0	44.2	72.1	32.2	37.8	209.6	173.9,17.0	73
111	N2-COOMe/N5-Me/N7-OH/10-=O		47.3	42.7	19.6	34.0	27.7	47.6	70.2	31.6	37.9	210.9	174.6,17.1	73,74,105
112	N2-COOMe/N7-OAc/10-=O		49.4	42.4	23.6	24.6	21.9	42.4	72.3	26.0	38.3	207.8	173.8	73
113	N2-COOMe/N7-OH/10-=O		49.8	42.6	24.0	24.8	21.2	45.7	70.4	29.2	38.5	209.5	174.6	73,74
114	X2-Me		43.8	29.9	35.2	22.0	22.2	36.5	28.4	26.4	22.3	29.5	20.1	4
115	N2-Me		42.6	36.8	29.4	27.4	25.7	38.3	33.2	21.9	27.4	19.8	19.9	4,2
116	N2-Me/3-=O 4-=/6-Me/N9-CMe2OH		48.6	48.6	201.9	128.5	156.7	38.2	41.0	23.9	42.1	24.5	12.0,27.1,72.7	31
117	X2-Me/N10-Me		50.2	28.7	38.8	21.6	33.4	40.3	26.7	27.6	29.8	40.4	23.6,23.3	5
118	N2-Me/N10-Me		43.7	33.9	33.9	22.5	32.9	39.1	32.1	22.5	33.9	33.9	19.8,19.8	5
119	X2-Me/N3-Me		43.3	34.8	39.8	31.2	32.9	37.7	27.1	27.6	20.8	29.2	16.7,21.0	102
120	N2-Me/N3-Me		43.0	39.3	33.5	34.2	22.7	39.3	33.5	21.8	28.1	24.6	17.3,16.1	102
121	X2-Me/N4-Me		42.3	35.2	36.7	27.6	35.6	31.3	32.8	21.6	27.5	28.2	20.1,23.3	102
122	N2-Me/N4-Me		41.9	36.5	38.0	33.6	34.8	38.4	32.9	21.8	27.3	19.6	19.9,22.9	102
123	X2-Me/N5-Me		45.2	27.5	36.8	29.2	37.0	43.4	20.7	27.4	21.3	30.1	19.9,19.6	5
124	N2-Me/N5-Me		40.9	36.0	30.4	30.4	36.0	40.9	27.4	25.4	25.4	27.4	20.3,20.3	5
125	X2-Me/N7-Me		44.4	34.6	27.7	21.3	25.6	36.7	36.9	29.7	27.4	27.7	19.2,20.1	5
126	N2-Me/N7-Me		44.2	36.9	29.9	27.1	19.3	44.2	36.9	29.9	27.1	19.3	20.0,20.0	5
127	X2-Me/N8-Me		43.1	28.0	36.8	21.9	33.2	37.4	36.0	33.8	29.6	28.6	20.3,23.1	102
128	N2-Me/N8-Me		42.4	36.6	30.2	27.1	30.2	38.3	38.6	28.0	32.6	16.2	19.9,22.4	102
129	X2-Me/N9-Me		42.7	34.8	27.3	21.5	30.4	38.2	32.7	30.6	33.9	37.1	19.4,23.1	102
130	N2-Me/N9-Me		42.3	36.4	29.3	27.2	25.6	37.5	32.8	30.6	33.6	28.7	19.8,23.3	102
131	X2-Me/X10-Me	E	49.3	28.3	37.0	21.5	23.1	30.4	26.3	20.8	27.0	29.9	20.7,19.2	5
132	N2-Me/X3-Me		43.6	38.9	31.7	29.7	27.1	32.3	31.0	22.9	26.6	28.3	14.2,18.6	102
133	N2-Me/X3-Me		43.1	43.8	32.9	36.8	26.2	38.4	32.7	21.8	27.5	20.7	17.5,20.6	102
134	X2-Me/X4-Me		43.4	28.2	45.9	28.0	42.5	38.2	28.0	27.7	20.9	28.6	20.3,23.1	102
135	N2-Me/X4-Me		42.7	29.9	35.1	28.5	31.0	31.2	32.8	21.5	27.2	19.2	19.8,19.1	102
136	X2-Me/X7-Me		44.2	29.3	36.4	21.6	29.3	44.2	29.3	36.4	21.6	29.3	20.3,20.3	5
137	X2-Me/X8-Me		42.9	33.8	28.9	21.7	28.3	31.6	40.7	27.5	34.9	27.5	19.6,22.3	102
138	N2-Me/X8-Me		42.1	36.4	29.3	27.3	26.5	38.7	42.2	27.5	36.1	19.9	19.9,23.2	102
139	X2-Me/X9-Me		44.1	28.9	37.1	22.0	32.9	36.8	28.0	36.3	26.6	37.9	20.3,23.1	102
140	N2-Me/X9-Me		35.3	36.4	29.2	27.2	25.3	38.3	26.8	27.0	28.1	25.2	19.8,18.2	102

cis-BICYCLO[4.4.0]DECANES (cont.)

Str	Substituents	A/E	C1	C2	C3	C4	C5	C6	C7	C8	C9	C10	Substituent Shifts	References
141	N2-OH		43.1	73.5	29.4	24.5	24.5	35.9	32.0	21.7	26.5	19.0		72
142	X2-OH N2-Me/X5-C(=CH$_2$)COOMe/7= 8-Me/X9-OAc		40.6	71.5	34.5	27.6	36.5	39.6	131.2	131.2	70.6	28.4	29.2,144.1,21.2	119
143	N2-OH X2-Me/X5-C(=CH$_2$)COOMe/7= 8-Me		49.5	71.2	42.2	29.2	41.1	45.5	128.6	133.9	73.8	31.1	18.9,143.3,21.0	119
144	X2-OH/N3-iPr/6-Me/10-==CH$_2$		60.3	67.4	48.9	18.5	39.8	35.5	30.8	22.7	30.2	146.8	26.3,28.2,112.3	113
145	2= 4=		34.8	131.7	123.5	123.5	131.7	34.8	27.8	23.4	23.4	27.8		21
146	2=/6-Me	E	42.4	132.6	124.5	22.8	36.9	31.4	31.6	22.0	21.8	26.7	27.5	9
147	2=/6-Me	A	40.5	132.0	126.9	23.5	28.2	31.0	41.0	22.3	26.9	29.8	27.0	9
148	3-==0		38.6	45.3	211.3	39.3	28.8	35.0	28.4	22.0	23.8	28.9		106,22,27
149	3-==0	E	37.6	42.2	215.5	37.3	32.2	34.7	24.8	26.5	19.9	30.9		9+22
150	3-==0	A	39.4	48.2	214.7	41.4	27.1	34.5	30.7	20.4	26.8	26.3		9+22
151	3-==0/6-Me	A	44.3	44.9	213.7	38.1	30.3	32.6	39.1	21.6	26.7	29.3	27.5	9
152	3-==0/6-Me/N10-Me		47.6	36.9	212.4	37.5	41.1	33.2	28.7	21.6	27.7	29.9	26.2,19.3	11
153	3-==0/6-Me/N9-t-Bu		44.7	44.5	210.5	37.5	30.4	32.0	39.8	22.2	48.2	30.3	27.0,32.0	10,9
154	3-==0/6-Me/X10-Me		51.6	39.6	211.6	37.6	31.4	33.0	39.6	21.4	35.9	31.4	27.7,20.1	11
155	3-==0/6-Me/X7-Me		43.2	42.5	212.8	37.3	36.3	34.9	30.1	30.6	20.6	27.4	21.2,15.5	3
156	3-==0/6-Me/X7-OCH$_2$CH$_2$O-7 N		42.7	44.2	212.2	37.9	29.7	41.3	112.5	29.1	22.3	28.3	17.8	111
157	X3-Me		36.9	41.8	27.3	36.3	26.3	36.0	32.4	21.6	27.5	27.3	22.9	4+102
158	N3-Me		36.6	35.3	33.9	30.2	32.9	36.1	26.3	27.5	21.4	32.9	23.1	4,2
159	X3-Me/N4-Me		37.2	42.6	33.5	40.2	35.6	36.7	32.5	21.2	27.5	27.0	20.4,20.4	5
160	N3-Me/N4-Me		37.4	35.1	34.7	34.7	35.1	37.4	31.7	25.1	25.1	31.7	18.4,18.4	5
161	X3-Me/N8-Me		29.3	32.1	28.4	26.9	26.9	36.6	34.9	33.5	30.2	32.1	18.4,23.1	5
162	N3-Me/N8-Me		35.8	35.1	33.7	30.1	32.6	35.8	35.1	33.7	30.1	32.6	23.0,23.0	5
163	X3-Me/N9-Me		37.0	41.9	26.9	36.2	26.2	35.5	32.3	30.1	33.8	36.0	23.0,23.0	5
164	N3-Me/N9-Me		36.2	38.9	30.7	31.6	27.2	36.2	27.2	31.6	30.7	38.9	22.7,22.7	5
165	X3-Me/X8-Me		36.8	41.7	27.1	36.3	27.1	36.8	41.7	27.1	36.3	27.1	23.1,23.1	5
166	X3-OH		35.6	39.8	66.9	34.5	25.0	34.5	30.2	22.3	28.1	25.9		101
167	N3-OH		34.8	35.5	71.6	30.5	30.1	35.0	25.9	26.8	21.1	31.8		22,9,46,101

17.2 trans – Bicyclo [4.4.0] decanes

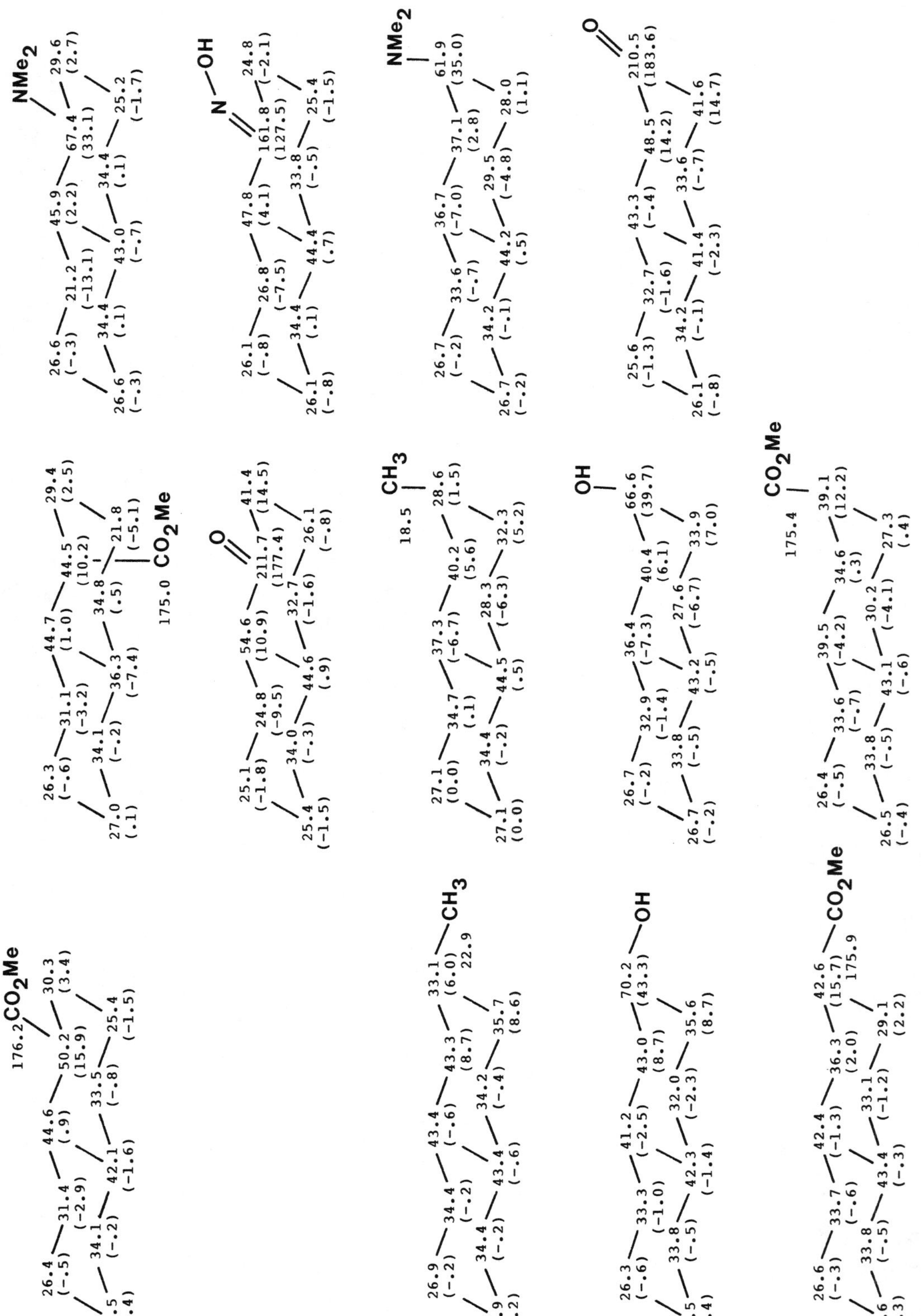

NMe₂ (structure 1):
- 29.6 (2.7)
- 67.4 (33.1)
- 45.9 (2.2)
- 25.2 (-1.7)
- 21.2 (-13.1)
- 34.4 (.1)
- 43.0 (-.7)
- 26.6 (-.3)
- 34.4 (.1)
- 26.6 (-.3)

N—OH (structure 2):
- 24.8 (-2.1)
- 161.8 (127.5)
- 47.8 (4.1)
- 25.4 (-1.5)
- 33.8 (-.5)
- 26.8 (-7.5)
- 44.4 (.7)
- 26.1 (-.8)
- 34.4 (.1)
- 26.1 (-.8)

NMe₂ (structure 3):
- 61.9 (35.0)
- 37.1 (2.8)
- 36.7 (-7.0)
- 28.0 (1.1)
- 29.5 (-4.8)
- 33.6 (-.7)
- 44.2 (.5)
- 26.7 (-.2)
- 34.2 (-.1)
- 26.7 (-.2)

O (structure 4):
- 210.5 (183.6)
- 48.5 (14.2)
- 43.3 (-.4)
- 41.6 (14.7)
- 33.6 (-.7)
- 32.7 (-1.6)
- 41.4 (-2.3)
- 25.6 (-1.3)
- 34.2 (-.1)
- 26.1 (-.8)

CO₂Me, 175.0 (structure 5):
- 29.4 (2.5)
- 44.5 (10.2)
- 44.7 (1.0)
- 21.8 (-5.1)
- 34.8 (.5)
- 31.1 (-3.2)
- 36.3 (-7.4)
- 26.3 (-.6)
- 34.1 (-.2)
- 27.0 (.1)

O, 211.7 (structure 6):
- 41.4 (14.5)
- 211.7 (177.4)
- 54.6 (10.9)
- 26.1 (-.8)
- 32.7 (-1.6)
- 24.8 (-9.5)
- 44.6 (.9)
- 25.1 (-1.8)
- 34.0 (-.3)
- 25.4 (-1.5)

CH₃, 18.5 (structure 7):
- 28.6 (1.5)
- 40.2 (5.6)
- 37.3 (-6.7)
- 32.3 (5.2)
- 28.3 (-6.3)
- 34.7 (.1)
- 44.5 (.5)
- 27.1 (0.0)
- 34.4 (-.2)
- 27.1 (0.0)

OH (structure 8):
- 66.6 (39.7)
- 40.4 (6.1)
- 36.4 (-7.3)
- 33.9 (7.0)
- 27.6 (-6.7)
- 32.9 (-1.4)
- 43.2 (-.5)
- 26.7 (-.2)
- 33.8 (-.5)
- 26.7 (-.2)

CO₂Me, 175.4 (structure 9):
- 39.1 (12.2)
- 34.6 (.6)
- 39.5 (-4.2)
- 27.3 (.4)
- 30.2 (-4.1)
- 33.6 (-.7)
- 43.1 (-.6)
- 26.4 (-.5)
- 33.8 (-.5)
- 26.5 (-.4)

CO₂Me, 176.2 (structure 10):
- 30.3 (3.4)
- 50.2 (15.9)
- 44.6 (.9)
- 25.4 (-1.5)
- 33.5 (-.8)
- 31.4 (-2.9)
- 42.1 (-1.6)
- 26.4 (-.5)
- 34.1 (-.2)
- 26.5 (-.4)

CH₃, 22.9 (structure 11):
- 33.1 (6.0)
- 43.3 (8.7)
- 43.4 (-.6)
- 35.7 (8.6)
- 34.2 (-.4)
- 34.4 (-.2)
- 43.4 (-.6)
- 26.9 (-.2)
- 34.4 (-.2)
- 26.9 (-.2)

OH, 70.2 (structure 12):
- 70.2 (43.3)
- 43.0 (8.7)
- 41.2 (-2.5)
- 35.6 (8.7)
- 32.0 (-2.3)
- 33.3 (-1.0)
- 42.3 (-1.4)
- 26.3 (-.6)
- 33.8 (-.5)
- 26.5 (-.4)

CO₂Me, 175.9 (structure 13):
- 42.6 (15.7)
- 36.3 (2.0)
- 42.4 (-1.3)
- 29.1 (2.2)
- 33.1 (-1.2)
- 33.7 (-.6)
- 43.4 (-.3)
- 26.6 (-.3)
- 33.8 (-.5)
- 26.6 (-.3)

Structure 1:

26.7 (−.2) — 33.4 (−.9) — 34.1 (−9.6) — 38.5 (4.2) — 126.8 (99.9)
‖
26.7 (−.2) — 33.4 (−.9) — 34.1 (−9.6) — 38.5 (4.2) — 126.8 (99.9)
H (left), H (right)

Structure 2:

CH₂ ‖
26.4 (−.5) — 29.0 (−5.3) — 42.4 (−1.3) — 147.7 (113.4) — 130.5 (103.6)
‖
26.1 (−.8) — 34.0 (−.3) — 38.9 (−4.8) — 34.3 (0.0) — 128.5 (101.6)
H (left), H (right)

Structure 3:

27.9 (1.0) — 35.3 (1.0) — 140.9 (97.2) — 119.1 (84.8) — 25.5 (−1.4)
‖
26.9 (0.0) — 35.6 (1.3) — 37.4 (−6.3) — 31.2 (−3.1) — 21.6 (−5.3)

Structure 4:

O ‖
27.0 (.1) — 35.6 (11.3) — 167.4 (123.7) — 124.3 (90.0) — 36.6 (9.7)
‖
25.6 (−1.3) — 34.5 (.2) — 37.9 (−5.8) — 29.2 (−5.1)

Structure 5:

26.8 (−.1) — 32.6 (−1.7) — 40.9 (−2.8) — 127.8 (93.5) — 131.7 (104.8) — 133.4 (106.5) — CH₃ 20.9
‖
26.7 (−.2) — 32.2 (−2.1) — 41.1 (−2.6) — 128.0 (93.7)
H (left), H (right)

Structure 6:

23.5 (−3.4) — 30.6 (−3.7) — 127.6 (83.9) — 30.6 (−3.7) — 23.5 (−3.4)
‖
23.5 (−3.4) — 30.6 (−3.7) — 127.6 (83.9) — 30.6 (−3.7) — 23.5 (−3.4)

Structure 7:

O ‖
22.4 (−4.5) — 21.6 (−12.7) — 146.0 (102.3) — 201.5 (167.2) — 38.3 (11.4)
‖
38.3 (11.4) — 201.5 (167.2) — 146.0 (102.3) — 21.6 (−12.7) — 22.4 (−4.5)
‖ O

Structure 8:

27.0 (.1) — 29.6 (−4.7) — 43.4 (−.3) — 140.0 (105.7) — 120.6 (93.7) — 124.1 (97.2)
‖
26.7 (−.2) — 32.9 (−1.4) — 40.8 (−2.9) — 131.1 (96.8) — CH₃ 19.9
H (left), H 19.4

trans–BICYCLO[4.4.0]DECANES

Str	Substituents	C1	C2	C3	C4	C5	C6	C7	C8	C9	C10	Substituent Shifts	References
1	PARENT	43.7	34.3	26.9	26.9	34.3	43.7	34.3	26.9	26.9	34.3		1,2,4,6,12-14,23,28, 52,72,82,103
2	1-Br	80.6	43.3	23.1	26.4	30.4	47.9	30.4	26.4	23.1	43.3		23
3	1-CH2OH/A3-OCH2CH2O-3E/4=/A7-Me E7-Me	42.0	45.3	105.8	127.2	132.5	53.8	32.2	41.4	18.4	34.3	60.8,22.0,32.2	120
4	1-CH2OTs/A3-OH E3=C(SMe)3/4=/A7-Me E7-Me	38.6	44.2	80.1	129.7	130.7	52.0	32.0	40.9	17.6	34.4	69.5,80.2,22.2,32.6	120
5	1-CH2OTs/E3-COOMe A3-OH/4=/A7-Me E7-Me	36.9	44.0	72.0	128.1	129.5	52.0	32.2	41.0	17.5	34.0	68.8,0,22.2,32.5	120
6	1-CHO/A3-OCH2CH2O-3E/4=/A7-Me E7-Me	51.4	47.9	104.1	129.7	130.7	52.4	32.8	41.2	18.8	32.2	207.1,21.0,30.8	120
7	1-Cl	77.4	42.2	22.2	26.4	29.2	52.4	35.6	42.2	22.2	32.2		23
8	1-Cl/6-Me	85.4	36.5	21.7	20.5	35.6	38.5	35.6	20.5	21.7	36.5	20.1	49
9	1-Cl/A2-Cl/6-Me	85.8	64.1	30.8	16.4	34.9	38.5	38.7	20.4	22.0	34.0	22.8	44
10	1-Cl/E2-Cl/6-Me	82.0	68.3	32.8	21.3	34.6	37.0	37.0	20.2	21.6	28.6	24.6	44
11	1-CN/3==O	43.3	50.6	205.6	40.4	29.0	42.4	29.7	24.8	22.0	36.5	120.1	121
12	1-CN/3==O/A7-Me E7-Me	40.8	53.2	205.5	41.2	24.2	50.8	33.4	41.2	19.5	38.4	122.0,20.0,31.8	120
13	1-CN/A3-OCH2CH2O-3E/4=/A7-Me E7-Me	38.6	47.2	104.1	128.8	130.4	51.4	32.8	40.6	19.3	37.4	123.0,20.2,31.4	120
14	1-CN/A3-OCH2CH2O-3E/A4=Br/A7-Me E7-Me	37.8	41.2	106.3	54.8	31.0	45.5	32.8	41.4	19.3	39.1	123.4,20.3,31.6	120
15	1-CN/A3-OCH2CH2O-3E/A7-Me E7-Me	38.0	46.4	107.0	36.2	22.1	52.4	33.2	41.4	19.4	39.4	123.5,19.9,32.1	120
16	1-COOH	48.8	38.8	23.6	26.8	29.4	45.7	29.4	26.8	23.6	38.8	183.3	70
17	1-F	93.3	37.7	22.1	26.2	29.2	43.9	29.2	26.2	22.1	37.7		23
18	1-I	75.6	45.8	24.9	26.5	32.6	48.8	32.6	26.5	24.9	45.8		23
19	1-Me	33.9	42.1	22.1	27.2	29.3	45.8	29.3	27.2	22.1	42.1	15.7	1,2,3,7,23
20	1-Me/2==CH2/3==CH2/A7-Me E7-Me	40.2	161.6	149.9	37.6	22.7	52.5	33.8	42.3	19.2	35.9	20.7,103.2,109.0,22.1,33.5	60
21	1-Me/2==O	48.3	215.6	37.4	26.2	27.7	46.1	28.0	26.2	21.4	32.5	15.7	15
22	1-Me/2==O/6-O-7E/8==O	48.5	211.1	37.6	22.0	28.6	70.0	61.6	205.2	32.6	23.4	20.7	110
23	1-Me/2==O/6-O-7E/E8==CHCOOMe	48.6	211.7	37.7	21.6	28.6	67.8	62.9	153.1	22.5	23.6	20.4,120.9	110
24	1-Me/2==O/6-O-7E/Z8==CHCOOMe	48.6	212.0	37.6	21.5	28.9	67.8	56.3	152.8	20.2	24.7	20.9,121.9	110
25	1-Me/2==O/6-OH/A7-OH/E8==CHCOOMe	51.0	213.8	35.5	19.0	28.4	77.5	80.5	161.8	20.1	28.9	20.1,116.4	110
26	1-Me/2==O/6= 8==O	50.6	210.9	37.7	22.9	29.7	165.9	125.8	198.1	33.6	31.7	23.3	111
27	1-Me/2==O/6+/A8-SCH2CH2S-8E	49.5	212.6	37.8	24.7	30.7	141.2	127.9	64.7	37.5	30.7	24.7	19
28	1-Me/2==O/8==O	47.3	213.8	37.4	25.9	45.1	45.1	43.3	209.4	37.4	32.3	15.0	111
29	1-Me/3==O	37.6	56.8	211.2	41.3	28.0	43.7	29.4	26.6	21.1	41.6	16.5	15
30	1-Me/3==O 4==CMe2/E7-OH A7-Me/A8-OH	36.3	60.2	202.0	131.2	25.7	45.6	73.1	74.3	25.7	32.9	18.6,144.1,21.4	117
31	1-Me/3==O/A7-Me E7-Me	37.4	58.6	209.2	41.5	32.4	51.2	32.4	41.5	18.0	41.0	18.5,20.6,32.4	83
32	1-Me/A2-Me/3==O 4= 6==/A9-C(Me)=CH2	39.7	55.2	202.7	122.9	145.8	138.8	135.4	31.2	41.4	41.4	30.7,14.8,148.3	3
33	1-Me/A2-O-3E/E4==C(Me)COOMe/7==CH2	36.4	62.7	51.9	139.7	26.7	38.9	148.7	36.4	23.3	36.4	15.1,131.0,106.8	112
34	1-Me/A2-O-3E/Z4==C(Me)COOMe/7==CH2	36.4	63.2	52.0	139.6	26.0	38.3	149.0	35.9	23.0	36.4	15.1,132.0,106.4	112
35	1-Me/A2-OH	38.3	75.4	28.6	20.5	28.9	38.6	28.9	26.8	21.9	34.9	16.1	7,1
36	1-Me/A2-OH E2-Me	40.7	74.3	35.9	22.2	29.2	38.6	28.9	26.7	21.8	31.4	14.8,24.3	7
37	1-Me/A2-OH/4==CMe2/7==CH2	41.2	79.6	34.6	126.9	27.9	48.3	149.5	37.3	23.3	36.8	10.1,123.1,106.9	64
38	1-Me/A3-C(Me)2OH/5= 7==O/E10-Me	40.6	35.3	36.5	27.2	128.5	148.2	205.7	40.1	29.1	41.6	21.2,71.9,15.3	116
39	1-Me/A3-C(Me)2OH/6=/E10-Me	37.8	35.4	34.3	26.4	28.1	143.9	120.6	25.2	27.2	42.0	21.0,72.6,16.0	116
40	1-Me/A3-COOEt/4==7==O/E10-Me	39.8	35.7	38.9	124.5	123.6	56.1	209.1	40.7	31.4	41.2	11.3,174.9,14.6	114
41	1-Me/A3-COOEt/5= 7==O/E10-Me	39.0	36.9	36.3	26.9	128.7	146.5	203.4	39.9	28.5	38.4	19.8,174.9,15.4	114
42	1-Me/A3-COOEt/6= E10-Me	38.0	40.0	38.0	27.8	29.1	143.0	120.4	27.5	27.0	40.6	18.7,175.8,15.7	116
43	1-Me/A3-OH	33.5	47.8	67.6	34.1	24.0	45.9	28.7	27.0	21.2	41.9	17.9	1
44	1-Me/A3-OH/A7-Me E7-Me	34.2	51.2	67.6	33.0	17.0	54.2	33.0	42.4	18.2	42.4	21.3,21.3,33.0	83
45	1-Me/A4-t-Bu	33.6	39.3	21.2	42.7	29.1	41.7	29.0	26.9	22.2	39.4	18.7,32.2	10
46	1-Me/A5-Me E5-Me	34.3	42.6	18.7	42.9	33.1	54.1	22.1	28.1	22.1	45.8	19.2,21.4,33.1	83
47	1-Me/E2-CH2OH/3-=CH2/A7-Me E7-Me	39.1	59.2	147.9	39.1	24.3	55.3	33.5	42.1	19.3	37.9	15.3,58.8,106.4,21.8,33.7	65
48	1-Me/E2-CHO A2-OH/3-CHO 3=/7-=CH2/E8-Me	42.2	77.4	139.6	157.6	27.7	40.1	155.7	38.1	30.8	31.7	14.9,201.3,192.7,106.1,18.4	91
49	1-Me/E2-COOMe/4-=O 5=	38.8	51.7	36.6	197.1	124.3	168.8	32.8	26.7	21.8	39.8	18.3	112
50	1-Me/E2-COOMe/4-=O/7-=CH2	38.6	53.9	40.5	208.6	40.7	49.6	147.7	36.1	23.1	38.9	12.2,0,107.9	112

214

trans-BICYCLO[4.4.0]DECANES (cont.)

Str	Substituents	C1	C2	C3	C4	C5	C6	C7	C8	C9	C10	Substituent Shifts	References
51	1-Me/E2-COOMe/A4-OCH2CH2O--4E/6=	36.7	51.1	34.1	---	37.1	138.9	123.9	25.4	18.8	41.7	19.4	112
52	1-Me/E2-COOMe/A4-OCH2CH2O--4E/7-==CH2	38.4	52.1	34.0	---	33.9	47.8	149.1	36.4	23.3	39.1	12.0,0,106.8	112
53	1-Me/E2-COOMe/A4-OCH2CH2O--4E/7-==O	41.0	51.8	34.0	---	30.2	55.5	210.1	37.7	22.3	40.9	13.2	112
54	1-Me/E2-Me	36.8	43.6	31.3	27.3	29.6	47.0	29.3	27.0	22.4	39.1	10.5,15.2	7
55	1-Me/E2-OAc/5=	39.0	80.3	24.3	23.9	118.8	141.9	31.9	27.8	22.0	38.5	18.3	44
56	1-Me/E2-OAc/6=	39.3	80.4	27.6	24.9	31.5	141.1	122.6	25.7	18.6	35.7	18.6	44
57	1-Me/E2-OAc/6=/E7-OH	39.5	80.3	27.4	24.4	31.2	143.3	126.8	67.2	28.4	33.2	18.1	110
58	1-Me/E2-OH	39.2	79.6	30.4	24.4	29.1	44.2	28.1	26.8	21.7	37.3	9.8	1,7
59	1-Me/E2-OH A2-Me	41.5	75.2	37.6	23.5	29.1	41.2	28.8	26.7	21.9	32.0	12.2,22.8	7
60	1-Me/E2-OH/5=	40.0	78.3	27.1	24.8	118.6	142.3	32.2	27.9	22.2	38.8	17.3	44
61	1-Me/E2-OH/6-O-7E/E8-==CHCOOMe	39.4	76.3	30.4	21.0	29.0	66.8	62.3	154.6	22.9	27.9	13.8,120.3	110
62	1-Me/E2-OH/6-O-7E/Z8-==CHCOOMe	39.4	76.5	31.0	20.1	29.3	66.9	55.7	153.8	26.4	28.9	14.2,121.6	110
63	1-Me/E2-OH/6=	40.5	79.0	30.9	25.2	31.9	142.3	121.9	25.9	19.0	36.2	17.5	44
64	1-Me/E2-OMs/6=/A8-SCH2CH2S-8E	39.1	88.6	28.7	23.9	30.6	140.9	126.2	64.8	37.1	35.4	17.7	18
65	1-Me/E3-C(Me)2OH/6=/E10-Me	37.7	40.6	44.5	29.0	32.7	143.2	119.8	37.4	27.3	41.2	18.5,72.3,15.7	116
66	1-Me/E3-COOMe/6=/E10-Me	37.6	41.3	47.8	30.0	31.9	141.7	121.1	25.9	27.2	40.9	18.2,211.5,15.6	116
67	1-Me/E3-COOEt/6=/E10-Me	37.4	41.8	39.8	30.4	31.7	141.7	120.9	25.8	27.2	40.8	18.1,176.1,15.6	116
68	1-Me/E3-OH	34.7	50.9	66.9	36.4	20.9	44.9	28.1	26.9	21.2	41.6	16.6	1,3
69	1-Me/E4-OH/A7-Me E7-Me	35.1	54.1	66.8	36.7	20.9	53.0	32.7	41.9	18.5	42.4	19.8,20.9,33.0	83
70	1-Me/E4-C(Me)/7-==CH2	35.9	42.0	27.0	45.8	29.6	49.8	149.4	36.9	23.8	41.2	16.7,149.4,105.5	66
71	1-O-2E/3-=O/6-Me	68.9	62.8	206.9	37.6	31.5	34.0	33.3	21.3	23.9	29.8	20.2	7
72	1-O-2E/3-=O/6-Me/E10-Me	70.8	59.1	206.4	33.3	35.1	35.1	38.0	21.5	35.2	30.1	20.3,13.8	110
73	1-O-2E/3-=O/6-Me/E7-OAc	69.0	61.3	206.0	32.8	27.8	38.7	78.1	26.8	21.0	28.8	15.1	110
74	1-O-2E/3-=O/6-Me/E7-OH	69.5	61.3	207.0	33.1	27.9	39.7	76.4	30.2	21.4	29.7	14.0	110
75	1-O-2E/3-=O/6-Me/E7-OTs	68.6	61.3	204.6	32.7	27.8	39.4	86.7	27.9	21.0	28.5	14.8	110
76	1-O-2E/A3-CH2COOMe/6-Me/7-==O	67.2	61.9	22.7	23.9	25.5	48.3	213.1	37.4	21.2	28.7	38.6,21.7	110
77	1-O-2E/E3-CH2COOMe/6-Me/7-==O	66.8	63.0	30.7	23.9	25.5	48.3	212.9	37.4	21.2	28.7	37.2,21.4	110
78	1-OAc/6-Me	87.2	28.1	21.5	20.7	35.0	37.5	35.0	20.7	21.5	28.1	20.0	48
79	1-OAc/A2-OAc/6-Me	86.3	69.9	26.2	19.5	33.9	36.8	35.3	20.5	21.4	24.4	20.8	48
80	1-OAc/E2-OAc/6-Me	88.2	73.2	27.5	19.5	33.9	39.6	35.3	20.2	21.6	24.7	20.2	48
81	1-OH	70.2	39.9	21.7	26.4	28.7	44.3	28.7	26.4	21.7	39.9		7,23,99
82	1-OH/2-=O/6-Me	78.8	214.6	36.7	21.9	33.6	41.1	33.5	20.4	20.4	26.4	18.7	48
83	1-OH/2-/6-Me	70.7	133.8	129.5	23.2	31.1	36.0	34.8	20.7	20.8	26.4	19.5	7
84	1-OH/6-Me	73.0	34.3	21.1	20.8	35.1	36.8	35.1	20.8	21.1	34.3	20.4	7,44,48
85	1-OH/A2-C1/6-Me	75.4	64.3	30.6	16.4	34.6	36.9	37.6	20.5	21.3	32.1	21.8	44
86	1-OH/A2-Me/6-Me	75.4	40.9	28.3	17.1	38.2	37.1	35.4	21.4	20.9	32.3	16.7,21.7	7
87	1-OH/A2-OAc/6-Me	73.2	76.3	26.2	17.0	34.5	36.4	36.6	20.6	20.8	30.8	20.3	44,48
88	1-OH/A2-OH/6-Me	73.9	75.9	29.3	16.5	35.0	36.3	36.9	20.6	20.9	30.8	20.6	44,7,48
89	1-OH/A2-OH/E3-==CHCOOMe/6-Me/E10-Me	76.7	78.1	160.8	21.4	36.4	37.5	36.5	20.7	30.0	30.7	118.8,20.4	110
90	1-OH/A4-OH/6-Me/E10-Me	74.8	25.4	29.7	67.3	32.9	36.8	34.6	19.9	30.4	33.7	22.9,14.9	7
91	1-OH/E2-Me/3-=O/6-Me	78.7	49.6	211.9	37.9	35.6	37.4	35.2	20.7	20.4	30.8	6.7,20.8	7,64
92	1-OH/E2-Me/6-Me	74.5	34.3	29.9	20.7	35.7	37.3	35.1	21.4	20.8	30.5	14.9,20.3	7,64
93	1-OH/E2-OAc/6-Me	74.0	75.0	26.8	19.8	33.8	38.0	34.8	20.8	20.8	29.4	19.8	48
94	1-OH/E2-OH/6-Me	74.9	71.2	30.5	19.5	34.0	37.5	35.0	20.6	21.0	28.6	19.8	48
95	1-OH/E4-OH/6-Me	70.7	33.3	30.8	65.7	44.2	37.6	33.7	20.0	20.8	34.5	21.0	7
96	1-OH/E4-OH/6-Me/E10-Me	73.6	29.8	31.2	66.8	45.0	38.8	34.7	20.3	30.3	33.9	21.1,15.2	7,8
97	1=	140.9	119.1	25.5	20.7	35.6	35.6	35.6	26.9	27.9	35.3		29,61
98	1= 2-C1/6-Me	138.1	125.0	26.8	19.8	39.5	37.2	42.1	22.2	27.1	34.8	24.4	44
99	1= 2-Me 3-=O/6-Me	162.8	128.0	199.3	33.6	37.5	36.1	42.0	21.3	26.7	27.6	10.7,22.3	83,90
100	1= 3-=O	167.4	124.3	36.6	36.6	29.2	37.9	34.5	25.6	27.0	35.6		76

Str	Substituents	C1	C2	C3	C4	C5	C6	C7	C8	C9	C10	Substituent Shifts	References
101	1-==0/6-Me	170.2	123.9	199.1	34.0	38.1	36.0	41.6	21.8	27.2	32.8	22.1	3,47,76
102	3-==0/6-Me/A7-Me	170.0	126.0	199.2	34.3	32.1	39.4	39.5	28.6	20.8	31.8	23.8,16.4	3
103	1-==0/6-Me/A7-OCH2CH2O-7E	167.7	125.6	199.0	33.9	31.5	45.0	112.4	30.1	21.8	26.9	20.6	111
104	1-==0/6-Me/E7-OH	169.5	125.1	199.9	33.8	34.3	41.9	77.9	30.3	23.3	32.2	15.5	110
105	1-==0/6-Me/E7-OTs	165.6	126.2	198.6	33.4	33.9	41.0	87.8	28.0	22.9	31.4	16.5	110
106	1-==0/6-Me/A7-SCH2CH2S-7E	166.2	126.0	198.7	34.4	31.3	45.6	79.7	39.0	24.3	30.8	22.2	19
107	1-==0/6-Me/A8-t-Bu	172.1	124.8		34.4	39.6	36.6	34.7	44.8	25.9	29.1		76
108	3-==0/6-Me/E10-Me	173.8	121.6	199.9	38.4	38.4	36.4	41.9	41.2	36.4	34.2	23.0,18.1	110
109	3-==0/6-Me/E7-Me	171.0	123.9	199.4	34.0	35.5	39.0	43.1	30.5	26.5	33.3	16.0,15.3	3
110	3-==0/6-Me/E7-OAc	166.6	125.7	198.5	33.5	34.0	40.4	79.1	26.8	23.0	31.7	16.6	110
111	3-==0/E4-OAc/E5-Me/6-OH/E8-C(Me)=CH2	162.7	122.7	193.2	75.1	43.0	71.6	43.0	39.1	31.2	32.2	10.0,148.4	58
112	1-==0/E4-OAc/E5-Me/E8-C(Me)=CH2	164.2	122.3	192.9	77.5	44.0	44.6	44.0	39.3	31.3	35.1	15.4,148.2	58
113	3-==0/E7-C(Me)=CH2	165.5	124.0	199.0	39.2	31.5	37.5	36.4	44.1	29.1	35.1	148.1	115
114	1-==0/E8-t-Bu	167.2	124.0		36.6	29.5	37.9	35.6	47.2	27.2	35.6		76
115	2-=OAc/6-Me	139.9	129.4	22.8	18.6	39.4	34.8	41.8	22.1	27.0	27.9	24.4	71
116	3-==0/6-Me/E9-t-Bu	172.3	125.5	197.5	33.9	37.4	36.0	35.7	21.3	44.8	33.9	24.5,31.7	10
117	1-/6-Me	143.6	119.2	26.0	19.1	40.2	34.8	42.3	22.5	28.7	32.8	24.4	3,44
118	1-/6-Me/A10-Cl	141.6	128.2	26.0	18.3	41.1	34.2	41.4	17.0	36.6	65.5	27.2	44
119	1-/6-Me/E10-Cl	141.1	121.0	26.0	18.7	40.5	36.6	41.0	22.4	40.0	61.5	24.9	44
120	1-/6-Me/E7-Cl	141.5	123.0	26.0	18.9	37.2	41.0	72.3	33.3	27.3	31.6	18.4	44
121	1-/A3-SCH2CH2S-3E/6-Me/A7-SCH2CH2S-7E	141.7	128.0	65.3	38.2	33.2	44.1	80.6	39.0	25.1	30.5	23.8	19
122	1-/E5-Cl/6-Me	142.4	118.8	25.9	29.3	71.7	40.4		22.2	27.7	32.6	18.5	44
123	1-/E5-Me/6-Me/8-==CMe2		119.7	25.9	27.6	40.2		42.5		33.9	31.5	15.6,17.6	116
124	1-/E5-Me/6-Me/A7-(C=CH)CO	144.5	122.0	29.1	26.6	34.0	37.7	41.1	41.1	29.2	25.5	19.5,15.4,174.9	80
125	1-/E5-Me/6-Me/A8-C(Me)=CH2	143.9	120.5	25.5	27.3	38.6	38.1	39.9	37.0	28.5	30.1	16.0,21.4,150.0	116
126	1-/E5-Me/6-Me/E8-C(Me)=CH2	142.9	120.1	25.9	27.2	41.0	37.8	45.0	41.0	33.2	32.8	15.7,18.4,150.4	116
127	1-Me/E2-OH/6-/A8-SCH2CH2S-8E	39.5	78.2	30.3	24.2	31.0	143.1	126.4	65.3	37.4	35.5	16.5	18
128	(6)1-	127.6	30.6	23.5	23.5	30.6	127.6	30.6	38.3	23.5	30.6		29,26
129	(6)1- 2-=0 7-=0	146.0	201.5	38.3	23.5	21.6	146.0	201.5	38.3	22.4	21.6		108
130	(6)1-/E2-Me/E3-OH/E4-OH/E9-C(CH2OH)=CH2	130.2	43.1	79.8	72.0	39.2	125.9	30.2	27.7	37.0	32.7	17.0,153.5	63
131	(6)1-/E2-Me/E3-OH/E4-OH/E9-C(Me)=CH2	129.1	41.7	79.2	71.5	38.4	124.9	39.7	26.6	40.5	31.2	16.5,148.9	30,32,33,55
132	(6)1-/E3-SePh/E4-Cl	125.5	33.8	44.5	59.2	37.2	123.9	29.6	22.8	22.8	29.8		42
133	(6)1-/E3-SePh/E4-OMe	126.3	35.7	42.6	78.7	34.7	124.9	29.5	22.6	22.6	29.7		42
134	2-=CH2 3-=	42.4	147.7	130.5	128.5	34.3	38.9	34.0	26.1	26.4	29.0	108.6	36
135	A2-=NOH	47.8	161.8	24.8	25.4	33.8	44.4	34.4	26.1	26.1	26.8		34
136	2-==0	54.6	211.7	41.4	26.1	32.7	39.6	34.0	25.4	25.1	24.8		78,15,24,25,72
137	2-==0 3-/A5-Me/6-Me/A8-C(Me)=CH2/10=	139.6	189.3	127.4	154.0	44.7	39.5	41.5	40.0	30.5	135.4	16.7,31.0,148.3	3
138	2-==0/5-=0	49.8	209.2	36.8	36.8	209.2	49.8	26.4	24.8	24.8	26.4		73
139	2-==0/5-=0/E7-Me	51.2	208.7	38.3	37.6	208.7	56.0	31.3	34.0	34.2	26.7	21.0	73
140	2-==0/5-=0/E8-Me	49.5	208.9	36.8	36.8	208.9	49.7	34.8	31.4	33.4	26.4	22.2	73
141	2-==0/6-Me	57.6	210.9	40.8	20.5	40.8	39.6	41.3	21.2	25.6	22.6	16.9	15
142	2-==0/7-=0	55.5	209.9	41.2	24.9	24.4	55.5	209.9	41.2	24.9	24.4		108
143	2-==0/8-Me 8=	50.0	211.7	41.8	25.9	32.4	40.2	38.1	131.9	119.6	24.4	23.0	78
144	2-==0/8-=	49.8	211.1	41.3	25.6	32.2	39.7	32.9	124.6	125.3	24.1		78
145	2-==0/8=/E10-Me	58.0	211.4	42.8	27.0	32.9	40.8	32.9	123.2	132.6	29.0	20.9	78
146	2-==0/A4-C(Me)=CH2/6-Me/E7-Me/10=	147.6	203.1	41.5	39.2	43.3		38.9	26.6	25.6	134.9	144.4,24.8,16.0	116
147	2-==0/A4-Me	54.7	212.1	47.5	30.6	38.4	38.9	34.0	25.4	25.1	24.6	19.0	78
148	2-==0/A4-Me E4-Me	54.0	211.6	54.4	35.6	46.2	40.0	34.1	25.6	25.2	24.6	25.8,32.0	78
149	2-==0/A4-Me E4-Me/8=	49.1	210.9	54.4	35.4	45.5	35.4	33.2	125.0	125.5	24.1	25.2,31.8	78
150	2-==0/A4-Me/8=	49.8	211.7	47.5	30.3	37.7	34.3	33.0	124.9	125.3	23.8	18.4	78

trans-BICYCLO[4.4.0]DECANES (cont.)

Str	Substituents	C1	C2	C3	C4	C5	C6	C7	C8	C9	C10	Substituent Shifts	References
151	2=0/A5-Br	48.7	210.5	37.2	35.5	58.8	48.5	32.7	25.3	24.7	24.9	19.2,28.6	108
152	2=0/A5-Me E5-Me	49.2	212.2	38.1	41.7	32.6	52.2	27.5	25.9	25.1	25.6	19.0,28.7,23.2	78
153	2=0/A5-Me E5-Me/8-Me 8=	45.0	211.9	37.9	41.1	32.2	47.3	31.5	132.1	119.1	25.2	18.9,28.5	78,88
154	2=0/A5-Me E5-Me/8=	44.8	210.9	37.6	40.8	32.0	46.7	26.3	124.6	124.8	24.9	19.6,28.9,21.4	78
155	2=0/A5-Me E5-Me/8=/E10-Me	52.9	211.6	39.1	42.6	32.8	48.0	26.1	123.3	131.9	29.4		78
156	2=0/A5-Me E5-Me/E10-Me	55.8	213.0	39.8	43.7	33.2	53.2	27.4	25.2	34.3	30.2	19.5,28.8,21.2	78
157	2=0/A5-OH	47.6	213.7	36.0	33.8	68.5	48.5	29.6	25.6	25.0	25.0		108
158	2=0/A7-OH	47.8	214.6	41.5	26.0	28.4	48.2	69.3	32.9	18.6	24.8		108
159	2=0/E10-Me	61.7	212.5	43.0	27.7	33.5	45.8	34.0	25.1	34.5	30.0	20.8	78
160	2=0/E10-Ph	59.5	210.6	42.5	27.5	33.3	45.8	33.7	25.4	35.2	42.3	146.2	89
161	2=0/E3-COMe	55.8	209.4	63.5	28.6	31.7	45.8	34.3	25.6	25.2	25.0	206.4	46
162	2=0/E3-COMe A3-Me	51.1	213.2	61.7	33.7	28.4	44.3	34.3	25.3	25.5	25.6	208.5,20.0	47
163	2=0/E3-COOH	55.3	207.0	57.4	29.3	31.7	45.2	34.3	25.6	25.2	25.0	170.7	45
164	2=0/E4-C(Me)=CH2/6-Me/E7-Me/E9-OH	56.1	211.1	46.0	41.5	42.8	40.7	41.3	39.3	69.5	30.2	147.4,12.9,15.5	51,57
165	2=0/E4-OH/A5-Me/6-Me/A8-C(Me)=CH₂/10=	141.4	200.8	43.0	66.4	46.4	38.8	44.4	40.6	30.3	134.8	8.9,30.9,148.0	3
166	2=0/E5-Me/6-Me/A8-C(Me)=CH₂/10=	147.9	205.5	40.3	29.3	38.1	40.7	39.5	37.3	30.9	128.9	15.6,21.1,148.8	116
167	2=0/E7-Br	54.4	209.0	40.6	25.3	31.0	51.6	58.3	38.0	24.7	24.9		108
168	2=0/E7-OH	53.2	211.8	41.4	25.9	28.1	51.5	74.9	35.1	22.6	24.4		108
169	E2-Br/3=0	52.1	65.3	199.6	34.3	34.3	43.7	33.7	26.5	26.5	30.0		16
170	E2-Br/3=0/E4-Br	51.5	62.5	191.9	53.7	45.6	43.7	33.8	26.2	25.7	32.8		16
171	A2-CH₂OOC-3A E2-Me/5=	46.6	41.5	41.4	20.8	116.0	142.1	36.8	26.5	29.2	30.0	75.0,176.6,20.0	124
172	A2-CH₂OH/E3-COOH/5=	42.2	42.6	42.6	28.5	118.9	139.9	36.1	28.3	27.6	30.8	61.2,178.9	109
173	A2-CH₂OOC-3E/5=	34.9	36.5	38.5	21.8	116.2	138.2	35.2	27.4	26.6	30.2	69.1,179.6	109
174	E2-C1/3=0	51.8	70.6	202.0	40.8	33.8	42.7	33.2	26.3	26.3	32.4		16
175	E2-C1/3=0/E4-Me	52.3	70.8	202.0	44.4	42.9	42.3	33.1	26.4	25.8	32.4		16
176	E2-COEt/3=0	45.0	68.7	209.1	40.2	33.0	41.9	33.2	26.0	25.4	32.6	208.1	47
177	A2-COOCH₂-3A/5=	37.4	45.0	35.0	28.5	116.2	143.4	36.3	28.7	26.3	29.4	176.2,72.0	122
178	A2-COOCH₂-3E A3-Me	37.3	52.2	37.4	32.8	119.0	136.7	35.8	27.9	27.4	30.9	177.2,70.6,23.0	109,123
179	A2-COOCH₂-3E/5=	36.0	41.3	33.7	24.0	115.9	138.9	35.2	27.7	27.1	30.8	177.5,71.8	109
180	A2-COOH E2-Me/A3-COOH/5=	50.5	47.6	41.9	27.6	116.5	141.7	37.9	28.0	30.5	33.0	179.6,18.5,178.6	124
181	A2-COOH E2-Me/E3-COOH/5=	47.4	46.5	50.4	28.0	119.3	137.6	36.3	27.7	27.5	28.5	177.5,24.7,177.2	109,123,124
182	A2-COOH/A3-CH₂OH/5=	40.7	46.9	34.4	28.8	118.0	141.0	37.2	29.8	27.7	31.8	177.3,65.5	122
183	A2-COOH/A3-COOH/5=	40.3	47.1	39.1	29.9	116.5	142.2	37.0	29.0	27.6	31.8	179.4,177.0	122
184	A2-COOH/E3-CH₂OH/5=	39.8	46.6	40.2	26.5	119.9	138.1	36.1	28.0	27.4	31.2	176.8,65.6	109
185	A2-COOH/E3-COOH A3-Me/5=	36.6	51.0	43.8	31.8	118.9	135.6	35.6	27.6	27.2	31.3	180.5,177.0,24.5	109,123
186	A2-COOH/E3-COOH/5=	39.6	45.8	42.1	25.5	119.4	137.4	35.7	27.6	27.0	31.0	177.4,176.4	109
187	A2-COOMe	44.7	44.5	29.4	21.8	34.8	36.3	34.1	27.0	26.3	31.1	175.0	6
188	E2-COOMe	44.6	50.2	30.3	25.4	33.5	42.1	34.1	26.5	26.4	31.4	176.2	6
189	A2-COOMe E2-Me/A3-COOMe/5=	49.5	47.0	40.8	26.4	115.4	140.5	37.0	27.0	29.4	31.9	176.4,18.2,175.4	109,123,124
190	A2-COOMe E2-Me/E3-COOMe/5=	46.8	46.1	49.5	27.1	118.3	136.7	35.5	26.8	26.2	27.7	174.6,24.2,174.1	122
191	A2-COOMe/A3-COOMe/5=	39.6	46.1	38.0	29.0	115.7	141.2	36.3	29.0	26.7	30.9	176.4,174.4	109,123
192	A2-COOMe/E3-COOMe A3-Me/5=	35.9	49.8	43.5	30.9	117.9	134.5	34.6	26.6	26.3	30.2	177.4,173.8,23.7	109,123
193	A2-COOMe/E3-COOMe/5=	39.1	44.7	41.7	25.0	118.8	136.5	35.1	26.7	26.4	30.3	174.2,172.8	109
194	A2-COOOC-3A E2-Me/5=	45.4	47.7	42.9	20.3	114.9	142.1	36.1	26.1	29.3	29.6	172.9,170.5,20.6	124
195	A2-COOOC-3A/5=	37.8	46.7	39.2	25.2	116.3	143.4	36.2	28.5	26.0	29.2	170.8,169.3	122
196	A2-COOOC-3E A3-Me/5=	34.6	50.1	44.6	31.9	115.5	139.3	33.5	25.7	25.5	28.3	175.9,170.4,21.5	109,123
197	A2-COOOC-3E/5=	35.0	38.6	34.6	21.4	115.9	140.0	34.8	26.8	29.9	29.9	174.1,172.0	109
198	A2-Me	46.8	33.3	33.3	20.9	35.3	36.1	35.3	27.1	27.5	31.8	13.4	4
199	E2-Me	50.0	38.0	34.6	26.6	34.9	43.3	34.8	27.0	27.2	30.8	19.8	4+102,2
200	2-Me 2= 4=	43.4	140.0	124.1	131.1	125.6	40.8	32.9	26.7	27.0	29.6	19.9	36

217

trans-BICYCLO[4.4.0]DECANES (cont.)

Str	Substituents	C1	C2	C3	C4	C5	C6	C7	C8	C9	C10	Substituent Shifts	References
201	E2-Me A2-CH$_2$OOC-3E/5=	43.3	40.7	43.9	22.1	115.2	138.6	35.8	27.7	27.0	28.9	25.3,74.2,179.4	109,123,124
202	E2-Me A2-COOOC-3E/5=	44.1	47.9	44.0	19.0	114.3	138.6	36.7	28.3	27.0	32.7	22.9,174.9,172.3	109,123,124
203	A2-Me E2-Me/3-=/6-Me	52.9	47.2	215.3	34.3	40.3	33.5	43.6	21.2	26.5	22.4	20.7,24.7,17.7	83
204	A2-Me E2-Me/3-=0/6-Me/10=	147.6	48.0	214.6	33.3	34.7	33.8	38.1	17.5	24.7	123.4	28.3,26.6,23.7	83
205	A2-Me E2-Me/E3-OH/6-Me	52.2	38.4	78.6	27.4	39.8	33.8	44.8	27.3	21.5	21.4	27.4,14.9,19.0	67
206	A2-Me E2-Me/E3-OH/6-Me	52.7	38.8	79.4	27.4	40.3	34.1	45.2	21.7	27.4	21.7	15.2,27.4,19.4	83
207	A2-Me/3-=0 4=/6-Me/E9-C(Me)$_2$OH	44.5	45.3	204.0	125.9	160.4	36.2	40.0	22.4	49.1	26.1	13.4,20.8,72.6	54,31,50,53,56
208	E2-Me/3-=0 4=/6-Me/E9-C(Me)$_2$OH	48.3	42.7	202.3	126.3	160.4	36.2	37.9	21.9	48.5	25.2	11.8,17.2,72.6	54
209	E2-Me/3-=0/6-Me	51.5	45.6	213.0	38.2	41.0	33.9	41.9	21.4	26.3	25.9	11.3,16.3	83
210	A2-Me/A3-Me	40.2	39.3	35.8	26.4	28.0	36.2	35.1	27.0	27.6	31.6	15.5,19.5	102
211	E2-Me/A3-Me	42.8	40.4	35.1	33.8	28.8	44.2	34.6	27.0	27.3	30.9	17.5,12.6	102
212	E2-Me/A4-Me	50.7	31.6	42.2	28.3	40.5	36.7	34.9	27.0	27.2	30.6	20.0,19.2	102
213	E2-Me/A8-Me	50.6	37.8	36.6	26.6	35.0	36.6	34.6	28.0	32.4	24.7	19.8,18.3	102
214	E2-Me/A9-Me	43.5	37.6	36.6	26.6	34.7	43.7	28.2	32.1	28.7	36.5	19.8,18.5	102
215	A2-Me/E10-Me	52.5	28.6	34.1	20.8	35.5	35.5	35.5	26.5	36.8	34.6	12.6,19.2	5,2
216	A2-Me/E3-Me	48.1	39.3	37.4	28.7	35.2	35.2	35.2	27.0	27.4	31.6	6.9,20.3	102
217	E2-Me/E3-Me	49.2	44.3	39.2	35.8	34.5	35.2	34.8	26.7	27.2	31.0	16.1,21.1	102,2
218	A2-Me/E4-Me	46.1	33.8	43.5	26.5	44.2	35.9	35.1	26.9	27.5	31.4	14.1,23.1	102
219	E2-Me/E4-Me	49.3	37.6	45.4	32.5	43.6	42.9	34.6	26.8	27.1	30.6	19.8,22.9	102,2
220	A2-Me/E5-Me	46.0	33.6	34.0	30.3	38.9	42.1	31.7	27.3	28.3	32.0	13.4,19.9	5
221	E2-Me/E5-Me	49.3	37.9	36.0	36.0	37.9	49.3	30.9	26.9	26.9	30.9	20.0,20.0	5
222	A2-Me/E7-Me	46.0	33.5	34.4	20.9	31.7	42.3	38.9	36.6	27.0	32.3	13.3,20.0	5
223	E2-Me/E7-Me	49.5	38.0	36.4	26.7	31.0	49.5	38.0	36.4	26.5	31.0	20.3,20.3	5
224	A2-Me/E8-Me	46.0	32.9	36.4	20.7	35.1	35.5	44.0	28.0	36.0	31.4	13.3,22.9	102
225	A2-Me/E8-Me	49.3	37.7	36.5	26.5	34.7	42.8	43.5	32.9	35.8	30.6	20.0,22.8	102
226	A2-Me/E9-Me	46.1	33.2	34.4	20.8	35.0	35.4	34.9	35.6	33.6	40.6	13.3,23.0	102
227	E2-Me/E9-Me	49.4	37.7	36.4	26.4	34.4	42.6	34.4	35.4	33.1	39.4	19.8,23.0	102
228	E2-NMe$_2$	45.9	67.4	29.6	25.5	34.4	43.0	34.4	26.6	26.6	21.2		75
229	A2-OAc/6-Me/10=	139.6	76.3	32.2	17.3	41.4	34.0	40.7	18.6	26.0	128.8	26.0	44,71
230	E2-OAc/6-Me/10=	140.8	71.6	34.2	20.3	41.2	35.8	40.2	18.6	25.7	116.6	24.9	44,71
231	E2-OAc/E5-OAc/E7-Ph	45.1	74.2	28.1	29.3	74.7	47.3	49.1	37.0	25.2	28.7	146.7	89
232	A2-OH	47.4	74.3	34.3	20.0	33.8	35.6	33.9	26.4	26.8	29.6		1
233	E2-OH	50.4	74.6	35.8	24.1	33.7	41.2	33.5	26.2	26.4	29.1		1
234	E2-OH A2-Me	53.3	72.7	43.0	23.4	43.0	34.4	34.3	26.8	26.5	25.6	21.3	7
235	A2-OH E2-Me	51.0	40.9	40.9	21.5	34.7	37.3	37.3	34.3	26.4	25.3	28.4	7
236	A2-OH/6-Me/8-=/9-iPr	47.3	71.6	41.6	18.8	41.3	32.4	44.6	116.6	142.5	30.2	23.3,18.0,35.1	85,84
237	A2-OH E2-Me/6=	46.3	72.2	40.4	22.5	35.2	137.0	124.6	26.8	23.1	25.2	22.0	61
238	A2-OH E2-Me/A5-Me/6=	41.7	72.4	35.1	27.9	37.3	141.2	124.5	26.8	23.0	25.3	22.1,19.9	61
239	A2-OH E2-Me/E5-Br/6-Me/8=/9-iPr	48.2	72.2	42.7	24.3	68.4	38.4	124.5	116.2	141.9	30.2	29.8,14.0,34.8	85,84,86
240	A2-OH E2-Me/E5-Br/6-Me/A9-Cl E9-iPr	47.4	71.4	42.7	30.3	67.5	39.4	33.0	33.8	81.0	37.3	29.8,14.3,40.9	85
241	A2-OH E2-Me/E5-Br/6-Me/E9-iPr/E10-Cl	58.4	72.9	44.4	30.4	67.6	42.4	39.9	19.3	52.3	63.6	35.7,14.9,27.2	87
242	A2-OH E2-Me/E5-Me/6=	46.6	72.6	40.9	21.9	37.0	140.7	121.9	26.8	26.0	25.3	21.6,18.2	61
243	A2-OH/5-Me 5=	43.5	69.5	28.6	28.1	122.9	131.0	30.0	27.7	26.4	29.2	18.5	79
244	E2-OH/5-Me 5=	46.9	73.7	29.8	29.6	123.3	130.6	32.7	27.2	26.2	30.1	18.5	79
245	A2-OH/6-Me	48.5	71.8	34.1	16.9	43.7	33.7	41.7	21.9	27.3	26.0	19.1	61
246	E2-OH/6-Me	52.4	70.8	36.6	20.4	41.9	34.8	41.2	21.7	26.7	23.0	16.8	1,3
247	A2-OH/6-Me/10=	144.4	74.5	34.1	16.8	41.9	34.0	40.8	18.7	26.0	125.7	26.8	44
248	E2-OH/6-Me/10=	145.5	69.6	37.9	20.5	41.5	35.6	40.3	18.8	25.7	115.8	25.1	44
249	A2-OH/6=	42.1	71.3	25.2	20.9	34.8	136.0	124.3	26.4	21.9	25.2		61
250	A2-OH/6=/A9-Me	40.7	73.0	33.8	21.4	34.8	135.1	122.2	32.7	26.2	32.7	21.0	61

trans-BICYCLO[4.4.0]DECANES (cont.)

Str	Substituents	C1	C2	C3	C4	C5	C6	C7	C8	C9	C10	Substituent Shifts	References
251	A2-OH/6=/A9-t-Bu	40.7	74.1	34.0	21.5	35.0	135.3	123.3	28.6	41.6	25.1	32.1	61
252	A2-OH/6=/E9-Me	43.2	69.8	33.1	20.6	34.1	136.1	123.5	34.6	28.7	33.4	22.4	61
253	A2-OH/6=/E9-t-Bu	43.5	69.5	32.8	19.8	33.5	135.3	123.2	26.9	43.5	26.4	31.7	61
254	A2-OH/A3-Br/6=	35.5	74.5	53.8	28.9	29.3	134.2	125.1	25.7	21.7	25.0		61
255	A2-OH/A3-Me/6=	36.6	76.4	33.5	26.2	29.3		123.9	26.6	22.1	25.2	15.9	61
256	E2-OH/A5-OH/E7-Ph	41.9	73.7	28.6	30.9	63.6	44.4	48.9	34.6	25.0	28.6	145.2	89
257	A2-OH/E10-Me	53.1	66.0	32.5	19.9	34.0	36.2	34.8	26.0	35.2	34.1	19.1	107
258	E2-OH/E10-Ph	54.2	74.8	34.6	23.3	33.2	41.2	33.3	25.8	36.5	49.8	146.4	89
259	A2-OH/E3-Br/6=	41.8	78.4	58.7	31.6	35.3	132.5	125.0	26.5	21.6	25.0		61
260	A2-OH/E3-Br/E5-Me/6=	42.0	75.0	58.1	40.3	37.8	136.4	122.3	27.5	21.5	25.2	17.3	61
261	A2-OH/E3-Me/6=	42.7	75.6	37.4	28.7	34.6	135.5	124.3	26.4	21.9	25.2	18.5	61
262	A2-OH/E3-Me/E5-Me/6=	43.0	76.3	37.2	38.2	36.5		121.9	26.6	21.7	25.3	18.1,17.8	61
263	A2-OH/E4-OH/A5-Me/6=Me/A8-C(CH2OH)=CH2	141.4	37.3	199.8	126.8	160.6	40.3	47.1	37.1	31.9	129.3	9.6,32.5,154.7	63
264	A2-OH/E4-OH/A5-Me/6=Me/A8-C(Me)=CH2/10=	140.3	44.8	210.6	38.1	47.0	39.1	41.0	21.6	30.4	128.8	8.9,32.1,149.1	3,59,104
265	E2-OH/E4-OH/A5-Me/6=Me/A8-C(Me)=CH2/10=	140.6	67.2	39.8	67.0	47.1	38.7	44.3	39.4	29.9	118.9	9.4,30.1,149.4	3,59
266	A2-OH/E4-OH/A5-Me/6=Me/A8-iPr/10=	140.2	75.2	36.3	65.4	47.7	38.8	43.7	39.3	29.2	129.1	9.1,32.4,32.8	3
267	A2-OH/E5-Me/6=	42.4	71.3	30.2	30.2	37.0	139.6	121.6	26.9	21.7	25.4	18.0	61
268	E2-OH/E5-OAc/E7-Ph	47.7	73.0	32.9	28.8	75.4	48.4	49.7	37.6	25.9	30.1	147.5	89
269	E2-OH/E5-OCH2Ph/E10-Ph	52.0	74.1	31.7	28.3	80.5	46.4	28.5	25.6	36.4	50.1	146.0	89
270	E2-OH/E5-OH/E7-Ph	47.9	73.1	31.9	31.9	74.5	51.8	50.1	36.4	25.7	28.6	146.0	89
271	2= 3-Me 4=	40.9	127.8	131.7	133.4	128.0	41.1	32.2	26.7	26.8	32.6	20.9	36
272	3-=0	43.3	48.5	210.5	41.6	33.6	41.4	34.2	26.1	25.6	32.7		15+27,16,46
273	3-=0 4=/6-Me/10-=CH2	47.7	37.3	199.8	126.8	160.6	38.3	37.8	23.0	35.5	147.5	17.1,107.5	112
274	3-=0/6-Me	44.6	44.8	210.6	38.1	40.6	33.6	41.0	26.0	26.0	29.0	14.9	3,15,47,66
275	3-=0/6-Me/A10-Me	46.4	43.8	211.6	38.2	40.6	33.4	43.7	17.2	32.8	32.8	18.6,14.1	11
276	3-=0/6-Me/A7-OCH2CH2O-7 E	41.4	44.1	211.1	37.7	30.4	41.7	112.3	30.2	22.7	28.1	13.0	111
277	3-=0/6-Me/A9-t-Bu	41.0	44.9	210.1	38.2	39.7	33.7	37.8	21.2	41.4	29.2	17.2,31.7	10
278	3-=0/6-Me/E10-Me	50.6	40.9	212.0	40.3	40.9	33.2	41.2	21.4	35.6	32.3	15.9,19.6	11
279	3-=0/6-Me/E7-Me	46.2	44.6	211.6	38.5	38.1	36.0	42.5	30.6	26.3	29.4	9.8,15.6	3
280	3-=0/A4-Br	43.6	42.7	202.7	52.3	41.8	36.5	34.1	26.5	26.5	32.5		16
281	3-=0/A4-C CH	44.4	44.8	204.8	43.4	39.8	37.9	34.4	26.6	26.6	32.8	82.9	16
282	3-=0/A4-Cl	43.8	43.0	203.1	61.2	41.7	36.0	34.2	26.5	26.6	32.6		16
283	3-=0/A4-Cl E4-Me	44.1	43.7	203.2	70.9	49.7	38.3	34.0	26.5	26.1	32.6		16
284	3-=0/A4-COMe E4-Me/6=Me	45.5	44.7	210.3	62.5	52.2	34.3	40.2	21.2	25.9	28.3	209.7,22.7,16.0	47
285	3-=0/A4-F	44.6	45.1	205.4	93.7	40.3	36.4	34.4	21.6	26.6	32.9		16
286	3-=0/A4-Me	43.6	44.9	212.4	44.9	40.3	36.6	34.5	26.8	26.7	33.6	17.7	16
287	3-=0/A4-OAc	44.7	45.3	205.3	77.0	39.1	37.1	34.5	26.6	26.6	33.0		16
288	3-=0/A4-OMe	44.7	44.9	209.9	84.4	40.3	36.6	34.5	26.7	26.7	33.1		16
289	3-=0/A4-Ph	44.1	46.5	209.6	54.5	37.1	46.7	34.2	26.6	26.6	33.4	139.2	16
290	3-=0/E4-Br	43.7	47.6	199.4	56.4	46.3	43.6	34.0	26.3	26.4	32.6		16
291	3-=0/E4-Cl	43.8	47.8	200.1	64.0	45.4	42.5	34.0	26.3	26.4	32.6		16
292	3-=0/E4-Cl A4-Me	43.7	45.3	202.6	74.1	51.7	33.3	34.0	26.4	26.0	32.8		16
293	3-=0/E4-COMe A4-Me	43.0	45.4	212.0	62.4	41.7	37.1	33.7	26.1	25.6	32.8	208.0,20.5	47
294	3-=0/E4-F	43.7	46.8	203.2	92.4	40.8	40.2	34.0	26.4	26.5	32.9		47
295	3-=0/E4-Me	44.6	48.7	210.0	44.9	43.4	42.7	34.5	26.7	26.7	33.1	14.6	16
296	3-=0/E4-NO2	43.4	47.2	198.2	92.2	38.0	40.0	33.9	26.3	26.3	32.6		16
297	3-=0/E4-O-5A/6-=Me/10-=CH2	37.5	36.3	204.8	55.5	63.8	35.9	36.5	22.9	36.2	147.0	15.0,107.4	112
298	3-=0/E4-OAc	44.0	47.0	202.8	76.2	39.6	40.7	34.1	26.4	26.6	33.0		16
299	3-=0/E4-OMe	44.3	47.7	207.2	83.9	41.3	34.2	34.2	26.4	26.6	33.0		16
300	3-=0/E4-Ph	44.5	49.1	207.3	57.3	42.8	42.9	34.5	26.7	26.7	33.1	140.0	16

trans-BICYCLO[4.4.0]DECANES (cont.)

Str	Substituents	C1	C2	C3	C4	C5	C6	C7	C8	C9	C10	Substituent Shifts	References
301	A3-Br/A4-Br	35.6	36.9	53.9	53.9	36.9	35.6	32.7	26.4	26.4	32.7		46
302	E3-C(Me)2OH/6-Me/A10-Me	47.0	28.0	50.1	22.7	41.6	33.6	44.6	17.4	33.6	33.7	72.8,19.5,14.8	64
303	E3-C(Me)2OH/6-Me/E10-Me	51.3	24.9	49.6	22.3	41.7	33.4	41.9	21.6	36.7	31.5	73.0,16.7,20.1	64
304	E3-CN/4-=O	43.3	38.3	41.5	199.6	47.5	43.5	32.4	26.4	26.5	34.2	117.2	16
305	A3-COOMe	39.5	34.6	39.1	27.3	30.2	43.1	33.8	26.5	26.4	33.6	175.4	6
306	E3-COOMe	42.4	36.3	42.6	29.1	33.1	43.4	33.8	26.6	26.6	33.7	175.9	6
307	A3-Me	37.3	40.2	28.6	32.3	28.3	44.5	34.4	27.1	27.1	34.7	18.5	4+102
308	E3-Me	43.4	43.3	33.1	35.7	34.2	43.4	34.4	26.9	26.9	34.4	22.9	4,2
309	3-Me 3=/7-=CH2/E10=iPr	41.6	36.4	132.7	120.1	29.7	49.2	152.6	36.6	25.8	28.9	21.8,104.6,26.2	39
310	A3-Me/A8-Me	37.9	40.0	28.7	32.3	28.3	37.9	40.0	28.7	32.3	28.3	18.5,18.5	5
311	A3-Me/E4-Me	36.8	42.0	34.2	35.5	36.7	44.3	34.4	27.0	27.0	33.9	12.6,20.2	5
312	E3-Me/E4-Me	43.6	43.6	39.4	39.4	43.6	43.6	34.0	26.8	26.8	34.0	20.2,20.2	5,2
313	A3-Me/E8-Me	36.7	39.8	28.4	32.1	28.2	44.0	43.2	33.1	35.7	34.4	18.4,22.9	5
314	E3-Me/E8-Me	43.0	43.0	33.1	35.6	34.2	43.0	43.0	33.1	35.6	34.2	22.8,22.8	5
315	A3-Me/E9-Me	36.9	40.2	28.2	32.3	28.2	44.0	34.3	35.8	33.3	43.6	18.5,22.9	5
316	E3-Me/E9-Me	43.3	43.3	33.1	35.7	34.0	42.9	34.0	35.7	33.1	43.3	22.8,22.8	5,2
317	A3-NMe2	36.7	37.1	61.9	28.0	29.5	44.2	34.2	26.7	26.7	33.6		75
318	A3-OH	36.4	40.4	66.6	33.9	27.6	43.2	33.8	26.7	26.7	32.9		1
319	E3-OH	41.2	43.0	70.2	35.6	32.0	42.3	33.8	26.5	26.3	33.3		1,46
320	A3-OH/6-Me	37.8	36.0	66.6	28.6	35.6	33.7	41.5	21.9	27.0	28.8	14.7	1
321	E3-OH/6-Me	43.1	38.1	71.0	31.2	40.0	33.0	41.1	21.9	26.7	28.8	15.7	1,3,7
322	E3-OH/6-Me/A7-Me	38.5	38.5	71.1	31.4	35.0	35.3	35.6	29.3	20.9	29.3	18.5,14.1	7
323	E3-OH/6-Me/E7-Me	44.5	38.2	71.3	31.5	37.2	35.9	43.1	30.9	26.7	29.1	10.6,15.5	7,3
324	E3-OH/A4-NMe2	36.5	35.5	67.2	66.9	30.5	36.8	33.5	26.6	26.6	33.7		75
325	E3-OH/A4-NMe2	40.6	38.4	69.8	62.0	31.3	39.1	33.6	26.4	26.3	34.0		75
326	A3-OH/A4-NMe3 I	35.5	38.3	64.6	74.8	29.4	37.0	33.6	26.4	26.6	34.0		75
327	E3-OH/A4-NMe3 I	35.4	40.5	65.9	73.3	27.6	36.3	34.7	26.4	26.4	34.9		75
328	A3-OH/E4-NMe2	35.3	38.1	65.1	66.6	32.5	41.5	33.3	26.6	26.7	33.9		75
329	E3-OH/E4-NMe2	40.6	40.9	69.4	69.1	27.5	42.2	33.4	26.4	26.4	33.6		75
330	A3-OH/E4-NMe3 I	35.8	42.0	66.1	75.5	28.2	40.8	32.8	26.4	26.9	33.6		75
331	E3-OH/E4-NMe3 I	40.4	43.1	70.9	78.2	32.6	41.7	32.9	26.3	26.1	32.9		75
332	A3-OH/E4-OH	35.0	38.8	68.7	70.9	36.0	40.8	33.3	25.9	26.3	32.8		46
333	3=	34.1	38.5	126.8	126.8	38.5	34.1	34.1	26.7	26.7	33.4		46
334	E4-C(Me)=CH2/6-Me/E7-Me	144.3	122.9	31.3	37.7	43.3	38.7	44.1	31.2	27.9	32.6	150.3,18.1,15.7	3

220

References to Bicyclo[4.4.0]decanes and related* systems

1. Grover, S. H. and Stothers, J. B., *Can. J. Chem.* **52**, 870 (1974).
2. Dalling, D. K., Grant, D. M. and Paul, E. G., *J. Am. Chem. Soc.* **95**, 3718 (1973).
3. Birnbaum, G. I., Stoessl, A., Grover, S. H. and Stothers, J. B., *Can. J. Chem.* **52**, 993 (1974).
4. Pehk, T. I., Laht, A. K., Musaev, I. A., Kurashova, E. K. and Sanin, P. I., *Neftekhimiya* **14**, 541 (1974).
5. Pehk, T. I., Laht, A. K., Musaev, I. A., Kurashova, E. K. and Sanin, P. I., *Neftekhimiya* **15**, 329 (1975).
6. Gordon, M., Grover, S. H. and Stothers, J. B., *Can. J. Chem.* **51**, 2092 (1973).
7. Ayer, W. A., Browne, L. M., Fung, S. and Stothers, J. B., *Org. Magn. Reson.* **11**, 73 (1978).
8. Ayer, W. A. and Paice, M. G., *Can. J. Chem.* **54**, 910 (1976).
9. Browne, L. M., Klinck, R. E. and Stothers, J. B., *Org. Magn. Reson.* **12**, 561 (1979).
10. House, H. O. and Lusch, M. J., *J. Org. Chem.* **42**, 183 (1977).
11. Caine, D. and Smith, Jr. T. L., *J. Org. Chem.* **43**, 755 (1978).
12. Metzger, P., Cabestaing, C., Casadevall, E. and Casadevall, A., *Org. Magn. Reson.* **19**, 144 (1982).
13. Dalling, D. K. and Grant, D. M., *J. Am. Chem. Soc.* **96**, 1827 (1974).
14. Boaz, H., *Tetrahedron Lett.* 55 (1973).
15. Grover, S. H., Marr, D. H., Stothers, J. B. and Tan, C. T., *Can. J. Chem.* **53**, 1351 (1975).
16. Metzger, P., Casadevall, E., Casadevall, A. and Pouet, M.-J., *Can. J. Chem.* **58**, 1503 (1980).
17. Heng, K. K. and Smith, R. A. J., *Tetrahedron* **35**, 425 (1979).
18. Smith, R. A. J. and Hannah, D. J., *Tetrahedron* **35**, 1183 (1979).
19. Smith, R. A. J. and Hannah, D. J., *Synth. Commun.* **9**, 301 (1979).
20. Blunt, J. W., Coxon, J. M., Lindley, N. B. and Lane, G. A., *Aust. J. Chem.* **29**, 967 (1976).
21. Dauben, W. G. and Michno, D. M., *Tetrahedron* **37**, 3263 (1981).
22. Pehk, T., Laht, A. and Lippmaa, E., *Org. Magn. Reson.* **19**, 21 (1982).
23. Schneider, H.-J. and Gschwendtner, W., *J. Org. Chem.* **47**, 4216 (1982).
24. Oppolzer, W., Snowden, R. L. and Simmons, D. P., *Helv. Chim. Acta* **64**, 2002 (1981).
25. Lewis, P. H., Middleton, S., Rosser, M. J. and Stock, L. E., *Aust. J. Chem.* **32**, 1123 (1979).
26. Benkeser, R. A., Belmonte, F. G. and Kang, J., *J. Org. Chem.* **48**, 2796 (1983).
27. Abraham, R. J., Bergen, H. A. and Chadwick, D. J., *Tetrahedron* **38**, 3271 (1982).
28. Onopchenko, A., Cupples, B. L. and Kresge, A. N., *Macromolecules* **15**, 1201 (1982).
29. Becker, K. B., *Helv. Chim. Acta* **60**, 68 (1977).
30. Stoessl, A., Stothers, J. B. and Ward, E. W. B., *Can. J. Chem.* **56**, 645 (1978).
31. Stoessl, A., Stothers, J. B. and Ward, E. W. B., *Can. J. Chem.* **53**, 3351 (1975).
32. Masamune, T., Murai, A., Takasugi, M., Matsunaga, A., Katsui, N., Sato, N. and Tomiyama, K., *Bull. Chem. Soc. Japan* **50**, 1201 (1977).
33. Stoessl, A., Ward, E. W. B. and Stothers, J. B., *Tetrahedron Lett.* 3271 (1976).
34. Geneste, P., Durand, R., Kamenka, J.-M., Beierbeck, H., Martino, R. and Saunders, J. K., *Can. J. Chem.* **56**, 1940 (1978).
*35. Dauben, W. G., Michno, D. M. and Olsen, E. G., *J. Org. Chem.* **46**, 687 (1981).

36. Dauben, W. G. and Michno, D. M., *J. Am. Chem. Soc.* **103**, 2284 (1981).
*37. Warner, P. M., Ah-King, M. and Palmer, R. F., *J. Am. Chem. Soc.* **104**, 7166 (1982).
*38. Gunther, H., Schmickler, H., Bremser, W., Straube, F. A. and Vogel, E., *Angew. Chem., Int. Ed. Engl.* **12**, 570 (1973).
39. Sakurai, H., Hosomi, A., Saito, M., Sasaki, K., Iguchi, H., Sasaki, J. and Araki, Y., *Tetrahedron* **39**, 883 (1983).
*40. Paquette, L. A., Jendralla, H., DeLucca, G. and , *J. Am. Chem. Soc.* **106**, 1518 (1984).
41. Davies, S. G. and Whitham, G. H., *J. Chem. Soc. Perkin Trans. II* 861 (1975).
42. Garratt, D. G. and Kabo, A., *Can. J. Chem.* **58**, 1030 (1980).
43. Kelly, D. P., Underwood, G. R. and Barron, P. F., *J. Am. Chem. Soc.* **98**, 3106 (1976).
44. Coxon, J. M. and Gibson, J. R., *Aust. J. Chem.* **32**, 2223 (1979).
45. Lewis, P. H., Middleton, S., Rosser, M. J. and Stock, L. E., *Aust. J. Chem.* **33**, 1049 (1980).
46. Galteri, M., Lewis, P. H., Middleton, S. and Stock, L. E., *Aust. J. Chem.* **33**, 101 (1980).
47. Middleton, S. and Stock, L. E., *Aust. J. Chem.* **33**, 2467 (1980).
48. Coxon, J. M. and Gibson, J. R., *Aust. J. Chem.* **34**, 1451 (1981).
49. Blunt, J. W., Coxon, J. M. and Gibson, J. R., *Aust. J. Chem.* **34**, 2469 (1981).
50. Murai, A., Abiko, A., Ono, M. and Masamune, T., *Bull. Chem. Soc. Japan* **55**, 1191 (1982).
51. Katsui, N., Yagihashi, F., Murai, A. and Masamune, T., *Bull. Chem. Soc. Japan* **55**, 2428 (1982).
52. Lippmaa, E. and Pehk, T., *Eesti NSV Tead. Akad. Toim., Keem., Geol.* **17**, 287 (1968).
53. Stoessl, A., Stothers, J. B. and Ward, E. W. B., *J. Chem. Soc. Chem. Commun.* 709 (1974).
54. Kelly, R. B., Alward, S. J., Murty, K. S. and Stothers, J. B., *Can. J. Chem.* **56**, 2508 (1976).
55. Stoessl, A. and Stothers, J. B., *Can. J. Chem.* **61**, 1766 (1983).
56. Murai, A., Abiko, A., Ono, M., Katsui, N. and Masamune, T., *Chem. Lett.* 1209 (1978).
57. Katsui, N., Yagihashi, F., Murai, A. and Masamune, T., *Chem. Lett.* 1455 (1980).
58. Murai, A., Taketsuru, H., Yagihashi, F., Katsui, N. and Masamune, T., *Bull. Chem. Soc. Japan* **53**, 1045 (1980).
59. Stillman, M. J., Stothers, J. B. and Stoessl, A., *Can. J. Chem.* **59**, 2303 (1981).
60. Berger, J., Yoshioka, M., Zink, M. P., Wolf, H. R. and Jeger, O., *Helv. Chim. Acta* **63**, 154 (1980).
61. Snider, B. B. and Deutsch, E. A., *J. Org. Chem.* **48**, 1822 (1983).
*62. Fringuelli, F., Hagaman, E. W., Moreno, L. N., Taticchi, A. and Wenkert, E., *J. Org. Chem.* **42**, 3168 (1977).
63. Ward, E. W. B., Stoessl, A. and Stothers, J. B., *Phytochem.* **16**, 2024 (1977).
64. Gerber, N. C. and Denney, D. Z., *Phytochem.* **16**, 2025 (1977).
65. Toyota, M., Asakawa, Y. and Takemoto, T., *Phytochem.* **20**, 2359 (1981).
66. Wenkert, E., Buckwalter, B. L., Burfitt, I. R., Gasic, M. J., Gottlieb, H. E., Hagaman, E. W., Schell, F. M. and Wovkulich, P. M., "Topics in Carbon-13 NMR Spectroscopy", G.C. Levy, Ed., 1976, vol. 2, 81ff.
67. Alewood, P. F., Benn, M., Wong, J. and Jones, A. J., *Can. J. Chem.* **55**, 2510 (1977).
*68. Warner, P. M., Palmer, R. F. and Lu, S.-L., *J. Am. Chem. Soc.* **99**, 3773 (1977).
*69. Olah, G. A., Liang, G., Ledlie, D. B. and Costopoulos, M. G., *J. Am. Chem. Soc.* **99**, 4196 (1977).

70. Iwasaki, S., *Helv. Chim. Acta* **61**, 2831 (1978).
71. Coxon, J. M. and Gibson, J. R., *Aust. J. Chem.* **35**, 1165 (1982).
72. "Sadtler Standard Carbon-13 NMR Spectra", Sadtler Research Laboratories, Inc., Philadelphia, Pa.
73. Irikawa, H. and Okumura, Y., *Bull. Chem. Soc. Japan* **51**, 2086 (1978).
74. Irikawa, H. and Okumura, Y., *Bull. Chem. Soc. Japan* **51**, 657 (1978).
75. Laffite, C., Wylde, R. and Mavel, G., *Org. Magn. Reson.* **12**, 77 (1979).
76. Loomes, D. J. and Robinson, M. J. T., *Tetrahedron* **33**, 1149 (1977).
77. Park, H., King, P. E. and Paquette, L. A., *J. Am. Chem. Soc.* **101**, 4773 (1979).
78. Fringuelli, F., Pizzo, F., Taticchi, A., Halls, T. D. J. and Wenkert, E., *J. Org. Chem.* **47**, 5056 (1982).
79. Danheiser, R. L., Martinez-Davila, C. and Sard, H., *Tetrahedron* **37**, 3943 (1981).
80. Bohlmann, F., Jakupovic, J., Muller, L. and Schuster, A., *Angew. Chem., Int. Ed. Engl.* **20**, 292 (1981).
*81. Gadek, T. R. and Vollhardt, K. P. C., *Angew. Chem., Int. Ed. Engl.* **20**, 802 (1981).
82. Booth, H. and Griffiths, D. V., *J. Chem. Soc. Perkin Trans. II* 842 (1973).
83. Buckwalter, B. L., Burfitt, I. R., Nagel, A. A., Wenkert, E. and Naf, F., *Helv. Chim. Acta* **58**, 1567 (1975).
84. Rose, A. F. and Sims, J. J., *Tetrahedron Lett.* 2935 (1977).
85. Rose, A. F., Sims, J. J., Wing, R. M. and Wiger, G. M., *Tetrahedron Lett.* 2533 (1978).
86. Howard, B. M. and Fenical, W., *J. Org. Chem.* **42**, 2518 (1977).
87. Kazlauskas, R., Murphy, P. T., Daly, R. J. and Oberhansli, W. E., *Aust. J. Chem.* **30**, 2679 (1977).
88. Liu, H.-J. and Browne, E. N. C., *Can. J. Chem.* **59**, 601 (1981).
89. Whitesell, J. K. and Chen, H. H., University of Texas at Austin, unpublished results.
90. Hagaman, E. W., *Org. Magn. Reson.* **8**, 389 (1976).
91. Kubo, I., Miura, I., Pettei, M. J., Lee, Y.-W., Pilkiewicz, F. and Nakanishi, K., *Tetrahedron Lett.* 4553 (1977).
*92. Neidlein, R. and Wesch, K. F., *Helv. Chim. Acta* **66**, 891 (1983).
*93. Gunther, H. and Keller, T., *Chem. Ber.* **103**, 3231 (1970).
*94. Ghatak, U. R., Sanyal, B., Ghosh, S., Sarkar. M, Raju, M. S. and Wenkert, E., *J. Org. Chem.* **45**, 1081 (1980).
*95. Seebach, D., Hassig, R. and Gabriel, J., *Helv. Chim. Acta* **66**, 308 (1983).
96. Vogel, E., Breuer, A., Sommerfeld, C.-D., Davis, R. E. and Liu, L.-K., *Angew. Chem., Int. Ed. Engl.* **16**, 169 (1977).
*97. Benkeser, R. A., Laugal, J. A. and Rappa, A., *Tetrahedron Lett.* **25**, 2089 (1984).
*98. Calmes, D., Gorrichon-Guigon, L., Maroni, P., Accary, A., Barret, R. and Huet, J., *Tetrahedron* **37**, 897 (1981).
99. Crandall, J. K., Magaha, H. S., Henderson, M. A., Widener, R. K. and Tharp, G. A., *J. Org. Chem.* **47**, 5372 (1982).
*100. Halton, B. and Officer, D. L., *Aust. J. Chem.* **36**, 1291 (1983).
101. Oritani, T., Ichimura, M., Hanyu, Y. and Yamashita, K., *Agric. Biol. Chem.* **47**, 2613 (1983).
102. Pehk, T. I., Laht, A. K., Musaev, I. A., Kurashova, E. K. and Sanin, P. I., *Neftekhimiya* **16**, 663 (1976).
103. Johnson, L. F. and Jankowski, W. C., "Carbon-13 NMR Spectra", Wiley-Interscience, New York, 1972, spectrum 401.
104. Baker, F. C. and Brooks, C. J. W., *Phytochem.* **15**, 689 (1976).

105. Irikawa, H., Oda, K. and Okumura, Y., *Rep. Fac. Sci. Shizuoka Univ.* **13**, 29 (1979).
106. Rizvi, S. Q. A. and Williams, J. R., *J. Org. Chem.* **46**, 1127 (1981).
107. Hagishita, S. and Kuriyama, K., *J. Chem. Soc. Perkin Trans. I* 950 (1980).
108. Hamon, D. P. G. and Richards, K. R., *Aust. J. Chem.* **36**, 2243 (1983).
109. Denisenko, V. A., Novikov, V. L. and Elyakov, G. B., *Bull. Acad. Sci. USSR, Div. Chem. Sci.* 721 (1979).
110. Bensel, N., Marschall, H., Weyerstahl, P. and Zeisberg, R., *Liebigs Ann. Chem.* 1781 (1982).
111. Bensel, N., Klinkmuller, K.-D., Marschall, H. and Weyerstahl, P., *Liebigs Ann. Chem.*. 1572 (1977).
112. Bokel, H.-H., Marschall, H. and Weyerstahl, P., *Liebigs Ann. Chem.* 73 (1982).
113. Thomas, A. F., Ozainne, M., Decorzant, R., Naf, F. and Lukacs, G., *Tetrahedron* **32**, 2261 (1976).
114. Naf, F., Decorzant, R. and Thommen, W., *Helv. Chim. Acta* **62**, 114 (1979).
115. Bessiere, Y., Barthelemy, M., Thomas, A. F., Pickenhagen, W. and Starkemann, C., *Nouv. J. Chim.* **2**, 365 (1978).
116. Naf, F., Decorzant, R. and Thommen, W., *Helv. Chim. Acta* **65**, 2212 (1982).
117. Nakanishi, K., Crouch, R., Miura, I., Dominguez, X., Zamudio, A. and Villarreal, R., *J. Am. Chem. Soc.* **96**, 609 (1974).
118. Irikawa, H., Koyama, T. and Okumura, Y., *Bull. Chem. Soc. Japan.* **52**, 637 (1979).
119. de Pascual Teresa, J., Moreno Valle, M. A., Gonzalez, M. S. and Bellido, I. S., *Tetrahedron* **40**, 2189 (1984).
120. Dailey, Jr., O. D. and Fuchs, P. L., *J. Org. Chem.* **45**, 216 (1980).
121. Seuron, N., Ourevitch, M., Seyden-Penne, J. and Guilhem, J., *Bull. Soc. Chim. Fr. II* 124 (1984).
122. Denisenko, V. A., Novikov, V. L., Kamernitskii, A. V. and Elyakov, G. B., *Bull. Acad. Sci. USSR, Div. Chem. Sci.* 931 (1980).

Chapter 18

Spirocyclics

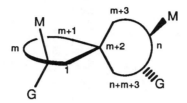

Substantial problems exist with all previous systems for specifying the relative orientations of groups between the two rings of spirocyclics. As with the bicyclo[2.2.2]octanes where literature schemes were found to be inadequate, we have defined a new system. Quite simply, groups in each ring are defined as M (mit) if they are oriented towards the lower numbered of the two atoms of the other ring immediately adjacent to the junction; otherwise they are G (gegen).

Whitesell, J.K. and Minton, M.A. *J. Org. Chem.*, **50**, 509 (1985).

18.1 Spirocyclo[3.4]octanes

Str	Substituents	C1	C2	C3	C4	C5	C6	C7	C8	Substituent Shifts	References
1	1=O/M5-Me	215.4	42.7	22.8	74.2	43.4	34.7	22.9	33.6	15.3	34
1	1=O/G5-Me	215.3	42.9	20.0	74.1	38.5	35.4	22.9	33.5	15.7	34

18.2 Spirocyclo[3.5]nonanes

Str	Substituents	C1	C2	C3	C4	C5	C6	C7	C8	C9	Substituent Shifts	References
1	1=O/M5-Me	215.4	41.5	23.0	69.6	36.5	31.2	24.7		33.1	16.7	34
2	1=O/G5-Me	215.9	42.3	18.5	70.4	34.0	30.7	25.2	21.9	32.6	16.8	34
3	1=O/M5-Me/M9-Me	216.4	42.8	19.0	74.5	37.9	32.2	25.0	32.2	37.9	17.5,17.5	34
4	1=O/G5-Me/G9-Me	218.2	44.1	12.9	75.2	36.1	30.7	25.8	30.7	36.1	17.1,17.1	34
5	5=/7-COOH 7-Ph/8=	34.7	15.8	34.7	39.3	133.7	124.1	52.8	124.1	133.7	179.0,141.9	36

18.3 Spirocyclo[4.4]nonanes

Str	Substituents	C1	C2	C3	C4	C5	C6	C7	C8	C9	Substituent Shifts	References
1	1=O	221.3	37.3	20.2	38.7	56.7	36.9	26.4	26.4	36.9		2
2	1=O 2=	213.9	132.9	161.7	45.9	53.9	37.8	25.3	25.3	37.8		41
3	1=O/G2-(CH₂)₄-2M	225.8	59.1	36.2	36.2	59.1	37.9	26.6	26.6	37.9	37.9,37.9	2
4	1=O/G2-(CH₂)₅-2M	224.7	49.9	32.0	35.3	57.2	37.7	26.4	26.4	37.7	33.7,33.7	2
5	G1-Me M1-Me/2=O/G6-Me M6-Me	56.2	223.1	33.3	25.0	51.6	44.0	42.0	18.7	31.1	21.9,20.0,26.3,25.5	24
6	G1-Me M1-Me/G6-Me M6-Me	43.6	43.4	18.4	31.7	57.5	43.6	43.4	18.4	31.7	26.4,25.6,26.4,25.6	24
7	C1-Me M1-Me/M2-OH/G6-Me M6-Me	56.2	80.5	28.4	28.3	46.0	44.4	42.2	18.4	32.3	22.8,17.2,26.0,25.6	24
8	1= 3=	143.9	127.9	127.9	143.9	64.1	32.4	26.0	26.0	32.4		21,22,23
9	1= 3=/6= 8=	150.5	151.0	151.0	150.5	77.0	150.5	151.0	151.0	150.5		20,21
10	1= 3=/7=	144.8	127.9	127.9	144.8	62.0	36.6	130.4	130.4	36.6		21

18.4 Spirocyclo [4.5] decanes

Str	Substituents	C1	C2	C3	C4	C5	C6	C7	C8	C9	C10	Substituent Shifts	References
1	PARENT	38.7	24.1	24.1	38.7	42.6	38.7	24.5	26.8	24.5	38.7		1,2
2	1-=0	221.0	38.0	20.0	35.3	49.7	33.3	23.5	27.1	23.5	33.3		2,42
3	1-=0/4-==CH2/M8-t-Bu	219.7	36.4	27.0	155.5	50.8	34.0	34.0	47.6	22.4	34.0	106.1,27.5	26
4	1-=0/7= 8-Me	223.5	37.6	18.7	33.9	47.3	32.3	118.6	133.0	28.3	26.8	23.2	18
5	1-==0/E4-==CHMe		35.9	22.2	145.7	51.8	33.2	21.7	25.7	21.7	33.2	116.9	26
6	1-==0/G2-=(CH2)4-2M	224.7	57.2	35.3	32.0	49.9	33.7	22.9	26.5	22.9	33.7	37.7,37.7	2
7	1-==0/G2-=(CH2)5-2M	223.9	50.5	31.2	31.2	50.5	33.6	22.8	26.5	22.8	33.6	33.6,33.6	2
8	1-==0/Z4-==CHMe	222.1	37.4	29.2	144.8	50.2	31.2	21.6	25.6	21.6	31.2	119.0	26
9	G1-C(Me)=CH2/G4-Me/7-=0	55.2	27.4	32.3	41.5	51.9	48.1	212.1	41.0	21.6	31.2	145.3,17.2	11
10	M1-C(Me)=CH2/G4-Me/7-=0	52.5	27.1	32.1	39.3	52.3	48.1	212.7	41.3	22.2	31.5	145.2,16.3	11,30
11	G1-iPr/G4-Me/7-=0	58.5	30.3	30.3	46.2	49.4	53.0	213.3	40.9	22.9	24.4	28.9,15.7	1
12	G1-iPr/G4-Me/7-==0 8-Me 8=	57.1	25.4	29.9	46.2	48.5	49.6	200.8	135.6	144.4	26.0	29.3,17.1,15.6	3
13	M1-iPr/G4-Me/7-==0 8-Me 8=	52.5	24.8	31.0	40.8	50.2	49.6	200.9	136.0	144.7	33.5	28.5,17.8,15.7	3
14	G1-iPr/G4-Me/7-==0 8=	57.3	25.5	30.0	46.7	48.2	45.6	200.9	136.0	144.7	26.0	29.4,17.2	1
15	G1-iPr/M4-Me/7-==0	53.4	24.8	32.2	41.7	51.4	48.9	212.8	41.5	22.7	30.7	27.8,17.1	1
16	M1-iPr/M4-Me/7-==0 8-Me 8=	57.2	25.7	30.1	47.5	48.3	39.0	201.0	135.2	144.9	38.0	29.3,16.4,15.8	3,35
17	G1-iPr/M4-Me/7-==0 8-Me 8=	51.1	23.9	31.3	41.5	50.5	46.1	200.9	136.0	144.5	32.4	29.2,16.6,15.6	3
18	G1-iPr/M4-Me/7-==0 8=	51.3	23.3	31.3	41.6	50.4	46.1	201.0	130.2	150.0	32.4	29.4,16.6	1
19	G1-iPr/M4-Me/G7-OCH2CMe2CH2O-7M	49.4	23.3	30.3	41.3	48.3	36.6	99.6	32.3	19.9	30.2	27.8,16.5	1
20	M1-Me/G4-Me	38.7	30.3	30.3	38.7	45.9	30.6	23.2	26.9	23.2	30.6	15.6,15.6	1
21	G1-Me/G4-Me/7-=0	45.3	30.8	30.8	45.3	49.6	52.5	213.3	40.7	23.1	24.4	16.2,16.2	1
22	M1-Me/G4-Me/7-=0	40.1	31.2	30.5	39.9	51.6	47.2	213.2	41.7	23.1	26.0	15.8,15.8	1
23	G1-Me/G4-Me/7-==0 8=	45.5	30.2	30.2	45.5	48.5	48.5	200.9	130.5	149.8	30.5	16.7,16.7	1
24	M1-Me/G4-Me/7-==0 8=	40.6	31.2	30.8	40.0	49.6	45.0	200.3	130.5	149.5	25.5	17.4,16.3	1
25	G1-Me/G4-Me/G7-OCH2CH2O-7M	43.6	31.9	31.9	43.6	46.7	46.9	110.1	35.4	21.5	32.3	17.5,17.5	1
26	M1-Me/G4-Me/G7-OCH2CH2O-7M	39.4	30.7	30.0	38.2	47.6	37.6	110.7	36.0	20.5	29.2	16.3,15.7	1
27	G1-Me/G4-Me/G7-OCH2CH2O-7M	44.2	32.0	32.0	44.2	46.1	42.6	98.9	34.6	20.4	27.1	17.3,17.3	1
28	M1-Me/G4-Me/G7-OCH2CH2O-7M	40.0	30.7	30.5	37.7	47.0	35.4	99.4	32.6	19.5	27.1	16.2,15.8	1
29	G1-Me/G4-Me/M7-Me	46.5	32.2	31.4	41.4	45.7	49.0	29.7	35.4	24.1	30.1	18.8,17.2,23.4	1
30	G1-Me/G4-Me/M7-OH	46.4	32.2	31.2	42.7	47.0	48.6	48.6	35.4	22.4	27.5	18.7,16.6	1
31	M1-Me/M4-Me	43.4	31.9	31.9	31.9	45.0	28.4	24.1	26.7	24.0	26.2	17.9,17.9	1
32	1= 2-COOEt/6-=0 7= 8-OEt/M10-Me	140.5	139.4	29.0	31.1	64.0		100.8	174.9	34.9	40.0	164.2,16.1	13
33	1=/6-=0	133.4	132.5	31.3	32.3	64.1	211.5	118.7	27.7	23.0	36.4		19
34	2-=0 3=/7= 8-Me	47.3	195.6	131.7	171.8	43.5	36.4	39.9	133.7	27.8	39.9	23.1	15
35	2-=0 3=/G8-t-Bu	45.0	209.0	132.5	169.5	50.7	38.2	23.8	47.7	23.8	38.2	32.4	25
36	2-=0/6-Me 6=/8-=0/G10-Me	38.5	216.7	36.8	27.4	47.2	163.5	127.2	197.6	42.0	48.1	21.1,15.8	16
37	2-=0/6-Me 6=/8-=0/M10-Me	36.7	216.7	39.0	33.3	48.0	163.4	127.0	197.6	42.1	45.9	21.6,16.7	16
38	M2-C(Me)=CH2 / 6-Me 6=-=0/G10-Me	40.9	46.6	32.8	34.4	50.2	166.5	125.6	198.8	43.0	43.0	147.2,20.9,15.9	6,4
39	M2-C(Me)=CH2/6-Me 6= 8-=0 9= 10-Me	41.4	49.3	33.5	36.5	52.8	125.6	198.9	186.3	126.0	164.4	146.1,20.7,21.4	4
40	M2-C(Me)=CH2/6-Me 6= 8-=0/M9-OH/G10-Me	40.7	47.4	32.8	32.8	51.7	164.2	126.0	126.3	126.0	48.0	147.0,22.6,12.2	5
41	M2-C(Me)=CH2/G6-CHO/G8-OH/G10-Me	40.5	47.3	32.6	25.2	46.7	172.3	122.0	199.4	74.0	48.0	147.9,66.3,16.6	7
42	M2-C(Me)=CH2/G6-CH2OAc/G8-OAc/M9-OAc/G10-Me	41.1	47.3	32.9	26.1	48.0	44.8	33.0	72.1	36.8	40.9	147.4,65.7,11.4	7
43	M2-C(Me)=CH2/G6-CH2OH/8-=0/G10-Me	40.0	47.2	32.5	25.0	46.8	45.8	31.4	73.7	47.3	47.3	147.6,64.3,16.7	6,10
44	M2-C(Me)=CH2/G6-CH2OH/G8-OH/G10-Me	40.6	47.4	32.7	25.4	46.7	50.1	42.4	211.6	46.7	43.0	148.1,64.1,16.7	6,7,17
45	M2-C(Me)=CH2/G6-CH2OH/M8-OH/G10-Me	41.0	47.3	32.7	24.1	47.3	48.5	36.7	69.9	40.8	41.2	148.3,64.4,16.6	9
46	M2-C(Me)=CH2/G6-CHO/G8-OH/G10-Me	41.8	47.4	32.6	25.9	46.9	58.4	34.1	66.1	38.3	35.6	147.3,205.0,16.4	6,17
47	M2-C(Me)=CH2/G6-CHO/G8-OH/M9-OH/G10-Me	42.2	47.3	32.7	26.8	48.3	57.6	30.5	74.1	76.8	47.1	147.0,204.3,11.2	6,17
48	M2-C(Me)=CH2/G6-CHO/M8-OH/G10-Me	42.5	47.2	31.2	25.1	47.6	54.2	32.2	65.0	37.9	35.2	147.5,206.3,16.2	9
49	M2-C(Me)=CH2/M6-CH2OH/G8-OH/G10-Me	42.1	47.9	31.2	31.3	46.4	49.1	34.0	66.6	40.4	35.4	148.4,62.1,17.2	8,10
50	M2-C(Me)=CH2/M6-CH2OAc/G8-OAc/G10-Me	41.3	47.5	31.1	31.5	46.2	44.0	30.4	69.1	36.0	35.2	147.3,63.9,16.8	8,10

18.4 Spirocyclo [4.5] decanes (cont)

Str	Substituents	C1	C2	C3	C4	C5	C6	C7	C8	C9	C10	Substituent Shifts	References
51	M2-C(Me)=CH2/M6-CHO/G8-OH/G10-Me	42.4	48.0	30.8	30.7	46.3	58.6	32.9	66.6	39.9	36.9	147.6,205.5,17.0	6
52	M2-C(Me)=CH2/M6-CHO/G8-OH/M9-OH/G10-Me	42.8	47.9	30.9	31.2	46.7	57.9	30.0	71.4	76.9	43.1	147.1,204.7,11.7	10
53	M2-CMe2OH/6-Me 6=/M10-Me	28.5	50.9	28.5	36.0	49.0	140.3	121.9	25.2	33.8	37.2	79.4,20.7,16.7	12,16
54	G2-CMe2OH/6-Me 6=/M10-Me	30.1	54.8	28.6	37.2	33.8	141.4	122.2	22.2	31.4	26.4	72.5,24.2,17.1	16
55	G2-COMe/6-Me 6=/M9-BR/G10-Me M10-Me	35.6	54.4	31.3	35.1	54.6	140.7	119.8	35.8	63.0	41.8	211.1,24.0,17.1	14
56	6==0	35.9	25.7	25.7	35.9	57.4	212.1	39.6	27.9	23.5	40.5		2
57	6==0/9= 10-Me	34.9	27.2	27.2	34.9	58.6	203.2	36.5	25.5	121.8	139.6	18.7	37
58	6==0/G7-(CH2)4-7M	37.8	25.8	25.8	37.8	56.5	216.1	56.5	38.7	20.5	38.7	37.8,37.8	2
59	6==0/G7-(CH2)5-7M	38.2	26.2	26.2	38.2	56.2	216.9	48.5	36.4	18.9	39.3	34.9,34.9	2
60	M6-0Ac/9= 10-Me	33.3	27.4	26.9	36.7	49.1	76.9	24.5	22.3	121.2	138.3	19.3	37
61	M6-OH/9= 10-Me	32.7	27.6	27.0	37.2	50.8	74.1	27.1	21.8	121.0	138.5	19.6	37
62	G7-(CH2)5-7M	41.2	24.8	24.8	41.2	43.2	50.2	34.3	38.0	19.9	38.7	39.4,39.4	2
63	7==0	38.4	24.5	24.5	38.4	47.7	53.5	211.9	41.5	24.0	37.2		1
64	M7-Me	43.0	25.4	24.2	35.2	43.4	48.0	30.0	35.6	24.0	38.1	23.2	1
65	G7-0CH2CH2O-7M	39.1	24.3	24.3	39.1	44.0	45.6	109.9	35.4	21.3	36.8		1
66	G7-0CH2CMe2CH2O-7M	39.1	24.1	24.1	39.1	43.3	41.1	98.6	34.9	20.1	37.3		1
67	M7-OCHO	42.0	25.2	24.1	35.9	43.8	43.5	72.3	32.0	21.6	36.9		1
68	M7-OCOCF3		24.7	23.9		43.4	42.6	77.4	31.3	21.2	36.6		1
69	M7-OH	42.4	25.3	24.0	35.6	44.1	47.8	69.1	36.1	22.0	37.3		1
70	M7-OH G7-Me	37.7	23.4	24.8	42.1	42.8	50.8	71.2	39.9	20.0	36.7	31.6	1

228

18.5 Spirocyclo [5.5] undecanes

Str	Substituents	C1	C2	C3	C4	C5	C6	C7	C8	C9	C10	C11	Substituent Shifts	References
1	PARENT	37.8	22.4	27.8	22.4	37.8	33.1	37.8	22.4	27.8	22.4	37.8		2,27
2	1-=0	214.0	38.7	28.0	21.2	39.5	49.2	34.4	22.6	26.9	22.6	34.4		2
3	1-=0 M2-Me 3=/G4-Br/G5-Me M5-Me/G8-Br/M9-C1 G9-Me	209.9	42.6	130.0	136.6	48.7	58.4	36.3	60.6	70.8	40.1	27.1	14.2,24.8,23.8,24.0	38
4	1-=0/G2-(CH2)4-2M	216.9	56.2	39.3	18.9	36.4	48.5	34.9	26.8	26.8	22.5	34.9	38.2,38.2	2
5	1-=0/G2-(CH2)5-2M	218.3	48.7	34.8	18.0	34.8	48.7	35.3	22.2	26.5	22.2	35.3	35.3,35.3	2
6	1-Me 1= 3-=0/G5-Me M5-Me/G8-C1/M9-Br G9-Me	166.2	128.2		48.8	42.2	42.0	41.9	68.6	66.8	36.1		26.5,0,23.7	31
7	1-Me 1= 3-=0/G5-Me M5-Me/M8-C1/G9-Br M9-Me	167.0	127.4		49.1	42.4	43.1	39.6	65.4	68.3	36.3	29.7	25.7,24.1,31.5,24.3	31
8	1-Me 1= 3-=0/M4-Br/G5-Me M5-Me/G8-C1/M9-Br G9-Me	166.5	126.5	189.3	65.7	42.3	49.2	41.5	68.4	66.0	37.5	30.9	26.5,24.6,19.0,23.7	31
9	1-Me 1= 3-=0/M4-Br/G5-Me M5-Me/M8-C1/G9-Br M9-Me	156.1	126.8		64.2	44.8	49.6	40.1	65.7	68.0	35.4	25.5	24.8,19.1,32.3,24.2	31
10	1-Me 1=/G3-OAc/M4-Br/G5-Me M5-Me/M8-Br/M9-C1 G9-Me	144.4	121.6	75.1	62.2	45.8	47.7	40.0	62.9	70.5	39.0	31.2	25.6,24.9,18.0,24.0	39
11	1-Me 1=/G3-OH/M4-Br/G5-Me M5-Me/G8-Br/M9-C1 G9-Me	142.7	124.3	73.0	70.7	45.5	47.9	40.1	62.4	70.6	39.1	31.4	25.7,25.1,17.9,24.1	39
12	1-Me 1=/G3-OH/M4-Br/G5-Me M5-Me/G8-C1/M9-Br G9-Me	143.1	127.6	73.2	71.0	45.8	47.6	42.1	69.2	67.1	37.8	31.7	25.8,25.2,17.9,24.0	31
13	1-Me 1=/M3-OH/M4-Br/G5-Me M5-Me/G8-C1/M9-Br G9-Me	143.2	124.0	68.4	66.2	42.8	47.9	42.0	69.2	67.2	38.3	31.0	26.2,24.8,19.5,23.9	31
14	1-Me 1=/M4-Br/G5-Me M5-Me/8= 9-Me	140.8	122.6	36.6	63.3	41.8	44.0	31.1	121.5	134.0	28.8	29.8	23.4,24.9,17.4,23.4	39
15	1-Me 1=/M4-Br/G5-Me M5-Me/G8-Br/M9-C1 G9-Me	139.6	122.6	36.3	60.8	42.8	47.7	40.3	62.9	71.0	39.5	31.6	26.0,24.6,17.1,24.1	39
16	M1-Me G1-0-2G/3= 4-Br/G5-Me M5-Me/G8-Br/M9-C1 G9-Me	58.1	56.2	124.1	143.6	45.8	49.1	39.5	63.1	71.1	39.0	25.8	24.3,23.5,23.8,24.7	39
17	M1-Me G1-0-2G/G3-OH/M4-Br/G5-Me M5-Me/M9-C1 G9-Me	62.9	65.4	72.2	67.2	43.9	46.2	40.2	62.3	70.8	39.6	25.6	25.4,24.9,18.3,24.5	39
18	G1-Me M1-Me/M2-Br/5-=CH2/M8-C1/G9-Br M9-Me	43.9	63.4	35.9	33.5	145.8	50.4	40.5	67.9	68.2	37.2	25.5	23.6,17.5,114.7,23.9	31,39
19	G1-Me M1-Me/M2-Br/M3-OAc/5-=CH2/8-C1 8= 9-Me	43.1	62.9	73.3	36.5	140.4	48.8	38.4	127.7	123.8	29.2	25.4	24.0,19.9,115.4,19.2	39
20	G1-Me M1-Me/M2-Br/M3-OAc/5-=CH2/G8-C1/M9-Br G9-Me	43.6	66.2	71.4	35.1	146.7	44.2	33.9	65.0	70.9	33.2	25.5	24.5,33.0,114.5,25.0	32,31
21	G1-Me M1-Me/M2-Br/M3-OAc/5-=CH2/M8-C1/G9-Br M9-Me	44.5	62.5	73.3	37.2	141.3	50.2	40.6	67.5	67.9	37.2	25.7	24.2,20.2,117.8,23.9	31
22	G1-Me M1-Me/M2-Br/M3-OH/5-=CH2/8-C1 8= 9-Me	42.9	70.7	72.0	37.9	140.6	49.0	38.5	127.8	124.0	29.2	25.5	24.1,20.6,115.6,19.3	39
23	G1-Me M1-Me/M2-Br/M3-OH/5-=CH2/G8-C1/M9-Br G9-Me	43.7	76.2	69.7	39.2	147.5	44.0	34.0	65.2	71.0	33.4	25.7	24.3,33.1,113.8,25.3	31
24	G1-Me M1-Me/M2-Br/M3-OH/5-=CH2/M8-C1/G9-Br M9-Me	44.3	70.4	71.8	38.6	141.4	50.3	40.6	67.6	68.0	37.2	25.6	24.2,20.8,117.7,23.9	31
25	G1-Me M1-Me/M3-OAc/5-=CH2/8-C1 8= 9-Me	37.1	37.4	70.7	36.3	141.2	50.3	40.0	127.8	124.5	29.7	25.7	25.0,25.2,116.3,19.4	39
26	G1-Me M1-Me/M3-OAc/5-=CH2/M8-C1/G9-Br M9-Me	38.3	41.4	70.6	36.0	144.5	48.3	40.7	67.9	68.7	37.9	25.8	24.5,23.7,116.3,23.9	31
27	G1-Me M1-Me/M3-OH/5-=CH2/8-C1 8= 9-Me	37.1	43.5	68.0	37.3	143.8	47.1	39.8	127.8	124.3	29.7	25.6	25.1,25.9,114.0,19.3	39
28	M1-OH/G3-C1 M3-Me/M4-Br/G7-Me M7-Me/8=/11-=CH2	70.6	35.2		62.2	47.2			139.2	122.4	34.9		28.2,23.8,26.3,112.7	39
29	1= 3-=0	158.7	127.4	199.4	33.8	32.9	35.5	36.0	21.7	26.0	21.7	36.0		29,43
30	1= 3-=0/8=	157.6	127.8	199.2	33.8	33.0	34.0	35.3	126.7	124.3	32.1	22.2		29,43
31	1= 3-=0/M7-OH G7-C CH	157.3	130.3	210.5	33.9	31.9	43.8	86.4	35.4	22.5	20.8	25.7	73.4	33
32	1= 3-=0/M9-t Bu	155.4	127.9	198.6	34.7	33.9	38.1	37.7	22.8	48.2	22.8	37.7	32.4	25
33	G2-(CH2)4-2M	50.2	43.2	38.7	19.9	38.0	34.3	39.4	22.5	27.6	22.5	39.4	41.2,41.2	2
34	G2-(CH2)5-2M	49.5	34.1	38.2	18.8	38.2	34.1	40.2	22.6	27.4	22.6	40.2	40.2,40.2	2
35	G2-C1/M3-OH G3-Me/7-Me 7=/M10-Br/G11-Me M11-Me	39.1	66.1	71.9	35.9	31.6	48.1	139.9	122.0	36.2	61.2	42.7	24.7,25.8,22.1,17.3	39,40
36	M2-0-3M G3-Me/7-Me 7=/M10-Br/G11-Me M11-Me	29.4	58.7	55.5	26.5	25.6	43.0	141.6	119.6	35.7	61.8	42.0	22.3,25.4,20.6,17.6	39,40
37	2= 3-Me/G7-Me M7-Me/G8-Br/11-=CH2	30.4	119.7	132.7	26.6	25.0	50.8	42.8	66.1	33.1	35.7	145.6	23.1,17.4,23.6,112.6	39
38	3-=0	35.0	36.2	210.7	36.2	35.0	31.2	35.3	21.2	25.9	21.2	35.3		28

References

1. Kutschan, R., Ernst, L. and Wolf, H., *Tetrahedron* **33**, 1833 (1977).
2. Zimmermann, D., Ottinger, R., Reisse, J. , Cristol, H. and Brugidou, J., *Org. Magn. Reson.,* **6**, 346 (1974).
3. Kutschan, R., Schiebel, H. -M., Schroder, N. and Wolf, H., *Chem. Ber.* **110**, 1615 (1977).
4. Coxon, D. T., Price, K. R., Howard, B., Osman, S. F., Kalan, E. B. and Zacharius, R. M., *Tetrahedron Lett.* 2921 (1974).
5. Anderson, R. C., Gunn, D. M., Murray-Rust, J., Murray-Rust, P. and Roberts, J. S., *J. Chem. Soc. Chem. Commun.* 27 (1977).
6. Stoessl, A., Stothers, J. B. and Ward, E. W. B., *J. Chem.* **56**, 645 (1978); Katsui, N., Yajihashi, F., Murai, A. and Masamune, T. *Chem. Lett.,* 1205 (1978).
7. Katsui, N., Matsunaga, A., Kitahara, H., Yagihashi, F., Murai, A., Masamune, T. and Sato, N., *Bull Chem. Soc. Japan* **50**, 1217 (1977).
8. Katsui, N., Yagihashi, F., Matsunaga, A., Orito, K., Murai, A. and Masmune, T., *Chem. Lett.* 723 (1977).
9. Stoessl, A. and Stothers, J. B., *Can. J. Chem.,* **58**, 2069 (1980).
10. Katsui, N., Yagihashi, F., Murai, A. and Masmune, T., *Bull Chem. Soc. Japan,* **55**, 2424 (1982).
11. Manh, D. D. K., Ecoto, J., Fetizon, M., Colin, H. and Diez-Masa, J. -C., *J. Chem. Soc. Chem. Commun.* 953 (1981).
12. Rustaiyan, A., Behjati, B. and Bohlmann, F., *Chem. Ber.* **109**, 3953 (1976).
13. Dauben, W. G. and Hart, D. J., *J. Am. Chem. Soc.* **99**, 7307 (1977).
14. Suzuki, M., Kowata, N. and Kurosawa, E. *Tetrahedron* **36**, 1551 (1980).
15. Wenkert, E., Buckwalter, B. L., Craveiro, A. A., Sanchez, E. L. and Sathe, S. S., *J. Am. Chem. Soc.* **100**, 1267 (1978).
16. Lafontaine, J., Mongrain, M., Sergent-Guay, M., Ruest, L. and Deslongchamps, P., *Can. J. Chem.* **58**, 2460 (1980).
17. Stoessl, A. and Stothers, J. B.,*Can. J. Chem.* **61**, 1766 (1983).
18. Garst, M. E., McBride, B. J. and Johnson, A. T., *J. Org. Chem.* **48**, 8 (1983).
19. Piers, E., Lau, C. K. and Nagakura, I., *Can. J. Chem.* **61**, 288 (1983).
20. Smolinski, S., Balazy, M., Iwamura, H., Sugawara, T., Kawada, Y. and Iwamura, M., *Bull Chem. Soc. Japan* **55**, 1106 (1982).
21. Semmelhack, M. F., Foos, J. S. and Katz, S., *J. Am. Chem. Soc.* **95**, 7325 (1973).
22. van deVen, L. J. M. and deHaan, J. W., *J. Magn. Reson.* **19**, 31 (1975).
23. Holder, R. W., Daub, J. P., Baker, W. E., Gilbert, III, R. H. and Graf, N. A., *J. Org. Chem.* **47**, 1445 (1982).
24. Berger, J., Yoshioka, M., Zink, M. P., Wolf, H. R. and Jeger, O., *Helv. Chim. Acta* **63**, 154 (1980).
25. Martin, S. F., University of Texas at Austin, personal communication.
26. Shimada, J., Hashimoto, K., Kim, B. H., Nakamura, E. and Kuwajima, I., *J. Am. Chem. Soc.* **106**, 1759 (1984).
27. Dalling, D. K., Grant, D. M. and Paul, E. G., *J. Am. Chem. Soc.* **95**, 3718 (1973).
28. McMurry, J. E. and Andrus, A., *Tetrahedron Lett.* **21**, 4687 (1980).
29. Gramlich, W., *Liebigs Ann. Chem.* 121 (1979).
30. Oppolzer, W., Zutterman, F. and Battig, K., *Helv. Chim. Acta* **66**, 522 (1983).
31. Gonzalez, A. G., Martin, J. D., Martin, V. S. and Norte, M., *Tetrahedron Lett.* 2719, (1979).

32. Schmitz, F. J., Michaud, D. P. and Schmidt, P. G., *J. Am. Chem. Soc.* **104**, 6415 (1982).

33. Rao, C. S. S., Kuwar, G., Rajagopalan, K. and Swaminathan, S., *Tetrahedron* **38**, 2195 (1982).

34. Trost, B. M. and Scudder, P. H., *J. Am. Chem. Soc.* **99**, 7605 (1977).

35. Baldwin, S. W. and Fredericks, J. E., *Tetrahedron Lett.* **23**, 1235 (1982).

36. Grovenstein, Jr., E. and Lu, P. -C., *J. Org. Chem.* **47**, 2928 (1982).

37. Coxon, J. M. and Gibson, J. R., *Aust. J. Chem.* **35**, 1165 (1982).

38. Selover, S. J. and Crews, P., *J. Org. Chem.* **45**, 69 (1980).

39. Sims, J. J., Rose, A. F. and Izac, R. R., *Marine Natural Products, Chemical and Biological Perspectives*, vol. 2, Scheuer, P. J., Ed., 1978, p. 297ff.

40. Suzuki, M., Furusaki, A. and Kurosawa, E., *Tetrahedron* **35**, 823 (1979).

41. Koller, M., Karpf, M. and Dreiding, A. S., *Helv. Chim. Acta* **66**, 2760

42. Crandall, J. K., Magaha, H. S., Henderson, M. A., Widener, R. K. and Tharp, G. A., *J. Org. Chem.* **47**, 5372 (1982).

43. *Sadtler Standard Carbon-13 NMR Spectra*, Sadtler Research Laboratories, Inc., Philadelphia, Pa.